# Biochemistry and Molecular Biology: Concepts, Techniques and Applications

# Biochemistry and Molecular Biology: Concepts, Techniques and Applications

Editor: Sydney Marsh

R CALLISTO
REFERENCE

www.callistoreference.com

**Callisto Reference,**
118-35 Queens Blvd., Suite 400,
Forest Hills, NY 11375, USA

Visit us on the World Wide Web at:
www.callistoreference.com

ISBN: 978-1-64116-103-9 (Hardback)

**Cataloging-in-Publication Data**

Biochemistry and molecular biology : concepts, techniques and applications / edited by Sydney Marsh.
    p. cm.
Includes bibliographical references and index.
ISBN 978-1-64116-103-9
1. Biochemistry. 2. Molecular biology. 3. Biomolecules. I. Marsh, Sydney.
QH345 .B56 2019
572--dc23

# Table of Contents

# Preface

The world is advancing at a fast pace like never before. Therefore, the need is to keep up with the latest developments. This book was an idea that came to fruition when the specialists in the area realized the need to coordinate together and document essential themes in the subject. That's when I was requested to be the editor. Editing this book has been an honour as it brings together diverse authors researching on different streams of the field. The book collates essential materials contributed by veterans in the area which can be utilized by students and researchers alike.

The field of biochemistry studies the structural and functional aspects of biomolecules like lipids, proteins, carbohydrates and nucleic acids as well as their interactions with each other. It helps in understanding the processes occurring within living organisms. The applications of biochemistry are in medicine, nutrition and agriculture. Molecular biology is a branch of biochemistry. It studies the interactions of DNA, proteins and RNA as well as their biosynthesis. The study of molecular biology is mostly quantitative and a lot of emphasis is placed on the understanding of gene structure and function. There has been rapid progress in these fields in recent years and their applications are finding their way across multiple industries. This book presents the complex subjects of biochemistry and molecular biology in a comprehensible manner. It will serve as a reference text for students as well as biologists, researchers, geneticists and other experts working on this field.

Each chapter is a sole-standing publication that reflects each author's interpretation. Thus, the book displays a multi-facetted picture of our current understanding of application, resources and aspects of the field. I would like to thank the contributors of this book and my family for their endless support.

Editor

# Histone deacetylase inhibitor sodium butyrate suppresses DNA double strand break repair induced by etoposide more effectively in MCF-7 cells than in HEK293 cells

Liping Li[1,2*†], Youxiang Sun[1,2†], Jiangqin Liu[1], Xiaodan Wu[1,2], Lijun Chen[1,2], Li Ma[1] and Pengfei Wu[1,2]

## Abstract

**Background:** Histone deacetylase inhibitors (HDACi's) are emerging as promising anticancer drugs alone or in combination with chemotherapy or radiotherapy agents. Previous research suggests that HDACi's have a high degree of selectivity for killing cancer cells, but little is known regarding the impact of different cellular contexts on HDACi treatment. It is likely that the molecular mechanisms of HDACi's involve processes that depend on the chromatin template, such as DNA damage and repair. We sought to establish the connection between the HDACi sodium butyrate and DNA double-strand break (DSB) damage in human breast cancer MCF-7 and non-cancerous human embryonic kidney293 (HEK293) cells.

**Results:** Sodium butyrate inhibited the proliferation of both HEK293 and MCF-7 cells in a dose- and time- dependent manner, but the effects on MCF-7 cells were more obvious. This differential effect on cell growth was not explained by differences in cell cycle arrest, as sodium butyrate caused an arrest in $G_1/G_2$ phase and a decrease in S phase for both cell lines. At high doses of sodium butyrate or in combination with etoposide, MCF-7 cells formed fewer colonies than HEK293 cells. Furthermore, sodium butyrate enhanced the formation of etoposide-induced γ-H2AX foci to a greater extent in MCF-7 than in HEK293 cells. The two cells also displayed differential patterns in the nuclear expression of DNA DSB repair proteins, which could, in part, explain the cytotoxic effects of sodium butyrate.

**Conclusions:** These studies suggest that sodium butyrate treatment leads to a different degree of chromatin relaxation in HEK293 and cancerous MCF-7 cells, which results in differential sensitivity to the toxic effects of etoposide in controlling damaged DNA repair.

**Keywords:** Double strand breaks, Histone deacetylase inhibitor, MCF-7, HEK293, Etoposide, Sodium butyrate

## Background

Eukaryotic DNA is bound by histones and organized into chromatin, which serves as the true in vivo substrate of transcription, replication and DNA repair. Post-translational modification of histones alters chromatin structure; for example, histone acetylation plays a central role in the unwinding of DNA. Histone deacetylase inhibitors (HDACi's) globally increase histone acetylation, relaxing chromatin structure and leading to reversible decondensation of chromatin regions [1]. These inhibitors of chromatin-modifying enzymes are emerging as a promising anticancer drug and already have shown anticancer effects in both pre-clinical and clinical settings [2,3]. HDACi's are gaining increasing attention because of their therapeutic effectiveness in selectively killing cancer cells and their mild toxicity profile [3-5].

Double strand breaks (DSBs) in DNA occur naturally in the genome during replication and are increased by exogenous DNA damaging agents. Many anti-cancer therapeutics, including radiotherapy and chemotherapy agents, kill tumor cells by inducing DSBs. DSB repair is

* Correspondence: lipinggdmc@126.com

†Equal contributors

¹Key Laboratory for Medical Molecular Diagnostics of Guangdong Province, Guangdong Medical College, Xincheng Road, Dongguan 523808, P R China

²Department of Biochemistry, School of Basic Medicine, Guangdong Medical College, Dongguan 523808, P R China

essential for cell viability and normal growth because a single unrepaired DSB can lead to programmed cell death. DSBs can be repaired through several pathways including homologous recombination, non-homologous end-joining and single strand annealing [6]. Understanding the relationship between DSB repair and HDACi anticancer effects is important for elucidating mechanistic details of DSB repair within chromatin that have the potential to be exploited in the clinic.

Sodium butyrate, a naturally occurring short-chain fatty acid that is a by-product of carbohydrate metabolism in the gut, is one of the most widely studied HDACi's [7]. We studied the effect of sodium butyrate alone and combination with the DNA damaging agent, etoposide. Etoposide, a classical chemotherapeutic drug of cancer, interrupts the normal function of topoisomerase II (Topo II) during DNA replication and generates DSBs [1]. We treated both healthy human embryonic kidney 293 (HEK293) and breast cancer MCF-7 cells with sodium butyrate, and our results demonstrate that sodium butyrate treatment increases sensitivity to the cytotoxic effect of etoposide and reduces DSB repair capacity in MCF-7 but not in HEK293 cells.

## Methods
### Ethics statement
All results of this research were based on the use of cultured human (MCF-7) cell lines. Neither human (human subjects or human derived material) nor animals (vertebrates or any regulated invertebrates) were used in this experimental research.

### Cell lines and reagents
Human breast cancer cell MCF-7 and human embryonic kidney 293 cells were obtained from Dr. Fen Xia (The Ohio State University College of Medicine, Columbus, Ohio.). The cells were maintained in DMEM supplemented with 10% fetal bovine serum, 50 units/mL penicillin, and 50 µg/mL streptomycin (Invitrogen, Gibco) at 37°C under 5% $CO_2$.

Sodium butyrate was purchased from Sigma-Aldrich, and etoposide from Selleck Chemicals. The subcellular protein fraction kit was purchased from Thermo Scientific, and the Cell Counting Kit (CCK-8) was from Beyotime (P.R. China). Rabbit antibodies for Rad51, Rad52 and CtIP were purchased from Abcom, and rabbit antibodies for Ku80 and RPA70, mouse antibodies for acetyl histone H4 and γ-H2AX, and secondary antibody for anti- rabbit, anti-mouse and Alexa-fluor488-conjugated anti-mouse were purchased from Cell Signaling Technology. All other reagents were of analytic grade and purchased from standard suppliers.

### Cell proliferation assays
HEK293 and MCF-7 cells were seeded in a 96-well plate at a density of $3 \times 10^3$ cells/well and then were treated with DMSO vehicle or various concentrations of sodium butyrate in 100 µl medium for the indicated times. After the treatment period, 10 µl CCK-8 mixture was added to each well, and the plates were incubated for 40 min at 37°C. The absorbance was measured in a microplate reader (Biotek, USA) at a wavelength of 450 nm.

### Cell cycle analysis
HEK293 and MCF-7 cells were treated with either DMSO vehicle or various concentrations of sodium butyrate for 24 h, and then cell cycle distribution was determined using standard ethanol fixation and propidium iodide staining followed by flow cytometry (Bio-Rad, USA) as previously described [8].

### Colony forming assay
HEK293 and MCF-7 cells were treated with either DMSO vehicle or various concentrations of sodium butyrate for 4 h followed by the addition of DMSO vehicle or etoposide to a final concentration of 0.5 µM for another 20 h incubation. After the treatment period, cells were counted and reseeded in duplicate in 60 mm dishes at a density of 300 cells per dish. Colonies were stained after 15 d incubation and counted as positive if >50 cells were visible. The survival fraction was calculated as follows: (number of colonies for sodium butyrate/number of cells plated)/(number of colonies for corresponding control/number of cells plated).

### Immunofluorescence assay
Immunohistochemistry was performed as previously described [8]. Staining patterns were visualized via fluorescence microscopy (Leica, German). Cells with >5 foci were recorded, and100 to 300 cells were counted per slide. The average foci in one cell and the percentage of foci-positive cells in the whole cell population were calculated.

### Western blot analysis
Soluble nuclear and chromatin-bound protein fractions were extracted using a subcellular protein fraction kit. 50 µg protein per well were separated by SDS/polyacrylamide gel electrophoresis and electroblotted onto nitrocellulose membranes. The transblotted membranes were washed twice with TBS containing 0.1% Tween20 (TBST). After blocking with TBST containing 5% nonfat milk for 40 min, the membranes were incubated with an appropriate primary antibody in TBST containing 1% nonfat milk at 4°C overnight. After treatment with the primary antibodies, the membranes were washed twice with TBST for a total of 20 min, followed by incubation with anti-rabbit or anti-mouse IgG-HRP conjugates for 1

h at room temperature and then washed twice with TBST for a total of 1 h. The immunoblots were visualized by enhanced chemiluminescence.

## Statistical analysis

The data were analyzed via ANOVA followed by a Bonferroni post test using GraphPad Prism version 4.02 for Windows (GraphPad Software).

## Results

### Sodium butyrate suppresses the proliferation of HEK293 and MCF-7 cells to different extents

To verify the effects of sodium butyrate on the proliferation of HEK293 cells and cancerous MCF-7 cells, we performed CCK-8 assays. Sodium butyrate suppressed both HEK293 and MCF-7 cell growth in a dose- and time- dependent manner (Figure 1A and B). No significant effect was found with 0.1mM sodium butyrate for either cell line; however, increasing concentrations of sodium butyrate increasingly reduced the proliferation over time. The effect appeared more dramatic for MCF-7 cells: at 24 h, no difference could be seen at any concentration for HEK293, whereas for MCF-7 cells, the 2.0 mM ($P<0.05$) and 8.0 mM ($P<0.01$) groups had significantly reduced $A_{450nm}$ values compared with the vehicle control. Additionally, at 48 h with 0.1, 0.5, 2.0 and 8.0 mM sodium butyrate concentrations, the inhibitory rate of HEK293 (5.5%, 17.9%, 22.5%, and 22.5%) was lower

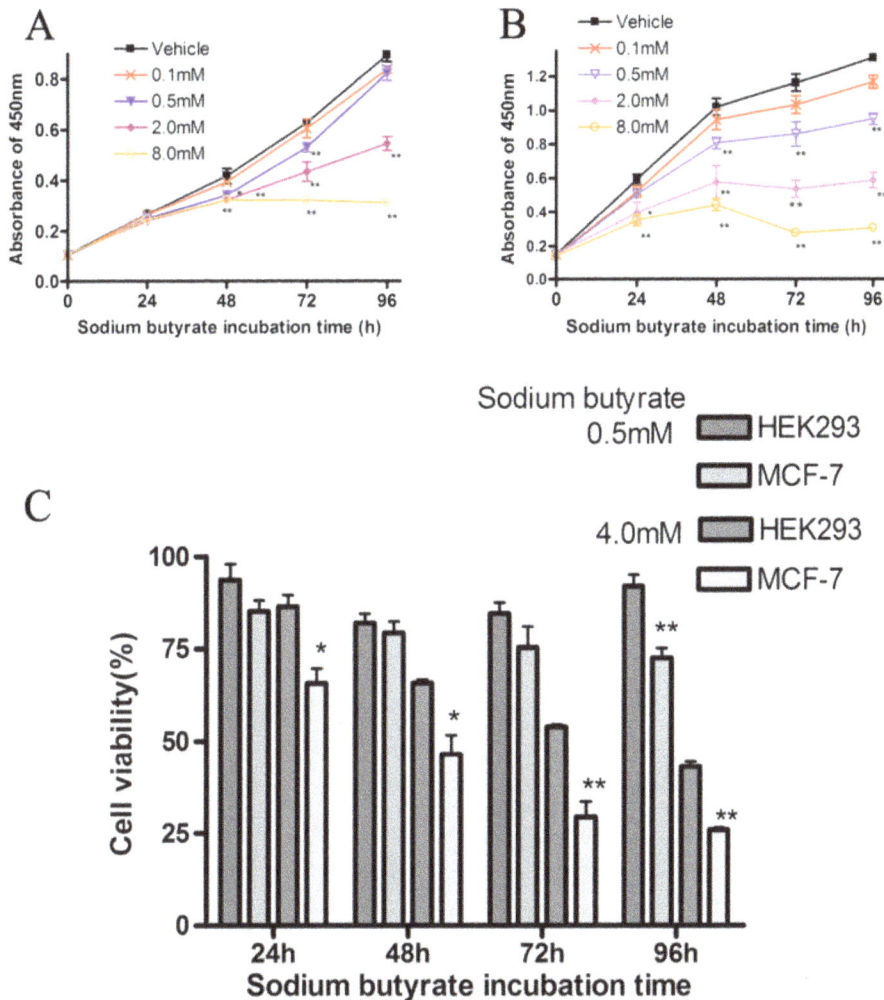

Figure 1 Sodium butyrate suppresses the proliferation of HEK293 and MCF-7 cells to different extents. (A) HEK293 cells were incubated with DMSO vehicle, 0.1, 0.5, 2.0, or 8.0 mM sodium butyrate for the indicated times. After the treatment period, cells were assessed for cell proliferation by CCK-8 assay. The $A_{450nm}$ value corresponds to the amount of cells. *$P< 0.05$ and **$P< 0.01$ versus the corresponding time for vehicle-treated cells. (B) MCF-7 cells were incubated with DMSO vehicle or sodium butyrate and were assessed by CCK-8 assay as in panel A. (C) MCF-7 cell growth inhibition was compared for HEK293 versus MCF-7 cells after treatment with 0.5 or 4.0 mM sodium butyrate for the indicated times. Results represent the CCK-8 assay values at each respective drug treatment relative to that of the DMSO vehicle control. *$P< 0.05$ and **$P< 0.01$ for MCF-7 cells versus the corresponding treatment for HEK293 cells. All data represent the means +/– SD of 3 experiments performed in triplicate.

than that of MCF-7 cells (7.2%, 20.7%, 43.2%, and 56.5%). These findings suggest that MCF-7 cell growth is inhibited more strongly than HEK293 by sodium butyrate.

To directly compare the effects of sodium butyrate on MCF-7 versus HEK293, we calculated the % viability for 0.5 mM and 4.0 mM sodium butyrate treatment at different times. MCF-7 cells were more greatly inhibited than HEK293 were upon 0.5 mM sodium butyrate treatment for 96 h (72.5% versus 92.0%, $P<0.01$) and upon 4.0 mM sodium butyrate treatment for 24 h (65.7% versus 86.6%, $P<0.05$), 48 h (46.5 versus 65.8% %, $P<0.05$), 72 h (29.5% versus 53.9%, $P<0.01$), and 96 h (26.0% versus 43.1%, $P<0.01$) (Figure 1C). These finding verify that HEK293 cells are more resistant than MCF-7 cells to the cytotoxic effects of sodium butyrate.

### Sodium butyrate decreases the proportion of cells in S phase for both HEK293 and MCF-7 cells

Cell proliferation is closely associated with the cell cycle, which is regulated by checkpoints that are activated by the DNA damage response pathway. To determine whether the differential effects of sodium butyrate on proliferation in HEK293 and MCF-7 cells can be explained by differential redistribution of cell cycle progression, we treated each cell line for 24 h with 0.5, 2.0, or 8.0 mM butyrate. Our results demonstrate that for both cell lines, sodium butyrate robustly induces the accumulation of cells in $G_1$ and $G_2$ phase with a concomitant decrease of cells in S phase (Figure 2). These results suggest that sodium butyrate triggers cell cycle checkpoints in both cell lines, indicating that the differences in growth response to

sodium butyrate are not caused by differential control of the cell cycle.

### Sodium butyrate suppresses cell growth synergistically with etoposide, and the effect is more dramatic for MCF-7 cells than for HEK293 cells

To further verify the growth inhibitory effects of sodium butyrate in HEK293 and MCF-7 cells, an equivalent number of each type of cell were seeded for colony forming assay after 24 h treatment with vehicle or 2.0 mM butyrate in the absence or presence of 0.5 μM etoposide, a classic DNA damage reagent. Because co-treatment with HDACi and Topo II inhibitor only has a synergistic effect if HDACi is administrated before Topo II inhibitor [9], we exposed the cells to 2.0 mM butyrate before etoposide was added. While each drug alone had minimal effect, the two drugs together decreased the number of colonies that grew after 15 days in both HEK293 and MCF-7 cells; however, the inhibitory effects of the two drugs in combination were more obvious for MCF-7 cells (Figure 3A). These results suggest that the two drugs may function synergistically to reduce cell viability and that the synergistic effects are more pronounced for cancerous MCF-7 cells.

To verify these findings, we treated cells with a range of doses of sodium butyrate with or without etoposide for 24 h prior to plating. In the absence of etoposide, the survival fraction for the 0.5 or 2.0 mM sodium butyrate doses, was significantly increased for MCF-7 cells compared to HEK293 cells, but the opposite trend was observed for the 8.0 mM sodium butyrate dose (Figure 3B). Additionally, in the presence of etoposide, 2.0 mM or 8.0 mM sodium

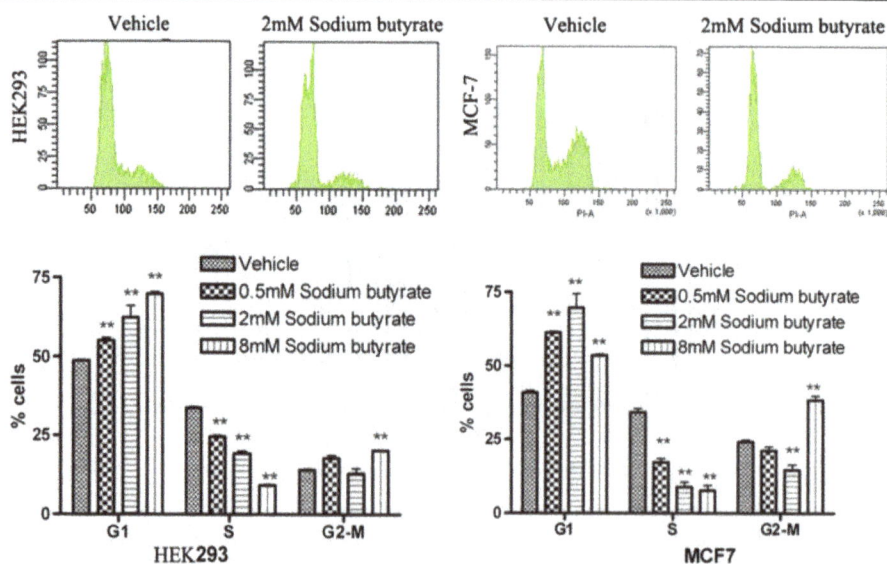

**Figure 2 Sodium butyrate decreases the proportion of cells in S phase for both HEK293 and MCF-7 cells.** HEK293 and MCF-7 cells were treated with DMSO vehicle, 0.5, 2.0, or 8.0 mM sodium butyrate for 24 h. Cell cycle analysis was performed by flow cytometry using propidium iodide staining. Representative histograms are shown above, and quantification of the cells in each phase of the cell cycle is provided below. The values represent the means + SD of triplicate experiments. **$P< 0.01$ versus vehicle-treated cells.

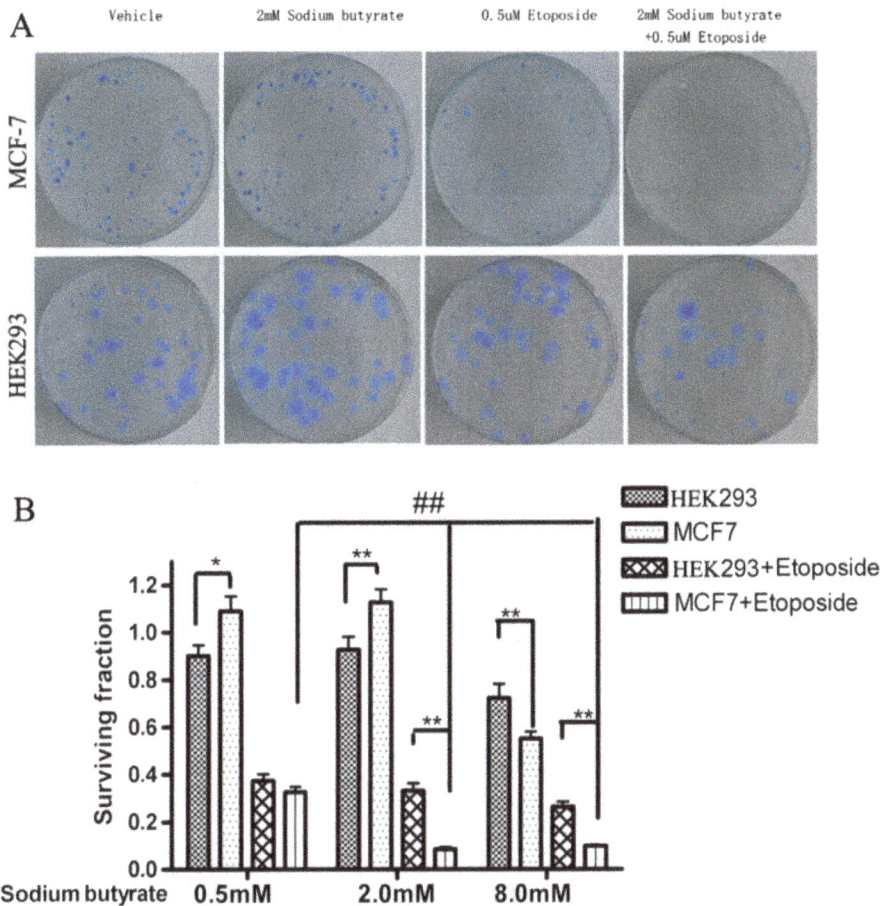

**Figure 3 Sodium butyrate enhances the cytotoxic effects of etoposide, and the effect is more dramatic in MCF-7 cells than in HEK293 cells. (A)** MCF-7 and HEK293 cells were pre-treated with DMSO vehicle or 2 mM sodium butyrate for 4 h, and then were treated with vehicle control or 0.5 μM etoposide for 20 h. After the treatment period, cells were re-seeded for colony forming assay and grown for 15 days. The colonies were stained with 0.5% methylene blue for visualization. Results are representative of 3 independent experiments. **(B)** HEK293 and MCF-7 cells were treated for 4 h with 0, 0.5, 2.0, or 8.0 mM sodium butyrate and then were exposed to DMSO vehicle or 0.5 μM etoposide for 20 h. After the treatment period, cells were re-seeded for colony forming assay. The fraction of surviving colonies was determined for the 0.5, 2.0, and 8.0 mM sodium butyrate groups relative to the corresponding 0 mM sodium butyrate group.*$P < 0.05$ and **$P < 0.01$ for MCF-7 versus HEK293 cells, ##$P < 0.01$ versus the corresponding co-treatment of 0.5 mM sodium butyrate plus etoposide in MCF-7 cells. Results represent the means + SD of quadruplicate experiments.

butyrate led to more obviously decreased colony formation for MCF-7 cells than for HEK293 cells. These results suggest that under lower, non-toxic doses of butyrate, MCF-7 cells may have intrinsic mechanisms for protection from cell death, but that under harsher drug conditions of either higher dose sodium butyrate or a combination of sodium butyrate and etoposide, cancerous MCF-7 cells are more sensitive than HEK293 cells to drug treatment.

**Sodium butyrate enhances etoposide-induced γ-H2AX accumulation to a greater extent in MCF-7 cells than in HEK293 cells**

To further investigate the potential contribution of DNA DSBs to the differential growth response of MCF-7 and HEK293 cells, we tested whether sodium butyrate can increase nuclear γ-H2AX foci, a marker of DNA DSBs. 2.0

mM butyrate alone did not significantly affect γ-H2AX foci in either cell line. However, the addition of 10 μM etoposide caused a difference in the number of foci per nucleus, which was statistical in MCF-7 cells, but not in HEK293 cells (Figure 4A, $P < 0.01$). Furthermore, the percentage of foci–containing cells was significantly increased in MCF-7, but not HEK293 cells (Figure 4B, $P < 0.05$). These results suggest that low dose butyrate treatment can heighten etoposide-induced DNA damage and γ-H2AX accumulation to a greater extent in MCF-7 cells than in HEK293 cells.

**Sodium butyrate suppresses nuclear expression of DNA DSB repair proteins induced by etoposide in MCF-7 and HEK293 cells**

To determine how sodium butyrate may affect the etoposide-induced DSB repair process, we examined

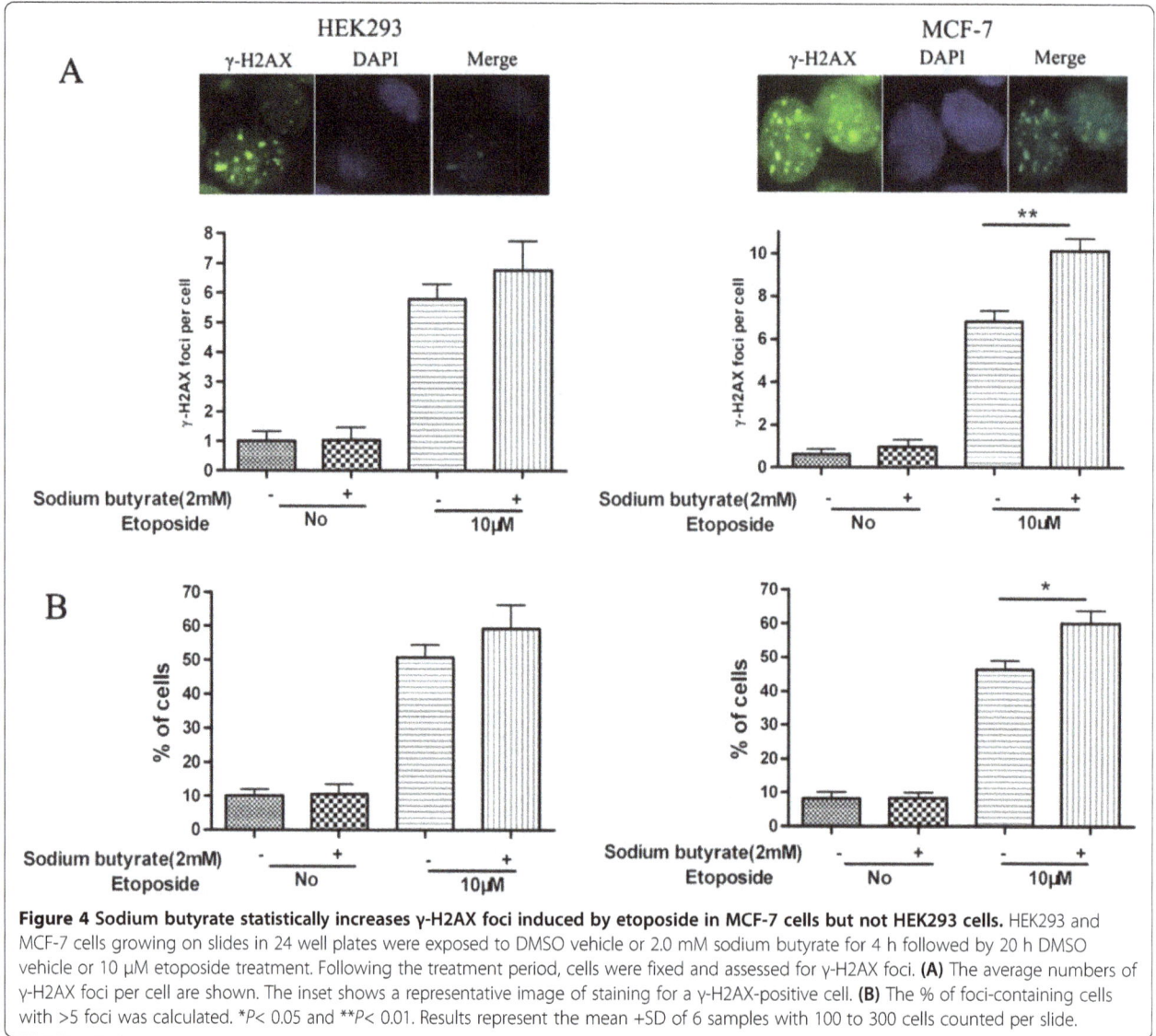

**Figure 4 Sodium butyrate statistically increases γ-H2AX foci induced by etoposide in MCF-7 cells but not HEK293 cells.** HEK293 and MCF-7 cells growing on slides in 24 well plates were exposed to DMSO vehicle or 2.0 mM sodium butyrate for 4 h followed by 20 h DMSO vehicle or 10 μM etoposide treatment. Following the treatment period, cells were fixed and assessed for γ-H2AX foci. **(A)** The average numbers of γ-H2AX foci per cell are shown. The inset shows a representative image of staining for a γ-H2AX-positive cell. **(B)** The % of foci-containing cells with >5 foci was calculated. *$P < 0.05$ and **$P < 0.01$. Results represent the mean +SD of 6 samples with 100 to 300 cells counted per slide.

the expression of proteins involved in the DNA damage response pathway. Because DSBs are generated in DNA, which is located in the nucleus and is characterized by surrounding chromatin architecture, we assessed the expression changes in chromatin-bound and nuclear soluble protein fractions. HEK293 and MCF-7 cells were pre-treated cells with sodium butyrate before exposure to 10 μM etoposide for inducing detectable DSBs and DNA damage response. As shown in Figure 5, the chromatin-bound protein of Acetyl-histone H4 (AceH4) is increased in both HEK293 and MCF-7 cells; however, the increase in MCF-7 cells is observed at a lower sodium butyrate dose. This is consistent with the greater sensitivity of MCF-7 cells as compared to HEK293 cells to the sodium butyrate/etoposide combination.

The results for the other proteins we tested are more complex. Chromatin-bound Rad52 did not show reproducible modulation upon treatment with sodium butyrate and etoposide. However, RPA70, Ku80 and CtIP showed variable patterns of modulation, with increase in chromatin-bound protein in HEK293 cells at certain doses of sodium butyrate, but decrease gradually in MCF-7 cells follow increasing doses. The soluble nuclear portion of these proteins also showed variable trends of modulation, high doses (2 mM and 8 mM) of sodium butyrate attenuate the upregulation of Rad51, RPA70 and CtIP induced by etoposide in MCF-7 cells but not in HEK293 cells. Collectively, these data support the idea that the HDACi sodium butyrate is involved in the mechanism of etoposide-induced DSB damage and repair to improve its anticancer activity.

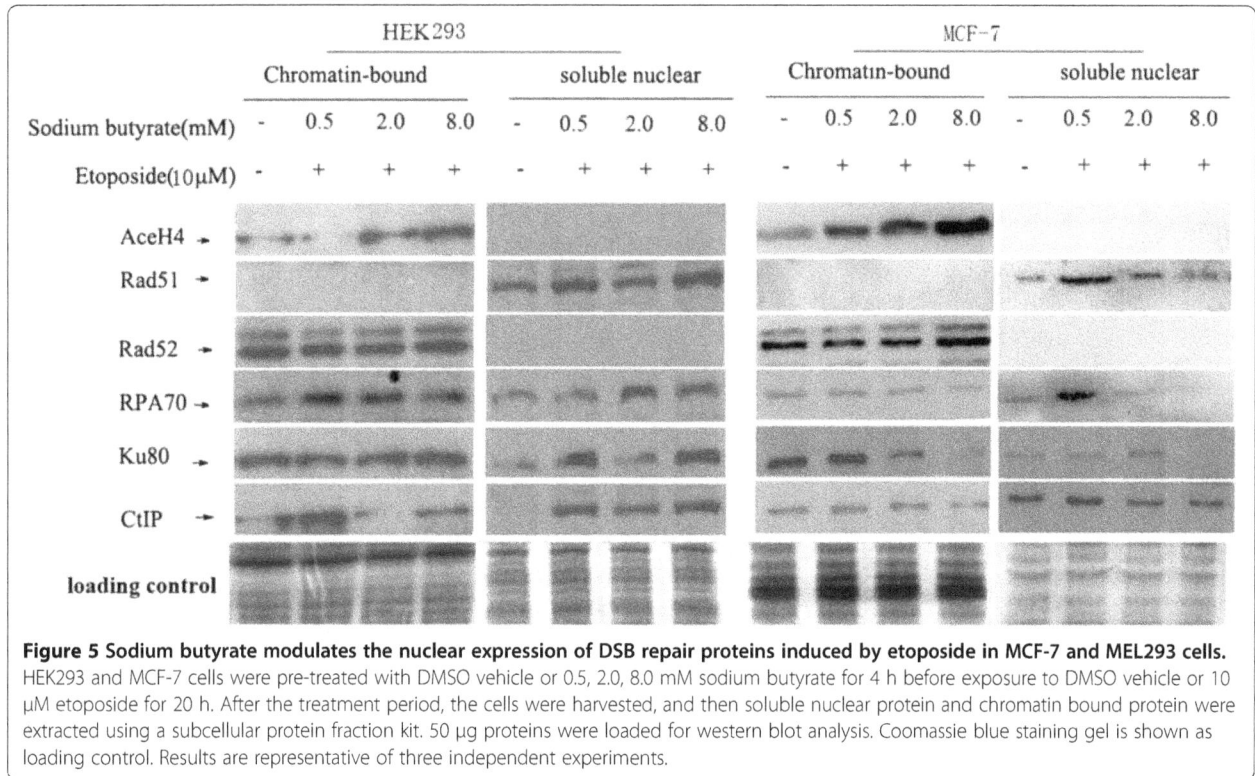

**Figure 5 Sodium butyrate modulates the nuclear expression of DSB repair proteins induced by etoposide in MCF-7 and MEL293 cells.**
HEK293 and MCF-7 cells were pre-treated with DMSO vehicle or 0.5, 2.0, 8.0 mM sodium butyrate for 4 h before exposure to DMSO vehicle or 10 μM etoposide for 20 h. After the treatment period, the cells were harvested, and then soluble nuclear protein and chromatin bound protein were extracted using a subcellular protein fraction kit. 50 μg proteins were loaded for western blot analysis. Coomassie blue staining gel is shown as loading control. Results are representative of three independent experiments.

## Discussion

Research has shown that HDACi's affect gene transcription, induction of cell cycle arrest, differentiation, apoptosis, and inhibition of cell survival, particularly in tumor cell lines [4,10]. Consistent with those data, we found that the growth of both HEK293 cells and cancerous MCF-7 cells are inhibited in a dose- and time-dependent manner by sodium butyrate treatment alone. Core histone acetylation and deacetylation are connected with checkpoint activation and repression [11], and HDACi also mediates acetylation-dependent changes in non-histone proteins involved in cell cycle regulation [12]. Our data show that the cell cycle of both HEK293 cells and cancer MCF-7 cells are arrested by sodium butyrate mostly in G1 but also in G2 phases and reduced in S phase simultaneously, and a similar result has been observed in Hela cells before [13]. So, cell cycle arrest might be one of the main reasons for the effects of sodium butyrate in promoting inhibition of proliferation.

$HDAC_{1/2}$ play redundant and essential roles in tumor cell survival. Their deletion leads to nuclear bridging, nuclear fragmentation, and mitotic catastrophe, mirroring the cytotoxic effects of HDACi's on cancer cells [14]. Human $HDAC_{1/2}$ function in the DNA damage response pathway to promote DNA repair, and their deletion leads to hypersensitivity to DNA-damaging agents and sustained DNA-damage signaling [15]. Consequently, various HDACi's with different HDAC inhibition profiles have

been reported to induce DNA damage [1,5,16]. Sodium butyrate exhibits radiosensitizing effects in several cancers, such as cervical cancer and melanoma cells [17]. We demonstrated that low dose and short times of sodium butyrate incubation led to protection from killing in MCF-7 cells, as measured by colony forming assay, whereas higher doses or a combination of sodium butyrate and etoposide had clear effects on the viability of MCF-7 cells. Higher concentrations of HDACi are believed to exert cell cycle redistribution, induction of apoptosis, and downregulation of surviving signals; whereas lower, nontoxic doses of HDACi might not be strong enough to produce DSBs, but might allow gene activation that sustains cell survival in MCF-7 cells. Consistently, 24 h treatment of 2.0 mM sodium butyrate alone did not affect cell viability, with no detectable increase in the γ-H2AX foci in either HEK293 or cancer MCF-7 cells line; however, when combined with etoposide, 2.0 mM butyrate sensitized MCF-7 cancer cells but not HEK293 cells to produce more γ-H2AX foci. This chemotoxic synergy is likely due to both increased numbers of DSBs and hyperactivation of the cytotoxic arm of the DNA damage response.

HDACi's have been shown to directly downregulate homologous recombination and non-homologous end-joining in many cancer cell lines [17-20]. For example, in prostate cells, treatment with HDACi downregulates the protein expression levels of BRCA1, RAD51 and DNA-PK, and the mRNA expression levels of ATM,

BRCA1, BRCA2, RAD51 and XRCC4 [4,21]. Synchronized HeLa cells in G1 phase have decreased non-homologous end-joining in the presence of butyrate [17]. Additionally, HDACi's attenuate the upregulation of Ku70, Ku80 and DNA-PK induced by ionizing radiation in several cancer cells [22,23]. On the other hand, the effects of HDACi on non-homologous end-joining and homologous recombination proteins to date have not been observed in normal human cells.

Interactions between chromatin and DNA damage response proteins are central to the cellular response to DSBs. Assembly factors and repair at most DNA breaks in mammalian cells occurs by recruiting complexes from the nucleoplasm. Therefore, the protein expression of chromatin-bound and soluble nuclear compartments reflects an underlying state of DSB repair. We found that sodium butyrate differentially affects the nuclear expression of some DSB repair proteins in MCF-7 and HEK293 cells. The chromatin-bound protein of Acetyl-histone H4 was more highly induced by sodium butyrate in the presence of etoposide for MCF-7 cells than for HEK293 cells. Differences were also observed for Rad51 (a key component of homologous recombination pathway, which resides only in the soluble nuclear fraction), Ku80 (which serves as a key initiator component of the non-homologous end-joining pathway), CtIP (which initiates DSB end resection and generates 3' ended single strand DNA overhangs necessary for homologous recombination), and RPA70 (which is coated onto the overhangs of single stranded DNA created during resection). However, Rad52 (which facilitates upstream and downstream sequences of DSB annealing during signal -strand annealing pathway) did not consistently change with increasing doses of sodium butyrate. These results indicate that the DSB repair pathway induced by etoposide is suppressed after butyrate pretreatment in both cell lines, but that the effect is more dramatic for cancer MCF-7 cells. These results suggest that sodium butyrate may enhance cancerous cell MCF-7 killing by inhibiting recruitment of repair factors to damaged DNA and reducing its repair capacity.

We have compared two different cell lines, one cancerous and one non-cancerous; however, we acknowledge that these cell lines also have differences in the species and tissue derivation. Therefore, further work is needed to determine whether the differences in response that we have observed in this study may extend to other cancerous and non-cancerous cell pairs. Undoubtedly, there is a range of response to sodium butyrate that is likely to vary according to the cancer type and stage of malignancy. However, our results provide important mechanistic information about how sodium butyrate may function differentially in different cells, which may explain the therapeutic efficacy and low toxicity of

HDACi's in selectively killing cancer cells [3-5]. Practically, the combination of sodium butyrate may enhance the efficacy of etoposide and permit lower concentrations. Additionally, the understanding of how these two drugs function synergistically may facilitate the development of future therapeutics to selectively treat cancer.

## Conclusions

In summary, our findings indicate that the cytotoxic effects of the HDACi sodium butyrate occur, in part, via down-regulation of DSB repair protein accessibility to the nucleus and the sites of damage, with the outcome of affecting the repair capacity. Differences in the chromatin compaction in HEK293 cells and cancer MCF-7 cells in response to HDACi treatment might determine distinct fates between survival and death by controlling the DSB repair pathway. Consequently, sodium butyrate sensitizes cancerous MCF-7 more highly than HEK293 cells to the cytotoxic effect of DNA damage agents. Further work is needed to elucidate the precise pathways and targets by which HDACi's exert these chemosensitizing effects in cancer cells.

## Abbreviations

HEK293: Human embryonic kidney 293; HDACi: Histone deacetylase inhibitor; HDAC: Histone deacetylase; γ-H2AX: Histone variant H2A phosphorylated at serine 139; DSBs: Double strand breaks; DNA-PK: DNA-dependent Protein Kinase; CtIP: C-terminal binding protein-interacting protein; ATM: Ataxia telangiectasia mutated; RPA: Replication protein A; BRCA1: Breast cancer 1; BRCA2: Breast cancer 2; XRCC4: X-ray repair cross complementing4; Topo II: Topoisomerase II; CCK-8: Cell Counting Kit; TBST: TBS containing 0.1% Tween20.

## Competing interests

The authors declare that they have no competing interests.

## Authors' contributions

LL, YS, and JL participated in the study design. YS, XW, and JL performed experiments. LL and LM drafted the manuscript and figures, LL, XW, PW and LC participated in manuscript revision. All authors read and approved the final manuscript.

## Acknowledgements

This work was supported by grants from the National Natural Science Foundation of China (31271436), the Science and Technology Planning Project of Guangdong Province, China (2010B060900109), the Natural Science Foundation of Guangdong Province, China (S2012010008225), and the Administration of Traditional Chinese Medicine of Guangdong Province, China (1050051).

## References

1.  Vashishta A, Hetman M: Inhibitors of histone deacetylases enhance neurotoxicity of DNA damage. *Neuromolecular Med* 2014,Epub ahead of print
2.  Iwahashi S, Utsunomiya T, Imura S, Morine Y, Ikemoto T, Arakawa Y, et al. Effects of valproic acid in combination with S-1 on advanced

pancreatobiliary tract cancers: clinical study phases I/II. Anticancer Res. 2014;34(9):5187–91.

3. Khabele D. The therapeutic potential of class I selective histone deacetylase inhibitors in ovarian cancer. Front Oncol. 2014;4:111.

4. Chao OS, Goodman OB Jr: Synergistic loss of prostate cancer cell viability by co-inhibition of HDAC and PARP. *Mol Cancer Res* 2014, Epub ahead of print

5. Lee JH, Choy ML, Ngo L, Foster SS, Marks PA. Histone deacetylase inhibitor induces DNA damage, which normal but not transformed cells can repair. Proc Natl Acad Sci USA. 2010;107(33):14639–44.

6. Thompson LH. Recognition, signaling, and repair of DNA double-strand breaks produced by ionizing radiation in mammalian cells: The molecular choreography. Mutation Research. 2012;751(2):158–246.

7. Dong Q, Sharma S, Liu H, Chen L, Gu B, Sun X, et al. HDAC inhibitors reverse acquired radio resistance of KYSE-150R esophageal carcinoma cells by modulating Bmi-1 expression. Toxicol Lett. 2014;224(1):121–9.

8. Li L, Wang H, Yang ES, Arteaga CL, Xia F. Erlotinib attenuates homology-directed recombinational repair of chromosomal breaks in human breast cancer cells. Cancer Res. 2008;68(22):9141–6.

9. Marchion DC, Bicaku E, Daud AI, Sullivan DM, Munster PN. In vivo synergy between topoisomerase II and histone deacetylase inhibitors: predictive correlates. Mol Cancer Ther. 2005;4(12):1993–2000.

10. Chao H, Wang L, Hao J, Ni J, Chang L, Graham PH, et al. Low dose histone deacetylase inhibitor, LBH589, potentiates anticancer effect of docetaxel in epithelial ovarian cancer via PI3K/Akt pathway in vitro. Cancer Lett. 2013;329(1):17–26.

11. Thurn KT, Thomas S, Raha P, Qureshi I, Munster PN. Histone deacetylase regulation of ATM-mediated DNA damage signaling. Mol Cancer Ther. 2013;12(10):2078–87.

12. Gabrielli B, Brown M. Histone deacetylase inhibitors disrupt the mitotic spindle assembly checkpoint by targeting histone and nonhistone proteins. Adv Cancer Res. 2012;116:1–37.

13. Koprinarova M, Markovska P, Iliev I, Anachkova B, Russev G. Sodium butyrate enhances the cytotoxic effect of cisplatin by abrogating the cisplatin imposed cell cycle arrest. BMC Mol Bio. 2010;11:49.

14. Haberland M, Johnson A, Mokalled MH, Montgomery RL, Olson EN. Genetic dissection of histone deacetylase requirement in tumor cells. Proc Natl Acad Sci USA. 2009;106(19):7751–5.

15. Miller K, Tjeertes J, Coates J, Legube G, Polo S, Britton S, et al. Human HDAC1 and HDAC2 function in the DNA-damage response to promote DNA nonhomologous end-joining. Nat Struct Mol Biol. 2010;17(9):1144–51.

16. Conti C, Leo E, Eichler GS, Sordet O, Martin MM, Fan A, et al. Inhibition of histone deacetylses in cancer cells slows down replication forks, activates dormant origins, and induces DNA damage. Cancer Res. 2010;70(11):4470–80.

17. Koprinarova M, Botev P, Russev G. Histone deacetylase inhibitor sodium butyrate enhances cellular radiosensitivity by inhibiting both DNA nonhomologous end joining and homologous recombination. DNA Repair (Amst). 2011;10(9):970–7.

18. Xiao W, Graham PH, Hao J, Chang L, Ni J, Power CA, et al. Combination therapy with the histone deacetylase inhibitor LBH589 and radiation is an effective regimen for prostate cancer cells. PLoS One. 2013;8(8):e74253.

19. Tang J, Cho NW, Cui G, Manion EM, Shanbhag NM, Botuyan MV, et al. Acetylation limits 53BP1 association with damaged chromatin to promote homologous recombination. Nat Struct Mol Biol. 2013;20(3):317–25.

20. Ladd B, Ackroyd JJ, Hicks JK, Canman CE, Flanagan SA, Shewach DS. Inhibition of homologous recombination with vorinostat synergistically enhances ganciclovir cytotoxicity. DNA Repair (Amst). 2013;12(12):1114–21.

21. Kachhap SK, Rosmus N, Collis SJ, Kortenhorst MS, Wissing MD, Hedayati M, et al. Downregulation of homologous recombination DNA repair genes by HDAC inhibition in prostate cancer is mediated through the E2F1 transcription factor. Plos One. 2010;5(6):e11208.

22. Blattmann C, Oertel S, Ehemann V, Thiemann M, Huber PE, Bischof M, et al. Enhancement of radiation response in osteosarcoma and rhabdomyosarcoma cell lines by histone deacetylase inhibition. Int J Radiat Oncol Biol Phys. 2010;78(1):237–45.

23. Kuribayashi T, Ohara M, Sora S, Kubota N. Scriptaid, a novel histone deacetylase inhibitor, enhances the response of human tumor cells to radiation. Int J Mol Med. 2010;25(1):25–9.

# The laforin/malin E3-ubiquitin ligase complex ubiquitinates pyruvate kinase M1/M2

Rosa Viana, Pablo Lujan and Pascual Sanz[*]

## Abstract

**Background:** Lafora disease (LD, OMIM 254780) is a fatal neurodegenerative disorder produced mainly by mutations in two genes: *EPM2A*, encoding the dual specificity phosphatase laforin, and *EPM2B*, encoding the E3-ubiquitin ligase malin. Although it is known that laforin and malin may form a functional complex, the underlying molecular mechanisms of this pathology are still far from being understood.

**Methods:** In order to gain information about the substrates of the laforin/malin complex, we have carried out a yeast substrate-trapping screening, originally designed to identify substrates of protein tyrosine phosphatases.

**Results:** Our results identify the two muscular isoforms of pyruvate kinase (PKM1 and PKM2) as novel interaction partners of laforin.

**Conclusions:** We present evidence indicating that the laforin/malin complex is able to interact with and ubiquitinate both PKM1 and PKM2. This post-translational modification, although it does not affect the catalytic activity of PKM1, it impairs the nuclear localization of PKM2.

**Keywords:** Laforin, Malin, Ubiquitination, Pyruvate kinase, Nuclear localization

## Background

Lafora disease (LD, OMIM 254780) is a rare fatal type of progressive myoclonus epilepsy that initially manifests during adolescence. Recessive mutations in two main loci, *EPM2A* or *EPM2B*, are known to produce LD ([1–3]). The *EPM2A* gene encodes the dual-specificity phosphatase laforin and the *EPM2B* gene encodes the E3-ubiquitin ligase malin. Although it is known that both proteins form a functional complex ([4–6]), the underlying molecular mechanisms of this pathology are still far from being understood. For this reason the search for substrates of the laforin/malin complex has been a good strategy to understand the molecular basis of this pathology. Different techniques have been used with this aim, from regular yeast two-hybrid screening, using either laforin or malin as bait, to co-immunoprecipitation/pull down analyses and functional interactions. All these approaches have identified several putative substrates of the laforin/malin complex that have enriched our knowledge on the different cellular pathways where laforin and malin may participate (see [7] for a review).

In order to identify novel phospho-protein substrates of laforin, we have used the yeast substrate-trapping system described by Fukada and Noda [8], designed to identify substrates of protein tyrosine phosphatases. This technique is based on the use of phosphatase mutants which harbor a substitution in invariant amino acid residues in the catalytic domain, making them catalytically inactive, but retaining the capability to recognize the phosphorylated substrate. In addition, the yeast model co-expresses a wide spectrum tyrosine kinase (i.e., v-src), which will phosphorylate putative target substrates. In this way, the possibilities to identify putative phospho-tyrosine containing substrates of a particular phosphatase are enhanced. Here, we have identified the two muscular isoforms of pyruvate kinase (PKM1 and PKM2) as novel interaction partners of laforin. Pyruvate kinase regulates the rate-limiting final step of glycolysis, which catalyzes the transfer of a high-energy phosphate group from phosphoenolpyruvate (PEP) to adenine diphosphate (ADP), yielding one molecule of pyruvate and one molecule of adenosine triphosphate (ATP) (see [9] for review). Four

* Correspondence: sanz@ibv.csic.es

Instituto de Biomedicina de Valencia, CSIC, and Centro de Investigación Biomédica en Red de Enfermedades Raras (CIBERER), Jaime Roig 11, 46010 Valencia, Spain

isoforms of pyruvate kinase have been described in mammals: PKR, present mainly in erythrocytes; PKL, present in liver, kidney, small intestine and pancreas; PKM1, present in muscle, heart and brain; and PKM2, widely distributed and also present in fetal tissues and cancer cells ([9, 10]). PKM1 and PKM2 result from an alternative splicing of two mutually exclusive exons of the *PKM* gene (exon 9 is present in PKM1 whereas exon 10 is present in PKM2). This difference in exon usage results in proteins that differ only in 22 amino acids distributed along a small region of 56 amino acids [from residue 378 to 434 [11], (Fig. 1a, black box)]. These changes confer distinct kinetic and regulatory properties to PKM1 and PKM2: although both PKM1 and PKM2 are proteins of 531 aa, with a similar molecular weight (58 kDa) and similar structural domains ([9, 10]) (Fig. 1a), only PKM1 is fully

active whereas PKM2 may form inactive dimeric complexes ([9, 10]).

In this work we present evidence of the interaction between laforin and both PKM1 and PKM2. As a result of this interaction, we show that the laforin/malin complex ubiquitinates PKM1 and PKM2 and that this post-translational modification, although it does not affect the catalytic activity of PKM1, it impairs the nuclear localization of PKM2.

## Results

### The muscular isoform of Pyruvate Kinase (PKM) interacts with laforin

As indicated at the Background section, in this work we have used the yeast substrate-trapping system [8], designed to identify substrates of protein tyrosine phosphatases, to

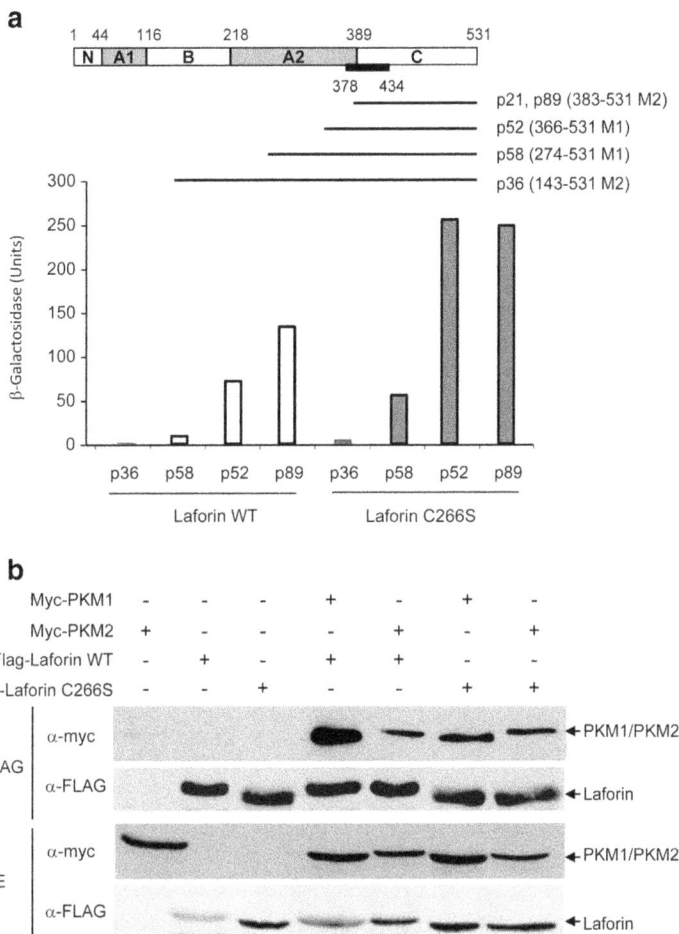

**Fig. 1** PKM1 and PKM2 interact with laforin. **a** Yeast substrate-trapping system with LexA-laforin C266S was used to identify putative partners from a human brain library. Five clones corresponding to different lengths of PKM1 and PKM2 were selected (in brackets the amino acids included in the clones corresponding either to PKM1 of PKM2). A diagram of the different domains in PKMs is presented (black bar indicates the region that is different between PKM1 and PKM2). The strength of the interaction of these clones with either LexA-laforin C266S or LexA-laforin wild type was measured (β-galactosidase activity). **b** Co-immunoprecipitation analysis of PKM1 and PKM2 with laforin. HEK293 cells were transfected with the indicated plasmids. Crude extracts and co-immunoprecipitation analyses with anti-FLAG antibody was performed as indicated in Methods. Proteins in the immunoprecipitates and in the crude extracts (CE, 25 μg) were analyzed by western blotting using the indicated antibodies

screen for possible phospho-protein substrates of laforin. Following the conditions required for this type of screening, we initially constructed a catalytically inactive laforin mutant, Laf-C266S, which affects the key Cys residue in the catalytic site [7]. This mutant retained the capability to recognize laforin binding partners (i.e. malin, R5/PTG) (not shown). A yeast strain co-expressing LexA-Laforin C266S and the protein tyrosine kinase v-src was transformed with a human brain cDNA library, and after screening more than 300,000 independent transformants, we recovered 97 putative positive clones. Unfortunately, 58 of them were false positives since when we purified the corresponding plasmids they were not able to reproduce the two-hybrid interaction. Among the other 39 positive clones, we identified already known laforin binding partners, such as laforin itself and R5/PTG, what validated the technique (Table 1). The rest of the clones interacted with laforin with different strength (in Table 1 we describe only the 17 clones that presented the strongest interaction). Among them, we focused our attention to those that coded for the two muscular isoforms of pyruvate kinase (PKM1 and PKM2), as we recovered five PKM-related clones which encoded for different fragments of both PKM1 and PKM2 (Fig. 1a). In these clones the interaction with laforin was much stronger when only the C-terminal part of PKMs was present in the clones [clones p21 and p89 (aa 383–531 of PKM2) and p52 (aa 366–531 of PKM1)], and decreased when larger parts of the protein were present in the clones [clone p58 (aa 274–531 of PKM1), clone p36 (aa 143–531 of PKM2)] (Fig. 1a). We repeated the two-hybrid interaction of these clones using LexA-Laforin WT as bait and found that the interaction was maintained although at lower intensity (Fig. 1a). We also noticed that all these interactions were maintained under growth conditions that prevented the expression of v-src, indicating that, contrary to the experimental design, they were not dependent on the phosphorylation of the substrate by this protein kinase.

In order to confirm the interaction between laforin and the PKM1 and PKM2 isoforms, we carried out a co-immunoprecipitation analysis in mammalian cells. First, we obtained by PCR the full length cDNAs of PKM1 and PKM2 from the human brain cDNA library and subcloned them into plasmid pCMV-myc, to allow their expression in mammalian HEK293 cells. We observed that the expression of the PKM1 isoform yielded a protein with a slightly faster mobility in SDS-PAGE, probably due to differences in protein sequence in comparison to PKM2 (Fig. 1b. third panel). We co-expressed in HEK293 cells the PKM1 and PKM2 isoforms and laforin (tagged with a FLAG epitope) either wild type or C266S, immunoprecipitated the crude extracts with anti-FLAG antibody and analyzed the immunoprecipitates by western blotting using anti-FLAG and anti-myc antibodies. As shown in Fig. 1b (top panel) both laforin-WT and laforin-C266S were able to co-immunoprecipitate both myc-PKM1 and myc-PKM2 isoforms. Taken together, these results indicate that laforin interacts with both PKM1 and PKM2, and that the phosphatase activity of laforin is dispensable for the interaction with the PKMs isoforms. We tried to repeat the co-immunoprecipitation experiment with endogenous levels of proteins but unfortunately we did not find any reliable commercial antibody which recognized endogenous levels of laforin.

## The laforin/malin complex polyubiquitinates PKM isoforms, attaching K63-linked ubiquitins

It is known that laforin forms a functional complex with the E3-ubiquitin ligase malin, in which laforin recognizes the substrates that are eventually ubiquitinated by malin. In order to check if the interaction between PKM1 and PKM2 and laforin resulted in the ubiquitination of the PKM isoforms, we co-expressed in HEK293 cells myc-PKM1 or myc-PKM2, laforin, malin and a 6xHis-tagged ubiquitin. The 6xHis tag present in the ubiquitin allows the recovery of polyubiquitinated forms by metal affinity

**Table 1** Positive clones identified in substrate-trap screening of a human brain cDNA library using LexA-laforin C266S as bait. Yeast THY-AP4 cells transformed with plasmid pBridge-laforin C266S were co-transformed with a human brain cDNA library in pACT2

| No. of positive clones | Encoded protein | Uniprot number | Intensity of interaction |
|---|---|---|---|
| 5 | Pyruvate kinase PKM1/PKM2 | Q504U3 | From ++ to +++ |
| 3 | E3 ubiquitin ligase TOPORS | Q9NS56 | ++ |
| 3 | Cytochrome C oxidase subunit 5A | P20674 | ++ |
| 1 | Laforin (EPM2A) | O95278 | +++ |
| 1 | PP1 regulatory subunit R5/PTG (PPP1R3C) | Q9UQK1 | ++ |
| 1 | G/T mismatch-specific thymine DNA glycosylase | Q13569 | +++ |
| 1 | 26S proteasome regulatory subunit 6A | P17980 | ++ |
| 1 | HMG domain-containing protein 4 | Q9UGU5 | ++ |
| 1 | Translation initiation factor eIF-2B subunit alpha | Q14232 | ++ |

Positive clones were confirmed, subjected to BLAST analysis and protein interaction estimated by β-galactosidase filter lift assays (+++, when the blue color appeared in less than 5 min; ++, when the blue color appeared between 5 and 20 min)

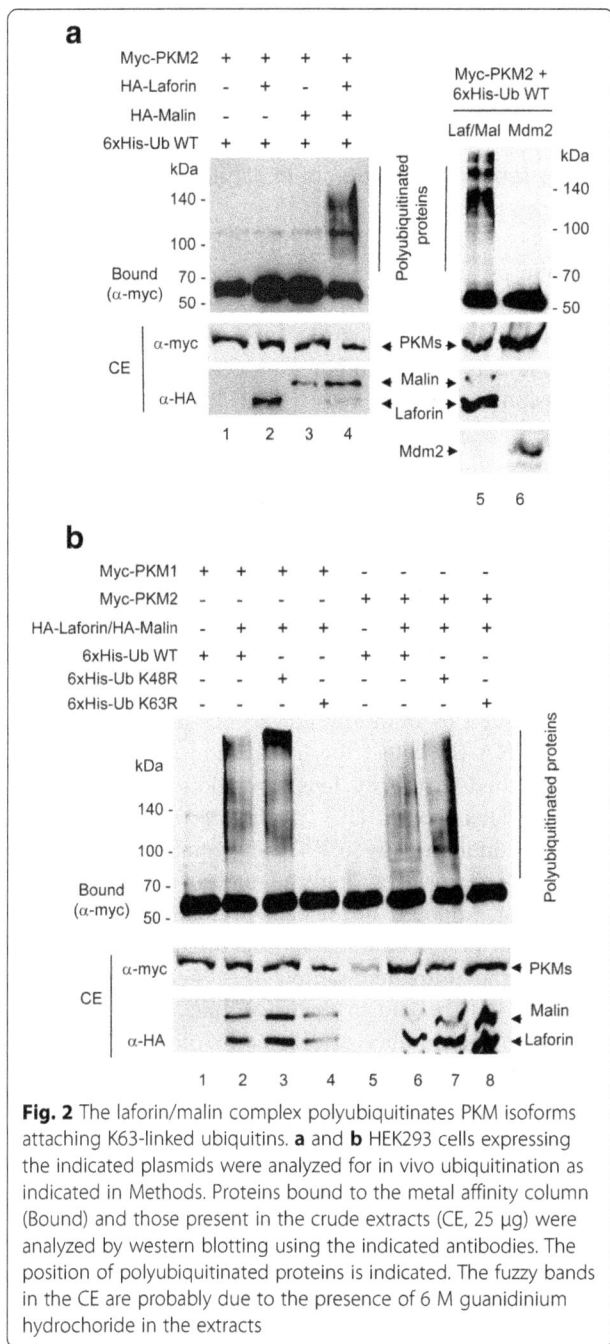

**Fig. 2** The laforin/malin complex polyubiquitinates PKM isoforms attaching K63-linked ubiquitins. **a** and **b** HEK293 cells expressing the indicated plasmids were analyzed for in vivo ubiquitination as indicated in Methods. Proteins bound to the metal affinity column (Bound) and those present in the crude extracts (CE, 25 μg) were analyzed by western blotting using the indicated antibodies. The position of polyubiquitinated proteins is indicated. The fuzzy bands in the CE are probably due to the presence of 6 M guanidinium hydrochloride in the extracts

modification when we co-expressed Mdm2, an unrelated E3-ubiquitin ligase [lane 6 in Fig. 2a for PKM2; similar results were obtained for PKM1 (not shown)].

In order to analyze the topology of the laforin/malin-induced polyubiquitinated chains in PKMs, we co-expressed in HEK293 cells mutated forms of ubiquitin that prevented the formation of ubiquitin chains in either K48 (K48R mutant) or K63 (K63R mutant) (K48-linked polyubiquitin chains signal proteins to be degraded by the proteasome, whereas K63-linked polyubiquitin chains play a role in cell signaling and other physiological processes). As shown in Fig. 2b, in the presence of K48R-ubiquitin, a similar pattern of polyubiquitination of PKM1 and PKM2 by the laforin/malin complex was obtained (lanes 3 and 7, respectively), whereas in the presence of K63R-ubiquitin, no polyubiquitinated forms were obtained (lanes 4 and 8, respectively). These results indicate that the laforin/malin complex promotes the polyubiquitination of PKM1 and PKM2 through the attachment of K63-linked ubiquitin moieties. This topology was similar to the one found in other substrates of the laforin/malin complex, such as R5/PTG and AMPK subunits [13], and suggests that the modified PKMs may be involved in cellular functions not related to proteasomal degradation.

### Consequences of the action of the laforin/malin complex on the biochemical properties of PKM1 and PKM2

We analyzed next the effect of the laforin/malin complex on the enzymatic activity of PKMs. With this aim, the pyruvate kinase activity present in HEK293 cells expressing PKM1 or PKM2 and co-expressing or not the laforin/malin complex was measured by standard methods. As expected, the overexpression of PKM1 increased the pyruvate kinase activity over endogenous levels (expressed as Units/mg total protein in the extract) (Fig. 3a, white bars). The activity in the extract was increased further if the cells co-expressed the laforin/malin complex (Fig. 3a, black bars) but the co-expression of laforin alone (Fig. 3a, grey bars) did not have this effect. On the contrary, the overexpression of PKM2 did not enhance the pyruvate kinase activity over endogenous levels (Fig. 3a) (what was in agreement with the reported reduced activity of this isoform [10]), and the laforin/malin complex had no effect on its activity (Fig. 3a). Although at first glance these results could suggest that the laforin/malin complex increased the intrinsic enzymatic activity of PKM1, we think that this is a misleading conclusion, since the steady-state levels of PKM1 were higher in cells expressing the laforin/malin complex (Fig. 3b, compare lane 4 with lane 6). So, we think that the increase in the total pyruvate kinase activity detected in cells co-expressing PKM1 and the laforin/malin complex is just a reflection of more amount of the PKM1 protein being present in the extracts. We also observed that the laforin/malin complex did not

chromatography in the presence of guanidinium hydrochloride, an inhibitor of the action of deubiquitinases [12]. As shown in Fig. 2a, the laforin/malin complex was able to promote the polyubiquitination of PKM2 (lane 4). This modification was dependent on the presence of the laforin/malin complex, since it was drastically reduced in its absence (lane 1) or when only laforin or malin were expressed in the cells (lanes 2 and 3 respectively). Similar results were obtained when PKM1 was assayed in the same way (not shown). This ubiquitination was specific of the laforin/malin complex since we did not observe this

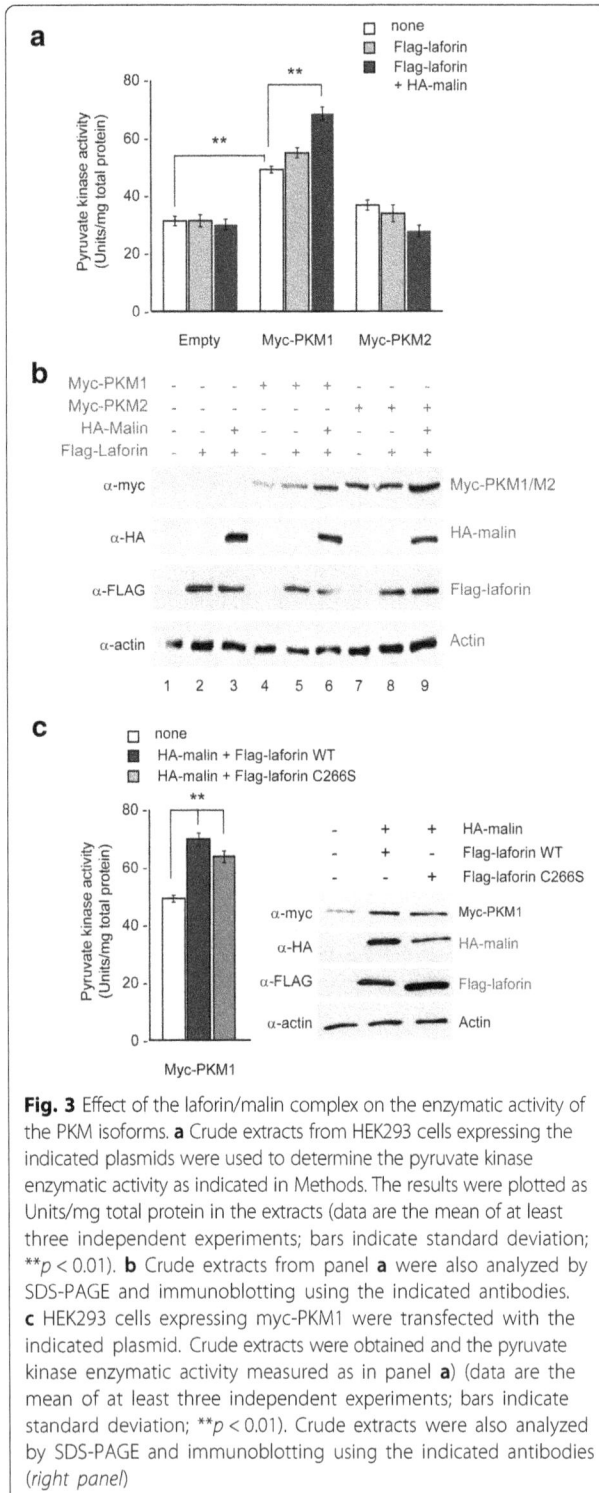

**Fig. 3** Effect of the laforin/malin complex on the enzymatic activity of the PKM isoforms. **a** Crude extracts from HEK293 cells expressing the indicated plasmids were used to determine the pyruvate kinase enzymatic activity as indicated in Methods. The results were plotted as Units/mg total protein in the extracts (data are the mean of at least three independent experiments; bars indicate standard deviation; **$p < 0.01$). **b** Crude extracts from panel **a** were also analyzed by SDS-PAGE and immunoblotting using the indicated antibodies. **c** HEK293 cells expressing myc-PKM1 were transfected with the indicated plasmid. Crude extracts were obtained and the pyruvate kinase enzymatic activity measured as in panel **a**) (data are the mean of at least three independent experiments; bars indicate standard deviation; **$p < 0.01$). Crude extracts were also analyzed by SDS-PAGE and immunoblotting using the indicated antibodies (*right panel*)

modify the total endogenous enzymatic activity of PKM isoforms present in Hek293 cells (transformed with an empty vector; Fig. 3a, empty).

We repeated the assay with cells expressing the catalytically inactive laforin C266S form. As shown in Fig. 3c, the

laforin C266S/malin complex was also able to increase the total enzymatic activity and the steady-state levels of PKM1 in the cells, indicating that this effect was not dependent on the phosphatase activity of laforin.

### Altered subcellular location of PKM2 due to the laforin/malin complex

It has been described that both PKM1 and PKM2 are cytosolic proteins, what is consistent with their role in the glycolytic pathway ([14, 15]). However, PKM2 but not PKM1 is also able to translocate to the nucleus, where it acts as a protein kinase and regulates gene expression in different ways ([16–18]). In order to check whether the ubiquitination of PKM2 by the laforin/malin complex could affect its translocation to the nucleus, we transfected human osteosarcoma U2OS cells [which are larger than HEK293 cells, so the different subcellular compartments (especially nucleus) are better observed] with myc-PKM2 in the presence or absence of FLAG-laforin and HA-malin. Transfected cells were treated or not with UV light (120 mJ/cm$^2$; 120 s) to induce the accumulation of PKM2 into the nucleus [19]. Cells were then analyzed by immunofluorescence using anti-PKM2 and anti-HA (to detect the localization of malin, a protein enriched in the nucleus) antibodies. In agreement with previous reports [19], treatment with UV light of the cells containing myc-PKM2 and transformed with an empty HA-vector promoted the nuclear accumulation of PKM2 (Fig. 4a); 60 % of the cells presented a nuclear accumulation of PKM2 after UV-treatment in contrast to 30 % in the untreated sample (Fig. 4c). However, the expression of the laforin/malin complex produced a reduction in the number of cells containing PKM2 at the nucleus in both UV treated and untreated cells [Fig. 4b; compare cells expressing malin (white arrows) with other cells that do not express malin]: in untreated cells, only 18 % of cells expressing the laforin/malin complex displayed a nuclear accumulation of PKM2, in contrast to 30 % in cells containing an empty plasmid, and in UV-treated cells, only 47 % of cells expressing the laforin/malin complex had a nuclear PKM2, in contrast to 60 % in cells containing an empty plasmid (Fig. 4c).

Taken all these results together, we suggest that the laforin/malin dependent ubiquitination of PKM2 partially impairs its nuclear localization.

### Discussion

In this work we describe additional substrates of the laforin/malin complex, namely the muscular isoforms of pyruvate kinase (PKM1 and PKM2). We present evidence of the interaction between laforin and both PKM1 and PKM2, and as a result of this interaction, we show that the laforin/malin complex ubiquitinates both PKM1 and PKM2 by attaching K63-linked polyubiquitin chains.

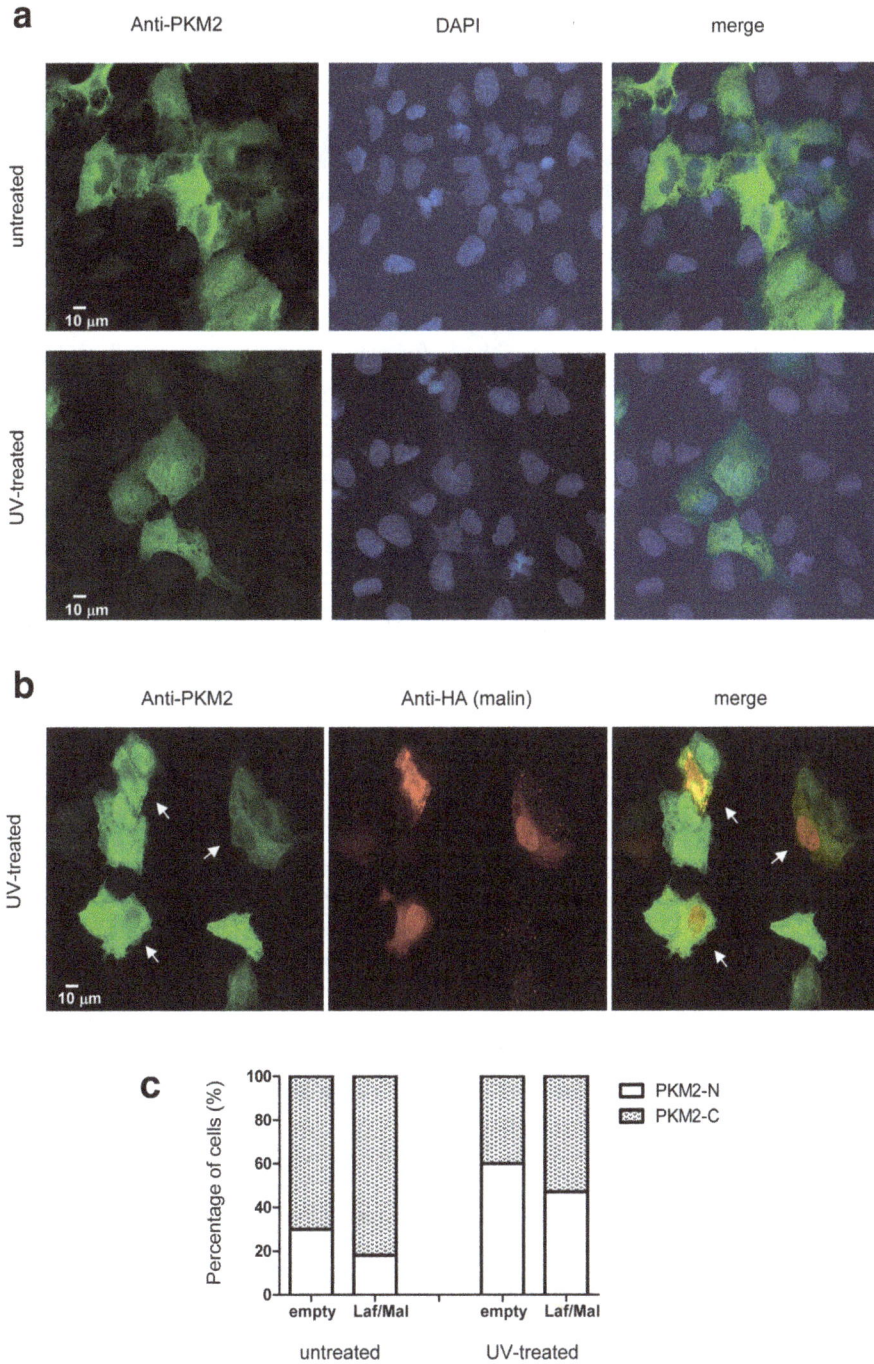

**Fig. 4** The laforin/malin complex alters the subcellular localization of PKM2. **a** U2OS cells were transfected with plasmid pCMV-myc-PKM2. Cells were then treated or not with UV light (120 mJ/cm$^2$; 120 s) to induce the accumulation of PKM2 into the nucleus. The localization of PKM2 was determined by immunofluorescence using anti-PKM2 antibodies (see Methods). DAPI staining was also performed to define the position of the nuclei. A merge image is presented. Scale bars are indicated. **b** U2OS cells expressing plasmid pCMV-myc-PKM2 were co-transfected with plasmids pFLAG-laforin and pCMV-HA-malin. Then, cells were treated of not with UV light as above. Panel shows the result in UV-treated cells. The localization of malin, a nuclear protein, and PKM2 was determined by immunofluorescence using anti-HA and anti-PKM2 antibodies, respectively. **c** The percentage of cells having a nuclear localization of PKM2 was determined in 100 cells from all the conditions. PKM-N: cells that present PKM2 inside the nucleus; PKM-C: cells that present PKM2 outside of the nucleus

These results are in agreement with recent data from the GGBase indicating that PKM2 was ubiquitinated at different sites (https://gygi.med.harvard.edu/ggbase/) and indicate for the first time that in the ubiquitination reaction participates, at least, the laforin/malin E3-ubiquitin ligase complex. Therefore, ubiquitination of PKMs should be added to the list of post-translational modifications that suffer these proteins [14]. The laforin/malin dependent ubiquitination of PKM1 and PKM2 did not modify their corresponding catalytic activities. However, we noticed that this post-translational modification increased the steady-state levels of the proteins, probably because the modified forms were more resistant to their proteolytic turnover. In this way, the effect of the laforin/malin dependent modification of PKM1 and PKM2 resembles that of AMPKβ subunits (other substrates of the laforin/malin complex) since also higher steady-state levels of these subunits were observed upon the action of the laforin/malin complex [13]. It is worth to point out that stabilization of PKM1 and PKM2 levels also occurs when a phosphatase inactive form of laforin (laforin C266S) is expressed, indicating that the phosphatase activity of laforin may not be required for inducing the malin-mediated polyubiquitination of PKM1 and PKM2, as already demonstrated in other laforin/malin substrates, such as R5/PTG or R6, two regulatory subunits of protein phosphatase type1 ([6, 20]).

We next studied the effect of the laforin/malin dependent polyubiquitination on the subcellular distribution of PKMs. It has been described that PKM2 but not PKM1 is able to translocate to the nucleus, where it displays specific functions ([14, 15]). This is probably due to the presence of a nuclear localization signal (NLS) in the specific region (aa 378 to 434) of PKM2, not present in PKM1. This NLS allows PKM2 to interact with members of the importin family, who mediate in the translocation of PKM2 to the nucleus [16]. Recently, it was reported that PKM2 interacted with the E3-SUMO ligase PIAS3 and this resulted in the sumoylation of PKM2 and in an enhancement of its nuclear localization [21]. On the contrary, our results suggest that the laforin/malin dependent ubiquitination of PKM2 interferes with the localization of PKM2 to the nucleus. Perhaps ubiquitination of PKM2 affects its recognition by the importin machinery, resulting in an impairment of nuclear translocation. Alternatively, ubiquitination of PKM2 could interfere with its sumoylation since both processes affect similar Lys residues and, in some cases, the introduction of one modification (i.e. ubiquitination) prevents the alternative post-translational alteration by the other (sumoylation) ([22, 23]). This negative relationship between ubiquitination and sumoylation might affect nuclear translocation of PKM2, as it has already been documented in other cases [24].

## Conclusions

We present evidence indicating that the laforin/malin complex is able to interact with and ubiquitinate both muscular isoforms of pyruvate kinase (PKM1 and PKM2). This post-translational modification, although it does not affect the catalytic activity of PKM1, it impairs the nuclear localization of PKM2.

## Methods

### Microorganisms, culture conditions and genetic methods

*Escherichia coli* DH5α strain was used as the host strain for plasmid constructions and protein production. It was grown in LB (1 % peptone, 0.5 % yeast extract, 1 % NaCl, pH 7.5) medium supplemented with 50 mg/l ampicillin. Yeast strain used in this work was THY-AP4 (*MATa, ura3, leu2, lexA::lacZ::trp1, lexA::HIS3, lexA::ADE2*). Yeast transformation was carried out using the lithium acetate protocol [25]. Yeast cultures were grown in synthetic complete (SC) medium lacking the corresponding supplements to maintain selection for plasmids [26].

### Plasmids

pCMV-HA-laforin, pFLAG-laforin, pFLAG-laforin C266S and pcDNA3-HA-malin plasmids were described previously [6]. Plasmids pBridge-laforin and pBridge-laforin C266S were obtained by subcloningBamHI/SalI fragments containing either laforin WT or C266S in pBridgeLexA/v-src [8]. Plasmids pCMV-myc-PKM1 and pCMV-myc-PKM2 were obtained by subcloning an EcoRI/XhoI fragment containing the corresponding ORF in pCMV-myc vector. Other plasmids used in this study were pCMV-6xHisUbiq (from Dr. M. Rodriguez, Proteomics Unit, CIC-BioGUNE, Vizcaya, Spain); pCMV-6xHisUbiq-K48R and pCMV-6xHisUbiq-K63R (from Dr. Ch. Blattner, Institute of Toxicology and Genetics, Karlsruhe Institute of Technology, Karlsruhe, Germany); pCMV-Mdm2 (from Dr. M. Gentry, University of Kentucky, Lexington, KY; [27]).

### Two-hybrid screening and analysis

A modified form of yeast two-hybrid screening named yeast substrate-trapping system [8] was used to identify proteins that interacted with LexA-laforin C266S (plasmid pBridge-laforin C266S). Plasmid pBridge-laforin C266S also contains the ORF corresponding to the mammalian protein kinase v-src under the control of the MET25 promoter. *Saccharomyces cerevisiae* THY-AP4 strain containing pBridge-laforin C266S plasmid was transformed with a commercial human brain cDNA library in pACT2 vector (Clontech, Madrid, Spain). Transformants were selected in SC + 2 % glucose plates lacking tryptophan, leucine, methionine and histidine and were subsequently screened for β-galactosidase activity using a filter lift assay [28]. pACT2-plasmids were recovered from positive clones and used to transform again *S.*

cerevisiae THY-AP4 strain containing pBridge-laforin C266S or laforin WT, to confirm the interaction. The strength of the interaction was determined by measuring β-galactosidase activity in permeabilized yeast cells and expressed in Miller units as described by Ludin and collaborators [29].

### Cell culture, transfection and preparation of crude extracts

Human embryonic kidney (HEK293) and human osteosarcoma (U2OS) cells were grown in Dulbecco's modified Eagle's medium (Lonza, Barcelona, Spain), supplemented with 100 units/ml penicillin, 100 μg/ml streptomycin, 2 mM glutamine and 10 % of inactivated fetal bovine serum (Lonza, Barcelona, Spain) in a humidified atmosphere at 37 °C with 5 % $CO_2$. When indicated, cells were transfected with 1 μg of each plasmid using X-treme GENE HP transfection reagent (Roche Diagnostics, Barcelona, Spain) according to the manufacturer's instructions. Cell extracts were prepared using lysis buffer A [25 mMTrisHCl at pH 7.4, 15 mM EDTA at pH 8, 50 mMNaF, 0.6 M sucrose, 15 mM $Na_4P_2O_7$, 1 % nonidet P40, 10 mMNaCl, 1 mM PMSF, and a complete protease inhibitor mixture (Roche Diagnostics, Barcelona, Spain)]. Cells were lysed by repeated passage through 24G × 5/8″ needle and whole lysates were centrifuged at 10,000 × g for 15 min. The supernatants were collected and 25 μg of total protein subjected to SDS-PAGE, transferred into PVDF membrane and revealed with the appropriated antibodies.

### Co-immunoprecipitation and western blotting

Mammalian HEK293 cells were transfected with the corresponding plasmids. Twenty-four hours after transfection, cells were harvested with lysis buffer B [50 mMTrisHCl pH 7.5, 10 mMNaCl, 2 mM EDTA, 15 % glycerol, 1 % nonidet P40, complete protease inhibitor cocktail (Roche Diagnostics, Barcelona, Spain) and 1 mM PMSF]. Cells were lysed by repeated passage through 24G×5/8″ needle, whole lysates were centrifuged at 10,000 × g for 15 min and the soluble fraction was collected for immunoprecipitation. Co-immunoprecipitation experiments were carried out with 500 μg of soluble protein extracts, in a final volume of 500 μl of lysis buffer, adding 1 μl of anti-FLAG monoclonal antibodies (Sigma, Madrid, Spain) as in [30]. For western blot analysis, proteins (25 μg) were denatured using sample buffer (125 mMTrisHCl pH 6.8, 20 % Glicerol, 0.01 % bromophenol blue, 4 % SDS) and heating to95 °C for 5 min. The samples were subjected to SDS-PAGE and transferred onto PVDF membranes (Millipore, Madrid, Spain). Membranes were blocked with 5 % milk in TBS-tween for 1 h and incubated with the following specific antibodies: anti-myc, anti-FLAG, anti-HA and anti-actin from Sigma (Madrid,

Spain); anti-Mdm2 from Santa Cruz Biotechnology (Madrid, Spain). Thereafter, blots were washed with TBS-tween and further incubated for 1 h with the corresponding secondary antibody conjugated with horseradish peroxidase. Finally, membranes were washed (3× 15 min) with TBS-tween and analyzed by chemiluminiscence (ECL Western Blotting Detection Reagents, GE Healthcare, UK) using an image reader LAS-4000 (GE Healthcare, UK).

### Analysis of ubiquitination

For ubiquitination assays, HEK293 cells were co-transfected with pCMV-myc-PKM1 or pCMV-myc-PKM2, 6xHis-tagged ubiquitin (WT, K48R or K63R) plasmids and, when indicated, with pcDNA3-HA-malin and pCMV-HA-laforin, or pCMV-Mdm2 plasmids, using X-treme GENE transfection reagent, according to the manufacturer's instructions (Roche Diagnostics, Barcelona, Spain). After 18 h of transfection, cells were lysed in guanidinium hydrochloride to inhibit the action of deubiquitinases and ubiquitinated proteins purified by metal affinity chromatography [12]. Bound proteins and clarified extracts were analyzed by immunoblotting with the appropriated antibodies.

### Determination of pyruvate kinase enzymatic activity

HEK293 cells transfected with the indicated plasmids were resuspended in lysis buffer C [50 mMTrisHClpH 7.5, 1 mM EDTA, 150 mMNaCl, 1%NP-40, 1 mM DTT, 10 μM fructose 1,6 bisphosphateand complete protease inhibitor mixture (Roche Diagnostics, Barcelona, Spain)]. Cells were lysed by repeated passage through 24G × 5/8″ needle and whole lysates were centrifuged at 10,000 × g for 15 min. Clarified extracts were used to determine the pyruvate kinase activity by standard methods (reaction buffer: 40 mM $K_2HPO_4$ pH 7.6, 0.58 mM phosphoenolpyruvate, 0.11 mM NADH, 6.8 mM $MgSO_4$, 1.5 mM ADP, 10 units lactate dehydrogenase, 10 μM fructose 1,6 bisphosphate). The disappearance of NADH absorbance was measured spectrophotometrically at 340 nm. One unit of enzymatic activity is defined as the amount of enzyme that is able to oxidize 1 μmol of NADH per 1 min at 25 °C. Enzymatic activity was normalized by the total amount of protein in the sample.

### Immunofluorescence analysis

Human osteosarcoma U2OS cells transfected with the indicated plasmids were grown on 12-well plates containing coverslips. Cells were fixed with 4 % paraformaldehyde in phosphate-buffered saline (PBS) for 10 min. Then, cells were permeabilized with 0.2 % Triton X-100 in PBS for 15 min, blocked 1 h with 10 % fetal bovine serum, 5 % nonfat dried milk, 0.5 % BSA and 0.1 % Triton X-100 in PBS, and incubated overnight at 4 °C in the same buffer containing anti-HA and anti-PKM2 antibodies. Samples were washed three timeswith PBS and incubated with a 1/500 dilution of anti-mouse Texas Red and anti-rabbit Alexa-Fluor 488 (Invitrogen, Madrid, Spain). Then, cellswere washed three times with PBS and mounted on slices using Aqua-Poly/Mount coverslipping medium (Polysciences, Inc. Eppelheim, Germany). Images were acquired with an uprightLeica DM RXA2microscope using an PL APO 63× oil 1.4 N.A. immersion objective, and the Leica IM50 software. At least 100 cells were counted in each condition.

## Statistical data analysis

Data are expressed as means with standard deviation. Statistical significance of differences between the groups was evaluated by a paired Student's $t$-test with two-tailed distribution. The significance has been considered at ** $p < 0.01$, as indicated.

### Abbreviations

AMPK: AMP-activated protein kinase; PKM: Muscular isoform of pyruvate kinase; R5/PTG: Protein phosphatase 1 glycogen targeting subunit; PVDF: Polyvinylidene difluoride membranes; SC: Synthetic complete medium; SDS-PAGE: Sodium dodecylsulfate polyacrylamide gel electrophoresis.

### Competing interests

The authors declare that they have no competing interests.

### Authors' contributions

RV and PL carried out the experiments described in this work. RV and PS analyzed the data. PS wrote the manuscript. All authors read and approved the final manuscript.

### Acknowledgements

We want to thank Dr. Noda, Dr. Rodriguez, Dr. Blattner and Dr. Gentry for materials. This work has been supported by a grant from the Spanish Ministry of Education and Science (SAF2011-27442) and a grant from Generalitat Valenciana (PrometeoII/2014/029).

### References

1. Minassian BA, Lee JR, Herbrick JA, Huizenga J, Soder S, Mungall AJ, et al. Mutations in a gene encoding a novel protein tyrosine phosphatase cause progressive myoclonus epilepsy. Nat Genet. 1998;20(2):171–4.
2. Serratosa JM, Gomez-Garre P, Gallardo ME, Anta B, de Bernabe DB, Lindhout D, et al. A novel protein tyrosine phosphatase gene is mutated in progressive myoclonus epilepsy of the Lafora type (EPM2). Hum Mol Genet. 1999;8(2):345–52.
3. Chan EM, Young EJ, Ianzano L, Munteanu I, Zhao X, Christopoulos CC, et al. Mutations in NHLRC1 cause progressive myoclonus epilepsy. Nat Genet. 2003;35(2):125–7.
4. Lohi H, Ianzano L, Zhao XC, Chan EM, Turnbull J, Scherer SW, et al. Novel glycogen synthase kinase 3 and ubiquitination pathways in progressive myoclonus epilepsy. Hum Mol Genet. 2005;14(18):2727–36.
5. Vilchez D, Ros S, Cifuentes D, Pujadas L, Valles J, Garcia-Fojeda B, et al. Mechanism suppressing glycogen synthesis in neurons and its demise in progressive myoclonus epilepsy. Nat Neurosci. 2007;10(11):1407–13.
6. Solaz-Fuster MC, Gimeno-Alcaniz JV, Ros S, Fernandez-Sanchez ME, Garcia-Fojeda B, Criado Garcia O, et al. Regulation of glycogen synthesis by the laforin-malin complex is modulated by the AMP-activated protein kinase pathway. Hum Mol Genet. 2008;17(5):667–78.
7. Gentry MS, Roma-Mateo C, Sanz P. Laforin, a protein with many faces: glucan phosphatase, adapter protein, et alii. FEBS J. 2013;280(2):525–37.
8. Fukada M, Noda M. Yeast substrate-trapping system for isolating substrates of protein tyrosine phosphatases. Methods Mol Biol. 2007;365:371–82.
9. Gupta V, Bamezai RN. Human pyruvate kinase M2: a multifunctional protein. Protein Sci. 2010;19(11):2031–44.
10. Mazurek S. Pyruvate kinase type M2: a key regulator of the metabolic budget system in tumor cells. Int J Biochem Cell Biol. 2011;43(7):969–80.
11. Noguchi T, Inoue H, Tanaka T. The M1- and M2-type isozymes of rat pyruvate kinase are produced from the same gene by alternative RNA splicing. J Biol Chem. 1986;261(29):13807–12.
12. Kaiser P, Tagwerker C. Is this protein ubiquitinated? Methods Enzymol. 2005;399:243–8.
13. Moreno D, Towler MC, Hardie DG, Knecht E, Sanz P. The laforin-malin complex, involved in Lafora disease, promotes the incorporation of K63-linked ubiquitin chains into AMP-activated protein kinase beta subunits. Mol Biol Cell. 2010;21(15):2578–88.
14. Luo W, Semenza GL. Emerging roles of PKM2 in cell metabolism and cancer progression. Trends Endocrinol Metab. 2012;23(11):560–6.
15. Chaneton B, Gottlieb E. Rocking cell metabolism: revised functions of the key glycolytic regulator PKM2 in cancer. Trends Biochem Sci. 2012;37(8):309–16.
16. Yang W, Xia Y, Hawke D, Li X, Liang J, Xing D, et al. PKM2 phosphorylates histone H3 and promotes gene transcription and tumorigenesis. Cell. 2012;150(4):685–96.
17. Gao X, Wang H, Yang JJ, Liu X, Liu ZR. Pyruvate kinase M2 regulates gene transcription by acting as a protein kinase. Mol Cell. 2012;45(5):598–609.
18. Lv L, Xu YP, Zhao D, Li FL, Wang W, Sasaki N, et al. Mitogenic and oncogenic stimulation of K433 acetylation promotes PKM2 protein kinase activity and nuclear localization. Mol Cell. 2013;52(3):340–52.
19. Yang W, Lu Z. Regulation and function of pyruvate kinase M2 in cancer. Cancer Lett. 2013;339(2):153–8.
20. Rubio-Villena C, Garcia-Gimeno MA, Sanz P. Glycogenic activity of R6, a protein phosphatase 1 regulatory subunit, is modulated by the laforin-malin complex. Int J Biochem Cell Biol. 2013;45(7):1479–88.
21. Spoden GA, Morandell D, Ehehalt D, Fiedler M, Jansen-Durr P, Hermann M, et al. The SUMO-E3 ligase PIAS3 targets pyruvate kinase M2. J Cell Biochem. 2009;107(2):293–302.
22. Xing X, Bi H, Chang AK, Zang MX, Wang M, Ao X, et al. SUMOylation of AhR modulates its activity and stability through inhibiting its ubiquitination. J Cell Physiol. 2012;227(12):3812–9.
23. Rubio T, Vernia S, Sanz P. Sumoylation of AMPKbeta2 subunit enhances AMP-activated protein kinase activity. Mol Biol Cell. 2013;24(11):1801–11. S1801-1804.
24. Anderson DD, Eom JY, Stover PJ. Competition between sumoylation and ubiquitination of serine hydroxymethyltransferase 1 determines its nuclear localization and its accumulation in the nucleus. J Biol Chem. 2012;287(7):4790–9.
25. Ito H, Fukuda Y, Murata K, Kimura A. Transformation of intact yeast cells treated with alkali cations. J Bacteriol. 1983;153:163–8.
26. Rose MD, Winston F, Hieter P. Methods in yeast genetics, a laboratory course manual. Cold Spring Harbor, New York: Cold Spring Harbor Laboratory Press; 1990.
27. Worby CA, Gentry MS, Dixon JE. Malin decreases glycogen accumulation by promoting the degradation of protein targeting to glycogen (PTG). J Biol Chem. 2008;283(7):4069–76.
28. Yang X, Hubbard EJ, Carlson M. A protein kinase substrate identified by the two-hybrid system. Science. 1992;257(5070):680–2.
29. Ludin K, Jiang R, Carlson M. Glucose-regulated interaction of a regulatory subunit of protein phosphatase 1 with the Snf1 protein kinase in Saccharomyces cerevisiae. Proc Natl Acad Sci U S A. 1998;95(11):6245–50.
30. Solaz-Fuster MC, Gimeno-Alcaniz JV, Casado M, Sanz P. TRIP6 transcriptional co-activator is a novel substrate of AMP-activated protein kinase. Cell Signal. 2006;18(10):1702–12.

# The eukaryotic translation initiation factor 3f (eIF3f) interacts physically with the alpha 1B-adrenergic receptor and stimulates adrenoceptor activity

Mario Javier Gutiérrez-Fernández[1,2], Ana Edith Higareda-Mendoza[3], César Adrián Gómez-Correa[2,3] and Marco Aurelio Pardo-Galván[1*]

## Abstract

**Background:** eIF3f is a multifunctional protein capable of interacting with proteins involved in different cellular processes, such as protein synthesis, DNA repair, and viral mRNA edition. In human cells, eIF3f is related to cell cycle and proliferation, and its deregulation compromises cell viability.

**Results:** We here report that, in native conditions, eIF3f physically interacts with the alpha 1B-adrenergic receptor, a plasma membrane protein considered as a proto-oncogene, and involved in vasoconstriction and cell proliferation. The complex formed by eIF3f and alpha 1B-ADR was found in human and mouse cell lines. Upon catecholamine stimulation, eIF3f promotes adrenoceptor activity in vitro, independently of the eIF3f proline- and alanine-rich N-terminal region.

**Conclusions:** The eIF3f/alpha adrenergic receptor interaction opens new insights regarding adrenoceptor-related transduction pathways and proliferation control in human cells. The elf3f/alpha 1B-ADR complex is found in mammals and is not tissue specific.

**Keywords:** eIF3f, Alpha 1B-ADR, Gαq/11, Adrenoceptors, Protein-protein interaction

## Background

The eukaryotic translation initiation factor 3f (eIF3f) is an ancient and conserved gene reported to be present in most eukaryotic organisms studied so far [1]. It was originally identified as a subunit of the protein synthesis-related eIF3 complex [2], where it is suggested to function as a protein synthesis inhibitor [3, 4]. In contrast, eIF3f acts as a translational enhancer by increasing protein synthesis efficiency in muscle hypertrophy [5]. Shut-off experiments in *Schizosaccharomyces pombe* showed that in a long-term period, eIF3f is essential for viability, and that depleting the expression of this gene markedly decreases global protein synthesis [6]. In accordance to this, cell viability was also compromised when eIF3f

expression was decreased in proliferating human A549 cells [7].

The eIF3f protein is a member of the Mov-34 family. Members of this family contain an MPN (Mpr1/Pad N-terminal) motif, which is found in subunits of other macromolecular complexes such as the proteasome and the COP9 signalosome; it has been related to complex assembly promotion and to mediating protein-protein interactions [8–10]. The MPN domain of human eIF3f protein is flanked by a proline- and alanine-rich N-terminal region and by the C-terminal region.

During terminal muscle differentiation, eIF3f interacts with hypophosphorylated S6K1 through its MPN domain and with the mTOR/Raptor complex (mTORC1) by interacting with a TOS site contained in its C-terminal region [11]. As a consequence of this interaction, mTOR/Raptor phosphorylates S6K1, and thus

* Correspondence: mapardo@umich.mx
[1]Instituto de Investigaciones Químico-Biológicas, Universidad Michoacana de San Nicolás de Hidalgo, Edificio B-3 Ciudad Universitaria Avenida Francisco J. Múgica S/N, Morelia, Michoacán 58030, México
Full list of author information is available at the end of the article

regulates downstream effectors of mTOR and cap-dependent translation initiation [5, 11].

eIF3f shows a remarkable ability to interact with many other proteins involved in a variety of cellular functions which are not directly involved in protein synthesis. For instance, it has been reported that eIF3f participates in the deubiquitination and activation of the development-related transmembrane protein Notch 1 [12]. In addition, it has been shown that HIV-1 replication is inhibited by eIF3f, through its proline- and alanine-rich N-terminal region [13, 14]. This inhibition was observed with the full-length eIF3f protein and with the 91 amino acid N-terminal region of the protein. In both cases, HIV-1 mRNA levels were reduced as a result of eIF3f interfering with the HIV-1 mRNA 3'-end processing [13, 14]. Furthermore, a recent report indicates that eIF3f is capable of interacting with the DNA repair-related protein hMSH4, facilitating hMSH4 stabilization [15]. These authors demonstrate that the eIF3f-hMSH4 interaction is through the N-terminal regions of both proteins.

Considering the ubiquitous nature of eIF3f, the aim of this work was to investigate other stable eIF3f - protein interactions in native cellular conditions. In this study, we found a novel physical interaction between human eIF3f and the alpha 1B-adrenergic receptor (alpha 1B-ADR) and that eIF3f stimulates adrenoceptor activity.

## Materials and methods
### Chemicals and materials
All chemicals were purchased from Sigma Aldrich [St. Louis, MO, USA], unless otherwise noted. Materials were mainly purchased from Corning [Corning, NY, USA], Bio-Rad [Hercules, CA, USA], EMD Millipore [Billerica, MA, USA], and Eppendorf [Hauppauge, NY, USA], unless otherwise noted.

### Cell cultures
All cell lines were purchased from the American Type Culture Collection [ATCC, Manassas, VA, USA]. Human lung carcinoma A549 cells (ATCC CCL185), human hepatocellular carcinoma HepG2 cells (ATCC HB-8065), Burkitt's Lymphoma Ramos cells (ATCC CRL-1596), and murine pre-osteoblast MC3T3-E1 Subclone 4 cells (ATCC CRL-2593) were thawed every month and routinely passaged twice per week into 75 cm$^2$ flasks (Corning, Corning, NY, USA) to maintain them in a logarithmic growth phase at 37 °C in a humidified atmosphere with 5 % $CO_2$ (NuAire, Plymouth, MN, USA). A549, HepG2, and Ramos were cultured in MEM medium supplemented with 10 % heat-inactivated FBS (Invitrogen, Carlsbad, CA, USA), 2 mM glutamine, 10 mM HEPES, and 1.5 g/L sodium bicarbonate. MC3T3-E1 cells were cultured in MEMalpha medium

(Invitrogen) without ascorbic acid, supplemented with 10 % heat-inactivated FBS. At 85 % confluence, cells were harvested using 0.25 % Trypsin-EDTA solution and were sub-cultured or collected for subsequent experimental analysis. Since this study does not involve humans, human data or animals, an ethics committee approval was not required.

### Native western blot analysis
Cells were lysed with ProteoJET Mammalian Cell Lysis Reagent (Fermentas, Hanover, MD, USA), following the manufacturer's protocol, to obtain total native conformation proteins. Protein concentrations were determined using a Bio-Rad Protein Assay Kit. A molecular weight (kDa) standard (NativeMark - Novex, Life Technologies, Grand Island, NY, USA) and equal amounts of protein extracts (50 µg) were subjected to electrophoresis on a 8 % SDS-free polyacrylamide gel (PAGE) and electrophoretically transferred to a polyvinylidene difluoride (PVDF) membrane following the manufacturer's instructions (Bio-Rad, Hercules, CA, USA). After blocking non-specific binding sites with 5 % skimmed milk, blots were incubated with primary rabbit polyclonal antibody specific to eIF3f (Biolegend, San Diego, CA, USA) or primary goat antibody specific to alpha 1B-ADR (Santa Cruz Biotechnology, Santa Cruz, CA, USA), and horseradish peroxidase-conjugated goat anti-rabbit (Biolegend) or rabbit anti-goat secondary antibody (Santa Cruz Biotechnology). The bound antibody was detected by enhanced chemiluminescence (ECL) on an X-ray film (GE Healthcare Life Sciences, Piscataway, NJ, USA).

### Complex member identification
After native electrophoresis, the 120 kDa region of the gel was excised and eluted overnight at 4 °C in 1 mL of ProteoJet Mammalian Cell Lysis Reagent. For non-specific bound protein removal, the elution was incubated 1 h at 4 °C with a non-specific IgG and 100 µL of 50 % Protein A-Sepharose 4B beads (Invitrogen). After centrifugation, the supernatant was incubated at 4 °C with the antibody against eIF3f and 100 µL of 50 % Protein A-Sepharose 4B beads, washed 4 times with lysis buffer, resuspended in sample buffer (2 % SDS, 20 % glycerol, and 0.5 % bromophenol blue in 62 mM Tris HCl buffer, pH 6.8), boiled for 5 min, and subjected to electrophoresis on a 10 % SDS-polyacrylamide gel (10 % SDS-PAGE). A PageRuler Plus Prestained Protein Ladder (Thermo Scientific, Rockford, IL, USA) was included in all SDS-PAGEs. The resolved proteins were detected by Coomassie blue R250 staining (Bio-Rad); the unknown protein bands were excised and sent for N-terminal protein/peptide sequencing (Iowa State University of Science and Technology, Ames, IA, USA). After sequencing service, the partial amino acid sequences

were subjected to a NCBI Blastp search to identify possible protein partners of eIF3f.

## Immunoprecipitate western blot analyses

As describe above, after a native electrophoresis, the 120 kDa region of the gel was excised, eluted, cleared with a non-specific IgG and Protein A-Sepharose 4B beads and centrifuged. The supernatant was incubated with the corresponding primary antibody (anti-eIF3f or anti-alpha 1B-ADR) and Protein A-Sepharose 4B beads, washed, resuspended in sample buffer, boiled, and subjected to electrophoresis (10 % SDS-PAGE). Proteins from the gel were electrophoretically transferred to a PVDF membrane and blotted with anti-eIF3f or anti-alpha 1B-ADR antibodies, and the corresponding secondary horseradish peroxidase-conjugated antibodies. The bound antibody was detected by ECL on an X-ray film.

## Plasmids

To express human eIF3f, we used the previously described pSK11F plasmid [7]. To express alpha 1B-ADR, we used plasmid AR0A1B0000 (Missouri S&T cDNA Resource Center, Rolla, MO, USA). To obtain plasmid eIF3fΔ91 (eIF3f lacking first 91 AA of the N-terminal region), template pSK11F and the forward 5'-CCCTTCCCCGGCGGCAGCATGGTC-3' and reverse 5'-CAGGTTTACAAGTTTTTCATTG-3' oligonucleotides were used to amplify the eIF3f coding sequence corresponding to amino acids 92–357. The forward oligo was designed to contain a classic Kozak consensus sequence (see underlined nucleotides), by modifying only 2 nucleotides. The amplicon was cloned in pGEM vector (Promega, Fitchburg, WI, USA) and verified by DNA sequencing (Elim Biopharmaceuticals, Hayward, CA, USA).

## [gamma-32P]GTP Binding Assay

Membranes from A549 human cells were obtained using ProteoExtract Subcellular Proteome Extraction Kit (Calbiochem, La Jolla, CA, USA), as described by the manufacturer. To obtain membranes with over expressed alpha 1B-ADR, before membrane extraction, cells were transiently transfected (48 h) with plasmid AR0A1B0000 using LipofectAMINE 2000 (Invitrogen) according to the manufacturer's specifications and as described previously [7]. For in vitro translation of eIF3f and eIF3fΔ91, mRNA was synthesized *in vitro* using T3 and T7 RNA polymerase (Invitrogen), respectively, and according to the manufacturer's instructions. The respective proteins were synthesized *in vitro* with a Rabbit Reticulocyte Lysate System (Promega), using 1 μg of mRNA. For the [gamma-32P]GTP binding assay [16], 20 μg of membrane protein and 5 μL of the translation reaction were

resuspended in 55 μL of 50 mM Tris–HCl (pH 7.4), 2 mM EDTA, 100 mM NaCl, 1 μM GDP, 3 mM MgCl and 30 nM the [gamma-32P]GTP (Institute of Isotopes Co Ltd, Budapest, Hungary). As a control, the reaction was also performed using water instead of a mRNA. The reactions were incubated at 30 °C for 5 min in the presence or absence of agonist (100 nM adrenaline). The reaction was terminated by adding 600 μL of ice-cold stop solution (50 mM Tris–HCl pH 7.5, 20 mM MgCl2, 150 mM NaCl, 0.5 % Nonidet, 100 μM GDP, and 100 μM GTP) and incubating for 30 min in ice. To each reaction, non-specific IgG and 100 μl of Protein A-Sepharose 4B beads were added and further incubated on ice for 20 min. Non-specifically bound protein was removed by centrifugation. The supernatant was then incubated 1 h at 4 °C with 1 μg of Gαq/11 antibody (Santa Cruz Biotechnology) and immunoprecipitated with 100 μl of Protein A-Sepharose 4B beads for 1 h at 4 °C. Immunoprecipitates were collected, washed 4 times in buffer without detergent, and resuspended in a TE buffer. The samples were analyzed in a scintillation counter (Wallac, Oy, Turku, Finland).

## Statistical analysis

All experiments were independently repeated at least three times. Results of multiple experiments are expressed as mean ± standard error (S.E.). Analysis of Student's t test was used to assess the differences between means. A $p < 0.05$ was accepted as statistically significant.

## Results

### eIF3f is immunodetected in a protein complex of approximately 120 kDa in native electrophoretic conditions

In denaturing electrophoresis conditions, mammalian eIF3f shows an apparent electrophoretic mobility of 47 kDa, albeit its molecular mass deduced from its amino acid sequence is 38.5 kDa. eIF3f migrates anomalously in SDS-PAGE, possibly due to the high proline content in the N-terminal region [2]. Human A549, HepG2, and Ramos, as well as murine pre-osteoblasts (cell line MC3T3-E1 Subclone 4), were independently lysed with a buffer that respects native protein conformation, as well as protein-protein interactions. Total protein from each cell type was subjected to a SDS-free PAGE and immunoblotted with eIF3f antibody. Figure 1 shows one major immunodetected band, of approximately 120 kDa, in each cell type. Other bands were observed in over-exposed blots (data not shown) and slightly appear in the blot shown in Fig. 1, around the 150 and 250 kDa region. The lower band intensity of these other complexes may be due to lower concentration, complex instability, or a transitory event. The eIF3

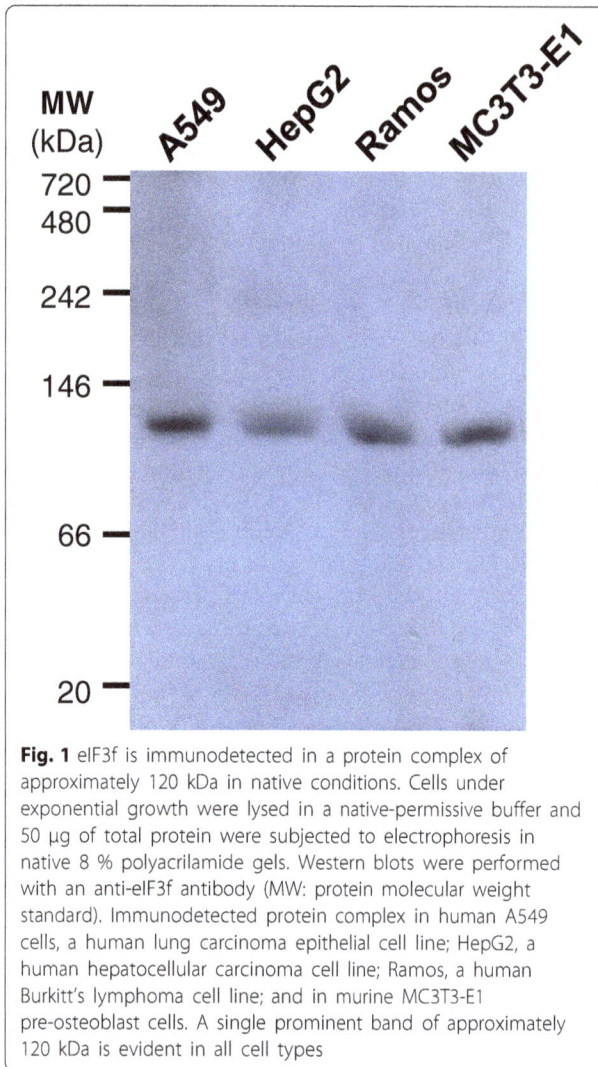

**Fig. 1** eIF3f is immunodetected in a protein complex of approximately 120 kDa in native conditions. Cells under exponential growth were lysed in a native-permissive buffer and 50 µg of total protein were subjected to electrophoresis in native 8 % polyacrilamide gels. Western blots were performed with an anti-eIF3f antibody (MW: protein molecular weight standard). Immunodetected protein complex in human A549 cells, a human lung carcinoma epithelial cell line; HepG2, a human hepatocellular carcinoma cell line; Ramos, a human Burkitt's lymphoma cell line; and in murine MC3T3-E1 pre-osteoblast cells. A single prominent band of approximately 120 kDa is evident in all cell types

two eluted proteins, a 47 kDa protein (eIF3f) and a second protein of approximately 60 kDa; similar results were obtained with the other cell lines (data not shown). The 60 kDa protein resolved from each cell type was excised from the denaturing gel and sent to N-terminal sequencing service. The partial amino acid sequence [MNPDLDT] (which was the same in A549, HepG2, Ramos, and MC3T3-E1), its protein position, and the expected molecular weight were the criteria used in a NCBI Blastp search. Results of this search showed that the human protein that fulfilled these criteria was the alpha-1B adrenergic receptor (alpha 1B-ADR), a 56.836 kDa (520 amino acids) plasma membrane protein. To verify this, the 120 kDa complex was immunoprecipitated with eIF3f antibody and then immunodetected using an alpha 1B-ADR antibody and with eIF3f antibody (Fig. 2b). In addition, the 120 kDa complex was immunoprecipitated with alpha 1B-ADR antibody and then immunodetected with eIF3f antibody and alpha 1B-ADR antibody (Fig. 2c). Both assays were positive. These results show that alpha 1B-ADR physically and stably interacts with eIF3f. Furthermore, under native conditions the 120 kDa complex from A549 cell extracts was immunodetected with alpha 1B-ADR antibody (Fig. 2d), confirming that this complex is composed by eIF3f and alpha 1B-ADR.

### eIF3f promotes adrenoceptor activity upon catecholamine stimulation

The stable interaction between eIF3f and alpha 1B-ADR raised the question if this interaction had a functional consequence. That is, if eIF3f stimulates or inhibits Alpha 1B-ADR activation. Alpha 1B-ADR is a member of the G protein-coupled receptor family of alpha 1-adrenergic receptors (alpha 1-ADRs), which is composed by the subtypes 1A, 1B, and 1D; all of which signal through the Gq/11 family of heterotrimeric G proteins [17]. The Gq/11 G proteins are membrane bound GTPases that are linked to 7-TM receptors and are formed by an alpha-, beta- and gamma- subunit [18]. In its inactive form, GDP is bound to the alpha subunit; when catecholamine binds to the receptor it causes a conformational change, which is recognized by the inactive form of the G protein complex and binds to it. The receptor triggers the exchange of bound GDP for GTP on the alpha subunit of the G-protein, which induces the GTP-alpha subunit to dissociate from the beta and gamma subunits. The GTP-alpha subunit (active form) then associates to downstream proteins involved in second messenger signaling cascades [18].

complex remained in the stacking gel. We tested several cell types to rule out the possibility that this immunodetected band may correspond to a specific cell line or organism. These results suggest that the 120 kDa complex may be present in mammalian organisms, and is not tissue specific.

### The 120 kDa complex is composed by eIF3f and the alpha 1B-adrenergic receptor

To identify the protein partners in the 120 kDa complex, total protein extracts under native conditions were used. Each of the 120 kDa regions were excised directly from the native gel, eluted in a protein-protein interaction permissive buffer, and immunoprecipitated with a specific antibody against eIF3f. After exhausting washes to eliminate unspecific protein binding, the immunoprecipitates were eluted in a non-reducing buffer, and run in a SDS-polyacrylamide gel. For proteins extracts from A549, Fig. 2a shows

To determine adrenoceptor activity, *in vitro* experiments were performed using radiolabeled [gamma-

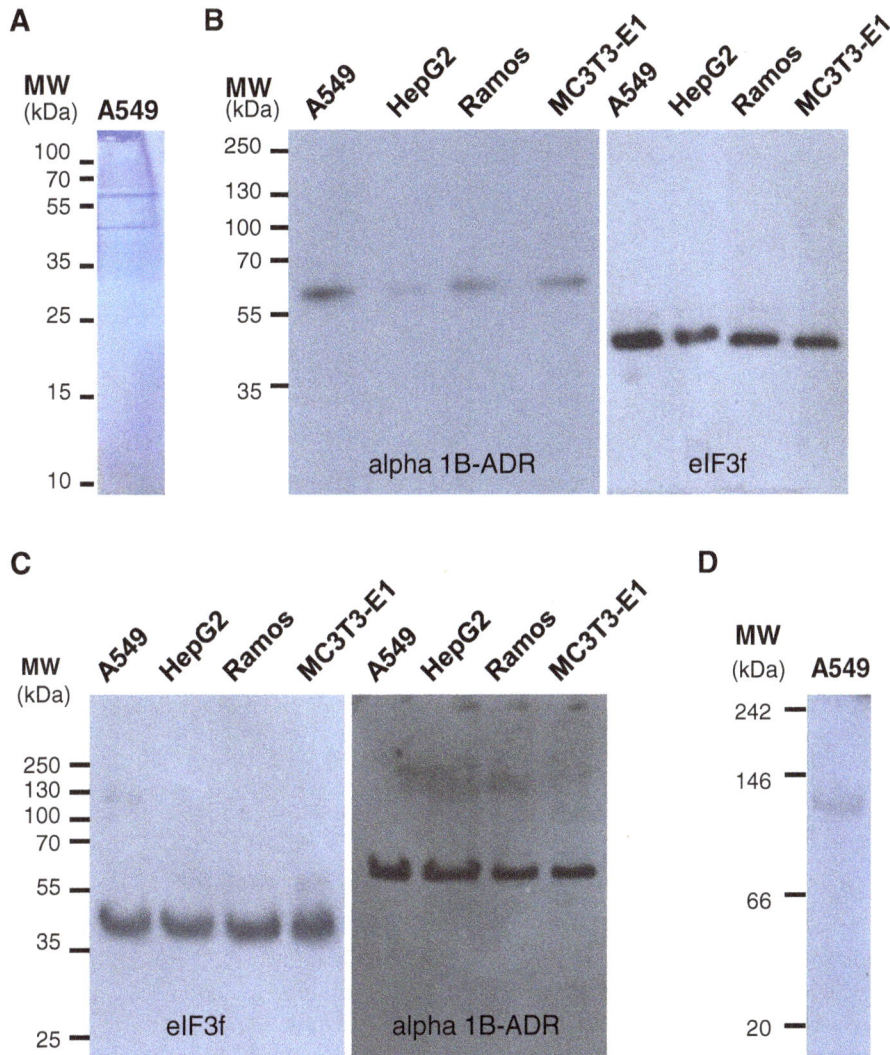

**Fig. 2** The 120 kDa complex is composed by eIF3f and the alpha-1B adrenergic receptor. **a** Resolved proteins in a SDS-PAGE from anti-eIF3f immunoprecipitate of the 120 kDa region: the expected 47 kDa corresponding to eIF3f, and an unknown ~60 kDa protein. The ~60 kDa protein was sent for N-terminal sequencing and a Blastp showed that it could be the alpha 1B-ADR. MW, prestained protein ladder; A549, resolved proteins in the 120 kDa complex obtained from total native protein extracts. **b** The 120 kDa complex was immunoprecipitated in native conditions with anti-eIF3f, and immunodetected with anti-alpha 1B-ADR or anti-eIF3f in denaturing conditions. **c** The 120 kDa complex was immunoprecipitated in native conditions with anti-alpha 1B-ADR, and immunodetected with anti-eIF3f or anti-alpha 1B-ADR in denaturing conditions. **d** Immunodetection of alpha 1B-ADR in native conditions using A549 cell extracts (MW: native protein standard), where the 120 kDa band is present. Results show that alpha 1B-ADR is the eIF3f partner in the 120 kDa complex

32P]GTP, and its transfer to Gαq/11 upon catecholamine stimulation, with or without the presence of eIF3f protein, and with or without over expressed alpha 1B-ADR membranes from A549. Reticulocyte-based translated eIF3f protein was added where indicated or the reticulocyte extract alone where eIF3f protein was not present. Individual assays were further distinctively immunoprecipitated with a Gαq/11 antibody and the [gamma-32P]GTP binding to Gαq/11 was quantified. This assay was based on a previous report [16], where the selectivity and significance in coupling of receptor to GTP-binding regulatory proteins (receptor

activation) was demonstrated. Figure 3 shows that the presence of eIF3f promotes adrenoceptor activation.

### The proline- and alanine-rich N-terminal region of eIF3f is not essential for adrenoceptor activation

eIF3f appeared early in Eukaryota, being identified as early as in protists [1]. eIF3f is present in most eukaryotic organisms, except in the budding yeast *Saccharomyces cerevisiae* [1, 6]. All eIF3fs reported to date include the MPN domain [1, 8–10] and a relatively conserved C-terminal region (Fig. 4). However, by

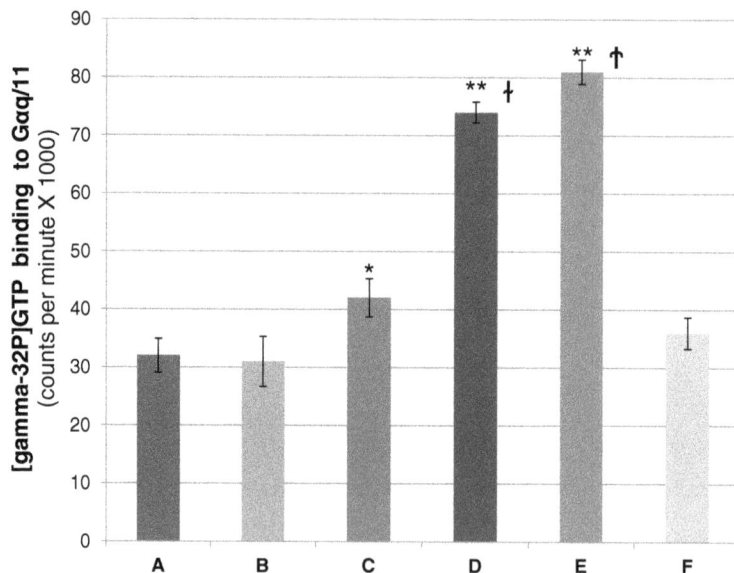

**Fig. 3** eIF3f promotes adrenoceptor activity upon catecholamine stimulation. Adrenoceptor activity is measured on the basis of *in vitro* agonist-promoted binding of [gamma-32P]GTP to G protein alpha subunits [16], in the presence of cell membrane fractions, and isolated subsequently by immunoprecipitation. In each condition, Gαq/11 immunoprecipitates were analyzed by scintillation for radiolabeled GTP binding. **a** Standard Gαq/11 activity reaction without catecholamine stimulation. **b** Standard Gαq/11 activity reaction with catecholamine stimulation. **c** Gαq/11 activity reaction without catecholamine stimulation and added eIF3f protein. **d** Gαq/11 activity reaction with catecholamine stimulation and added eIF3f protein. **e** Gαq/11 activity reaction with catecholamine stimulation and added eIF3f protein, using membrane fractions obtained from alpha 1B-ADR over-expressed A549 cultures. **f** Standard Gαq/11 activity reaction with catecholamine stimulation, using membrane fractions obtained from over-expressed alpha 1B-ADR A549 cultures. Data and error bars represent means ± standard error (S.E.) for three independent experiments; $*p < 0.05$ and $**p < 0.005$ with respect to the controls (A and F), $^{†}p < 0.005$ between C and D, and $^{†}p < 0.005$ between D and E

comparing protein sequence alignments of eIF3f from the N-terminal region to the beginning of the MPN domain (Fig. 4), it is evident that the proline- and alanine-rich N-terminal region of human eIF3f is a relatively recent evolutionary acquired region. Figure 4 shows the eIF3f protein alignment of some selected organisms. The proline- and alanine-rich N-terminal is present in mammals, non-avian reptiles, and birds, but not in amphibians or lower organisms, which suggests that this region was possibly acquired during the amniote clade. The proline- and alanine-rich N-terminal region of eIF3f has been reported to be important in the functional relationship between eIF3f and other proteins (see introduction), so we then asked if this N-terminal region could be relevant for adrenoceptor activation. To answer this question, we constructed an eIF3f clone that was devoid of this N-terminal region (indicated in Fig. 4), and tested *in vitro* for its ability to promote adrenoceptor activity upon catecholamine stimulation. Figure 5 clearly shows that the N-terminal region of eIF3f is not essential for adrenoceptor activation.

## Discussion

As pointed out earlier, eIF3f has a remarkable ability to interact with many proteins involved in a variety of

cellular functions [5, 8–15]. In this work, we report that in native undisturbed conditions, eIF3f stably interacts with the alpha 1B-adrenergic receptor (alpha 1B-ADR) and prompts Gαq/11 activation upon catecholamine stimulation. Previously, eIF3f was located in different subcellular compartments, including the plasma membrane fraction [19], which suggested a different function for eIF3f in addition to the protein synthesis process. Protein immunodetection of eIF3f in native conditions showed a clear single band in A549, HepG2, Ramos, and MC3T3-E1 cells, which localized approximately in the 120 kDa region. Specific immunoprecipitation with anti-eIF3f resolved two clearly distinguished protein bands, a 47 kDa protein corresponding to eIF3f and a 57 kDa protein that, by protein sequencing and specific immunodetection (Fig. 2), was identified as the alpha 1B-ADR. Interestingly, in native conditions we found no free unbound eIF3f protein, which confirms its high ability to interact with other proteins. In fact, most eIF3f is found in the eIF3 complex (data not shown).

In the presence of catecholamine, eIF3f stimulates adrenoceptor activity. The eIF3f/ alpha 1B-ADR interaction represents a novel and fascinating event in the control of adrenoceptor transducing activity, and disclose the possibility of new insights regarding alpha 1B-

```
                      Proline- and alanine-rich N-terminal region
H.sapiens       MATPAVPVSAPPATPTPVPAAA-PASVPAPTPAPAAAPVPAAAPASSSDP---AAAAAAT  56
M.musculus      MASPAVPANVPPATAAAAPAPVVTAAPASAPTPSTPAPTPAATPAASPAPVSSDPAVAAP  60
G.gallus        MA-----------APAPAAAPAPPAAAPAPS--PT------AAPATP----------TAG  31
O.hannah        MA-----------AAVGVPGASLSIANPSPPPPAT------SSPNTS----------SIT  33
X.tropicalis    ------------------------------------------------------------  0
D.rerio         ------------------------------------------------------------  0
A.thaliana      --------------------------------------------------------MA--  2
D.discoideum    ------------------------------------------------------------  0

                                                  Start of eIF3fΔ91
H.sapiens       AAPGQTPASAQAPAQTPAPALPGPALPGPFPGGRVVRLHPVILASIVDSYERRNEGAARV  116
M.musculus      AAPGQTPASAPAPAQTPAPSQPGPALPGPFPGGRVVRLHPVILASIVDSYERRNEGAARV  120
G.gallus        APPSAAPPPPAAPAAPPPAAPLSAALSGPFPGGRVVRLHPVVLASIVDSFERRNEGAARV  91
O.hannah        AMPASL----TPTAAP--PPPTPSALAAPFPGGRAVRLHPVVLASVVDSFERRNEGAARV  87
X.tropicalis    ----------------MAAAGSLSAAGSAMLHGPVVKVHPVVLASIVDSYERRNEGAGRV  44
D.rerio         ----------------------------MAVHGPVVKIHPVVLASIVDSYERRNEGASRV  32
A.thaliana      ----------ATSEHTILQFVSPSSTASATTSVLTARIHPLVIFNVCDCFVRRPDSAERV  52
D.discoideum    ---------------------MQKETSIFLPSQVTVKVHPVVIFNILDHYIRRNVGQDRV  39
                                                  ..::**::: .: * : **  .  **
                          MPN
H.sapiens       IGTLLGTVD-KHSVEVTNCFSVPHNESEDEVAVDMEFAKNMYELHKKVSPNELILGWYAT  175
M.musculus      IGTLLGTVD-KHSVEVTNCFSVPHNESEDEVAVDMEFAKNMYELHKKVSPNELILGWYAT  179
G.gallus        IGTLLGTVD-KHSVEVTNCFSVPHNESEDEVAVDMEFAKNMYELHKKVSPSEIILGWYAT  150
O.hannah        IGTLLGTVD-KYSVEVTNCFSVPHNESEDEVAVDMEFAKNMYELHKKVSPSEIILGWYAT  146
X.tropicalis    IGTLLGTTD-KHSVEVTNCFSVPHNESEDEVAVDMEFAKNMYELHKKVTSTEQIVGWYAT  103
D.rerio         IGTLLGTAD-KHSVDVTNCFSVPHNESEDEVAVDMEFAKNMYELHKKVSPSEVIVGWYAT  91
A.thaliana      IGTLLGSILPDGTVDIRNSYAVPHNESSDQVAVDIDYHHNMLASHLKVNSKETIVGWYST  112
D.discoideum    IGTLLGFNN-DGVLEIRNCFPVVHSET-EQIAVEMEYQRKMLDLHLKSSPREPIIGWYAT  97
                ******    .  ::: *.: * *.*: :::**:::: ::*   * * .  * *:***:*

H.sapiens       GHDITEHSVLIHEYYSREAP--NPIHLTVDTSLQNGRMSIKAYVSTLMGV-PGRTMGVMF  232
M.musculus      GHDITEHSVLIHEYYSREAP--NPIHLTVDTGLQHGRMSIKAYVSTLMGV-PGRTMGVMF  236
G.gallus        GHDITEHSVLIHEYYSREAH--NPIHLTVDTSLQNSRMSIKAYVSAPMGV-PGKTMGVMF  207
O.hannah        GHDITEHSVLIHEYYSREAH--NPIHLTVDTSLQNGRMSIKAYVSTTMGV-PGKTMGVMF  203
X.tropicalis    GNDITEHSVLIHEYYSREAT--NPIHMTVDTSLQGNRMNIKAYISSPMGV-PGKTMGVMF  160
D.rerio         GFDITEHSVLIHEYYSREAP--NPIHLTVDTALQSNKMNIRAYVSSQMGV-PGKTVGVMF  148
A.thaliana      GAGVNGGSSLIHDFYAREVP--NPIHLTVDTGFTNGEGTIKAFVSSNLSL-GDRQLVAHF  169
D.discoideum    GNDINENSVHINNFYRDEMGNSTPIHLTVDTGLTNDTMGIHAYMAHNLSLNPESSLGSYF  157
                *  :.  *  *.::*  *    .***:****.:      *:*:::  :.:     :   *

H.sapiens       TPLTVKYAYYDTERIGVDLIMKTCFSPNRVIGLSSDLQQVGGASARIQDALSTVLQYAED  292
M.musculus      TPLTVKYAYYDTERIGVDLIMKTCFSPNRVIGLSSDLQQVGGASARIQDALSTVLQYAED  296
G.gallus        TPLTVKYVYYDTERIGVDLIMKTCFSPNRVIGLSSDLQQVGSASARIQDTLGMVLQYAED  267
O.hannah        TPLTVKYAYYDTERIGVDLVMKTCFSPNRVINLSSDLQQVGTASAQIQDTLSTVLQYAED  263
X.tropicalis    TPLTVQYVYYDTERIGVDLITKTCANPNRSIGLTTDLQQVGVAANRLQDSLSTVLQYAED  220
D.rerio         TPLSVKYIYYDTERIGVDLLQRTRASPGRTNGLTTDLAQVAGAAGRVQEMLATVLSYIED  208
A.thaliana      QEIPVDLRMVDAERVGFDVLKATSV-----DKLPNDLEGMELTMERLLTLINDVYKYVDS  224
D.discoideum    SQLPLEILTFEAENAGLESIAQTKYDQQ-STSLLSELESLQGSLTKLDEMLESITSYIES  216
                  : :.    ::*. *.: :  *         * .:*  :  :  ::   :  : .* :.

H.sapiens       VLSGKVSADNTVGRFLMSLVNQVPKIVPDDFETMLNSNINDLLMVTYLANLTQSQIALNE  352
M.musculus      VLSGKVSADNTVGRFLMSLVNQVPKIVPDDFETMLNSNINDLLMVTYLANLTQSQIALNE  356
G.gallus        VLSGKVAADNTVGRFLMDLINQVPKISPEDFETMLNSNINDLLMVTYLANLTQSQIALNE  327
O.hannah        VLSGKVAADNTVGRFLMDLVNQVPKISPEDFETMLNSNINDLLMVTYLANLTQSQIALNE  323
X.tropicalis    VLSGKVTADNTVGRFLMDLVTQVPKIDPEDFEAMLNSNINDLLMVTYLANLTQSQIALNE  280
D.rerio         VLSGKVMADNSVGRFLMDLVNKVPKIPAEDFENMLNSNINDLLMVTYLANLTQAQIALNE  268
A.thaliana      VVGGQIAPDNNIGRFIADAVASLPKLPPQVFDNLVNDSLQDQLLLLYLSSITRTQLSLAE  284
D.discoideum    VEKGEIQGDPRLGRFLAKTIQALPKANAQVMDKVINNSVKDLLMIVYLSSLTRSQLAVAT  276
                *  *::  *  :***: . :  :**   : :: ::*..::* *:: **:.:*::*:::
                                                  Conserved C-terminal region
H.sapiens       KLVNL----     357
M.musculus      KLVNL----     361
G.gallus        KLLSL----     332
O.hannah        KILNL----     328
X.tropicalis    KLLNL----     285
D.rerio         KLVVL----     273
A.thaliana      KLNTAAQML     293
D.discoideum    KISHSLSN-     284
                *:
```

**Fig. 4** Protein sequence alignment of eIF3f orthologues. eIf3f sequences were obtained from NCBI (*Homo sapiens* gi:6685511, *Mus musculus* gi:341940488, *Gallus gallus* gi: 50749406, *Ophiophagus hannah* gi:565315948, *Xenopus tropicalis* gi:62859127, *Danio rerio* gi:317108137, *Arabidopsis thaliana* gi:23396614, *Dictyostelium discoideum* gi:74850733); and Clustal Omega (EMBL-EBI) was used for multiple sequence alignment. The red box represents the MPN domain. The proline- and alanine-rich N-terminal is present in mammals, non-avian reptiles, and birds, but not in amphibians or lower organisms

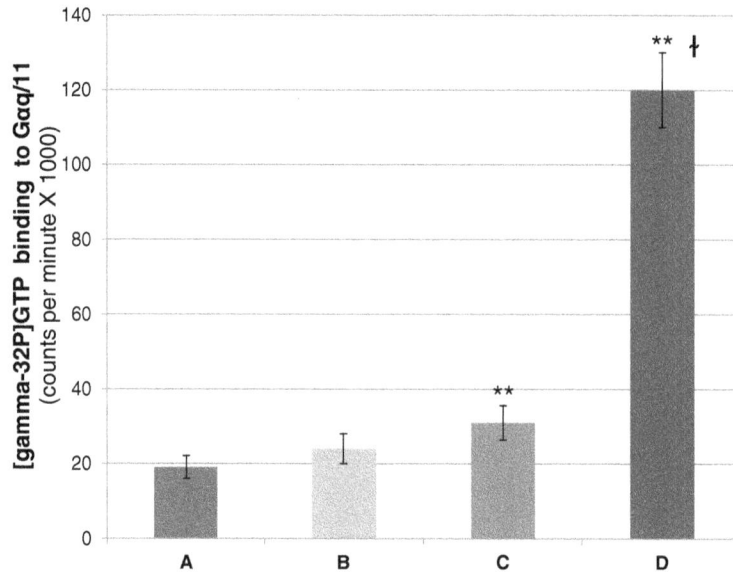

**Fig. 5** The proline-alanine rich amino terminal of eIF3f is not essential for adrenoceptor activation. The experimental conditions were the same as in Fig. 3. eIF3fΔ91 is a genetic construct where the first 91 amino acids of eIF3f were removed (proline- and alanine-rich N-terminal region). **a** Standard Gαq/11 activity reaction without catecholamine stimulation. **b** Standard Gαq/11 activity reaction with catecholamine stimulation. **c** Gαq/11 activity reaction without catecholamine stimulation and added eIF3fΔ91 protein. **d** Gαq/11 activity reaction with catecholamine stimulation and added eIF3fΔ91 protein. Data and error bars represent mean ± S.E. for three independent experiments; ** $p < 0.005$ with respect to the control (**a**), and † $p < 0.0005$ between C and D

ADR function in different cellular processes. Our results demonstrate that the proline- and alanine-rich N-terminal region of eIF3f is not required for adrenoceptor activation. Moreover, comparing the GTP binding to Gqα/11 in the presence of native eIF3f (Fig. 3 e) or truncated eIF3f (Fig. 5 D) under catecholamine stimulation, we observed that truncated eIF3f stimulates more GTP binding. Our interpretation is that the N-terminal region of eIF3f lowers its affinity for the adrenoceptor, possibly due to a steric impediment. Other studies should be conducted to elucidate if this has a functional cellular consequence. According to reported eIF3f amino acid sequences, this N-terminal region appeared during the amniote clade (Fig. 4). This was relevant to investigate, since this region was found important for other eIF3f-protein interactions [13, 14]. Taking in account that alpha 1B-ADR is present in vertebrates (Blastp in NCBI and UniProtKB EMBL-EBI) it is possible that the eIF3f/alpha 1B-ADR interaction would have arisen since then.

To explain the possible cellular function(s) derived from the eIF3f/alpha 1B-ADR interaction, we considered the following facts. The plasma membrane is an organized biological system that serves as a structural barrier and communication interface with the extracellular environment, and the alpha1-adrenergic receptors are embedded in this membrane. Alpha 1- adrenergic receptors bind to and are activated by endogenous catecholamine hormones, which are mainly involved in vasoconstriction [17]. They are coupled to phospholipase C, c-Jun N-

terminal kinase, and the mitogen-activated protein kinase downstream signal transduction pathways [17, 20, 21]; and have an important function in stress response that affects lipid, carbohydrate, and amino acid metabolism [22]. In addition to these effects, there is substantial evidence indicating that stimulation of alpha 1-ADRs by catecholamines generally enhances growth-related gene expression and cell growth in a variety of cells, including cardiac myocytes [23], vascular smooth muscle cells [24–26], hepatocytes [20, 27], and adipocytes [28].

Alpha 1B-ADR mediates co-mitogenic effects with catecholamines in different cells. For instance, activation of alpha 1B-ADR increases DNA synthesis in primary cultures of hepatocytes [27], and promotes malignancy in alpha 1B-ADR transfected Rat-1 fibroblasts [29, 30]. Induction of neoplastic transformation by the alpha 1B-ADR, thus, identifies this normal cellular gene as a proto-oncogene. Also, in Rat-1 fibroblasts, alpha 1-ADRs affect the expression of cell cycle-related genes in a differential manner: the over expression of alpha 1A-ADR and alpha 1D-ADR downregulated genes ascribed to the G1/S transition phase, such as Cyclin E and DNA polymerase; this over expression upregulated p27 kip and induced G1/S cell cycle arrest. In contrast, over expressed alpha 1B-ADR transfected cells did not affect Cyclin E or DNA polymerase expression; they showed downregulated p27 kip and stimulated cell cycle progression [30]. On the contrary, CHO cells over expressing human alpha1-adrenergic receptors showed that upon catecholamine activation, alpha 1A-ADR or alpha 1B-ADR -transfected cells exhibit inhibition of serum-promoted cell proliferation and were arrested at G1/S phase, whereas alpha 1D-ADR did not show any effect [31].

eIF3f is also a cell division and proliferation-related gene. In human A549 cells, eIF3f exhibits a fluctuating expression pattern in cycling cells, with maximum expression peaks in G1/S and in G2/M; transient expression analysis showed that eIF3f deregulation compromises cell viability and induces apoptosis [7]. In addition, altered expression of eIF3f has been reported in several human tumors, being found downregulated in some cell lines [32, 33] and upregulated in others [19].

The fact that eIF3f expression is induced in G1/S, that alpha 1B-ADR affects functions related to G1/S, and that both gene products relate physically and functionally, establish the interesting possibility that this relationship might be involved in the regulation of G1/S functions. We are presently exploring this possibility. In addition, since alpha 1B-ADR is considered as a proto-oncogene [29], and the deregulation of eIF3f is frequently associated to oncogenesis, it would be interesting to investigate if and how the interaction of these two proteins affects the control of cell proliferation, and eventually use these gene products as potential targets for cancer therapy.

## Conclusions

In the present work, we report that eIF3f physically and stably interacts with the alpha 1B-ADR, and that eIF3f stimulates adrenoceptor activity. This novel protein-protein interaction may represent a regulatory link between adrenoceptor-related signal transduction and, for instance, cell proliferation and protein synthesis control.

### Abbreviations
alpha 1-ADRs: alpha 1-adrenergic receptors; alpha 1B-ADR: alpha 1B-adrenergic receptor; eIF3f: eukaryotic translation initiation factor 3f.

### Competing interests
The authors declare that they have no competing interests.

### Authors' contributions
MJGF participated in the design of the study, carried out most of the molecular studies, and participated in the draft of the manuscript. AEHM carried out the cellular studies, performed the statistical analysis, and participated in the draft of the manuscript. CAGC participated with molecular experiments. MAPG conceived the study, participated in its design and coordination, and helped to draft the manuscript. All authors read, revised, and approved the final manuscript.

### Acknowledgements
This research was partially financed by the Universidad Michoacana de San Nicolás de Hidalgo through its CIC Research Program in favor of MAPG and AEHM. MJGF was a recipient of a PROMEP Doctoral Scholarship and CAGC of a CONACYT Masters Scholarship. PROMEP and the Universidad Tecnológica de Morelia financed the publication fee.

### Author details
[1]Instituto de Investigaciones Químico-Biológicas, Universidad Michoacana de San Nicolás de Hidalgo, Edificio B-3 Ciudad Universitaria Avenida Francisco J. Múgica S/N, Morelia, Michoacán 58030, México. [2]Present address: Universidad Tecnológica de Morelia, Morelia, Michoacán 58200, México. [3]División de Estudios de Posgrado de la Facultad de Ciencias Médicas y Biológicas "Dr. Ignacio Chávez", Universidad Michoacana de San Nicolás de Hidalgo, Morelia, Michoacán 58020, México.

### References
1. Rezende AM, Assis LA, Nunes EC, da Costa Lima TD, Marchini FK, Freire ER, et al. The translation initiation complex eIF3 in trypanosomatids and other pathogenic excavates - identification of conserved and divergent features based on orthologue analysis. BMC Genomics. 2014;15:1175.
2. Asano K, Vornlocher HP, Richter-Cook NJ, Merrick WC, Hinnebusch AG, Hershey JW. Structure of cDNAs encoding human eukaryotic initiation factor 3 subunits. Possible roles in RNA binding and macromolecular assembly. J Biol Chem. 1997;272(43):27042–52.
3. Shi J, Kahle A, Hershey JW, Honchak BM, Warneke JA, Leong SP, et al. Decreased expression of eukaryotic initiation factor 3f deregulates translation and apoptosis in tumor cells. Oncogene. 2006;25:4923–36.
4. Higareda-Mendoza AE, Farias AE, Llanderal JM, Mendez AB, Pardo-Galvan MA. A cell cycle and protein synthesis-related factor required for G2/M transition. Miami Nat Biotech Short Rep. 2003;14:111.
5. Csibi A, Tintignac LA, Leibovitch MP, Leibovitch SA. eIF3-f function in skeletal muscles: to stand at the crossroads of atrophy and hypertrophy. Cell Cycle. 2008;7(12):1698–701.
6. Zhou C, Arslan F, Wee S, Krishnan S, Ivanov AR, Oliva A, et al. PCI proteins eIF3e and eIF3m define distinct translation initiation factor 3 complexes. BMC Biol. 2005;3:14.
7. Higareda-Mendoza AE, Pardo-Galvan MA. Expression of human eukaryotic initiation factor 3f oscillates with cell cycle in A549 cells and is essential for cell viability. Cell Div. 2010;5:10.
8. Aravind L, Ponting CP. Homologues of 26S proteasome subunits are regulators of transcription and translation. Protein Sci. 1998;7:1250–4.
9. Fu H, Reis N, Lee Y, Glickman MH, Vierstra RD. Subunit interaction maps for the regulatory particle of the 26S proteasome and the COP9 signalosome. EMBO J. 2001;20(24):7096–107.
10. Sanches M, Alves BS, Zanchin NI, Guimarães BG. The crystal structure of the human Mov34 MPN domain reveals a metal-free dimer. J Mol Biol. 2007;370(5):846–55.
11. Csibi A, Cornille K, Leibovitch MP, Poupon A, Tintignac LA, Sanchez AM, et al. The translation regulatory subunit eIF3f controls the kinase-dependent mTOR signaling required for muscle differentiation and hypertrophy in mouse. PLoS One. 2010;5(2):e8994.
12. Moretti J, Chastagner P, Gastaldello S, Heuss SF, Dirac AM, Bernards R, et al. The translation initiation factor 3f (eIF3f) exhibits a deubiquitinase activity regulating Notch activation. PLoS Biol. 2010;8(11):e1000545.
13. Valente ST, Gilmartin GM, Mott C, Falkard B, Goff SP. Inhibition of HIV-1 replication by eIF3f. Proc Natl Acad Sci U S A. 2009;106(11):4071–8.
14. Valente ST, Gilmartin GM, Venkatarama K, Arriagada G, Goff SP. HIV-1 mRNA 3' end processing is distinctively regulated by eIF3f, CDK11, and splice factor 9G8. Mol Cell. 2009;36(2):279–89.
15. Chu YL, Wu X, Xu Y, Her C. MutS homologue hMSH4: interaction with eIF3f and a role in NHEJ-mediated DSB repair. Mol Cancer. 2013;12:51.
16. Barr AJ, Brass LF, Manning DR. Reconstitution of receptors and GTP-binding regulatory proteins (G proteins) in Sf9 cells. A direct evaluation of selectivity in receptor.G protein coupling. J Biol Chem. 1997;272(4):2223–9.
17. Graham RM, Perez DM, Hwa J, Piascik MT. alpha 1-adrenergic receptor subtypes. Molecular structure, function, and signaling. Circ Res. 1996;78(5):737–49.
18. Oldham WM, Hamm HE. Heterotrimeric G protein activation by G-protein-coupled receptors. Nat Rev Mol Cell Biol. 2008;9(1):60–71.
19. Harvey S, Zhang Y, Landry F, Miller C, Smith JW. Insights into a plasma membrane signature. Physiol Genomics. 2001;5(3):129–36.
20. Spector MS, Auer KL, Jarvis WD, Ishac EJ, Gao B, Kunos G, et al. Differential regulation of the mitogen-activated protein and stress activated protein

kinase cascades by adrenergic agonists in quiescent and regenerating adult rat hepatocytes. Mol Cell Biol. 1997;17(7):3556–65.

21. Hawes BE, van Biesen T, Koch WJ, Luttrell LM, Lefkowitz RJ. Dinstinct pathways of Gi- and Gq- mediated mitogen-activated protein kinase activation. J Biol Chem. 1995;270(29):17148–53.

22. Kunos G, Ishac EJN, Gao B, Jiang L. Inverse regulation of hepatic alpha 1B- and beta 2-adrenergic receptors. Cellular mechanisms and physiological implications. Ann N Y Acad Sci. 1995;757:261–71.

23. Milano CA, Dolber PC, Rockman HA, Bond RA, Venable ME, Allen LF, et al. Myocardial expression of a constitutively active alpha 1B-adrenergic receptor in transgenic mice induces cardiac hypertrophy. Proc Natl Acad Sci U S A. 1994;91(21):10109–13.

24. Jackson CL, Schwartz SM. Pharmacology of smooth muscle cell replication. Hypertension. 1992;20(6):713–36.

25. Hu ZW, Shi XY, Lin RZ, Chen J, Hoffman BB. alpha1-Adrenergic receptor stimulation of mitogenesis in human vascular smooth muscle cells: role of tyrosine protein kinases and calcium in activation of mitogen-activated protein kinase. J Pharmacol Exp Ther. 1999;290(1):28–37.

26. Nakaki T, Nakayama M, Yamamoto S, Kato R. Alpha 1-adrenergic stimulation and beta 2-adrenergic inhibition of DNA synthesis in vascular smooth muscle cells. Mol Pharmacol. 1990;37(1):30–6.

27. Cruise JL, Houck KA, Michalopoulos GK. Induction of DNA synthesis in cultured rat hepatocytes through stimulation of alpha 1 adrenoreceptor by norepinephrine. Science. 1985;227(4688):749–51.

28. Thonberg H, Zhang SJ, Tvrdik P, Jacobsson A, Nedergaard J. Norepinephrine utilizes alpha 1- and beta-adrenoreceptors synergistically to maximally induce c-fos expression in brown adipocytes. J Biol Chem. 1994;269(52):33179–86.

29. Allen LF, Lefkowitz RJ, Caron MG, Cotecchia S. G-protein-coupled receptor genes as protooncogenes: constitutively activating mutation of the alpha 1B-adrenergic receptor enhances mitogenesis and tumorigenicity. Proc Natl Acad Sci U S A. 1991;88(24):11354–8.

30. Gonzalez-Cabrera PJ, Shi T, Yun J, McCune DF, Rorabaugh BR, Perez DM. Differential regulation of the cell cycle by alpha1-adrenergic receptor subtypes. Endocrinology. 2004;145(11):5157–67.

31. Shibata K, Katsuma S, Koshimizu T, Shinoura H, Hirasawa A, Tanoue A, et al. alpha 1-Adrenergic receptor subtypes differentially control the cell cycle of transfected CHO cells through a cAMP-dependent mechanism involving p27Kip1. J Biol Chem. 2003;278(1):672–8.

32. Doldan A, Chandramouli A, Shanas R, Bhattacharyya A, Cunningham JT, Nelson MA, et al. Loss of the eukaryotic initiation factor 3f in pancreatic cancer. Mol Carcinog. 2008;47(3):235–44.

33. Doldan A, Chandramouli A, Shanas R, Bhattacharyya A, Leong SP, Nelson MA, et al. Loss of the eukaryotic initiation factor 3f in melanoma. Mol Carcinog. 2008;47(10):806–13.

# 4

# Characterization of a cold-active and salt tolerant esterase identified by functional screening of Arctic metagenomic libraries

Concetta De Santi[1*], Bjørn Altermark[1], Marcin Miroslaw Pierechod[1], Luca Ambrosino[2], Donatella de Pascale[2] and Nils-Peder Willassen[1]

## Abstract

**Background:** The use of metagenomics in enzyme discovery constitutes a powerful approach to access to genomes of unculturable community of microorganisms and isolate novel valuable biocatalysts for use in a wide range of biotechnological and pharmaceutical fields.

**Results:** Here we present a novel esterase gene (*lip3*) identified by functional screening of three fosmid metagenomic libraries, constructed from three marine sediment samples. The sequenced positive fosmid revealed an enzyme of 281 amino acids with similarity to class 3 lipases. The 3D modeling of Lip3 was generated by homology modeling on the basis of four lipases templates [PDB ID: 3O0D, 3NGM, 3G7N, 2QUB] to unravel structural features of this novel enzyme. The catalytic triad of Lip3 was predicted to be Asp207, His267 and the catalytic nucleophile Ser150 in a conserved pentapeptide (GXSXG). The 3D model highlighted the presence of a one-helix lid able to regulate the access of the substrate to the active site when the enzyme binds a hydrophobic interface. Moreover an analysis of the external surface of Lip3 model showed that the majority of the surface regions were hydrophobic (59.6 %) compared with homologous lipases (around 35 %) used as templates. The recombinant Lip3 esterase, expressed and purified from *Escherichia coli*, preferentially hydrolyzed short and medium length *p*-nitrophenyl esters with the best substrate being *p*-nitrophenyl acetate. Further characterization revealed a temperature optimum of 35 °C and a pH optimum of 8.0. Lip3 exhibits a broad temperature stability range and tolerates the presence of DTT, EDTA, PMSF, β-mercaptoethanol and high concentrations of salt. The enzyme was also highly activated by NaCl.

**Conclusions:** The biochemical characterization and homology model reveals a novel esterase originating from the marine Arctic metagenomics libraries with features of a cold-active, relatively thermostable and highly halotolerant enzyme. Taken together, these results suggest that this esterase could be a highly valuable candidate for biotechnological applications such as organic synthesis reactions and cheese ripening processes.

**Keywords:** Metagenomics libraries, Cold-active esterase, Salt, Homology modeling, Biotechnological applications

## Background

Extreme environments represent a great microbial resource for novel enzymes, the majority of which remains to be discovered. Metagenomics, the technique to access the genome content resource of non-cultivated microbes, is a powerful tool used in the discovery of novel industrial enzymes for biotechnological and pharmaceutical applications [1–5]. More than 99 % of the microorganisms cannot be cultivated [6] but discovered, using an alternative metagenomics approach to the traditional microbial screening methods to isolate enzyme from extreme environments [7–9].

Based on the direct cloning of the metagenomic DNA [10] for the construction of large clone libraries, metagenomics gives access to new genes, complete pathways and their products by multiple screening methods. Despite that there are several limitations in screening of such libraries, such as the functional expression of genes in a

* Correspondence: concetta.d.santi@uit.no
[1]NorStruct, Department of Chemistry, Faculty of Science and Technology, UiT The Arctic University of Norway, Tromsø, Norway
Full list of author information is available at the end of the article

heterologous, screening host, the metagenome-approach has led to the discovery of many novel enzymes such as new esterases (carboxyl ester hydrolases, EC 3.1.1.1) and lipases (triacylglycerol lipases, EC 3.1.1.3) [11, 12]. Lipolytic enzymes are found in all living organisms and most of the commercially produced enzymes originate from microbial sources. Lipolytic enzymes can be grouped into 8 different families based on their sequence, structure and biological functions [13]. These enzyme families are all characterized by a catalytic triad consisting of a nucleophilic serine, a catalytic acid (aspartate or glutamate) and a histidine residue which is located in a conserved Gly-Xaa-Ser-Xaa-Gly pentapeptide that forms a sharp elbow in the center of the α/β-fold [14]. Lipases can be distinguished from esterases by exhibiting the interfacial activation [15]. Both enzymes have a secondary structural elements, called lids, that change conformation to accommodate the substrates [16, 17]. This lid moves to expose the catalytic cleft at the lipid-water interface according to the activation mechanism typical of lipases. However, there are exceptions such as the previously characterized *Candida antarctica* Lip B [18].

To date, numerous novel lipolytic enzymes have been identified by functional metagenomics analysis of various microbial habitats, such as soil [19–21], hot springs [22], lake water [23] and marine sediments [24, 25]. In particular, cold-active esterases and lipases have been studied because of their structural flexibility if compared to mesophilic and thermophilic counterparts. A reduced number of disulfide bridges and prolines in loop structure has been observed in several cold lipases with a high catalytic activity and stability at low temperatures. Thus, lipolytic enzymes have emerged as key enzymes in the growing biotechnology industry [26].

In this study, we screened three small metagenomic libraries constructed from marine sediment samples in order to identify new esterases for developing a cocktail, together with other lypolitic enzymes, with application in food industry [27, 28]. After sequencing of a positive clone, we found the gene responsible for the esterase activity seen on tributyrin plates. Afterwards recombinant expression in *Escherichia coli*, the enzyme was analyzed for its substrate specificity, optimal pH and temperature, thermal stability, and effect of different additives on its enzymatic activity. Moreover, homology modeling was performed to relate the biochemical future of the enzyme to structural properties.

## Methods

### Sampling in the marine Arctic

During two research cruises in the high Arctic samples of seawater, sediment and various biota were taken. For the sediment sampling, a Van-Veen grab was used and two 50 ml tubes of the top 10 cm layer were filled at each sampling location and frozen; first at −20 °C and later at −80 °C. Three of these sediment-samples, which are described in Table 1, were used to extract total DNA. The first sampling was conducted in the Barents Sea area in May 2010 and the second around Svalbard in October 2011.

Both research cruises, performed by the University of Tromsø, were conducted in areas regulated by the Norwegian government, and no special sampling permission was necessary.

### High molecular weight DNA extraction and purification

The frozen sediments were aliquoted using a solid mortar pre-chilled with liquid nitrogen. A soft lysis protocol [29] was followed with some modifications. Five grams of sediment was resuspended in 10 mL of DNA extraction buffer and 100 μl of proteinase K (10 mg/ml) was added. The sample was incubated in a 56 °C water bath for 2 h with an occasional, gentle mixing. Then 1.5 mL 20 % SDS was added and samples were incubated at 60 °C for another 2 h. After centrifugation at 5000 g for 20 min, the DNA-containing supernatant was extracted with a phenol: chloroform: isoamyl alcohol mixture (25:24:1 volume ratio). Next, the aqueous phase was precipitated with isopropanol (0.7 volumes). The pellet was then washed with 70 % EtOH, air-dried and dissolved in TE buffer (pH 8.0). At this stage the raw DNA had a brown color which indicates a high content of contaminants.

To purify the DNA further, two protocols were followed; for sediment CTD 241–861 an ion-exchange hydroxyapatite column was used. Dry HTP-hydroxyapatite (Bio-Rad, USA) was resuspended in TE-phosphate buffer (10 mM Tris pH 8.0, 1 mM EDTA 25 mM Na-phosphate, pH 8.0), swirled and decanted three to four times to get rid of ultrafine particles using a home-made HTP column in a syringe at 1000 g. The final volume of the resin was between 0.6 and 0.8 ml. The column was then equilibrated with 10

**Table 1** Sample description, number of positive hits and total number of screened clones

| Sample number | Grain size | GPS COORDINATES | | Depth (m) | Number of positive clones | Total number of screened clones |
|---|---|---|---|---|---|---|
| | | Latitude | Longitude | | | |
| CTD241-86 | Clay | N73 13.521 | E16 20.547 | 475 | 11 | 384 |
| CTD249-119 | Clay | N77 8.920 | E31 16.667 | 191 | – | 1000 |
| HH596-1 | Sand | N79 12.820 | E19 18.976 | 0 | 8 | 2500 |

volumes of TE-phosphate buffer. The DNA solution was loaded onto the column and washed with increasing concentrations of sodium phosphate in TE-phosphate buffer (25 mM, 50 mM, 100 mM and 200 mM Na-phosphate pH 8.0). DNA was then eluted with TE-phosphate buffer containing 300 mM of Na-phosphate. Buffer exchange was performed using a Centricon 4 ml spin cartridge with a 100 kDa cut-off (Millipore, Germany). Sediment samples CTD 249–119 and HH 596–1 were purified using the Aurora DNA purifier from Boreal genomics, USA, which utilizes the SCODA (synchronous coefficient of drag alteration) DNA extraction technology [30]. Raw DNA was diluted to 5 ml in milliQ water and applied to the sample well of a precast gel cassette (1 % 0.25X TBE agarose gel, and 0.25× TBE buffer). The run parameters were as stated in the AURORA_HMW_DNA_SOIL_PROTOCOL, provided by the manufacturer.

The purified metagenomic DNA was quality checked by performing standard PCR targeting the 16S rRNA gene using universal primers (27 F and 1492R) and Taq polymerase.

## Creation of fosmid library, storage of clones and functional screening

The purified DNA was used with the Copy Control Fosmid Library Production Kit and the pCC1FOS Vector (Epicentre, USA) according to the manufacturer's protocol, to obtain the three metagenomic fosmid libraries. Colonies were picked and grown in 400 µl LB containing 12.5 µg/ml chloramphenicol and 10 % glycerol using 1.2 ml deepwell blocks (square wells) and sealed with "breathable" film (BREATHseal, Greiner bio-one, USA). Incubation was done in a plate shaker at 37 °C and 300 rpm. After colony picking, plates were re-sealed with alumina sealing film (alumaseal, Sigma-aldrich, USA) and a lid was put on before the plates were transferred to -80 °C for storage.

For detection of esterase activity, EPI300TM-T1R *E. coli* fosmid clones were transferred to Omni trays (Thermo Scientific Nunc, USA) containing LB agar medium, 12.5 µg/ml chloramphenicol and 1 % tributyrin as synthetic substrate. The replication of fosmid libraries was made by a 96 pin library copier (Thermo Scientific Nunc, USA). The appearance of a clear halo zone around colonies within 4 days at 20 °C was considered a positive indication of esterase activity.

## Fosmid purification and sequencing

The fosmid from the clone showing strongest esterase/lipase activity (evaluated by halo size) was included together with 167 randomly selected fosmids to be sequenced. Deepwell blocks (2.2 ml square wells) containing 1.5 ml of LB medium with 12.5 µg/ml chloramphenicol, and supplemented with 1X autoinduction solution (Epicentre, USA) were inoculated with the 168 fosmid-bearing clones. The

plates were incubated with shaking at 37 °C for 16 h. The fosmid DNA was then purified using the Montage 96 well kit from Millipore following the vacuum suction protocol. The resulting fosmid DNA was resuspended in 100 µl Tris buffer pH 8.0. The DNA concentration was measured using a Nanodrop Spectrophotometer at 260 nm. The concentration of DNA in each well was then adjusted to 120 ng/µl by adding more buffer. Pools of 7 × 24 fosmids were made by pipetting 4 µl of each of the 24 fosmid into 7 separate tubes. 7 individually tagged libraries were made from the pooled DNA, pooled again, and sequenced on the 454 GS-FLX machine (Roche, USA) using one half of a picotiter plate. The remaining fosmid DNA in the 96 well plates was utilized in end-sequencing by the Sanger method using BigDye chemistry and the primers T7 or EpiFOSF (forward) and EpiFOSR (reverse). All sequencing was performed at the Norwegian Sequencing Centre (NSC) in Oslo.

## Assembly and analysis of fosmid sequences

The sequence reads were screened for vector- and *E. coli* DNA and assembled using the Newbler Assembler software (454 Life Sciences), accessed remotely through the Bioportal in Oslo (now changed to Lifeportal, https://lifeportal.uio.no/). The 7 pools of sequences were separated according to their MID (Multiplex Identifier) and assembled individually. The Sanger end-sequences were then used to distinguish, within each pool, which fosmid-clone each contig originated from. This was done by local nucleotide blast searches against the assembled fosmid DNA. The complete insert belonging to the fosmid-clone showing esterase activity was further annotated and analyzed using GeneMark [31] (http://opal.biology.gatech.edu/). The GC content profile of the fosmid-DNA was analyzed online using EMBOSS Isochore with default settings (http://www.ebi.ac.uk/Tools/seqstats/emboss_isochore/). The fosmid insert containing the *lip3* gene has been deposited [GenBank: KJ538549].

## Gene cloning strategy

The *lip3* gene was amplified from purified fosmid DNA using a cloning method termed *FastCloning* [32]. The following primer pairs were used to PCR amplify pET-26b vector and insert separately:

VecFw 5 -TGTCTTAAGAGCTTACTGCACCACCAC CACCACCAC -3 ,

VecRv 5 -CTATCTATTATGTAATTATTCATATGTAT ATCTCCTTCTTAAAGTT-3 ,

InsertFw 5 -AACTTTAAGAAGGAGATATACATATG AATAATTACATAATAGATAG-3 ,

InsertRv 5 -GTGGTGGTGGTGGTGGTGCAGTAAG CTCTTAAGACA-3 . The expression vector encodes an in-frame C-terminal 6xHis-Tag. The PCR reaction conditions used were: 1 cycle (98 °C for 3 min), 20 cycles

(98 °C 15 s, 55 °C 30 s, and 72 °C 1 min), and a final cycle at 72 °C for C 10 min. PCR reactions were performed in a MJ Research PTC 200 thermal cycler (MJ Research, Canada). DpnI (Sigma-Aldrich, USA) was added to the PCR insert- and vector product separately. Vector and insert were mixed at a ratio of 1:4 and incubated for 2 h at 37 °C. The mixture was then used to transform NovaBlue Giga Singles competent cells (Novagen, Germany). The DNA sequence of the resulting construct was verified by bidirectional DNA sequencing. The expression vector containing lip3 was then transformed into E. coli BL21 (DE3) cells.

### Recombinant production and purification of Lip3

E. coli BL21 (DE3) cells carrying pET-26b-Lip3 vector were cultivated in Luria Broth (LB) medium with 50 µg/mL kanamycin for 16 h at 37 °C. To induce protein expression, overnight culture was diluted to an OD 600 nm of 0.1 in 3-L shake flasks containing 600 ml LB medium and antibiotic (50 µg/ml kanamycin). Cultures were grown at 37 °C with an agitation rate of 140 rpm until the OD 600 nm reached 0.6. IPTG was then added to a concentration of 0.2 mM to induce the expression. The culture was incubated for a further 16 h at 20 °C. Cells were then harvested by centrifugation at 3200 g at 4 °C for 30 min and frozen at –20 °C. The pellet was resuspended in 50 mM Tris-HCl pH 8.0, 500 mM NaCl and 10 % glycerol, sonicated, and cleared by ultracentrifugation at 75,000 g for 40 min. The crude extract was filtered using a 0.45 µm membrane, and loaded on a HisTrap HP 1 ml column (GE Healthcare, England) equilibrated with 50 mM Tris-HCl pH 8.0, 500 mM NaCl, 30 mM imidazole, 10 % glycerol. Lip3 was eluted with a linear imidazole gradient (10 ml of 0–500 mM). Fractions of 1 mL were collected and analyzed by SDS-PAGE, with Lip3 being detected by the presence of a band at the expected molecular weight after Comassie staining. Fractions containing the recombinant enzyme, were dialyzed against 20 mM Tris-HCl pH 8.0, 10 mM NaCl and 5 % glycerol at 4 °C overnight.

The recombinant protein was further purified using a 1 ml HiTrap Q HP column (GE Healthcare, England) equilibrated with buffer A (20 mM Tris-HCl pH 8, 10 mM NaCl, 5 % glycerol) and eluted with a linear gradient of 0–100 % of buffer B (20 mM Tris-HCl pH 8, 1 M NaCl, 5 % glycerol) at a flow rate of 1 ml/min. The proteins containing the esterase activity eluted at approximately 50 % buffer B.

SDS-PAGE was performed using 5 % polyacrylamide-stacking gel and a 12 % polyacrylamide-resolving gel with a Bio-Rad Mini-Protean II cell unit, at room temperature essentially as described by Laemmli. Opti-Protein XL protein molecular mass marker (ABM, Canada) was used as molecular weight standard. The protein concentration was determined according to the Bradford method with bovine serum albumin as the standard [33]. The protein content was measured by monitoring the optical density at 595 nm.

### Lipolytic activity assays

The lipolytic activities of the purified enzyme were determined by measuring the hydrolysis of synthetic substrates labeled with p-nitrophenol (p-NP). The reaction progress was followed by monitoring the absorbance at 405 nm in 1-cm path-length cells with a Cary 100 spectrophotometer (Varian, Australia), equipped with a temperature controller. To check the linearity of the reaction, two different concentrations of enzyme were tested for each condition. Stock solutions of p-nitrophenyl (p-NP)-esters were prepared by dissolving the substrates in pure acetonitrile. Assays were performed in 1 mL mixture containing purified enzyme (2 µg/mL), 100 mM Tris-HCl buffer (pH 8.0), 3 % acetonitrile and p-NP esters at different concentrations.

One unit of esterase activity was defined as the amount of enzyme needed to release 1 µmol p-NP in 1 min. All experiments were performed in triplicate. Results are expressed as mean values ± SE of the mean.

### Substrate specificity and enzyme kinetics

The substrate specificity of the esterase was investigated by measuring enzymatic activity toward a series of p-NP esters with various carbon chain lengths: p-NP acetate (C2), p-NP butanoate (C4), p-NP pentanoate (C5), p-NP octanoate (C8), p-NP decanoate (C10). Assays were carried out in duplicate at 35 °C by following the absorbance at 405 nm, and the kinetic parameters were determined from the rates of hydrolysis by fitting the rates to a Lineweaver-Burk double reciprocal plot. All kinetic data were analyzed by linear regression using SigmaPlot 10.0.

### Effects of pH and temperature on Lip3 activity

Esterase activity was measured at different pHs by using the buffers 0.1 M MES (pH 5.0–6.0), 0.1 M Na-phosphate (pH 6.0–7.5), 0.1 M Tris-HCl (pH 7.5–9.5) and 0.1 M CAPS (pH 9.5–10.5). The esterase activity at 25 °C was monitored by the amount of p-nitrophenol released from p-nitrophenyl (p-NP) esters at 348 nm, which is the pH-independent isosbestic wavelength of p-nitrophenoxide and p-nitrophenol. A molar extinction coefficient of 5000 $M^{-1}$ $cm^{-1}$ at 25 °C was used in the calculations.

The activity was expressed as percent relative activity with respect to maximum activity, which was considered as 100 %.

Esterolytic activity as a function of temperature was determined in the range of 10–65 °C with 5 °C increments using p-NP pentanoate (100 µM) as a substrate. The reaction buffer was 0.1 M Tris-HCl (pH 8.0) as this was determined to be optimal from the pH screening, and contained 3 % acetonitrile.

## Thermal stability of the esterase

The thermostability of the enzyme was examined at temperatures ranging from 25 to 70 °C. The enzyme was incubated at different temperatures for a total time of 2 h, and the residual activity was measured at 20-min intervals under standard conditions.

## Effect of additives and NaCl on enzyme activity

The effect of various additives on esterase activity was tested by incubating the protein for 1 h at 4 °C in presence of β-mercaptoethanol, EDTA, DTT or PMSF at final concentrations of 1 mM and 10 mM. The residual activities were measured by comparison with activity from the standard assay, containing no compounds, and defined as 100 %.

The effect of NaCl on esterase activity was investigated by assaying with increased salt concentrations in a range from 0 to 4 M at 35 °C using standard assay conditions. Additionally, assays were carried out after incubation of the enzyme in presence of 0, 1, 2, 3 M NaCl for 24 h at 4 °C.

## Sequence analysis

The Lip3 sequence was investigated for protein homology by searching the complete non-redundant protein databases (www.ncbi.nlm.nih.gov) using the BLAST software [34]. A multiple sequence alignment was constructed using the JalView software (Fig. 6) [35]. The N-terminal signal peptide prediction was made using SignalP 3.0 (http://www.cbs.dtu.dk/services/SignalP/) [36]. Molecular weights were determined using Protein Calculator V3.3 (http://www.scripps.edu/~cdputnam/protcalc.html).

## Lip3 modeling

Lip3 was modeled using four templates, obtained by scanning the Protein Data Bank database with the HHpred server [37]: a triacylglycerol lipase from *Yarrowia lipolytica* [PDB: 3O0D], a lipase from *Gibberella zeae* [PDB: 3NGM], a lipase from *Penicillium expansum* [PDB: 3G7N] and a lipase from *Serratia marcescens* [PDB: 2QUB]. The atomic coordinates of the templates were obtained from the Protein Data Bank. In order to create the 3D-model, the multiple sequence alignment between Lip3 and the sequences of the four templates, obtained in PIR format by HHpred server [32], was submitted to the comparative protein structure modeling software Modeller 9v11 [38]. Modeller algorithm was set to generate 53-dimensional models. To select the best model, structure validation was carried out by PDBsum pictorial database. In order to evaluate the stereochemical quality of the generated structures, models were uploaded, in the standard PDB file format, to the PDBsum server, to carry out a full set of Procheck structural analyses [39]. Moreover, the Z-score of the Lip3 model was calculated using the WhatIf web server [40]. The Z-score expresses how well the backbone conformations of all

residues correspond to the known allowed areas in the Ramachandran plot. The Solvent Accessible Surface Area (SASA) of the templates and Lip3 model was calculated by POPS algorithm [41]. Furthermore the electrostatic potential of Lip3 external surface was performed by APBS (Adaptive Poisson-Boltzmann Solver) and PDB2PQR software packages [42, 43]. Finally the molecular graphics software VMD [44] was used to display the obtained model (Fig. 7).

## Results

### Construction of metagenomic libraries and screening for lipolytic enzymes

Three small fosmid libraries were created from marine sediment samples. The total number of fosmid clones and the number of positive hits in the screening are shown in Table 1, together with the exact GPS coordinates.

A total of 19 fosmid clones showed a clear halo zone indicative of putative lipolytic activity (see Additional file 1). The clone displaying the largest halo size was sequenced, and a gene encoding a putative class 3 lipase was found. As shown in Fig. 1, we could not detect any clear phylogenetic relatedness of the ORFs encoded by the fosmid. Each ORF is similar to a variety of unrelated bacteria and no phylogenetic marker gene is present. The presence of four transposase genes in the fosmid indicates that the region which is cloned comes from a region within the host DNA with rearrangements and/or insertions due to the transposases. This is consistent with its highly variable GC-content profile (data not shown). An 843-bp ORF encoding a putative esterase/lipase (designated Lip3) was identified by GeneMark. The low sequence conservation, between Lip3 and the eight lipolytic families described by Arpigny [13], did not allow the construction of a meaningful phylogenetic tree for the full dataset. Despite this a sequence analysis using the Pfam protein families identified a α/β-hydrolase family sequence (Pfam01764) belonging to class 3 lipases. The lipase catalytic triad (serine, aspartate and histidine) is indicated (Fig. 7). Lip3 contain no signal peptide as predicted by SignalP.

### Expression and purification of recombinant Lip3

In order to study the biochemical properties of the enzyme, the *lip3* gene was cloned into pET26b in frame with the C-terminal 6x His tag encoded by the vector.

High amounts of active protein were achieved when *E. coli* BL21 (DE3) was induced overnight with 0.2 mM IPTG at 20 °C. The expressed protein was purified to homogeneity with a yield of 1.5 mg of protein from 2 L of cell culture and a SDS-PAGE analysis shows that under denaturing condition the molecular weight (MW) is around 31.2 kDa (see Additional file 2).

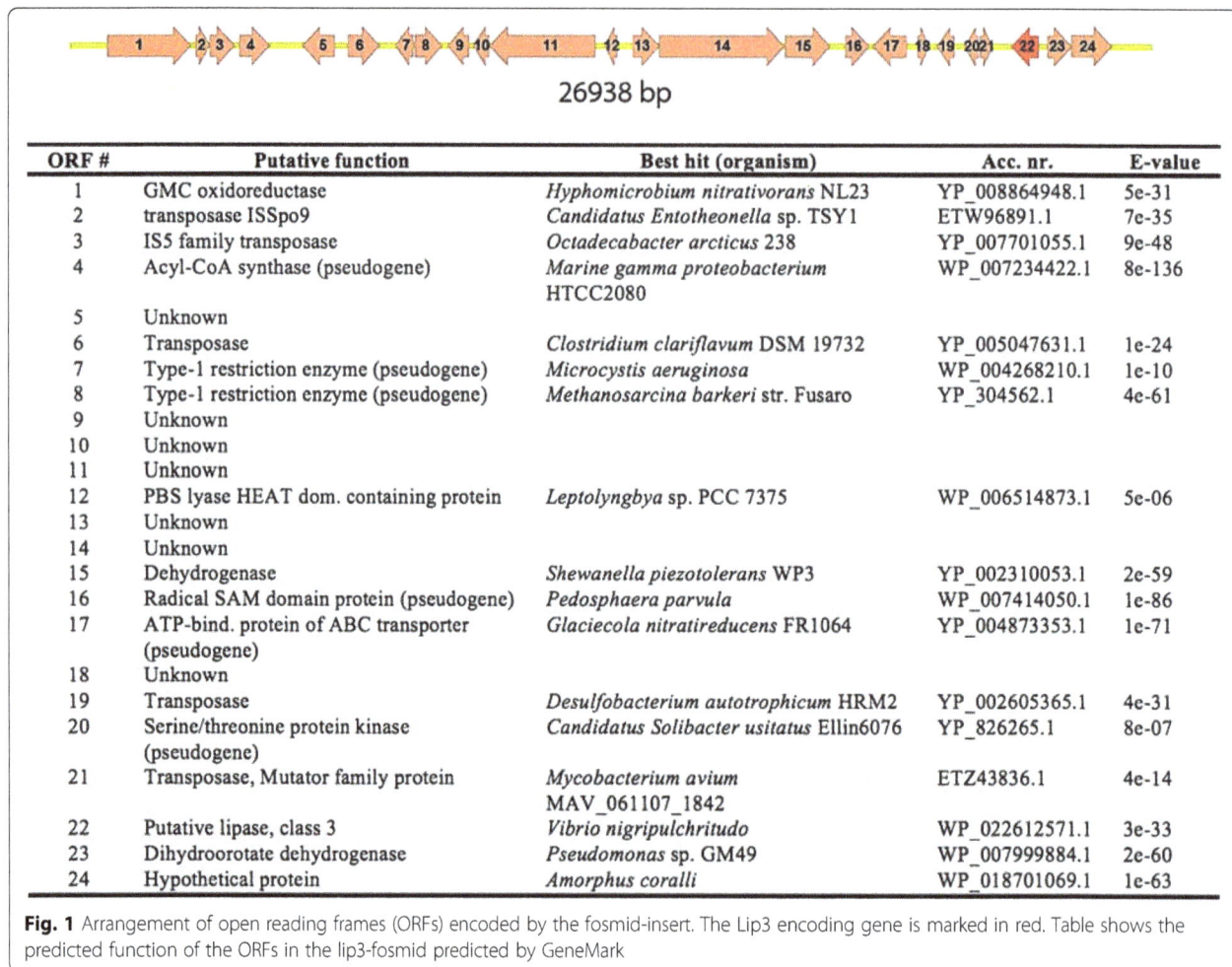

| ORF # | Putative function | Best hit (organism) | Acc. nr. | E-value |
|-------|-------------------|---------------------|----------|---------|
| 1 | GMC oxidoreductase | *Hyphomicrobium nitrativorans* NL23 | YP_008864948.1 | 5e-31 |
| 2 | transposase ISSpo9 | *Candidatus Entotheonella* sp. TSY1 | ETW96891.1 | 7e-35 |
| 3 | IS5 family transposase | *Octadecabacter arcticus* 238 | YP_007701055.1 | 9e-48 |
| 4 | Acyl-CoA synthase (pseudogene) | *Marine gamma proteobacterium* HTCC2080 | WP_007234422.1 | 8e-136 |
| 5 | Unknown | | | |
| 6 | Transposase | *Clostridium clariflavum* DSM 19732 | YP_005047631.1 | 1e-24 |
| 7 | Type-1 restriction enzyme (pseudogene) | *Microcystis aeruginosa* | WP_004268210.1 | 1e-10 |
| 8 | Type-1 restriction enzyme (pseudogene) | *Methanosarcina barkeri* str. Fusaro | YP_304562.1 | 4e-61 |
| 9 | Unknown | | | |
| 10 | Unknown | | | |
| 11 | Unknown | | | |
| 12 | PBS lyase HEAT dom. containing protein | *Leptolyngbya* sp. PCC 7375 | WP_006514873.1 | 5e-06 |
| 13 | Unknown | | | |
| 14 | Unknown | | | |
| 15 | Dehydrogenase | *Shewanella piezotolerans* WP3 | YP_002310053.1 | 2e-59 |
| 16 | Radical SAM domain protein (pseudogene) | *Pedosphaera parvula* | WP_007414050.1 | 1e-86 |
| 17 | ATP-bind. protein of ABC transporter (pseudogene) | *Glaciecola nitratireducens* FR1064 | YP_004873353.1 | 1e-71 |
| 18 | Unknown | | | |
| 19 | Transposase | *Desulfobacterium autotrophicum* HRM2 | YP_002605365.1 | 4e-31 |
| 20 | Serine/threonine protein kinase (pseudogene) | *Candidatus Solibacter usitatus* Ellin6076 | YP_826265.1 | 8e-07 |
| 21 | Transposase, Mutator family protein | *Mycobacterium avium* MAV_061107_1842 | ETZ43836.1 | 4e-14 |
| 22 | Putative lipase, class 3 | *Vibrio nigripulchritudo* | WP_022612571.1 | 3e-33 |
| 23 | Dihydroorotate dehydrogenase | *Pseudomonas* sp. GM49 | WP_007999884.1 | 2e-60 |
| 24 | Hypothetical protein | *Amorphus coralli* | WP_018701069.1 | 1e-63 |

**Fig. 1** Arrangement of open reading frames (ORFs) encoded by the fosmid-insert. The Lip3 encoding gene is marked in red. Table shows the predicted function of the ORFs in the lip3-fosmid predicted by GeneMark

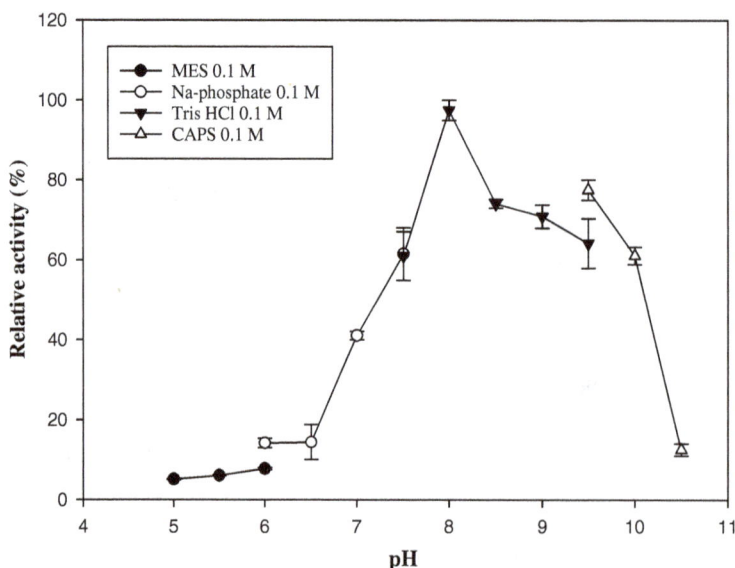

**Fig. 2** Effect of pH on Lip3 activity. Relative activity of *p*-NP-pentanoate (100 M) hydrolysis was performed in various pH buffers at 25 °C

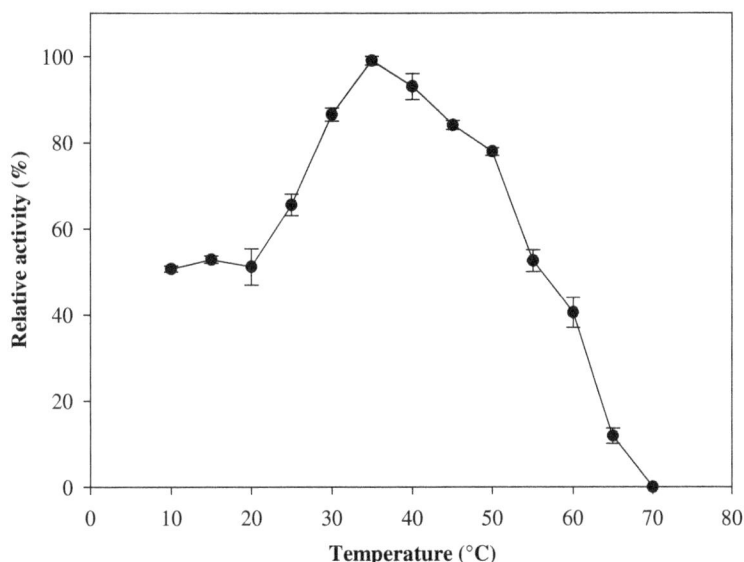

**Fig. 3** Effect of temperature on Lip3 esterase activity. The activity was determined at different temperatures in 0.1 M Tris-HCl buffer adjusted to pH 8.0 using p-NP-pentanoate (100 M) as substrate

## Effect of pH and temperature on enzyme activity and stability

The effect of pH on esterase activity was assessed at 25 °C (Fig. 2) using p-NP-C5 as substrate. The enzyme in the pH range of 5.0–10.5 showed maximal activity at pH 8.0 in Tris-HCl buffer.

The effect of temperature on esterase activity was determined using p-NP-C5 as substrate. Lip3 was active over a temperature range from 7 to 65 °C (Fig. 3), with an optimum temperature of 35 °C. No catalytic activity was detectable at 70 °C.

Lip3 hydrolysed p-NP esters with acyl-chain lengths from two to ten carbon atoms (C2–C10). All characterization was performed at 35 °C and in presence of 3 % acetonitrile using different stock solution in a range of 7-50 mM. In this condition, the enzyme displays highest activity towards p-NP acetate (Table 2) with a highest $k_{cat}$ and $k_{cat}/K_M$ value. Enzymatic activity was assayed in triplicate with an appropriate blank for the correction of the auto hydrolysis of the substrate. This result showed that the enzyme is an esterase and p-NP acetate was used as its preferred substrate for further studies.

To examine the thermal stability of Lip3 esterase, we pre-incubated the enzyme at different temperatures and measured the residual activity under standard assay conditions. The enzyme displayed a relatively high thermal stability at 60 °C, retaining approximately 80 % of its activity even after incubation for 120 min (Fig. 4). However, the stability of the enzyme decreased significantly after only 20 min at 70 °C.

## Lip3 activity in presence of additives and NaCl

The enzyme was unaffected by the presence of low concentrations of β-mercaptoethanol and DTT, while low concentrations of PMSF and EDTA gave a small but detectable decrease in activity. The relative activity was approximately halved when the additive concentration was tenfold higher (Table 3).

Lip3 activity was evaluated in presence of NaCl under the above mentioned assay conditions in 0.1 M Tris-HCl pH 8.0. Results show the activating effect of NaCl on Lip3, with the highest activity value being obtained in 3 M NaCl (Table 4). To test the stability of the enzyme with increasing amount of NaCl we measured the

**Table 2** Kinetic parameters for Lip3

| Substrate | $k_{cat}$ (s$^{-1}$) | $K_M$ (M·10$^{-3}$) | $s = k_{cat} / K_M$ (sec$^{-1}$·M$^{-1}$·10$^3$) |
|---|---|---|---|
| pNP-acetate | 1198 ± 200 | 2.38 ± 0.56 | 503.3 ± 214.5 |
| pNP-butanoate | 218.0 ± 15.5 | 1.05 ± 0.16 | 206.5 ± 48.6 |
| pNP-pentanoate | 152.8 ± 9.82 | 1.52 ± 0.20 | 100.2 ± 19.7 |
| pNP-octanoate | 100.0 ± 3.8 | 0.36 ± 0.04 | 271.0 ± 44.0 |
| pNP-decanoate | 29.9 ± 1.5 | 1.03 ± 0.12 | 29.3 ± 10.1 |

Esterase activity was measured at 35 °C for five minutes in presence of 0.1 M Tris-HCl pH 8.0

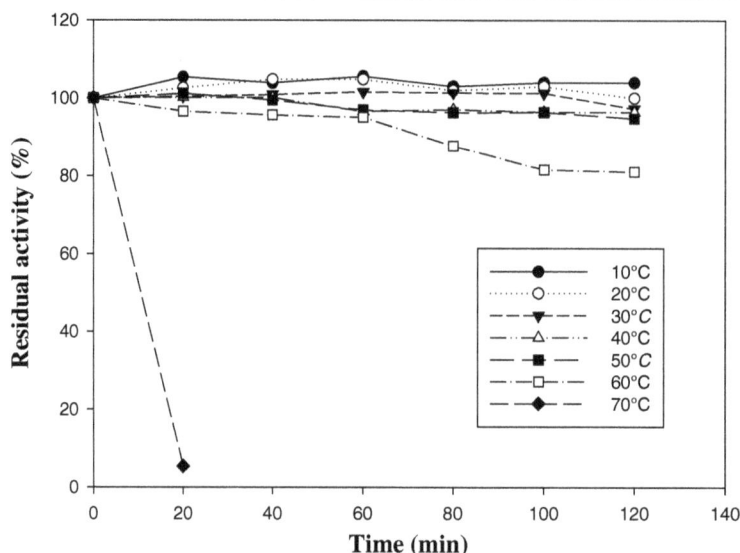

**Fig. 4** Thermal stability of Lip3 esterase. Activity was measured in the range from 10–70 °C using *p*-NP-acetate (100  M) as substrate

relative activity after incubation for 24 h at 4 °C. The stability approximately increased by a factor of four in presence of 3 M NaCl (Fig. 5) and a factor of approximately two in presence of 2 M NaCl.

### Analysis of the Lip3 sequence

A multiple sequence alignment, consisting of Lip3 together with the most similar amino acid sequences is shown in Fig. 6. Similarities were found with lipases from *Vibrio scophthalmi* and *Vibrio ichthyoenteri*, hypothetical proteins from *Flexthrix Dorotheae* and *Pseudanabaena sp.* PCC 6802, and a putative lipase from *Vibrio nigripulcritudo*. These sequences share only about 30 % identity and 50 % similarity with the Lip3 sequence. The alignment (Fig. 6) reveals that Lip3 contains the lipase-conserved catalytic triad residues, Asp207, His267 and the catalytic nucleophile Ser150, in the typical consensus pentapeptide G-X-S-X-G, also known the nucleophilic elbow.

### Lip3 modeling

The 3D-modeling of Lip3 was performed by homology modeling using the following structures as templates: a triacylglycerol lipase from *Yarrowia lipolytica* [PDB ID: 3OOD], a lipase from *Gibberella zeae* [PDB: 3NGM], a lipase from *Penicillium expansum* [PDB: 3G7N] and a lipase from *Serratia marcescens* [PDB: 2QUB]. These structures were chosen because they were predicted to have a significant structural homology with Lip3, calculated by HHpred server, despite a low percentage of sequence identity. Starting from the alignment of Lip3 sequence with the reference structures, a set of 50 all atoms models was generated. The best model (Fig. 7a) was selected in terms of energetic and stereochemical quality. In detail, it has 86.3 % of residues in the most favored regions and no residues in disallowed regions of the Ramachandran plot according to the PROCHECK program provided with PDBSum. Moreover this model has a WhatIf Z-score of –0.297, which is within expected ranges for well-refined structures. These values, compared with those of the template structures, indicated that a good quality model was created. The Lip3

**Table 3** Effect of different additives on Lip3 activity

| Additives | Relative activity % | |
| --- | --- | --- |
| | Concentration | Concentration |
| | 1 mM | 10 mM |
| Control | 100 ± 0.007 | 100 ± 0.004 |
| β-mercaptoethanol | 96.0 ± 0.005 | 56.0 ± 0.002 |
| EDTA | 91.0 ± 0.001 | 78.7 ± 0.003 |
| DTT | 106.0 ± 0.021 | 59.0 ± 0.005 |
| PMSF | 77.0 ± 0.011 | 55.0 ± 0.004 |

Esterase activities were measured toward various compounds at 35 °C in presence of 0.1 M Tris-HCl pH 8.0

**Table 4** Effect of NaCl on Lip3 activity

| NaCl (M) | Relative activity (%) |
| --- | --- |
| 0 | 100.0 ± 0.093 |
| 1 | 284.3 ± 0.107 |
| 2 | 334.3 ± 0.003 |
| 3 | 675.0 ± 0.021 |
| 4 | 528.0 ± 0.007 |

Esterase activity was measured at different NaCl concentrations using *p*-NP-acetate as substrate

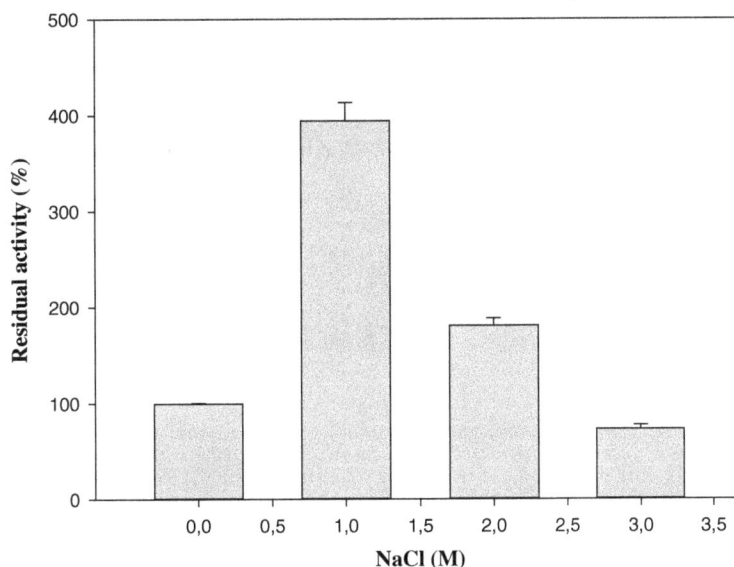

**Fig. 5** Stability profile of Lip3 with NaCl. Lip3 activity was evaluated after preincubation in presence of NaCl at 4 °C for 24 h

model displayed an alpha-beta structure characterized by 9 alpha-helices, three 310-helices and 11 beta-strands forming three sheets (Fig. 7), corresponding to 35, 5.4 and 16.8 % of sequence, respectively. Moreover the Lip3 structure seems to be stabilized by a disulfide bond (Cys179-Cys210). The SASA analysis of the Lip3 model carried out by the POPS algorithm revealed an exposed surface that is more hydrophobic than hydrophilic (59.6 % versus 40.4 %, see Additional file 3). On the contrary, the SASA analysis of the template structures always revealed an exposed surface that is less hydrophobic than hydrophilic (38.5 % versus 61.5 % for the *Penicillium expansum* lipase, 32.7 % versus 67.3 % for the *Gibberella zeae* lipase, 34.3 % versus 65.7 % for the *Yarrowia lipolytica* triacylglycerol lipase, 35.7 % versus 64.3 % for the *Serratia marcescens* lipase). Finally the analysis of the electrostatic potential of the external surface of Lip3 revealed the presence of positive and negative charged areas (see Additional file 3) around the active site.

## Discussion

In the current study we have identified a cold-active esterase (Lip3) originating from marine clay sediment, as result of functional screening of three small fosmid libraries. The choice of sampling area was based on the knowledge that marine habitats represent a vast resource of novel lipolytic genes due to the numerous microorganisms living there. Lipids from phytoplankton are one of the principal nutrition sources in marine food chains [45] and these can be removed from the surface layers by microbial activity when falling through the water column and lying on the benthos [46, 47]. The number of

clones positive for esterase activity was very high for sample number 1 (11 out of 384). We do not have an explanation for this, however the sample sites were not carefully inspected, and there are probably great variations of nutrients available at the respective sites. Additionally, the occurrence of false positives is a well-known phenomenon when using tributyrin as substrate [48]. As we did not sequence more of the positive fosmids, we can only speculate the reasons for this high hit-rate.

To address the biotechnological potential of Lip3 as a biocatalyst, we performed a biochemical and structural characterization of the recombinant enzyme. Lip3 showed activity at temperatures as low as 10 °C and has an optimum temperature of 35 °C, so it can be defined as a cold-active esterase. The highest relative activity was observed at pH 8.0 while a loss of activity was observed at pHs below 7.0 as reported for other identified esterases [49]. The sudden drop in activity at acidic pH can be explained by deprotonation of His267 (pKa 6.08). The drop at alkaline pH could be due to the protein denaturation rather than protonation changes in the active site residues.

The substrate specificity experiment revealed that in vitro the enzyme functions best as an esterase with a preference for $p$-NP esters of short- and medium chain fatty acids and could not hydrolyze substrates with long-chain fatty acids. In agreement with our results, another esterase from sea sediment metagenome was also found to have the highest hydrolytic activity with short- and middle- length $p$-NP esters [24].

In addition, Lip3 activity was evaluated in the presence of DTT, EDTA, β-mercaptoethanol and PMSF, at low and high concentrations. An inhibitory effect was observed at high concentrations additives. The inhibition

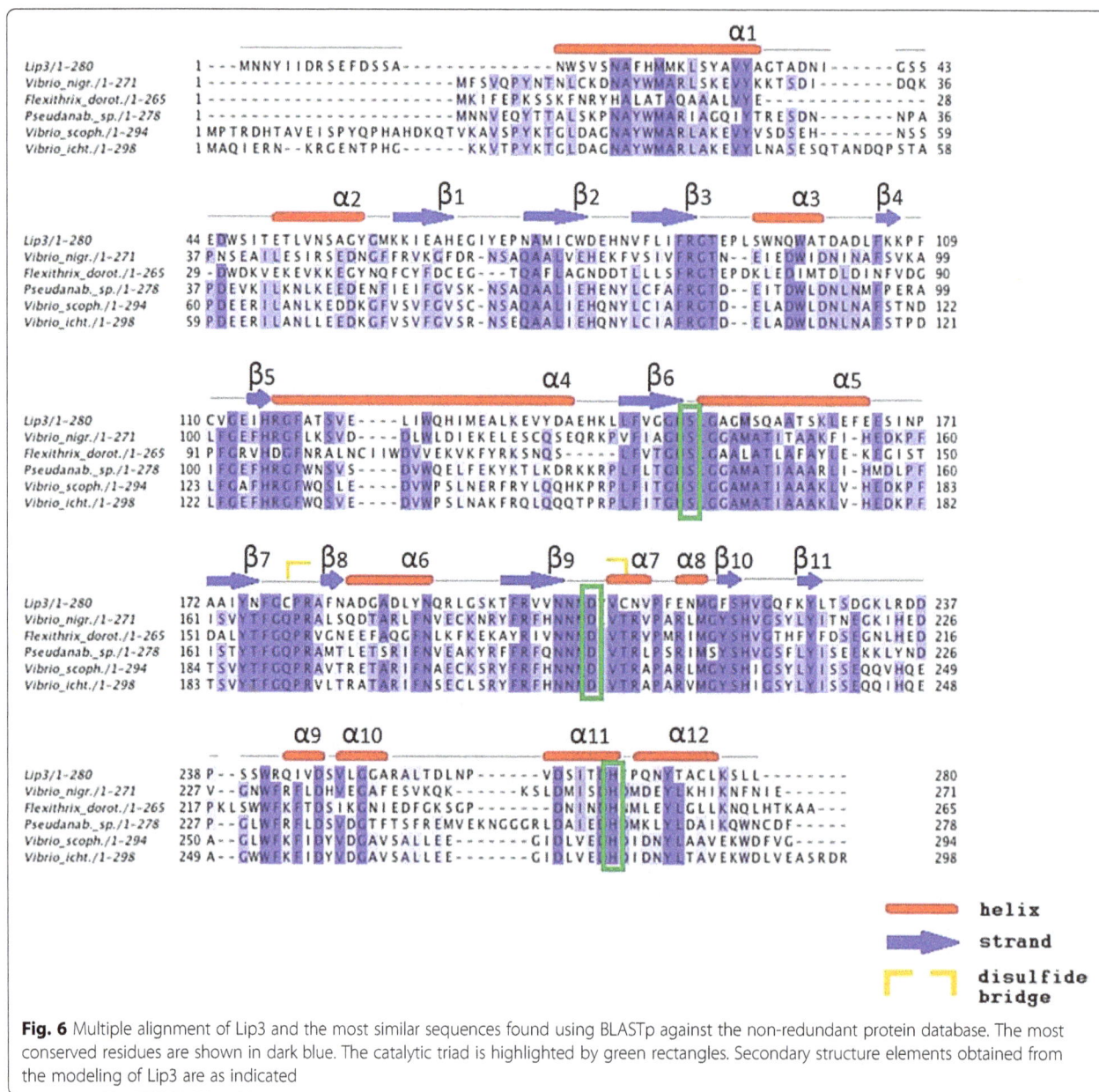

**Fig. 6** Multiple alignment of Lip3 and the most similar sequences found using BLASTp against the non-redundant protein database. The most conserved residues are shown in dark blue. The catalytic triad is highlighted by green rectangles. Secondary structure elements obtained from the modeling of Lip3 are as indicated

of esterase activity in the presence of EDTA can be attributed to its metal chelating effect. The weak inhibition of esterase activity caused by PMSF may be attributed to the attack of Ser150 responsible for the hydrolytic reaction in the active site since this inhibitor interacts selectively and irreversibly with the serine hydroxyl groups [50]. The relatively low reactivity could be explained by the low stability of PMSF in the assay conditions. A hypothetical disulfide bond between Cys179 and Cys210 was seen in the model, and the high concentrations of DTT and β-mercaptoethanol probably reduce this bond, thereby lowering the enzyme activity. Being an intracellular enzyme, the disulfide bonds rarely form due to the reducing environment. The surface exposure of the

disulfide bond might be low, so that it is more shielded from being naturally reduced.

The enzyme thermostability studies showed a quite stable behavior up to 60 °C for 1 h which decreased at higher temperatures. This is in contrast to the stability of a previously published cold-active lipase EstF, from deep-sea metagenomic library, which was stable up to 50 °C and had a dramatic decrease thereafter [51].

The Lip3 enzyme thus shows stability at moderate temperatures, and because of its high catalytic efficiency and specificity at low and moderate temperatures, it could be used in improving biotechnological or industrial processes. Applications may include the use of the enzyme as catalyst for organic synthesis of unstable

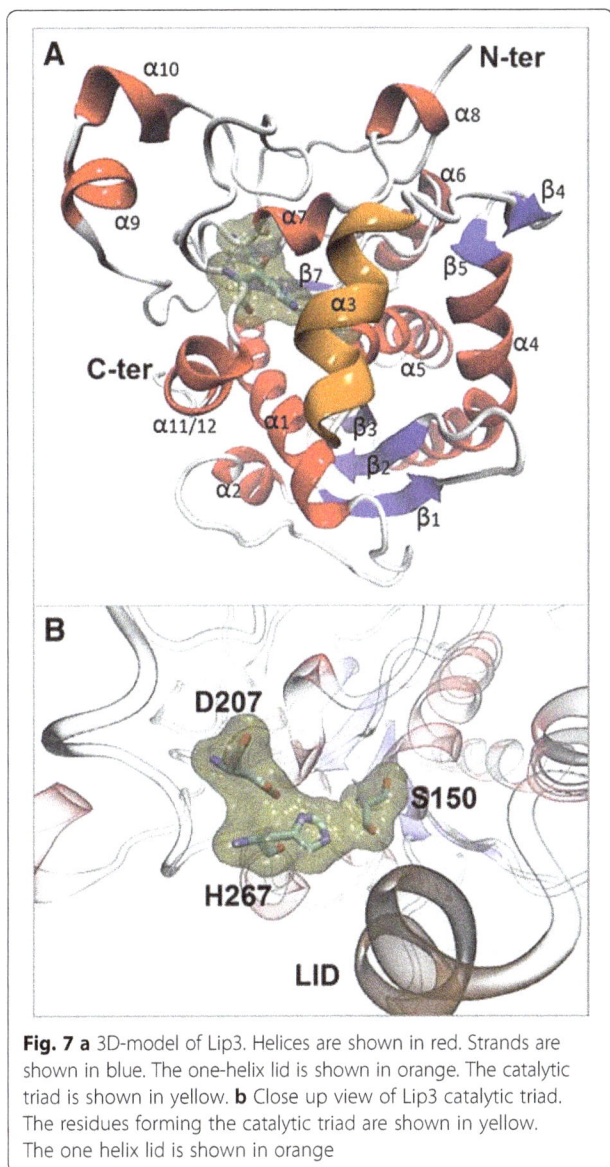

**Fig. 7 a** 3D-model of Lip3. Helices are shown in red. Strands are shown in blue. The one-helix lid is shown in orange. The catalytic triad is shown in yellow. **b** Close up view of Lip3 catalytic triad. The residues forming the catalytic triad are shown in yellow. The one helix lid is shown in orange

The solvent accessible surface area analysis suggests that this enzyme, thanks to its prevailing hydrophobic surface (59.6 %), could be adapted to hydrophobic environments, such as organic solvents, despite the opposite trend shown by the 4 templates (38.5 %, 32.7 %, 34.3 %, 35.7 %). In particular, the active site is covered by a mobile element, the lid (Fig. 7a), which opens when the enzyme binds a hydrophobic interface [54]. The surface exposed to the active site is hydrophobic, whereas the surface exposed to the outside of the enzyme is hydrophilic [55–57]. When the lid opens and exposes its hydrophobic surface, the SASA (solvent-accessible surface area) increases drastically [58]. Depending on the structure of the lid, different transition mechanisms have been proposed. In lipases with a simple one-helix lid, it is assumed that the transition is a fast rigid body movement [59]. In lipases with a more complex lid, instead, the secondary structure of this mobile element changes when it opens, undergoing a partial refolding [52] which might be a kinetic bottleneck [60]. Lip3 seems to belong to the first category, having a single-helix lid (Fig. 7a, α3) and it shows a higher catalytic activity if compared to other kinetic values of esterases and lipases described in the literature [61, 62].

It is already known that salt has a significant impact on the protein stability [63]. In our results, as shown in Fig. 5 and in the additional file 3, hydrate ions can have interacted with surface residues to stabilize the folded Lip3 conformation in presence of high NaCl concentrations. The theoretical isoelectric point (pI) of Lip3 is 4.98 which means that the global negative net charge of the esterase would be negative in a buffer solution of pH 8.0. Under these conditions, the substrates *p*-nitrophenyl esters are positively charged. The increase of the activity at high salt concentrations might be due to a salt effect on the hydrophobic interaction between the active site of Lip3 and the substrate. The decrease of Lip3 activity at really high salt concentrations might be due to a screening effect by salt of charge-charge amino acids interactions [64].

Crystallization of Lip3 will clearly be of value in future studies. Afterwards, structural studies of docking interactions between Lip3 and natural fat substrates will be a key to develop an improved mutant esterase which can ideally be extensively used in the dairy industry.

## Conclusions

In summary, a novel cold-active esterase, Lip3, was isolated from a metagenomic library constructed from an Arctic marine sediment sample. A homology model of the enzyme highlighted the presence of a mobile element (lid) covering the active site consisting in a catalytic triad (Ser150, Asp207 and His267). Moreover the 3D-model, despite the presence of charged areas in the

compounds at low temperature [52]. Due to the high stability in presence of NaCl, Lip3 could also be used in food technology applications [53] to accelerate cheese flavor during ripening. The advantage is to use this enzyme at the ripening temperature (7 to 53 °C) followed by an easy inactivation at higher temperatures to avoid the development of strong rancid flavor. Lip3 may be added individually or as a cocktail with other lipases.

Structural analysis was carried out focusing on the most conserved regions in the sequence alignment (Fig. 6). In Fig. 7b, the catalytic triad of Lip3, formed by Ser150, Asp207 and His267, is highlighted in yellow. The three amino acids forming the catalytic triad seem to be located in three different loops connecting three strands to three helices. In particular Ser150 is within a conserved motif called a "nucleophilic elbow".

external surface, revealed a surface area more hydrophobic than hydrophilic. Lip3 showed high activity both at low temperatures and in presence of high salt concentrations. These are very useful characteristics for biotechnological processes. The described esterase with its characteristics is a valuable contribution to the expanding enzymatic toolbox used by academia and the biotechnological industry.

## Additional files

> **Additional file 1:** Functional screening of a fosmid library on 1 % tributyrin (glycerol tributyrate) agar plates incubated at 20 °C. A clear halo zone is indicative of putative lipolytic activity. (PNG 480 kb)
>
> **Additional file 2:** SDS-PAGE analysis of the purified recombinant esterase, Lip3. Lane 1, purified esterase, Lip3 (31.2 kDa); Lane 2, Opti-Protein XL protein molecular mass marker. (PNG 224 kb)
>
> **Additional file 3:** Graphical representations of the electrostatic potential in the external surface of Lip3 model. Positive charges are shown in blue. Negative charges are shown in red. Uncharged areas are shown in white. Catalytic triad cloud is shown in yellow. (PNG 1699 kb)

## Abbreviations

DTT: Dithiothreitol; EDTA: Ethylene-Diamine-Tetra-Acetic acid; EtOH: ethanol; IPTG: Isopropyl β-D-1-thiogalactopyranoside; LB: Luria-Bertani; MW: Molecular Weight; OD: optical density; PCR: polymerase chain reaction; PDB ID: Protein Data Bank Identifier; PMSF: Phenylmethanesulfonylfluoride; pNP: *para*-Nitrophenol; RCF: relative centrifugal force; SASA: Solvent Accessible Surface Area; SDS-PAGE: Sodium dodecyl sulfate - polyacrylamide gel electrophoresis; TE: Tris EDTA.

## Competing interests

The authors declare that they have no competing interests.

## Authors' contributions

NPW and DdP conceived the study. BA, MP and CDS isolated the DNA and constructed the metagenomic libraries. CDS screened the libraries and expressed, purified and characterized the esterase. LA created the sequence alignments and carried out the molecular modeling studies. CDS, BA, MP and LA wrote the manuscript. All authors discussed the results, revised and approved the final manuscript.

## Acknowledgments

This work was supported by grants No. 219710 and 192123 from the Research Council of Norway.

## Author details

[1]NorStruct, Department of Chemistry, Faculty of Science and Technology, UiT The Arctic University of Norway, Tromsø, Norway. [2]Institute of Protein Biochemistry, National Research Council, Naples, Italy.

## References

1. Alcalde M, Ferrer M, Plou FJ, Ballesteros A. Environmental biocatalysis: from remediation with enzymes to novel green processes. Trends Biotechnol. 2006;24(6):281–7.
2. Daniel R. The soil metagenome—a rich resource for the discovery of novel natural products. Curr Opin Biotechnol. 2004;15(3):199–204.
3. Lorenz P, Schleper C. Metagenome—a challenging source of enzyme discovery. J Mol Catal B Enzym. 2002;19–20:13–9.
4. Schloss PD, Handelsman J. Biotechnological prospects from metagenomics. Curr Opin Biotechnol. 2003;14(3):303–10.
5. Streit WR, Daniel R, Jaeger KE. Prospecting for biocatalysts and drugs in the genomes of non-cultured microorganisms. Curr Opin Biotechnol. 2004;15(4):285–90.
6. Handelsman J, Rondon MR, Brady SF, Clardy J, Goodman RM. Molecular biological access to the chemistry of unknown soil microbes: a new frontier for natural products. Chem Biol. 1998;5(10):R245–249.
7. Daniel R. The metagenomics of soil. Nat Rev Microbiol. 2005;3(6):470–8.
8. Lorenz P, Eck J. Metagenomics and industrial applications. Nat Rev Microbiol. 2005;3(6):510–6.
9. Streit WR, Schmitz RA. Metagenomics—the key to the uncultured microbes. Curr Opin Microbiol. 2004;7(5):492–8.
10. Handelsman J. Metagenomics: application of genomics to uncultured microorganisms. Microbiol Mol Biol Rev. 2004;68(4):669–85.
11. Glogauer A, Martini VP, Faoro H, Couto GH, Muller-Santos M, Monteiro RA, et al. Identification and characterization of a new true lipase isolated through metagenomic approach. Microb Cell Fact. 2011;10:54.
12. Jeon JH, Kim JT, Kim YJ, Kim HK, Lee HS, Kang SG, et al. Cloning and characterization of a new cold-active lipase from a deep-sea sediment metagenome. Appl Microbiol Biotechnol. 2009;81(5):865–74.
13. Arpigny JL, Jaeger KE. Bacterial lipolytic enzymes: classification and properties. Biochem J. 1999;343(Pt 1):177–83.
14. Brault G, Shareck F, Hurtubise Y, Lepine F, Doucet N. Isolation and characterization of EstC, a new cold-active esterase from Streptomyces coelicolor A3(2). PLoS One. 2012;7(3). e32041.
15. Lopes DBFL, Fleuri LF, Macedo GA. Lipase and esterase - to what extent can this classification be applied accurately? Cienc Tecnol Aliment. 2011;31:608–13.
16. Fojan P, Jonson PH, Petersen MT, Petersen SB. What distinguishes an esterase from a lipase: a novel structural approach. Biochimie. 2000;82(11):1033–41.
17. Verger R. Interfacial activation of lipases: facts and artifacts. Tibtech. 1997;15:32–8.
18. Uppenberg J, Hansen MT, Patkar S, Jones TA. The sequence, crystal structure determination and refinement of two crystal forms of lipase B from Candida antarctica. Structure. 1994;2(4):293–308.
19. Berlemont R, Jacquin O, Delsaute M, La Salla M, Georis J, Verte F, et al. Novel cold-adapted esterase mhlip from an antarctic soil metagenome. Biology (Basel). 2013;2(1):177–88.
20. Hu XP, Heath C, Taylor MP, Tuffin M, Cowan D. A novel, extremely alkaliphilic and cold-active esterase from Antarctic desert soil. Extremophiles. 2012;16(1):79–86.
21. Wanga SWK, Lia L, Liua Y. Isolation and characterization of novel organic solvent-tolerant and halotolerant esterase from a soil metagenomic library. J Mol Catalysis B Enzymatic. 2013;95:1–8.
22. Tirawongsaroj P, Sriprang R, Harnpicharnchai P, Thongaram T, Champreda V, Tanapongpipat S, et al. Novel thermophilic and thermostable lipolytic enzymes from a Thailand hot spring metagenomic library. J Biotechnol. 2008;133(1):42–9.
23. Martinez-Martinez M, Alcaide M, Tchigvintsev A, Reva O, Polaina J, Bargiela R, et al. Biochemical diversity of carboxyl esterases and lipases from Lake Arreo (Spain): a metagenomic approach. Appl Environ Microbiol. 2013;79(12):3553–62.
24. Jiang X, Xu X, Huo Y, Wu Y, Zhu X, Zhang X, et al. Identification and characterization of novel esterases from a deep-sea sediment metagenome. Arch Microbiol. 2012;194(3):207–14.
25. Zhu Y, Li J, Cai H, Ni H, Xiao A, Hou L. Characterization of a new and thermostable esterase from a metagenomic library. Microbiol Res. 2013;168(9):589–97.
26. Joseph B, Ramteke PW, Thomas G. Cold active microbial lipases: some hot issues and recent developments. Biotechnol Adv. 2008;26(5):457–70.
27. Dherbécourt JFH, Jardin J, Maillard MB, Baglinière F, Barloy-Hubler F, Thierry A. Identification of a secreted lipolytic esterase in Propionibacterium freudenreichii, a ripening process bacterium involved in Emmental cheese lipolysis. Appl Environ Microbiol. 2010;76:1181–8.
28. Uno TIA, Miyamoto T, Kubo M, Kanamaru K, Yamagata H, Yasufuku Y, et al. Ferulic Acid Production in the Brewing of Rice Wine (Sake). J Inst Brew. 2012;115:116–21.

29. Zhou J, Bruns MA, Tiedje JM. DNA recovery from soils of diverse composition. Appl Environ Microbiol. 1996;62(2):316–22.

30. Pel J, Broemeling D, Mai L, Poon HL, Tropini G, Warren RL, et al. Nonlinear electrophoretic response yields a unique parameter for separation of biomolecules. Proc Natl Acad Sci U S A. 2009;106(35):14796–801.

31. Besemer J, Borodovsky M. GeneMark: web software for gene finding in prokaryotes, eukaryotes and viruses. Nucleic Acids Res. 2005;33(Web Server issue):W451–454.

32. Li C, Wen A, Shen B, Lu J, Huang Y, Chang Y. FastCloning: a highly simplified, purification-free, sequence- and ligation-independent PCR cloning method. BMC Biotechnol. 2011;11:92.

33. Kruger NJ. The Bradford method for protein quantitation. Methods Mol Biol. 1994;32:9–15.

34. Altschul SF, Gish W, Miller W, Myers EW, Lipman DJ. Basic local alignment search tool. J Mol Biol. 1990;215(3):403–10.

35. Waterhouse AM, Procter JB, Martin DM, Clamp M, Barton GJ. Jalview Version 2—a multiple sequence alignment editor and analysis workbench. Bioinformatics. 2009;25(9):1189–91.

36. Bendtsen JD, Nielsen H, von Heijne G, Brunak S. Improved prediction of signal peptides: SignalP 3.0. J Mol Biol. 2004;340(4):783–95.

37. Soding J, Biegert A, Lupas AN. The HHpred interactive server for protein homology detection and structure prediction. Nucleic Acids Res. 2005;33(Web Server issue):W244–248.

38. Sali A, Blundell TL. Comparative protein modelling by satisfaction of spatial restraints. J Mol Biol. 1993;234(3):779–815.

39. Laskowski RAMM, Moss DS, Thornton JM. PROCHECK: a program to check the stereochemical quality of protein structures. J Appl Crystal. 1993;26:283–91.

40. Vriend G. WHAT IF: a molecular modeling and drug design program. J Mol Graph. 1990;8(1):52–56, 29.

41. Cavallo L, Kleinjung J, Fraternali F. POPS: A fast algorithm for solvent accessible surface areas at atomic and residue level. Nucleic Acids Res. 2003;31(13):3364–6.

42. Baker NA, Sept D, Joseph S, Holst MJ, McCammon JA. Electrostatics of nanosystems: application to microtubules and the ribosome. Proc Natl Acad Sci U S A. 2001;98(18):10037–41.

43. Dolinsky TJ, Czodrowski P, Li H, Nielsen JE, Jensen JH, Klebe G, et al. PDB2PQR: expanding and upgrading automated preparation of biomolecular structures for molecular simulations. Nucleic Acids Res. 2007; 35((Web Server issue)):W522–525.

44. Humphrey W, Dalke A, Schulten K. VMD: visual molecular dynamics. J Mol Graph. 1996;14(1):33–8. 27-38.

45. Berge JP, Barnathan G. Fatty acids from lipids of marine organisms: molecular biodiversity, roles as biomarkers, biologically active compounds, and economical aspects. Adv Biochem Eng Biotechnol. 2005;96:49–125.

46. Kiriakoulakis KFE, Wolff GA, Freiwald A, Grehan A, Roberts JM, Freiwald A, et al. Lipids and nitrogen isotopes of two deep-water corals from the North-East Atlantic: initial results and implications for their nutrition. In: Freiwald ARJ, editor. Cold-water Corals and Ecosystems. Berlin: Springer; 2005. p. 715–29.

47. Russell NJ, Nichols DS. Polyunsaturated fatty acids in marine bacteria–a dogma rewritten. Microbiology. 1999;145(Pt 4):767–79.

48. Litthauer DAN, Piater LA, van Heerden E. Pitfalls using tributyrin agar screening to detect lipolytic activity in metagenomic studies. Afr J Biotechnol. 2010;9:4282–5.

49. Neves Petersen MT, Fojan P, Petersen SB. How do lipases and esterases work: the electrostatic contribution. J Biotechnol. 2001;85(2):115–47.

50. Ghati ASK, Pau G. Production and characterization of an alkalothermostable, organic solvent tolerant and surfactant tolerant esterase produced by a thermophilic bacterium geobacillus sp. agp-04 isolated from bakreshwar hot spring, India. J Microbiol Biotechnol Food Sci. 2013;3:55–162.

51. Fu C, Hu Y, Xie F, Guo H, Ashforth EJ, Polyak SW, et al. Molecular cloning and characterization of a new cold-active esterase from a deep-sea metagenomic library. Appl Microbiol Biotechnol. 2011;90(3):961–70.

52. Joseph BRP, Thomas G, Shrivastava N. Cold-active microbial lipases: a versatile tool for industrial applications. Biotech Mol Bio Rev. 2007;2:39–48.

53. Aravindan RAP, Viruthagiri T. Lipase applications in food industry. Indian J Biotechnol. 2007;6:141–58.

54. Brady L, Brzozowski AM, Derewenda ZS, Dodson E, Dodson G, Tolley S, et al. A serine protease triad forms the catalytic centre of a triacylglycerol lipase. Nature. 1990;343(6260):767–70.

55. Brzozowski AM, Derewenda U, Derewenda ZS, Dodson GG, Lawson DM, Turkenburg JP, et al. A model for interfacial activation in lipases from the structure of a fungal lipase-inhibitor complex. Nature. 1991;351(6326):491–4.

56. Brzozowski AM, Savage H, Verma CS, Turkenburg JP, Lawson DM, Svendsen A, et al. Structural origins of the interfacial activation in Thermomyces (Humicola) lanuginosa lipase. Biochemistry. 2000;39(49):15071–82.

57. Grochulski P, Li Y, Schrag JD, Cygler M. Two conformational states of Candida rugosa lipase. Protein Sci. 1994;3(1):82–91.

58. van Tilbeurgh H, Egloff MP, Martinez C, Rugani N, Verger R, Cambillau C. Interfacial activation of the lipase-procolipase complex by mixed micelles revealed by X-ray crystallography. Nature. 1993;362(6423):814–20.

59. Peters GH, Toxvaerd S, Olsen OH, Svendsen A. Computational studies of the activation of lipases and the effect of a hydrophobic environment. Protein Eng. 1997;10(2):137–47.

60. Rehm S, Trodler P, Pleiss J. Solvent-induced lid opening in lipases: a molecular dynamics study. Protein Sci. 2010;19(11):2122–30.

61. De Santi C, Tedesco P, Ambrosino L, Altermark B, Willassen NP, de Pascale D. A new alkaliphilic cold-active esterase from the psychrophilic marine bacterium Rhodococcus sp.: functional and structural studies and biotechnological potential. Appl Biochem Biotechnol. 2014;172(6):3054–68.

62. Sriyapai P, Kawai F, Siripoke S, Chansiri K, Sriyapai T. Cloning, Expression and Characterization of a Thermostable Esterase HydS14 from Actinomadura sp. Strain S14 in Pichia pastoris. Int J Mol Sci. 2015;16(6):13579–94.

63. Mao YJ, Sheng XR, Pan XM. The effects of NaCl concentration and pH on the stability of hyperthermophilic protein Ssh10b. BMC Biochem. 2007;8:28.

64. Perez-Jimenez R, Godoy-Ruiz R, Ibarra-Molero B, Sanchez-Ruiz JM. The efficiency of different salts to screen charge interactions in proteins: a Hofmeister effect? Biophys J. 2004;86(4):2414–29.

# 5

# Altered activity patterns of transcription factors induced by endoplasmic reticulum stress

Sheena Jiang[1], Eric Zhang[2], Rachel Zhang[2] and Xianqiang Li[1*]

## Abstract

**Background:** The endoplasmic-reticulum (ER) responds to the burden of unfolded proteins in its lumen by activating intracellular signal transduction pathways, also known as the unfolded protein response (UPR). Many signal transduction events and transcription factors have been demonstrated to be associated with ER stress. The process in which ER stress affects or interacts with other pathways is still a progressing topic that is not completely understood. Identifying new transcription factors associated with ER stress pathways provides a platform to comprehensively characterize mechanism and functionality of ER.

**Methods:** We utilized a transcription factor (TF) activation plate array to profile the TF activities which were affected by ER stress induced by pharmacological agents, thapsigargin (TG) and tunicamycin (TM) at 1 h, 4 h, 8 h and 16 h respectively, in MiaPACA2 cells. The altered activity patterns were analyzed and validated using gel shift assays and cell-based luciferase reporter assay.

**Results:** The study has not only confirmed previous findings, which the TFs including ATF4, ATF6, XBP, NFkB, CHOP and AP1, were activated by ER stress, but also found four newly discovered TFs, NFAT, TCF/LEF were activated, and PXR was repressed in response of ER stress. Different patterns of TF activities in MiaPaCa2 were demonstrated upon TM or TG treatment in the time course experiments. The altered activities of TFs were confirmed using gel shift assays and luciferase reporter vectors.

**Conclusion:** This study utilized a TF activation array technology to identify four new TFs, HIF, NFAT, TCF/LEF and PXR that were changed in their activity as a result of ER stress induced by TG and TM. The TF activity patterns were demonstrated to be diverse in response to the duration of TG or TM treatment. These new findings will facilitate further unveiling the complex mechanisms of the ER stress process and associated diseases.

**Keywords:** ER stress, UPR, TF, Plate array, Activation and Signal pathways

## Background

ER plays an important role in many biological functions such as folding and assembling the membrane and secreted proteins in eukaryotic cells [1]. Production process of these proteins in the lumen of the ER is believed to be led by the coordination between the extracellular and intracellular signals [2]. It has been previously reported [2–4] that, an imbalance between the protein-folding load and the capacity of the ER could happen due to either increase of protein-folding demand or disruption of protein-folding reactions, which will generate ER stress, and subsequently lead to accumulation of unfolded or misfolded proteins in the ER lumen. ER stress simultaneously activates Unfolded Protein Response (UPR) to reduce protein synthesis, degrade misfolded proteins, and produce molecular chaperones. The growing evidence suggests that ER stress (UPR) is an intricate molecular process, interacting with oxidative stress [5], $Ca^{2+}$ signal response [6], and the inflammatory response and other signal pathways. In addition, ER stress is associated with a variety of diseases caused by the accumulation of aggregated proteins such as neurodegenerative diseases and diabetes [7].

* Correspondence: jasonli@signosisinc.com
[1]Signosis Inc., 1700 Wyatt Drive, Suite #10-12, Santa Clara, CA 95054, USA
Full list of author information is available at the end of the article

It is well-known that in mammalian cells, the ER stress activates three distinct ER-localized transmembrane proteins, inositol-requiring enzyme 1 (IRE1), pancreatic ER kinase (PERK), and activating transcription factor 6 (ATF6) [8, 6]. The timing and duration of activation of these three proteins may be different from the previous speculation of parallel responses [9]. For examples, the activation of PERK could inhibit the global protein translation through phosphorylation of eIF2α, and a translational increase in the transcription factor ATF4 would promote UPR-specific gene expression [10, 11], and IRE1 would lead to the generation of a more potent form of XBP1 mRNA splicing version [12]. Furthermore, ATF6, an ER membrane transcription factor [13], undergoes proteolysis to release its cytoplasmic transactivation domain to become active [9]. Therefore, PERK, IRE1, and ATF6 are ultimately responsible for the activation of a set of transcription factors through a complicated and nonparallel process.

Cells respond to ER stress by inducing gene expression. Consequently, a signal transduced from the ER to the nucleus is required to activate transcription in response to the ER Stress. Activated TFs as the endpoints of the signal transduction pathway directly regulates the final gene expression in the nucleus. Therefore, TF activation is a measure of the effectiveness of ER stress in activating the three proteins mentioned above in the cells. A number of transcription factors have been found to take part in ER stress response, such as ATF6 [13], XBP1 [12], ATF4 [9], NFkB [14], AP1 [14], SREBP and CHOP (CÆBP homology) [15]. The activation of these TFs is believed to be associated with ER stress but through different mechanisms. The intrinsic ribonuclease activity of IRE1 also results in production and activation of XBP-1, inducing expression of genes involved in restoring protein folding or degrading unfolded proteins [12]. ATF4 is translationally up-regulated by eIF2a-mediated translational attenuation and PERK/eIF2α ~ P/ATF4 pathway is required not only for translational control, but also for activation of ATF6 [9] and CHOP and their target genes. Oligomerized Ire1 binds to TRAF2, TNF receptor associated factors, that activate NF-κB and c-Jun (AP-1), leading to expression of a set of genesassociated with host defense or alarm [14]. In addition, the different transcription factors may display different response time patterns during ER stress process and variable pathways.

Pharmacological agents are commonly used to treat cells to elevate unfolded proteins in most studies of ER stress, including dithiothreitol (DTT), which disrupts or prevents protein disulfide bonding; thapsigargin (TG), an inhibitor of the ER Ca2 dependent ATPase; or tunicamycin (TM), an inhibitor of protein glycosylation of newly synthesized proteins [16]. However, the concentration and duration of treatment vary from system to system. Typically only a few hours are sufficient to induce ER stress while a longer exposure often lead ER stress-mediated cell to death. Previous studies have indicated that three ER transmembrane components, IRE1, PERK and ATF6, displayed distinct sensitivities toward different forms of ER stress induced by these three agents, but it is not clear how ER stress is affected in downstream pathways and transcriptional regulation.

In this study, we employed a TF activation profiling array to systematically monitor ER stress-induced TF activity patterns with 1 h, 4 h, 8 h and 16 h of TM and TG treatment in pancreatic tumor cell MiaPaCa2, since this cell line has demonstrated a globally compromised ability to regulate the unfolded protein response and it has been widely used for studying ER stress process [17–19]. With the plate array assay, the activities of 48 TFs can be elucidated in a single experiment. Through a comparative study, it was observed that the activities of ATF4 and ATF6, XBP1, CHOP, AP1, NFkB, NFAT, TCF/LEF and HIF increased, while the activity of PXR decreased to different extents in response to TM and TG treatment. To our knowledge, the activation of NFAT, TCF/LEF, HIF and PXR under ER stress was observed for the first time. The altered TFs were further confirmed by conventional gel shift assays and luciferase reporter assays. Different patterns of TF activities in MiaPaCa2 were exhibited in response to different TM or TG treatment time, which may help to unveil the complicated mechanism of ER stress process.

## Methods
### Cell culture and nuclear extraction
MiaPaCa2 cells were seeded in 10 cm$^2$ culture plates in ATCC-formulated Dulbecco's Modified Eagle's Medium (ATCC), supplemented with 10 % FBS, 1 % nonessential minimal amino acids and 100 U/ml penicillin, 0.1 mg/ml streptomycin. The cells (about 80 % confluent) were then treated with 200nM TG and 10ug/ml TM for 1 h, 4 h, 8 h and 16 h respectively. Untreated cells were used as negative controls. Nuclear extracts were prepared with the nuclear extract kit (Signosis, Inc.) according to the user manual. The cells were washed twice in phosphate-buffered saline (PBS) and lysed on ice for 10 min in the extraction buffer I with gently shaking, and then were collected from the plates, and centrifuged at 15,000 rpm for 3 min at 4 °C. The supernatant (cytoplasmic fraction) was discarded; the pellet was then resuspended in 250 μl of extraction buffer II and incubated on ice for 2 h with gently shaking. After the mixture was centrifuged at 15,000 rpm for 5 min at 4 °C, the supernatant containing nuclear protein was collected and ready for assays. Protein concentrations were determined by the Bradford assay (Bio-Rad).

## TF activation profiling analysis

Each array assay was performed following the procedure described in the TF activation profiling plate array kit user manual (Signosis, Inc). 10 ug of nuclear extract was first incubated with the biotin labeled probe mix at room temperature for 30 min. The activated TFs were bound to the corresponding DNA binding probes. After the protein/DNA complexes were isolated from unbound probes, the bound probes were eluted and hybridized with the plate pre-coated with the capture oligos. The captured biotin-labeled probes were then detected with Streptavidin–HRP and subsequently measured with the chemiluminescent plate reader (Veritas microplate luminometer).

## Gel shift assay

The samples with 8 h of TG and TM treatment were chosen for gel shift assay analysis with EMSA kits (Signosis Inc). The TF DNA binding probe sequences are listed below.

1) ATF: CTGTCATGACGTCAAAAGTCG
2) NFkB: AGTTGAGGGGACTTTCCCAGGC
3) NFAT:
   ACGCCCAAAGAGGAAAATTTGTTTCATACA
4) AP1: CGCTTGATGACTCAGCCGGAA
5) CHOP: TTGCGGAGGATTGCGTTGACGA
6) TCF/LEF: ACGTTACTTTGATCTGATCAGGGC
7) XBP1:
   GATCTCCTAGCAACAGATGCGTCATCTC
8) HIF: GTGACTACGTGCTGCCTAG

The sequences that we used as probes for gel shift assay are identical to those we used as the probe mix for TF activation profiling array assay. 5ug nuclear extracts were incubated with 1× binding buffer and biotin-labeled probe for 30 min at room temperature to form protein/DNA complexes. The samples were then electrophoresed on a 6 % polyacrylamide gel in 0.5 % TBE at 120 V for 45 min and then transferred onto a nylon membrane in 0.5 % TBE at 300 mA for 1 h. After transfer and UV cross-linking, the membrane was detected with Streptavidin–HRP. The image was acquired using a FluorChem imager (Alpha Innotech Corp).

## Luciferase reporter assay

Luciferase reporter assay was carried out following the procedure in Luciferase reporter assay user manual (Signosis, Inc.). The reporter vectors contain 4 repeats of the corresponding DNA binding sequences shown in gel shift assay section. In order to distinguish ATF4 and ATF6 activation, we cloned the reporter vectors for ATF4 and ATF6 with 5 repeats of specific consensus sequences for ATF4 and ATF6 respectively, TGACGTAAG [20] for ATF4 and TGACGTGG [21] for ATF6. The cells were first transfected with luciferase reporter vectors (Signosis Inc) for 16 h with Fugene 6 (Promega) in a 96-well plate, and then treated without or with 200nM TG and 10ug/ml TM for 6 h, 8 h and 16 h respectively. After removing the culture media and rinsing the cells twice with PBS, 200 μl of 1× cell lysis buffer was added to lyse the cells. After dislodging the cells by scraping them off from the plate, we transferred the cells to a 1.5-ml microcentrifuge tube before being centrifuged at 14,000 rpm at room temperature for 1 min to remove cellular debris. 10 μl of the cell extract was mixed with 50 μl of substrate (Signosis Inc), and luminescence was measured using a luminometer.

## Statistical analyses

Data were analyzed by a method of two-sided and unpaired $t$-test using GraphPad Prism 6.0 software. The mean $\pm$ SD of multiple independent experiments were shown in data analysis. A $p$ value of $<0.05^*$ would be considered significant, $p < 0.01^{**}$ very significant, and $p < 0.001^{***}$ highly significant.

## Results and discussion

To examine TF activation patterns induced by ER stress, MiaPAC2 cells were treated with or without TG or TM for 1 h, 4 h, 8 h and 16 h prior to preparation of nuclear extracts for analysis with the TF activation profiling plate array I with slight modification (Table 1). The TFs were selected based on their important biological functions in crucial signal pathways which may associate with ER stress. The nuclear extracts were mixed with a biotin-labeled pool of DNA probe mix that correspond specifically to 48 TF response elements. After the probes were incubated with nuclear extracts, the complexes of TFs and probes were separated from the free probes. Through elution of bound probes, the composition and quantity of the bound probes were then determined using a plate array, which contained the pre-coated capture oligos in a 96-well white plate according to the position of the individual TFs indicated in Table 1, therefore, the plate would hybridize with any labeled probe that was present. After the hybridized signals were detected with a Streptavidin–HRP and HRP substrate, the resulting chemiluminescence was measured by a plate reader. The evolution of TF activity pattern in response to ER stress process was examined in a chronological sequence. The ATF and XBP1 activities were shown to increase significantly after only 1 h of TG and TM treatment. The activities of CHOP, AP1, NFkB, NFAT, TCF/LEF and HIF showed significant increases after 4 h of TG and TM treatment. All of TF activities reached to peak upon 8 h of TG and TM treatment. After 16 h treatment, only NFAT and TCF/LEF activities remained the same level as 8 h treatment, and the other TFs all decreased slightly (Fig. 1). In addition, we identified that the activity of

**Table 1** The diagram of TF Activation Plate Array I (revised). 48 TFs are included, locating in the column 1–6 and column 7–12 respectively

|   | 1 | 2 | 3 | 4 | 5 | 6 | 7 | 8 | 9 | 10 | 11 | 12 |
|---|---|---|---|---|---|---|---|---|---|----|----|----|
| A | AP1 | CDP | GATA | XBP | Pit | Stat3 | AP1 | CDP | GATA | XBP | Pit | Stat3 |
| B | AP2 | CREB | GR/PR | NFAT | PPAR | Stat4 | AP2 | CREB | GR/PR | NFAT | PPAR | Stat4 |
| C | AR | E2F-1 | HIF | CHOP | PXR | Stat5 | AR | E2F-1 | HIF | CHOP | PXR | Stat5 |
| D | ATF | EGR | HNF4 | NFkB | SMAD | Stat6 | ATF | EGR | HNF4 | NFkB | SMAD | Stat6 |
| E | Brn-3 | ER | IRF | Oct-4 | Sp1 | TCF/LEF | Brn-3 | ER | IRF | Oct-4 | Sp1 | TCF/LEF |
| F | C\EBP | Ets | MEF4 | p53 | SRF | TR | C\EBP | Ets | MEF2 | p53 | SRF | TR |
| G | CAR | FAST-1 | Myb | Pax-5 | SATB1 | YY1 | CAR | FAST-1 | Myb | Pax-5 | SATB1 | YY1 |
| H | CBF | GAS/ISRE | Myc-Max | Pbx1 | Stat1 | TFIID | CBF | GAS/ISRE | Myc-Max | Pbx1 | Stat1 | TFIID |

PXR decreased significantly after 4 h of TG treatment but only slightly decreased in TM-treated cells. Furthermore, HIF, TCF/LEF, NFAT and PXR were observed to be ER stress responsive TFs for the first time.

In order to validate the plate array results, the samples with optimal 8 h of TG and TM treatment were used for gel shift assay. As shown in Fig. 2, both TG- and TM-

activated ATF, XBP1, CHOP, AP1, NFkB, NFAT and HIF were able to be confirmed with gel shift assays. The decrease of PXR in DNA binding activity in TG-treated cells was also confirmed with the gel shift assay but the slight change in the activity of PXR in TM-treated cells identified by the plate array was not detectable with the gel shift assay. As the gel shift assay is considered to be

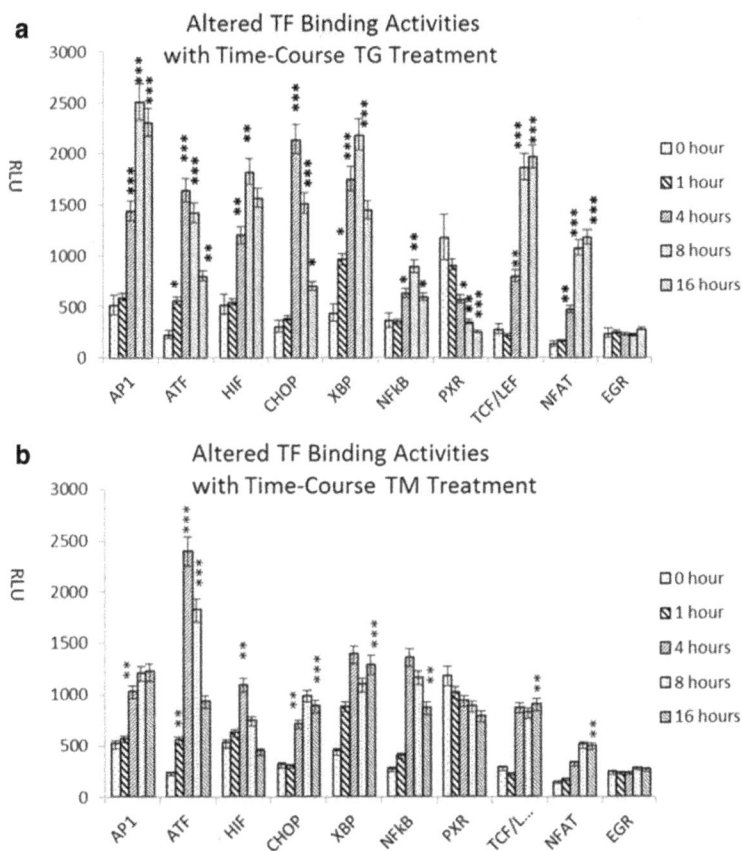

**Fig. 1** Plate array analysis of 48 TFs in MiaPAC2 cells treated without or with TG/TM treatment respectively. After 1 h, 4 h, 8 h and 16 h of treatment, the cells were subjected to nuclear extraction. The nuclear extracts were then used for TF activation plate assay. The data from control sample (without treatment), TG treated and TM treated samples were compared. Data were obtained from three independent experiments, $*P < 0.05$, $**P < 0.01$, $*** P < 0.001$; (**a**): TF activation DNA binding assay with TG treatment; (**b**). TF activation DNA binding assay with TM treatment

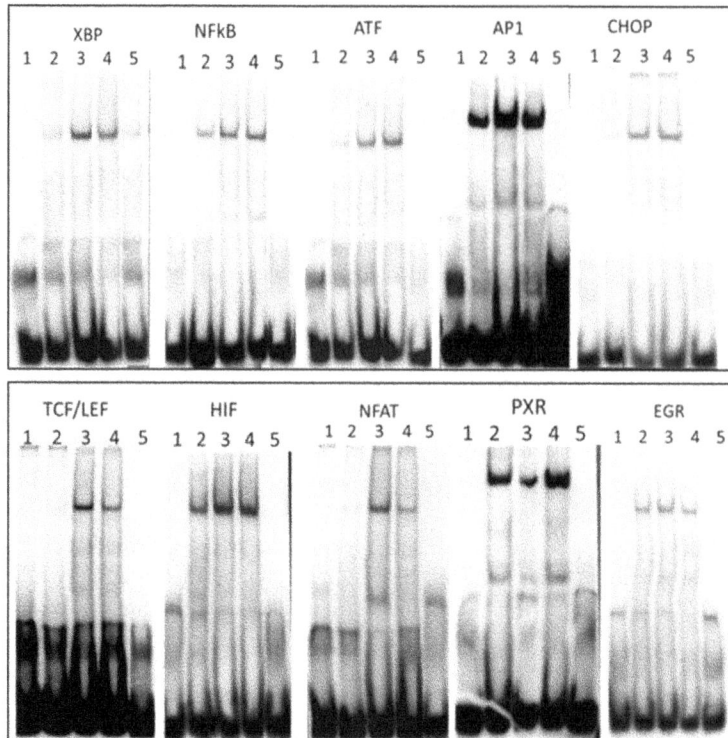

**Fig. 2** Nuclear extracts with 8 h of treatment were subjected to EMSA assay with different probes. **a**: TFs, XBP, NFkB, ATF, AP1 and CHOP, were reported to be associated with ER stress previously. **b**: TFs, TCF/LEF, HIF, NFAT and PXR, were the first time reported to be associated with ER stress in this study. EGR was used as a negative control. 1. Free probe only. 2. Without treatment; 3. TG treatment; 4. TM treatment; 5. Cold probe competition

a gold standard in analysis of DNA binding activities of TFs, we concluded that the activities of PXR decreased in TG-treated but not TM-treated cells. Furthermore, we introduced EGR as a control in gel shift assay. The array data showed no change of ERG in either TG- or TM-treated cells as compared to the untreated MiaPAC2 cells. The gel shift assay showed no difference in EGR between treated and untreated cells (data not shown). Through both the array and gel shift assays, we confirmed that ATF, XBP1, CHOP, AP1, NFkB, TCF/LEF, NFAT and HIF are indeed activated by TG and TM. The activity of PXR was down regulated by TG but not by TM.

In order to investigate whether the activation of these TFs can be quantitatively monitored with luciferase reporter assays, we employed a set of reporter vectors corresponding to these TFs to transfect into MiaPAC2 cells. The consensus sequence of ATF probe in the TF activation plate array assay and gel shift assay are for ATF family but cannot distinguish ATF family members, ATF4 and ATF6. We designed and cloned ATF4 and ATF6 reporter vectors with specific ATF4 and ATF6 DNA binding sequences respectively. After transfection of the vectors into the cells, the cells were treated with TG and TM treatment for 6 h, 8 h and 16 h before their luciferase activity was measured. We confirmed the

activation of ATF4, ATF6, XBP1, CHOP, AP1, NFkB, TCF/LEF, NFAT and HIF by TG and TM treatment, and repression of PXR by TG only but not by TM. The activation of ATF4 and XBP1 was observed to occur at the earlier stage during ER stress process. In addition, TG was shown to be a stronger inducer for CHOP, XBP, AP1, TCF/LEF, and PXR, whereas TM-activated ATF4, ATF6 and NFkB were shown to much more effective than TG. HIF responded equivalently to both TM and TG treatments (Fig. 3).

A possible mechanism underlying alteration of newly identified NFAT, HIF, TCF/LEF and PXR activities during ER stress process is presented here for further discussion. TM blocks the initial step of glycoprotein biosynthesis in the ER. Thus, treatment of TM causes accumulation of unfolded glycoproteins in the ER, effectively triggers eIF2α/ATF4 pathway and activates ATF4. ATF4 has demonstrated to be an early activated TF during early ER stress process and is the master regulator that plays a crucial role in the adaptation to stresses by regulating the transcription of many genes, such as CHOP and ATF6. These ATF4 target genes are themselves transcription factors that regulate the expression of a set of stress-induced target genes and amplify the signals by triggering other signaling pathways, such

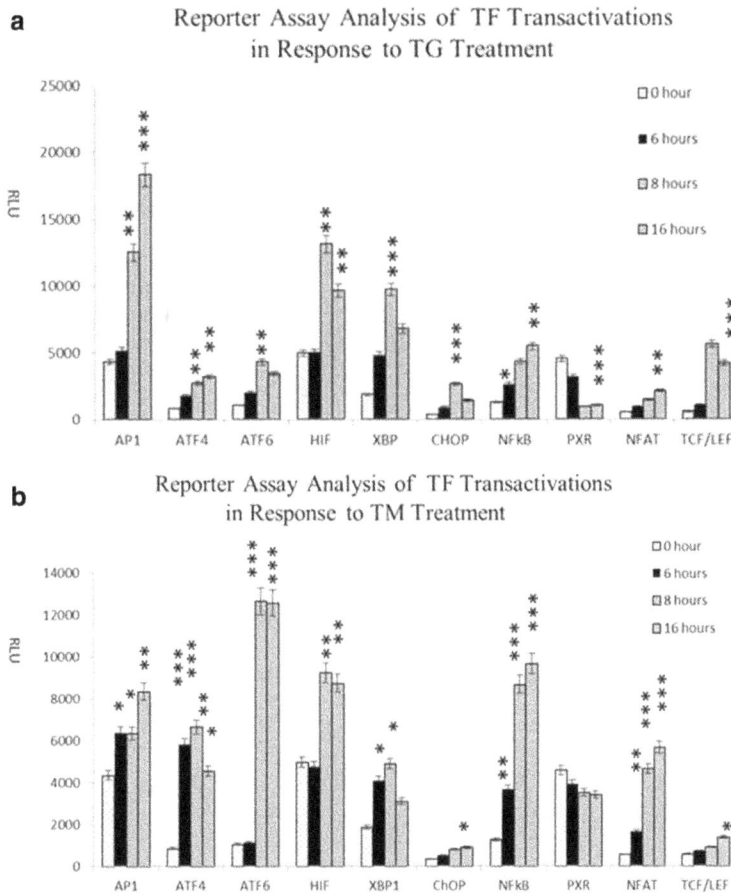

**Fig. 3** Transactivation of TFs in response to TG and TM treatment. The cells were transfected with different reporter vectors for 16 h, and treated with TG or TM respectively for 0 h (no treatment), 6 h, 8 h and 16 h The cells then were lysed and subjected to luciferase assay. Data were obtained from three independent experiments, *$P < 0.05$, **$P < 0.01$, *** $P < 0.001$; (**a**): Reporter assay with TG treatment; (**b**): Reporter assay with TM treatment

as inflammation and hypoxia via activating NFkB and HIF. Activation of these multiple TFs by the ER (UPR) may result in a complicated pattern of gene regulation through not only by target gene regulation but also by protein/protein interaction. It agrees with the observation from one of our previous studies that, the altered activities of TFs can be induced either by over expression TF or by interaction of TF with other proteins [22]. In addition, TG that can generate Cadysregulation and induce ER stress may result in significant increases in cytosolic $Ca^2$. $Ca^2$ disequilibrium releases beta-catenin from the plasma membrane, which subsequently leads to the accumulation of beta-catenin in the cytoplasm and formation of beta-catenin/TCF/LEF complex. This complex further translocates to the nucleus where it activates transcription [23]. Increased $Ca^2$ concentration in the cytoplasm could also directly activate NFAT. ER stress has been reported to lead to the phosphorylation of HIF-1α, which might result in an increase in the activity of HIF-1α [24, 25]. PXR is a nuclear receptor

recognized as a major regulator of xenobiotic metabolism and drug metabolism by regulating CYP3A4 [26]. The expression of PXR has been reported to be suppressed during ER stress by down-regulating HNF4 and up-regulating liver-enriched inhibitory protein (LIP) with TG treatment. The decrease in DNA binding activity of TG-induced PXR discovered in this study may be due to the pathway interactions with HNF4, ATF and CHOP [27]. With the array assay the different TF activity patterns were displayed in response to different ER stress pathways. Formation of homo- and heterodimers among these TFs families may build an integrated transcription factor network that determines precisely the initiation, magnitude, and length of the cellular response to ER stress in a fine-tuned and coordinated way. In spite of the exact mechanism how ER stress is regulated by TFs still remains not fully clear, the new findings in this study offer clues to dissect the cellular response to ER stress signaling pathways.

## Conclusion

We used TF activation plate array to profile the TF activities of TFs and reported four newly identified TFs whose activities were altered in response to ER stress. The activity patterns were shown to be distinctive with the different ER signal pathways.

### Abbreviations

AP1: activating protein; ATF: activating transcription factor; CHOP: DNA damage inducible transcript 3; ER: endoplasmic reticulum; HIF: hypoxia inducible factor; NFkB: nuclear factor of kappa light polypeptide gene enhancer in B-cells; PXR: pregnane X receptor; TCF/LEF: T-cell factor/lymphoid enhancer factor; TF: transcription factor; TG: thapsigargin; TM: tunicamycin; UPR: unfolded protein response; XBP: X-box binding protein .

### Competing interests

The authors declare that Sheena Jiang and Xianqiang Li are employed by Signosis Inc, and may gain financially from the company in future. Eric Zhang and Rachel Zhang don't gain financially from the company in the future.

### Authors' contributions

SJ and XL came up with the idea and wrote the manuscript. SJ conducted the tissue culture and luciferase reporter assay. SJ and EZ performed nuclear extraction and TF activation plate array. EZ and RZ performed EMSA. RZ edited the paper. All authors read and approved the final manuscript.

### Acknowledgements

This work is supported by Signosis Inc.

### Author details

[1]Signosis Inc., 1700 Wyatt Drive, Suite #10-12, Santa Clara, CA 95054, USA.
[2]Saratoga High School, 20300 Herriman Ave, Saratoga, CA 95070, USA.

### References

1. Gething MJ, Sambrook J. Protein folding in the cell. Nature. 1992;355:33–45.
2. Xu C, Bailly-Maitre B, Reed JC. Endoplasmic reticulum stress: cell life and death decisions. J Clin Invest. 2005;115:2656–64.
3. Sidrauski C, Chapman R, Walter P. The unfolded protein response: an intracellular signalling pathway with many surprising features. Trends Cell Biol. 1998;8:245–9.
4. Mori K. Tripartite management of unfolded proteins in the endoplasmic reticulum. Cell. 2000;101:451–4.
5. Zhang K. Integration of ER stress, oxidative stress and the inflammatory response in health and disease. Int J Clin Exp Med. 2010;3(1):33–40.
6. DuRose JB, Tam AB, Niwa M. Intrinsic capacities of molecular sensors of the unfolded protein response to sense alternate forms of endoplasmic reticulum stress. Mol Biol Cell. 2006;17:3095–310.
7. Lipson KL, Fonseca SG, Ishigaki S, Nguyen LX, Foss E, Bortell R, Rossini AA, Urano F. Regulation of insulin biosynthesis in pancreatic beta cells by an endoplasmic reticulum-resident protein kinase IRE1. Cell Metab. 2006;4:245–54.
8. Wu J, Kaufman RJ. From acute ER stress to physiological roles of the unfolded protein response. Cell Death Differ. 2006;13:374–84.
9. Teske BF, Wek SA, Bunpo P, Cundiff JK, McClintick JN, Anthony TG, Wek RC. The eIF2 kinase PERK and the integrated stress response facilitate activation of ATF6 during endoplasmic reticulum stress. Mol Biol Cell. 2011;22(22): 4390–405.
10. Harding HP, Novoa I, Zhang Y, Zeng H, Wek R, Schapira M, Ron D. Regulated translation initiation controls stress-induced gene expression in mammalian cells. Mol Cell. 2000;6:1099–108.
11. Ron D, Walter P. Signal integration in the endoplasmic reticulum unfolded protein response. Nat Rev Mol Cell Biol. 2007;8:519–29.
12. Yoshida H, Matsui T, Yamamotot A, Okada T, Mori K. XPB1 mRNA is induced by ATF6 and spliced by IRE1 in response to ER stress to produce a highly active transcription factor. Cell. 2001;107:881–91.
13. Yamamoto K, Yoshida H, Kokame K, Kaufman RJ, Mori K. Differential contributions of ATF6 and XBP1 to the activation of endoplasmic reticulum stress-responsive cis-acting elements ERSE, UPRE and ERSE-II. J Biochem. 2004;136:343–50.
14. Hummasti S, Hotamisligil GS. Endoplasmic reticulum stress and inflammation in obesity and diabetes. Circ Res. 2010;107:579–91.
15. Ozcan L, Tabas I. Role of endoplasmic reticulum stress in metabolic disease and other disorders. Annu Rev Med. 2012;63:317–28.
16. Oslowski CM, Urano F. Measuring ER stress and the unfolded protein response using mammalian tissue culture system. Methods Enzymol. 2011;490:71–92.
17. Verma G, Datta M. IL-1beta induces ER stress in a JNK dependent manner that determines cell death in human pancreatic epithelial MIA PaCa-2 cells. Apoptosis. 2010;15(7):864–76.
18. Cheng S, Swanson K, Eliaz I, McClintick JN, Sandusky GE, Sliva D. Pachymic acid inhibits growth and induces apoptosis of pancreatic cancer in vitro and in vivo by targeting ER stress. PLoS One. 2015;10(4), e0122270.
19. Cheng S, Swanson K, Eliaz I, McClintick JN, Sandusky GE, Sliva D. Imexon induces an oxidative endoplasmic reticulum stress response in pancreatic cancer cells. Mol Cancer Res. 2012;10(3):392–400.
20. Bouman L, Schlierf A, Lutz AK, Shan J, Deinlein A, Kast J, Galehdar Z, Palmisano V, Patenge N, Berg D, Gasser T, Augustin R, Trümbach D, Irrcher I, Park DS, Wurst W, Kilberg MS, Tatzelt J, Winklhofer KF. Parkin is transcriptionally regulated by ATF4: evidence for an interconnection between mitochondrial stress and ER stress. Cell Death Differ. 2011; 18(5):769–82.
21. Wang Y, Shen J, Arenzana N, Tirasophon W, Kaufman RJ, Prywes R. Activation of ATF6 and an ATF6 DNA binding site by the endoplasmic reticulum stress response. J Biol Chem. 2000;275:27013–20.
22. Jiang X, Norman M, Roth L, Li X. Protein-DNA array-based identification of transcription factor activities regulated by interaction with the glucocorticoid receptor. J Biol Chem. 2014;279:38480–5.
23. Carlisle RE, Heffernan A, Brimble E, Liu L, Jerome D, Collins CA, Mohammed-Ali Z, Margetts PJ, Austin RC, Dickhou JG. TDAG51 mediates epithelial-to-mesenchymal transition in human proximal tubular epithelium. Am J Physiol Renal Physiol. 2012;303:F467–81.
24. Pereira ER, Frudd K, Awad W, Hendershot LM. Endoplasmic reticulum (ER) stress and hypoxia response pathways interact to potentiate hypoxia-inducible factor 1 (HIF-1) transcriptional activity on targets like vascular endothelial growth factor (VEGF). J Biol Chem. 2014;289:3352–64.
25. Werno C, Zhou J, Brune B. A23187, ionomycin and thapsigargin upregulate mRNA of HIF-1alpha via endoplasmic reticulum stress rather than a rise in intracellular calcium. J Cell Physiol. 2008;215:708–14.
26. Vachirayonsti T, Ho KW, Yang D, Yan B. Suppression of the pregnane X receptor during endoplasmic reticulum stress is achieved by down-regulating hepatocyte nuclear factor-4α and up-regulating liver-enriched inhibitory protein. Toxicol Sci. 2015;144:382–92.
27. Desvergne B, Michalik L, Wahli W. Transcriptional regulation of metabolism. Physiol Rev. 2006;86:465–514.

# 6

# 1,2-Dichlorobenzene affects the formation of the phosphoenzyme stage during the catalytic cycle of the Ca²⁺-ATPase from sarcoplasmic reticulum

Javier Vargas-Medrano[3], Jorge A. Sierra-Fonseca[2] and Luis F. Plenge-Tellechea[1*]

## Abstract

**Background:** 1,2-Dichlorobenzene (1,2-DCB) is a benzene-derived molecule with two Cl atoms that is commonly utilized in the synthesis of pesticides. 1,2-DCB can be absorbed by living creatures and its effects on naturally-occurring enzymatic systems, including the effects on Ca²⁺-ATPases, have been poorly studied. Therefore, we aimed to study the effect of 1,2-DCB on the Ca²⁺-ATPase from sarcoplasmic reticulum (SERCA), a critical regulator of intracellular Ca²⁺ concentration.

**Results:** Concentrations of 0.05–0.2 mM of 1,2-DCB were able to stimulate the hydrolytic activity of SERCA in a medium-containing Ca²⁺-ionophore. At higher concentrations (0.25–0.75 mM), 1,2-DCB inhibited the ATP hydrolysis to ~80 %. Moreover, ATP hydrolysis and Ca²⁺ uptake in a medium supported by K-oxalate showed that starting at 0.05 mM,1,2-DCB was able to uncouple the ratio of hydrolysis/Ca²⁺ transported. The effect of this compound on the integrity of the SR membrane loaded with Ca²⁺ remained unaffected. Finally, the analysis of phosphorylation of SERCA by [γ-³²P]ATP, starting under different conditions at 0° or 25 °C showed a reduction in the phosphoenzyme levels by 1,2-DCB, mostly at 0 °C.

**Conclusions:** The temperature-dependent decreased levels of phosphoenzyme by 1,2-DCB could be due to the acceleration of the dephosphorylation mechanism – E₂P·Ca₂ state to E₂ and P$_i$, which explains the uncoupling of the ATP hydrolysis from the Ca²⁺ transport.

**Keywords:** Ca²⁺-ATPase, Sarcoplasmic reticulum, Dichlorobenzene, Phosphoenzyme

## Background

1,2-Dichlorobenzene (1,2-DCB) is a chlorinated benzene molecule used as precursor in the synthesis of pesticides. 1,2-DCB can be absorbed and accumulated by humans and wildlife, and it has been detected in different biological fluids. Moreover, 1,2-DCB has been linked to severe human health problems such as liver damage and anemia [1–6]. Oral administration of a radioactively-labeled 1,2-DCB to rats revealed that it can be distributed among all tissues, including skeletal muscle. In addition, it has been shown that DCB molecules can induce a rise in intracellular Ca²⁺ in human neuronal SH-SY5Y cells [7, 8]. Given this evidence, we hypothesized that 1,2-DCB can affect the functionality of skeletal muscle proteins, particularly the Ca²⁺-ATPase from sarcoplasmic reticulum (SERCA), a predominant protein of skeletal muscle, a hypothesis that has not previously been explored. SERCA is a large (110-kDa) sarcoplasmic reticulum protein with 10 transmembrane regions, and it plays the essential role of reducing the intracellular Ca²⁺ concentration by consuming ATP as a source of energy [9–12]. Each molecule of SERCA protein utilizes the energy of one ATP molecule to pump 2Ca²⁺ across the sarcoplasmic membrane, thus maintaining low cytoplasmic Ca²⁺ concentrations and high intravesicular Ca²⁺

* Correspondence: fplenge@uacj.mx
[1]Departamento de Ciencias Químico Biológicas, Laboratorio de Biología Molecular y Bioquímica (Edif. T-216), Instituto de Ciencias Biomédicas, Universidad Autónoma de Ciudad Juárez, Plutarco Elías Calles #1210 Fovissste Chamizal, Ciudad Juárez, Chihuahua C.P. 32310, Mexico
Full list of author information is available at the end of the article

concentrations [13, 14]. The catalytic cycle of SERCA involves two main conformational states, $E_1$ with high-$Ca^{2+}$ and $E_2$ with low-$Ca^{2+}$ affinities [15–17]. In addition, the three-dimensional structures of the two conformational stages of SERCA and phosphorylated intermediate (EP) have been successfully determined [18–22]. Furthermore, the coupling ratio of both transported $Ca^{2+}$ and hydrolyzed ATP occurs through a sequence of phosphorylated and non-phosphorylated enzymatic intermediates, with or without $Ca^{2+}$ bound to the enzyme, and it has been shown that EP accumulation can serve as an indicator of enzyme turnover, which can be affected by a variety of factors such as temperature, nature of the substrate, $Ca^{2+}$ concentration and presence of organic solvents [23].

Given the critical role of SERCA in maintaining appropriate cytosolic $Ca^{2+}$ levels, chemical agents that affect its functionality can have harmful effects to the overall cellular function. In this regard, a multitude of molecules have been described to act as inhibitors of SERCA [24]. Naturally-occurring compounds such as cyclopiazonic acid (a fungal toxin), and thapsigargin (a sesquiterpene lactone) are two of the most widely used inhibitors of SERCA [25–27]. Inorganic compounds such as vanadate, tungstate and molybdate have also been reported to inhibit SERCA activity [28–31]. Other natural compounds such as ellagic acid and gingerol, which have been described to stimulate SERCA activity, have been shown to be promising therapeutic targets for cardiovascular dysfunction [32, 33]. In addition, several molecules, including thapsigargin derivatives and flavonoids are being developed as potential cancer treatments based on their SERCA-mediated anti-proliferative and pro-apoptotic effects [34, 35]. Furthermore, a broad variety of organic solvents and hydrophobic compounds bearing nucleophilic groups have been extensively reported as inhibitors of SERCA activity [36–41]. The inhibitory effect of these compounds was related to their hydrophobicity, and notably, they were also able to stimulate SERCA activity at certain concentrations. In this study, we provide evidence, for the first time, that 1,2-DCB affects $Ca^{2+}$ homeostasis by affecting SERCA, a fundamental $Ca^{2+}$ modulator. We report here the effect of 1,2-DCB on several key features of SERCA, including ATP hydrolytic activity, the ratio of hydrolysis of ATP/$Ca^{2+}$ transported, and the phosphorylation stage (EP) during the catalytic cycle of this enzyme.

## Methods

### Chemicals and reagents

[$\gamma$-$^{32}$P]ATP and $^{45}$CaCl$_2$ were purchased from Perkin-Elmer, USA. [$^3$H]Glucose and $^{45}$CaCl$_2$ were products of DuPont NEN. The $Ca^{2+}$ ionophore A23187 (calcimycin), EGTA (ethylene glycol-bis (β-aminoethyl ether)-N,N,N',N'-tetra acetic acid), Mops (4-morpholinepropanesulfonic acid), liquid scintillation cocktail (Sigma-Fluor S-4023), ATP disodium salt, LaCl$_3$, K-oxalate, benzene, 1,2-DCB was obtained from Sigma, Co. (St. Louis, MO). CaCl$_2$ and other chemicals ware purchased from J.T. Baker (México). The $Ca^{2+}$ ionophore A23187 was dissolved in ethanol. Benzene and 1,2-DCB were dissolved in methanol. Ethanol and methanol never reached a concentration higher than 1 % (v/v) in the reaction media after the addition of 1,2-DCB or A23187.

### Preparation of sarcoplasmic vesicles

SR (sarcoplasmic reticulum) membranes rich in $Ca^{2+}$-ATPase (SERCA) were prepared from the low density SR of skeletal fast-twitch muscle of adult New Zealand rabbit hind limbs, and preserved in a buffer containing 10 mM Mops, pH 7.0 and 30 % sucrose and stored at –80 °C as described by Eletr and Inesi [42]. The sarcoplasmic membrane concentration refers to milligrams (mg) of total protein per milliliters (ml) and was measured by the colorimetric procedure described by Lowry et al. [43]. We used bovine serum albumin as standard protein.

### Free $Ca^{2+}$ concentration in the reaction media

Free $Ca^{2+}$ concentration was adjusted by adding the appropriate volume of CaCl$_2$ and EGTA to the enzymatic reaction media [44]. Free $Ca^{2+}$ concentration was calculated by the computer program *Calcium* which is based on the absolute constant stability for the $Ca^{2+}$-EGTA complex, the EGTA protonation equilibrium, $Ca^{2+}$ ligands and media pH [45, 46].

### $Ca^{2+}$-ATPase (SERCA) hydrolytic activity supported by A23187 or K-oxalate

Initial rates of ATP hydrolysis by SR vesicles were measured at 25 °C for 5 min (min) by following the liberation of inorganic phosphate ($P_i$). A typical assay medium was buffered at pH 7.0, and contained 20 mM Mops, pH 7.0, 80 mM KCl, 5 mM MgCl$_2$, 1 mM EGTA, 0.967 mM CaCl$_2$ (10 µM free $Ca^{2+}$), 0.01 mg/ml SR vesicles, 1.5 µM A23187 (to generate leaky vesicles to measure in min) or 5 mM K-oxalate (no leaks and to measure in min). A typical reaction was started after the addition of 1 mM of ATP. Any change in the concentration of SR protein added to the reaction medium is detailed in the figure captions. The compound 1,2-DCB was added at the indicated concentrations (0.05–1 mM). The appearance of $P_i$ from the hydrolysis of ATP mediated by SERCA was evaluated with a molybdovanadate reagent previously described by Lin and Morales [47]. Experimental details are indicated in the corresponding figure legends.

## Ca²⁺ uptake supported by K-oxalate

SERCA is a $Ca^{2+}$ pump that depends on $Mg^{2+}$ and the presence of ATP to transport $Ca^{2+}$ across of the SR membrane. For this reason, we studied the $Ca^{2+}$ transported by SERCA in the presence of 1,2-DCB. $Ca^{2+}$ transported by SERCA was measured at 25 °C using $^{45}CaCl_2$ as a radioactive tracer followed by sample filtration according with the methods described by Martonosi and Feretos [48]. The experiment was conducted in a reaction mixture containing 20 mM Mops, pH 7.0, 80 mM KCl, 5 mM $MgCl_2$, 1 mM EGTA, 5 mM K-oxalate instead of ionophore A23187 (to keep vesicles intact), 0.967 mM $CaCl_2$ (10 μM free $Ca^{2+}$) with 1000 cpm $^{45}CaCl_2$ per nmol of $Ca^{2+}$, and different concentrations of 1,2-DCB. The reaction contained 0.01 mg/ml of protein and was started after adding 1 mM of ATP. $Ca^{2+}$ uptake mediated by SERCA was stopped by filtering the vesicles in 0.45 μm HA type nitrocellulose membranes filters (Millipore, Milford, MA). The filters containing the $Ca^{2+}$-loaded SR vesicles were washed two times with 2 ml of a buffer containing 20 mM Mops pH 7.0, 80 mM KCl, 5 mM $MgCl_2$ and 1 mM $LaCl_3$. Finally, $^{45}Ca^{2+}$ transported was measured by scintillation spectroscopy using 3 ml of scintillation cocktail per vial. The use of radioactive standards allowed us to express the $Ca^{2+}$ uptake data as nmol $Ca^{2+}$/mg of protein.

## Membrane integrity test

In order to determine if 1,2-DCB was affecting the membrane integrity and therefore interfering with $Ca^{2+}$ measurements, SR vesicles (0.01 mg/ml of protein) were loaded with $^{45}Ca^{2+}$ by performing a $Ca^{2+}$ transport assay using the active transport of SERCA as described before for $Ca^{2+}$ uptake [48]. The reaction was started with 1 mM of ATP and the reaction ran for 5 s (s) until SR vesicles were loaded with detectable amounts of $Ca^{2+}$. In order to determine if DCB induced vesicle leakage, the $^{45}Ca^{2+}$-loaded vesicles were exposed to 1 mM of a 1,2-DCB right before stopping the reaction by filtration. In addition, two positive controls were incorporated in substitution for oxalate: $Ca^{2+}$ ionophore (1.5 μM of A23187) in combination with 1 mM 1,2-DCB, and $Ca^{2+}$ ionophore alone. Very importantly, A23187 does not allow $Ca^{2+}$ to be accumulated into SR vesicles. For more details of the experiment please review $Ca^{2+}$ uptake supported by oxalate above. The amount of $^{45}Ca^{2+}$ loaded into the vesicles was determined by scintillation spectroscopy.

## Ca²⁺ binding at equilibrium assay

The binding of $Ca^{2+}$ to SERCA was tested in the absence of ATP using $^{45}CaCl_2$. Unbound $Ca^{2+}$ was evaluated from the filter wet volume by using [³H]glucose as a marker. This approach was used to test the $Ca^{2+}$ high-affinity conformational state of SERCA ($E_1Ca_2$ stage). The effect of 1,2-DCB on $E_1Ca_2$ state was evaluated at 25 °C for 5 min in a $Ca^{2+}$-saturating medium containing 20 mM Mops, pH 7.0, 5 mM $MgCl_2$, 80 mM KCl, 0.1 mM EGTA, 0.105 mM [$^{45}Ca$]$CaCl_2$ (~5000 cpm/nmol) (10 μM free $Ca^{2+}$), 1 mM [³H]glucose (~10,000 cpm/nmol), and 0.2 mg/ml of SR protein. Subsequently, we added different concentrations of 1,2-DCB. In order to test the effect of 1,2-DCB on $E_2$ stage we used a $Ca^{2+}$-free media containing 20 mM Mops, pH 7.0, 5 mM $MgCl_2$, 80 mM KCl, 0.1 mM EGTA, 1 mM [³H]glucose (~10,000 cpm/nmol), and 0.2 mg/ml of SR protein, and different concentrations of 1,2-DCB. The incubation was performed at 25 °C for 5 min. Later, the reaction for $E_2$ stage was supplemented by adding a buffer containing 20 mM MOPS, pH 7.0, 5 mM $MgCl_2$, 80 mM KCl, 0.1 mM EGTA, 3.15 mM [$^{45}Ca$]$CaCl_2$ (~5000 cpm/nmol) and 1 mM [³H]glucose (~10,000 cpm/nmol) with a final free-$Ca^{2+}$ concentration of 10 μM. Following, the assay was prolonged for 1 min at 25 °C. Samples of 1 ml (0.2 mg of protein) were filtered under vacuum in 0.45 μm nitrocellulose filters (Millipore, HA type). Any further washing to determine $^{45}Ca$ and ³H associated with the filters and the amount of radioactivity in the filters was determined by scintillation spectroscopy. Specific binding of $Ca^{2+}$ to the SERCA was calculated by subtracting unbound $Ca^{2+}$ retained by the filter.

## Phosphorylation of SERCA by [γ³²P]ATP

Maximal levels of EP (phosphorylated enzyme intermediate) were determined using [γ-³²P]ATP as a substrate and starting from different incubation conditions: $E_1Ca_2$ -DCB, $E_2$ -DCB and $E_2$[γ-³²P]ATP-DCB in similar manner as we described in previous work [25, 26]. These conditions were studied during the enzyme turnover by mixing equal volumes (0.5 ml) of SR vesicles with the phosphorylating medium with or without $Ca^{2+}$ (after both solutions were mixed, the final volume was 1 ml and 0.1 mg/ml of SR protein). The SR vesicles (0.2 mg protein/ml) were suspended in a tube with 0.5 ml of 20 mM Mops, pH 7.0, 80 mM KCl, 5 mM $MgCl_2$, 1 mM EGTA, 0.967 mM $CaCl_2$ (10 μM free $Ca^{2+}$), and 15 μM A23187. To measure $E_1Ca_2$-DCB stage, the reaction mixture was preincubated with a defined concentration of 1,2-DCB and the phosphorylation reaction was started by mixing 0.5 ml of this medium with 20 μl of medium containing 20 mM Mops, pH 7.0, 80 mM KCl, 5 mM $MgCl_2$, 1 mM EGTA, 0.967 mM $CaCl_2$, and 1.25 mM [γ-³²P]ATP (50 μM final concentration, ~20,000 cpm/ nmol). When the enzyme (0.2 mg/ml) was in nominally $Ca^{2+}$-free medium ($E_2$ -DCB stage), the phosphorylating medium contained $CaCl_2$ to give 10 μM free $Ca^{2+}$ after mixing. In order to determine if 1,2-DCB was directly affecting the $E_2$ATP in a $Ca^{2+}$ free medium, the formation of the EP stage was started when $Ca^{2+}$ (to give 10 μM free) was added into the reaction medium. Phosphorylation time was 2 s when the enzyme turnover was studied at 0 °C, and 5 s when studied at 25 °C. The reaction was started under continuous vortexing and stopped by adding 1 ml of ice-cold 250 mM perchloric acid and 2 mM sodium phosphate. Stopped reactions were incubated on ice for 5 min before filtration through 0.45 µm nitrocellulose filters (Millipore, HA type). SR vesicles retained by the filters were washed 5 times with ice-cold 250 mM perchloric acid and 2 mM sodium phosphate. Finally, the amount of radioactivity in the filters was determined by scintillation spectroscopy. These results are expressed as nmol EP/milligram of protein.

### Statistical analysis

Statistical and kinetic analysis was performed using Sigma Plot 11 software. Unless otherwise stated, the experiment values are represented as the mean of at least three independent experiments, each performed in triplicate +/− average SE, with statistical significance of $p < 0.05$ determined by t-student's paired $t$-test.

### Results

#### Effect of 1,2-DCB on the hydrolytic activity of SERCA supported by A23187

In order to determine the effect of 1,2-DCB on the ATP hydrolytic activity of SERCA, we titrated different 1,2-DCB concentrations in a typical $Ca^{2+}$-ATPase assay where SERCA hydrolyzed ATP to ADP and $P_i$ as described before. In this experiment we incubated SR vesicles with A23187, a $Ca^{2+}$ ionophore, which is used to generate leaky vesicles (absence of gradient) in order to avoid saturation of SR vesicles with $Ca^{2+}$, which would otherwise stop $Ca^{2+}$-ATPase activity. The dose-response curve revealed a biphasic effect of 1,2-DCB on the ATP hydrolytic activity of SERCA: while lower concentrations of the compound highly stimulated ATP hydrolysis, 1,2-DCB concentrations higher than 0.3 mM inhibited SERCA hydrolytic activity (Fig. 1). In addition, 0.5 mM of 1,2-DCB reduced SERCA activity to 50 % ($IC_{50} = 0.5$ mM). To determine the importance of the Cl atoms on the effect of the compound on SERCA hydrolytic activity, we used the benzene molecule, which is the same ring base structure, but depleted of Cl atoms (Fig. 1, inset). A dose-response curve was performed by titrating several benzene concentrations, and we found that the benzene molecule by itself did not affect SERCA hydrolytic activity at all, indicating that Cl atoms are involved in the biphasic effect observed on SERCA hydrolytic activity.

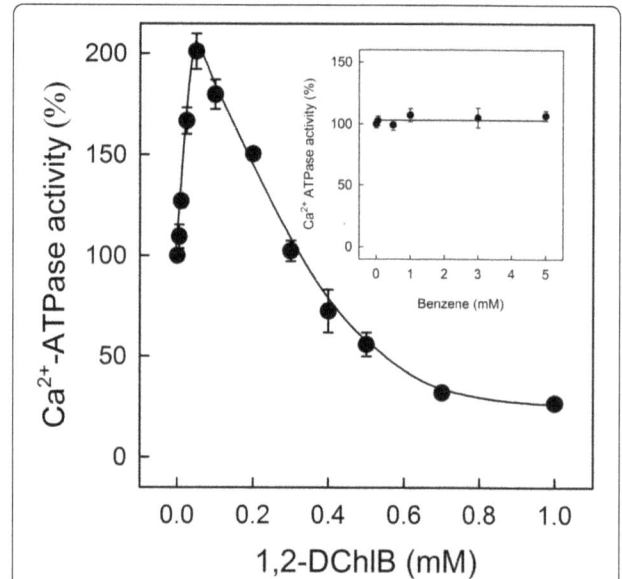

**Fig. 1** Effect of 1,2-DCB and benzene on the initial rates of ATP hydrolysis supported by A23187. The reaction was performed in a buffer containing 20 mM Mops, pH 7.0, 80 mM KCl, 5 mM MgCl2, 1 mM EGTA, 0.967 mM CaCl2, and 1.5 µM A23187. The reaction was conducted by adding 0.01 mg/ml of SR vesicles with different concentrations of 1,2-DCB (mM) or benzene (mM). First data point corresponds to $Ca^{2+}$-ATPase activity without DCB (control) and it was ~2.6 ± 0.05 µmol $P_i$/mg·min

#### Effect of 1,2-DCB on $Ca^{2+}$ uptake and ATP hydrolysis ratio mediated by SERCA and supported by K-oxalate

We next aimed to determine the effect of 1,2-DCB on $Ca^{2+}$ uptake mediated by SERCA. Vesicular fragments of SR exhibit linear rates of $Ca^{2+}$ transport upon addition of ATP when equilibrated in the presence of K-oxalate as a lumenal $Ca^{2+}$ regulator (non leaky vesicles). In order to avoid $Ca^{2+}$ leakage caused by the $Ca^{2+}$ ionophore (A23187), we used K-oxalate in both $Ca^{2+}$-transport and hydrolytic activity experiments in order to make them comparable to $Ca^{2+}$ transport. The dose-response curves showed that 1,2-DCB did not stimulate $Ca^{2+}$ uptake mediated by SERCA, but in contrast, $Ca^{2+}$ uptake was inhibited by DCB concentrations above 0.5 mM until achieving full inhibition at 1 mM (Fig. 2a). $Ca^{2+}$ transport cannot be measured in an assay containing a $Ca^{2+}$ ionophore (as used to measure the ATP hydrolysis) therefore these experimental conditions did not allow us to compare the $Ca^{2+}$ uptake assay with the previously conducted experiments of hydrolytic activity. The approach employed for $Ca^{2+}$ uptake measurement is less sensitive than the one using $Ca^{2+}$ ionophore and reduces enzyme activity by ~1 fold, thus showing less effect of 1,2-DCB (Fig. 2b), however, there were peaks of activation of approximately 15 and 20 % at DCB concentrations of 0.05 and 0.1 mM, respectively. Nevertheless, when we compared the ATP hydrolysis results with the

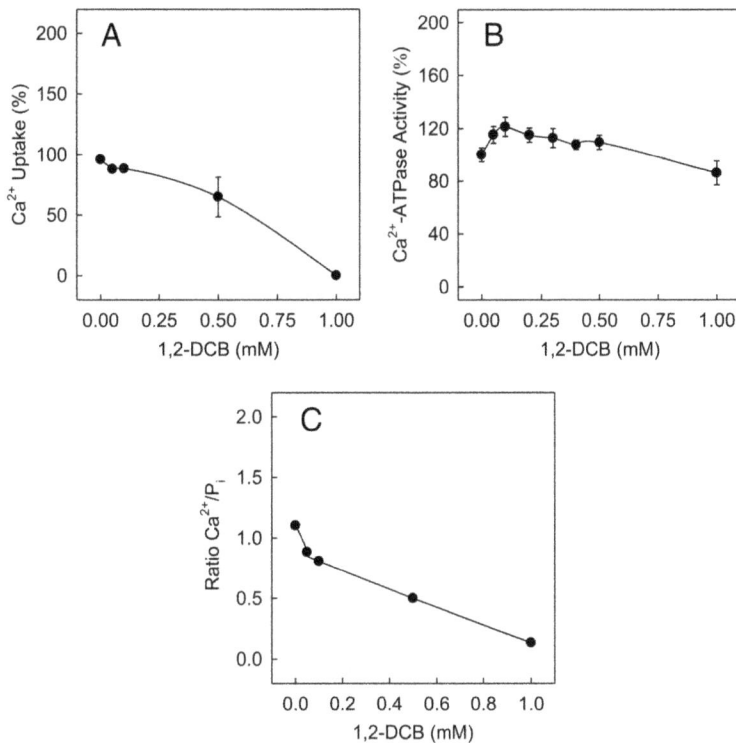

**Fig. 2** The $Ca^{2+}$-ATPase activity supported by oxalate was uncoupled by 1,2-DCB. The effect of 1,2-DCB on $Ca^{2+}/P_i$ coupling ratio was determined. The coupling ratio at each DCB concentration was calculated as the ratio between the $Ca^{2+}$ uptake rate and the ATP hydrolysis rate by SERCA. SR vesicles (0.01 mg/ml) were equilibrated at 25 °C in a medium containing 5 mM K-oxalate. Hydrolysis rate of ATP (**a**) or $^{45}Ca^{2+}$ uptake rates (**b**) were measured at same experimental conditions. **c** The $Ca^{2+}/P_i$ coupling ratio (calculated by dividing the $Ca^{2+}$ transport values by those of ATP hydrolysis), in presence of different concentrations of 1,2-DCB was uncoupled

ones conducted for $Ca^{2+}$ uptake, we were able to find a disruption in the $Ca^{2+}$ uptake/ATP hydrolysis ratio (calculated by dividing the $Ca^{2+}$ transport values by those of ATP hydrolysis), which indicates that the enzyme activity ratio was uncoupled, as opposed to normal physiological conditions during which the enzyme maintains a coupled ratio of $2Ca^{2+}$ for each mol of ATP hydrolyzed to ADP and $P_i$ (Fig. 2c). The observed ratio appears to confirm that 1,2-DCB concentrations greater than 0.05 mM cause a disturbance on the enzyme activity, an effect that is not observed only with the hydrolysis.

Given that 1,2-DCB affected $Ca^{2+}$ uptake mediated by SERCA, we aimed to determine if 1,2-DCB was also affecting the binding of $Ca^{2+}$ to SERCA. Direct measurements of $^{45}Ca^{2+}$ binding to the non-phosphorylated enzyme pre-incubated with the compound indicated that the $E_1$ conformation remained unaffected by treatment with 1,2-DCB (Fig. 3, open circles). In contrast with the $E_1$ form, treatment with 1,2-DCB starting from the $E_2$ form prior to binding of $Ca^{2+}$ affected only ~16 % of the total $Ca^{2+}$ bound to the enzyme (Fig. 3, closed circles), but the levels remained constant with all 1,2-DCB concentrations.

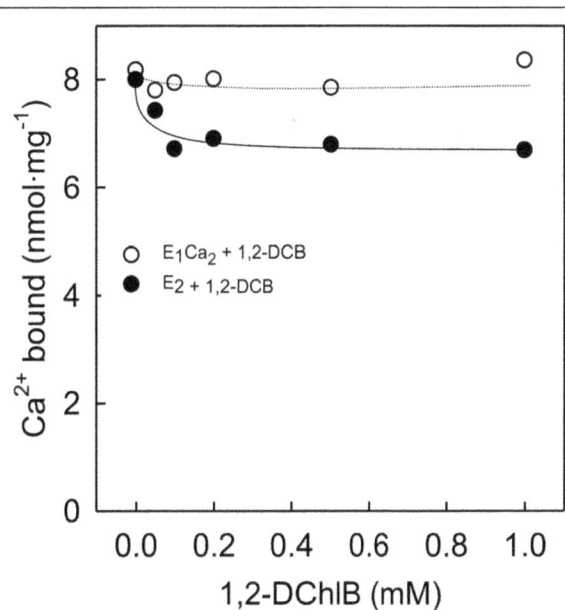

**Fig. 3** Binding of $Ca^{2+}$ to SERCA. To test if 1,2-DCB was affecting the binding of $Ca^{2+}$ to the enzyme we performed a binding assay as described in *Methods*. The complex $Ca^{2+}/SERCA$ ($E_1Ca_2$) (o) or enzyme prior binding to $Ca^{2+}$ ($E_2$) (●) were exposed to different concentrations of 1,2-DCB

## Membrane protein-dependent effect of 1,2-DCB and SR membrane integrity

1,2-DCB is a typical hydrophobic molecule and as such, it has affinity for hydrophobic environments such as biological membranes. For this reason, we used our model of study to investigate the possibility of 1,2-DCB targeting the SR membranes. First, we conducted a set of experiments using a high-dose of 1,2-DCB in reactions having 10-fold different amounts of SR vesicles (0.01–0.1 mg/ml total protein). In these experiments we tested hydrolysis of ATP mediated by SERCA and we found that the inhibitory effect of 1,2-DCB was significantly reduced with high amounts (5–10 folds higher) of SR vesicles (0.05–0.1 mg/ml total membrane protein), which strongly suggested that 1,2-DCB molecules were being diluted into the SR membranes (Fig. 4). These results indicate that DCB molecules were going from the aqueous environment of the reaction medium into the membranes, and when the number of SR vesicles was increased, there were less 1,2-DCB molecules per SR membrane. However, this does not confirm if 1,2-DCB was affecting the integrity of the SR membrane. In order to analyze if 1,2-DCB molecules were disrupting the integrity of the SR membrane, we loaded SR vesicles

with $Ca^{2+}$, using a typical $Ca^{2+}$ uptake assay as previously described, and 5 s before stopping the reaction we added 1 mM of 1,2-DCB into the reaction mix (Fig. 5). SR vesicles treated with 1,2-DCB (open circles) did not show any difference in levels of accumulated $Ca^{2+}$ inside of the SR vesicle when compared to controls (without 1,2-DCB, closed circles), thus demonstrating that 1,2-DCB did not disrupt membrane vesicles. In addition, we performed a positive control with a $Ca^{2+}$ ionophore (A23187) knowing that it produces leaky vesicles and therefore does not allow $Ca^{2+}$ accumulation (Fig. 5, closed and open triangles). As observed, vesicles with A23187 were only able to accumulate ~11 % of the total $Ca^{2+}$ accumulated in the control at 1 min of filling (100 %). This demonstrates that, at least in short time exposure, DCB does not disrupt the SR membrane integrity, which indicates that the effect of 1,2-DCB on SERCA hydrolytic activity was due to an effect on SERCA functionality.

## Formation of EP stage during the catalytic cycle of SERCA was inhibited by 1,2-DCB

The following step was to study the formation of EP stage by ATP during the turnover of SERCA, which is a crucial step in the catalytic cycle of this enzyme.

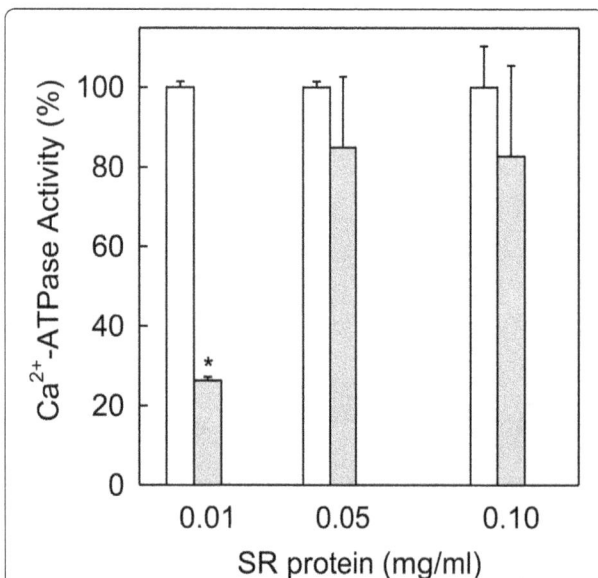

**Fig. 4** The effect of 1,2-DCB was SR vesicles-dependent. The assay medium contained 20 mM Mops, pH 7.0, 80 mM KCl, 5 mM MgCl₂, 1 mM EGTA, 0.967 mM CaCl₂, different SR protein concentrations (0.01 and 0.10 mg/ml), and 1.5–15 µM A23187 (A23187 was proportionally increased with respect SR total protein). SERCA activity was assayed at 25 °C in absence of compound as control (white bars) and with 1 mM 1,2-DCB (gray bars). The reactions started by the addition of 1 mM ATP. Controls were normalized to 100 % of $Ca^{2+}$-ATPase activity and the actual values were 2.6 ± 0.05, 2 ± 0.04, and 1.6 ± 1.3 µM P$_i$/min/mg. *, $p$ = <0.001, compared to their respectively control that does not have DCB

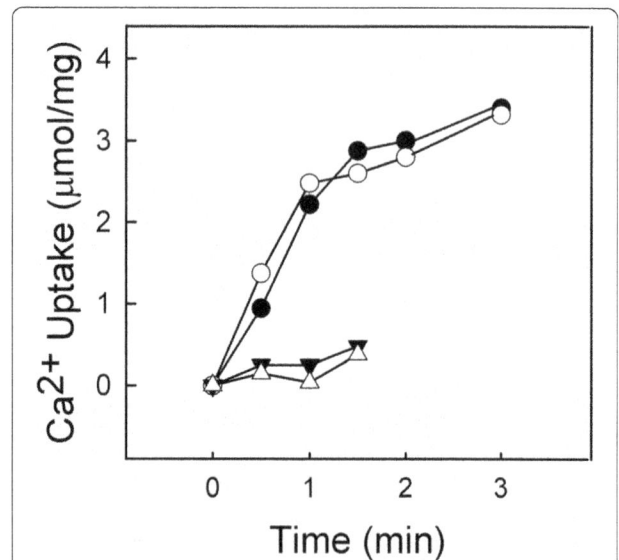

**Fig. 5** Time-dependent filling of SR vesicles and exposure to 1,2-DCB. In order to determine the effect of 1,2-DCB on the filling of the vesicles by SERCA protein, we exposed on the SR membrane, $^{45}Ca^{2+}$ uptake was measured under same conditions described for $Ca^{2+}$ transport. At different times in $Ca^{2+}$ uptake, 1,2-DCB (open circles) was added to the reaction media before stopping the reaction. The same reaction without DCB treatment was tested as a control (closed circles). Two additional positive controls were incorporated in substitution for oxalate, $Ca^{2+}$ ionophore (1.5 µM of A23187) was added to the assays with (closed triangles) or without 1 mM 1,2-DCB (open triangles). Very importantly, A23187 did not allow $Ca^{2+}$ to be accumulated into SR vesicles

Surprisingly, all the 1,2-DCB concentrations tested considerably reduced the formation of the $E_1P$ stage of SERCA (Fig. 6a). When starting from $E_1Ca_2$ plus 1,2-DCB, and the variation $E_2$ plus 1,2-DCB started with $Ca^{2+}$/ATP (inset), the EP levels were drastically reduced: $E_1Ca_2$ with 0.05 mM of 1,2-DCB barely reached 13.04 % of accumulated EP, with 0.1 and 0.2 mM of 1,2-DCB further reducing EP levels to 6.08 and 5.65 %, respectively and 0.5 mM of the compound completely abolishing EP accumulation. In the case of $E_2$, the EP levels were similarly affected: 0.05 mM of 1,2-DCB decreased EP to 17.5 %, with 0.1 mM further reducing it to 3.0 %, and the remaining concentrations showed no EP at all (Fig. 6a, inset). The results with $E_2ATP$ plus 1,2-DCB started with $Ca^{2+}$ at 0 or 25 °C were slightly different (Fig. 6b), and it appears that in this case the observed result could be more influenced by temperature. However, beside the temperature difference, a somewhat higher level of EP can still be observed at 0.05 mM of 1,2-DCB (Fig. 6b) when compared to the levels observed on Fig. 6a at the same DCB concentration, suggesting a possible protective effect by ATP, although experiments carried at zero degrees using DCB concentrations above 0.1 mM showed complete inhibition of EP formation (Fig. 6b, closed circles), but EP levels at 25 °C are greater than those observed at 0 °C. It is worth noting that the

data shown in both figures does follow a similar trend, that is a consistent reduction of EP levels as the 1,2-DCB concentration increases.

## Discussion

Our data show that 1,2-DCB exerts a dual response of stimulation and inhibition in SERCA hydrolytic activity supported by A23187 (Fig. 1). Similar effects of activation and inhibition of SERCA caused by other hydrophobic compounds have been reported in the literature, and possible explanations include alterations in the lipid environment of the enzyme, as well as accessing sensitive domains of the pump due to their hydrophobic properties [49–51]. The 1,2-DCB molecule possesses two Cl atoms in the benzene ring. In order to test the importance of the Cl atoms on the effect of this molecule on SERCA, we used benzene (which lacks Cl atoms) and we were able to avoid the inhibitory and stimulatory effects on SERCA activity produced by 1,2-DCB (Fig. 1). Our data supports a clear relation between the structure of 1,2-DCB and the effect produced on the ATP hydrolytic activity, since the absence of Cl atoms does not cause any effect on SERCA. Similar conclusions have been previously reported regarding the effects caused by the presence of electronegative groups [52–54]. Moreover, hydrophobic molecules can interact with hydrophobic

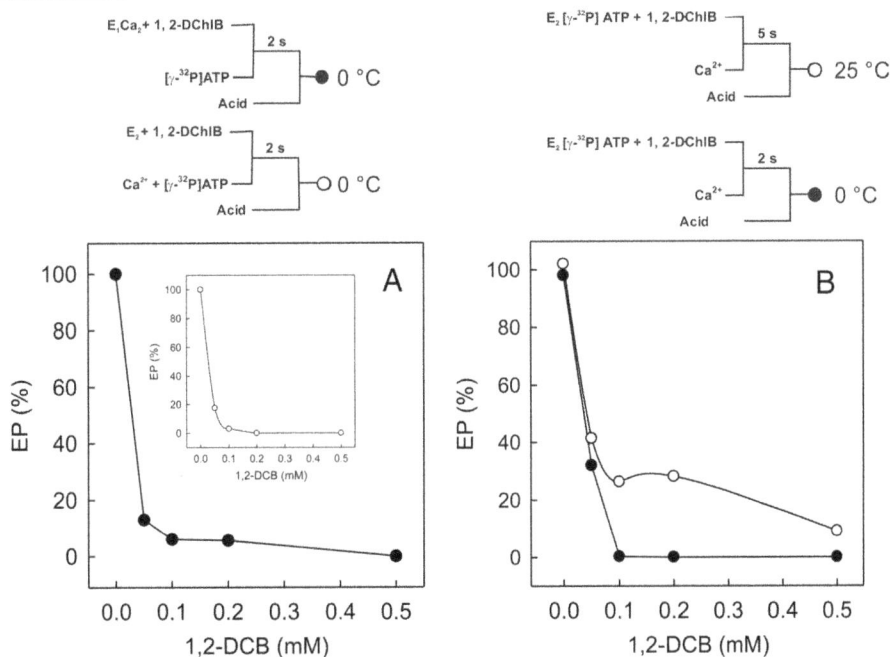

**Fig. 6** Effect of 1,2-DCB on functional states of SERCA as determined by ATP phosphorylation. SR vesicles at 0.1 mg of protein/ml were incubated at 25 °C in a medium containing: **a** 20 mM Mops, pH 7.0, 80 mM KCl, 5 mM MgCl$_2$, 15 μM A23187, 0.1 mM EGTA, 10 μM free Ca$^{2+}$, and a certain 1,2-DCB concentration starting with 50 μM [γ-$^{32}$P]ATP (●). *Inset*: The $E_2$ + 1,2-DCB prior phosphorylation was starting with 10 μM free Ca$^{2+}$ and 50 μM [γ-$^{32}$P]ATP (○). The phosphorylation in each case was maintained for 2 s at 0 °C. **b** For phosphorylation initiating from $E_2ATP$ Ca$^{2+}$ free, the enzyme (0.1 mg of protein/ml) was first treated with 1,2-DCB in a media containing 20 mM Mops, pH 7.0, 80 mM KCl, 5 mM MgCl$_2$, 15 μM A23187, 0.1 mM EGTA, followed by phosphorylation at 0 °C or 25 °C by adding [γ-$^{32}$P]ATP and 10 μM free Ca$^{2+}$. For more details see *Methods*

domains in the membrane proteins, thus affecting enzyme functionality [37, 53, 55].

We also addressed the possibility of 1,2-DCB affecting the activity of binding/translocation of $Ca^{2+}$ by SERCA. Surprisingly, a 1 mM concentration of 1,2-DCB fully inhibited $Ca^{2+}$ uptake without affecting $Ca^{2+}$-binding, which suggests that the binding of $Ca^{2+}$ is not a major problem for the inhibition of $Ca^{2+}$ transport. However, this is indicative that 1,2-DCB is affecting the $Ca^{2+}$ sensitive conformation of the enzyme to a lesser extent (Figs. 2a and 3 respectively). Nevertheless, our data of binding of $Ca^{2+}$ suggested that $Ca^{2+}$ was able to bind to the enzyme in turnover conditions almost at normal levels, even when $Ca^{2+}$ uptake was completely inhibited at the highest DCB concentration. Therefore, it was necessary to study the hydrolytic activity of SERCA under the same conditions used for the $Ca^{2+}$ transport assay, using K-oxalate instead of the $Ca^{2+}$ ionophore. When intravesicular $Ca^{2+}$ levels get higher (~2–3 mM), $Ca^{2+}$ reacts with oxalate to form calcium oxalate thus precipitating inside of the intact vesicle as crystals [13], which allows us to measure the enzyme activity in min. Surprisingly, we found that the hydrolytic activity of SERCA under these same conditions was still stimulated, but not inhibited at the same concentrations tested before (Fig. 3b). This may be due to the different experimental conditions such as the presence of ionophore, which allows the system to function more efficiently. The $Ca^{2+}$-ionophore acts as a freely mobile carrier that transports $Ca^{2+}$ across the membrane, and it is possible that it is acting simultaneously with the 1,2-DCB, stimulating the pump due to increased membrane permeability and absence of gradient [13, 56, 57]. This could explain why activation and inhibition are more pronounced when measured with the $Ca^{2+}$-ionophore than when measured with K-oxalate, where the vesicle interior is not gradient-free. We also found that the $Ca^{2+}/P_i$ ratio was partially uncoupled at 1,2-DCB concentrations between 0.05 and 0.5 mM, and fully uncoupled at 1 mM. This phenomenon of uncoupled $Ca^{2+}$-ATPase with a lower $Ca^{2+}/P_i$ ratio has been previously described in terms of the increased thermogenic activity that occurs when using intact vesicles [13–15], where it was found that after $Ca^{2+}$ saturation, ~82 % of the hydrolytic activity is uncoupled from the $Ca^{2+}$ transport, thus decreasing the $Ca^{2+}/P_i$ ratio.

Previous reports showed that non-polar organic solvents facilitate the incorporation of hydrophobic compounds to cell membranes, and that enzymatic stimulation or inhibition is related to both solvent and membrane protein quantity [38, 49]. In agreement with the literature, 1,2-DCB affects SERCA activity by sharing these features, as we further demonstrated that the inhibitory effect of 1,2-DCB was SR membrane-dependent (Fig. 4). This could be due to a partitioning effect of the DCB into the membrane fraction, which indicates that activation or inhibition is dependent on membrane composition and protein concentration. Organic solvents can reduce the inhibition of SERCA by certain drugs, which explains how hydrophobic drugs can move to a lipid environment in the membrane, thus causing enzyme inhibition [49, 50, 58]. Data from Lax et al. [59] showed that the addition of phosphatidylcholine liposomes to SERCA reactions caused a decrease in the inhibitory effect of the fungicide miconazol. 1,2-DCB shares similar features to those of miconazole and therefore, we believe that our data indicates that DCB is going into the SR vesicles because of its hydrophobic properties.

The $Ca^{2+}$ uptake assay employed in this study measured accumulated $Ca^{2+}$ inside the SR vesicles [48, 60]. However, if 1,2-DCB disrupts the time required to fill the vesicles with $Ca^{2+}$, this may be reflected as an inhibition on the accumulated $Ca^{2+}$ due to a loss of $Ca^{2+}$ caused by vesicle damage. Molecules such as 1,2-DCB can target SR membranes affecting the filling time of SR vesicles by the SERCA pump. For example, hexachlorocyclohexane, a hydrophobic compound, was shown to affect the activity of membrane proteins by disrupting the lipid bilayer [55]. For this reason, we filled out SR vesicles with $Ca^{2+}$ using the active transport of SERCA, followed by the addition of a high concentration of DCB (1 mM) 5 s before stopping the reaction. In agreement with our previous findings, 1,2-DCB was unable to decrease the levels of accumulated $Ca^{2+}$ when compared to controls (without DCB) (Fig. 5). Given that vesicle filling occurs in less than 1 s, which is the physiological time during which muscle relaxation occurs [15], we selected a longer vesicle filling time. This data clearly demonstrates that 1,2-DCB is not affecting the vesicle filling time by causing disturbances of the SR membrane. This was supported by our a positive control using $Ca^{2+}$ ionophore A23187 (absence of gradient) alone or with 1 mM 1,2-DCB, which did not allow accumulation of $Ca^{2+}$ inside the SR vesicles (Fig. 5). This phenomenon is referred to in the literature as "leaky vesicles", since this system allows the efflux of luminal $Ca^{2+}$, thus obtaining faster efflux [61, 62]. It was interesting to find that the SR membrane was not disrupted, which supports our findings that 1,2-DCB is causing uncoupling of the $Ca^{2+}/P_i$ ratio. Considering this phenomenon applied to a biological system, we can refer to the thermogenic activity previously proposed by Inesi and Tadini-Buoninsegni [15], where SERCA1 uncoupling interferes with the decrease of cytosolic $Ca^{2+}$ and the subsequent relaxation of muscle fibers. If we take into account that vesicles are filled with $Ca^{2+}$ in less than 1 s, a sufficient concentration of luminal $Ca^{2+}$ would not be reached, thus preventing the dissociation of the $E_2$-$P \cdot Ca_2$ form. This phenomenon could not lead to thermogenic activity due to a basal state of the pump. The $Ca^{2+}$ leakage from the pump could occur if the luminal $Ca^{2+}$ concentration reached such high levels that both its $E_2$-$P \cdot Ca_2$ dissociation constant and the cytosolic $Ca^{2+}$ concentrations would remain above the required levels to achieve maximal pump activation, which could occur under prolonged muscle activity. Even though the cytosolic $Ca^{2+}$ levels can remain high due to multiple action potentials, this does not appear to be our case, since we have no evidence of pump overload (Figs. 2a and 5).

In order to get more insights into the mechanism by which 1,2-DCB caused uncoupling of the SERCA activity, we studied the $E_1P$ step of the enzyme. Our results demonstrated that 1,2-DCB decreased the EP stage formation but was influenced by temperature when starting from $E_2ATP$, even when our previous experiments showed that 1,2-DCB produced a high rate of ATP hydrolysis. Our initial expectation was to observe elevated levels of EP by ATP, since it is common that hydrophobic molecules affect the $E_2$ state of SERCA which prevents or decreases the transition to the $E_1$ stage [25, 63]. Our data of EP inhibition suggests that 1,2-DCB (Fig. 6) could be affecting the $E_2$ step. Similar examples of this phenomenon are found in the literature [15, 49, 51, 64, 65]. Taken together, the inhibition of the EP stage and our previous results could explain why an uncoupled $Ca^{2+}$/$P_i$ ratio was observed, even when $Ca^{2+}$ is binding to the enzyme in the presence of 1,2-DCB. Phosphorylation of the enzyme is necessary for the appropriate $Ca^{2+}$ transport across the SR membrane. Therefore, it is possible that a $Ca^{2+}$ slippage process is taking place, where ATP hydrolysis is occurring without the subsequent accumulation of the transient $E_1P$ form and without $Ca^{2+}$ transport, with $Ca^{2+}$ "slipping" into the outside of the vesicles. This can be affected by several factors such as time, temperature, enzymatic stage and in our case the DCB concentration [66, 67] Our data shows that at 25 °C, EP levels were increased (Fig. 6b). However, the measurements followed the same pattern of EP decrease with increasing DCB concentration. Regarding the EP measurements conducted at 0 °C, which seem to go against the literature, it is well established that high levels of EP are obtained at low temperatures (0 °C) due to the rapid transition of the limiting $E_1P \cdot Ca_2$ step to $E_2P \cdot Ca_2$, and the subsequent $Ca^{2+}$ translocation that leads to the EP decomposition [68], in other words, the phosphoenzyme formed from ATP is more sensitive to ADP at 0 °C than at 30 °C, thus confirming that ATP hydrolysis coupled to $Ca^{2+}$ efflux and the ATP↔$P_i$ exchange reaction are inhibited at 0 °C [69]. However,

the experiments starting from $E_2ATP$ showed higher levels of accumulated EP, but these experiments were conducted at 0 and 25 °C, where A23187 is also present (although protein concentration, time and ATP are different). At 0.05 mM DCB, ~30–40 % of accumulated EP can be observed at both temperatures, indicating that ATP could be protecting the enzyme from inhibition. However, higher DCB concentrations (0.1 mM and subsequent) caused a drastic decrease in EP levels (Fig. 6b, filled circles). Other factors that could affect the $E_1P$ to $E_2P$ transition include detergents and the microviscosity of the lipid environment of the ATPase [70]. Temperature is another important factor that could potentially affect this transition, as it has been demonstrated that at 21 °C the $E_1P$ to $E_2P$ isomerization is rapidly affected at high KCl concentrations [71]. Temperature is also important for compound solubility in aqueous solutions, as it has been shown that the solubility of aroma compounds (such as methyl ketones, ethyl esters, aldehyde and alcohol) decreases when incubated on ice as the concentration of the compound is increased, which could contribute to explain the effect of temperature and DCB solubility at 0 and 25 °C in the EP measurements [72].

## Conclusions

Our data demonstrates that 1,2-DCB affects the ATP hydrolytic activity of SERCA, causing both stimulation and inhibition of ATP hydrolysis. This dual effect caused by 1,2-DCB fits it into a common pattern displayed by several other hydrophobic compounds. We also found that 1,2-DCB partially inhibited $Ca^{2+}$ transport and EP stage formation due to uncoupled enzyme activity, even without affecting the $Ca^{2+}$ bound to the enzyme. Finally, to the best of our knowledge, this is the first study designed to study the effect of 1,2-DCB, a chemical that is widely found in the environment, on the functionality of SERCA proteins, and thus the data presented here not only provide insights into the molecular mechanisms of the enzyme's catalytic cycle, but on the possible toxic effects caused by 1,2-DCB.

**Competing interests**
The authors of this manuscript declare that they have no competing interests.

**Authors' contributions**
JVM carried out the majority of the ATPase activity experiments, including preparation of SR vesicles, $Ca^{2+}$ transport experiments, also participating in data analysis and drafting the manuscript. JASF prepared SR vesicles used in different assays, conducted ATP hydrolysis measurements and participated in data analysis and manuscript preparation. LFPT conceived the study, prepared SR vesicles, measurement of $Ca^{2+}$ transport and phosphorylation assays experiments, analyzed and interpreted experimental data, participated in manuscript drafting and directed the study. All authors approved the final version of this manuscript.

## Acknowledgements

This project was supported by grant PIFI for CA-02 Diagnosis Academic Group from the Universidad Autónoma de Ciudad Juárez (UACJ), Chihuahua, México. Our recognition of the support provided by Dr. Eppie Rael through NIH grants S06 GM08012 and G12 RR08124 from the University of Texas at El Paso, TX, USA. We also wish to acknowledge the recommendations and comments made by Dr. Armando Gómez-Puyou and Dr. Marietta Tuena de Gómez from the Instituto de Fisiología Celular, UNAM, México.

## Author details

[1]Departamento de Ciencias Químico Biológicas, Laboratorio de Biología Molecular y Bioquímica (Edif. T-216), Instituto de Ciencias Biomédicas, Universidad Autónoma de Ciudad Juárez, Plutarco Elías Calles #1210 Fovissste Chamizal, Ciudad Juárez, Chihuahua C.P. 32310, Mexico. [2]Present address: Department of Biological Sciences, University of Texas at El Paso, El Paso, TX 79968, USA. [3]Present address: Department of Biomedical Sciences, Center of Emphasis for Neurosciences, Texas Tech University Health Science Center, El Paso, TX 79905, USA.

## References

1. Yoshida T, Andoh K, Kosaka H, Kumagai S, Matsunaga I, Akasaka S, et al. Inhalation toxicokinetics of p-dichlorobenzene and daily absorption and internal accumulation in chronic low-level exposure to humans. Arch Toxicol. 2002;76:306–15.

2. Bristol DW, Crist HL, Lewis RG, MacLeod KE, Sovocool GW. Chemical analysis of human blood for assessment of environmental exposure to semivolatile organochlorine chemical contaminants. J Anal Toxicol. 1982;6:269–75.

3. Kumagai S, Matsunaga I. Identification of urinary metabolites of human subjects exposed to o-dichlorobenzene. Int Arch Occup Environ Health. 1995;67:207–9.

4. Jan J. Chlorobenzene residues in human fat and milk. Bull Environ Contam Toxicol. 1983;30:595–9.

5. Mes J, Davies DJ, Turton D, Sun WF. Levels and trends of chlorinated hydrocarbon contaminants in the breast milk of Canadian women. Food Addit Contam. 1986;3:313–22.

6. Hsiao PK, Lin YC, Shih TS, Chiung YM. Effects of occupational exposure to 1,4-dichlorobenzene on hematologic, kidney, and liver functions. Int Arch Occup Environ Health. 2009;82:1077–85.

7. Hissink AM, Van Ommen B, Van Bladeren PJ. Dose-dependent kinetics and metabolism of 1,2-dichlorobenzene in rat: effect of pretreatment with phenobarbital. Xenobiotica. 1996;26:89–105.

8. Yan RM, Chiung YM, Pan CY, Liu JH, Liu PS. Effects of dichlorobenzene on acetylcholine receptors in human neuroblastoma SH-SY5Y cells. Toxicology. 2008;253:28–35.

9. MacLennan DH, Green NM. Structural biology. Pumping ions. Nat. 2000;405: 633–4.

10. Toyofuku T, Kurzydlowski K, Tada M, MacLennan DH. Identification of regions in the Ca(2+)-ATPase of sarcoplasmic reticulum that affect functional association with phospholamban. J Biol Chem. 1993;268:2809–15.

11. MacLennan DH, Klip A. Calcium transport and release by sarcoplasmic reticulum: a mini-review. Soc Gen Physiol Ser. 1979;33:61–75.

12. MacLennan DH. Purification and properties of an adenosine triphosphatase from sarcoplasmic reticulum. J Biol Chem. 1970;245:4508–18.

13. Barata H, de Meis L. Uncoupled ATP hydrolysis and thermogenic activity of the sarcoplasmic reticulum Ca2 + –ATPase: coupling effects of dimethyl sulfoxide and low temperature. J Biol Chem. 2002;277:16868–72.

14. Martonosi A, Feretos R. Sarcoplasmic reticulum. II. Correlation between adenosine triphosphatase activity and Ca++ uptake. J Biol Chem. 1964; 239:659–68.

15. Inesi G, Tadini-Buoninsegni F. Ca/H exchange, lumenal Ca release and Ca/ ATP coupling ratios in the sarcoplasmic reticulum ATPase. J Cell Commun Signal. 2014;8:5–11.

16. de Meis L, Vianna AL. Energy interconversion by the Ca2 + –dependent ATPase of the sarcoplasmic reticulum. Annu Rev Biochem. 1979;48:275–92.

17. de Meis L. Approaches to studying the mechanisms of ATP synthesis in sarcoplasmic reticulum. Methods Enzymol. 1988;157:190–206.

18. Henao F, Delavoie F, Lacapere JJ, McIntosh DB, Champeil P. Phosphorylated Ca2 + –ATPase stable enough for structural studies. J Biol Chem. 2001;276: 24284–5.

19. Toyoshima C, Nakasako M, Nomura H, Ogawa H. Crystal structure of the calcium pump of sarcoplasmic reticulum at 2.6 A resolution. Nature. 2000; 405:647–55.

20. Sorensen TL, Clausen JD, Jensen AM, Vilsen B, Moller JV, Andersen JP, et al. Localization of a K + – binding site involved in dephosphorylation of the sarcoplasmic reticulum Ca2 + – ATPase. J Biol Chem. 2004;279:46355–8.

21. Sorensen TL, Moller JV, Nissen P. Phosphoryl transfer and calcium ion occlusion in the calcium pump. Science. 2004;304:1672–5.

22. Toyoshima C, Iwasawa S, Ogawa H, Hirata A, Tsueda J, Inesi G. Crystal structures of the calcium pump and sarcolipin in the Mg2 + –bound E1 state. Nature. 2013;495:260–4.

23. Inesi G. Mechanism of calcium transport. Annu Rev Physiol. 1985;47:573–601.

24. Michelangeli F, East JM. A diversity of SERCA Ca2+ inhibitors. Biochem Soc Trans. 2011;39:789–97.

25. Plenge-Tellechea F, Soler F, Fernandez-Belda F. On the inhibition mechanism of sarcoplasmic or endoplasmic reticulum Ca2 + –ATPases by cyclopiazonic acid. J Biol Chem. 1997;272:2794–800.

26. Soler F, Plenge-Tellechea F, Fortea I, Fernandez-Belda F. Cyclopiazonic acid effect on Ca2 + –dependent conformational states of the sarcoplasmic reticulum ATPase. Implication for the enzyme turnover. Biochemistry. 1998; 37:4266–74.

27. Thastrup O, Cullen PJ, Drøbak BK, Hanley MR, Dawson AP. Thapsigargin, a tumor promoter, discharges intracellular Ca2+ stores by specific inhibition of the endoplasmic reticulum Ca2(+)-ATPase. Proc Natl Acad Sci. 1990;87: 2466–70.

28. Aureliano M, Tiago T, Gândara RM, Sousa A, Moderno A, Kaliva M, Salifoglou A, Duarte RO, Moura JJ. Interactions of vanadium(V)-citrate complexes with the sarcoplasmic reticulum calcium pump. J Inorg Biochem. 2005;99:2355–61.

29. Aureliano M, Henao F, Tiago T, Duarte RO, Moura JJ, Baruah B, Crans DC. Sarcoplasmic reticulum calcium ATPase is inhibited by organic vanadium coordination compounds: pyridine-2,6-dicarboxylatodioxovanadium(V), BMOV, and an amavadine analogue. Inorg Chem. 2008;47:5677–84.

30. Fraqueza G, Ohlin CA, Casey WH, Aureliano M. Sarcoplasmic reticulum calcium ATPase interactions with decaniobate, decavanadate, vanadate, tungstate and molybdate. J Inorg Biochem. 2012;107:82–9.

31. Aureliano M, Fraqueza G, Ohlin CA. Ion pumps as biological targets for decavanadate. Dalton Trans. 2013;42:11770–7.

32. Antipenko AY, Spielman AI, Kirchberger MA. Interactions of 6-gingerol and ellagic acid with the cardiac sarcoplasmic reticulum Ca2 + –ATPase. J Pharmacol Exp Ther. 1999;290:227–34.

33. Namekata I, Hamaguchi S, Wakasugi Y, Ohhara M, Hirota Y, Tanaka H. Ellagic acid and gingerol, activators of the sarco-endoplasmic reticulum Ca2 +-ATPase, ameliorate diabetes mellitus-induced diastolic dysfunction in isolated murine ventricular myocardia. Eur J Pharmacol. 2013;706:48–55.

34. Christensen SB, Skytte DM, Denmeade SR, Dionne C, Møller JV, Nissen P, Isaacs JT. A Trojan horse in drug development: targeting of thapsigargins towards prostate cancer cells. Anticancer Agents Med Chem. 2009;9:276–94.

35. Ogunbayo OA, Harris RM, Waring RH, Kirk CJ, Michelangeli F. Inhibition of the sarcoplasmic/endoplasmic reticulum Ca2 + –ATPase by flavonoids: a quantitative structure-activity relationship study. IUBMB Life. 2008;60:853–8.

36. Salama G, Scarpa A. Enhanced Ca2+ uptake and ATPase activity of sarcoplasmic reticulum in the presence of diethyl ether. J Biol Chem. 1980; 255:6525–8.

37. Almeida LM, Vaz WL, Stumpel J, Madeira VM. Effect of short-chain primary alcohols on fluidity and activity of sarcoplasmic reticulum membranes. Biochemistry. 1986;25:4832–9.

38. Bigelow DJ, Thomas DD. Rotational dynamics of lipid and the Ca-ATPase in sarcoplasmic reticulum. The molecular basis of activation by diethyl ether. J Biol Chem. 1987;262:13449–56.

39. Michelangeli F, Orlowski S, Champeil P, East JM, Lee AG. Mechanism of inhibition of the (Ca2(+)-Mg2+)-ATPase by nonylphenol. Biochemistry. 1990; 29:3091–101.

40. Ishida Y, Honda H. Inhibitory action of 4-aminopyridine on Ca(2+)-ATPase of the mammalian sarcoplasmic reticulum. J Biol Chem. 1993;268:4021–4.

41. Carfagna MA, Muhoberac BB. Interaction of tricyclic drug analogs with synaptic plasma membranes: structure-mechanism relationships in inhibition of neuronal Na+/K(+)-ATPase activity. Mol Pharmacol. 1993; 44:129–41.

42. Eletr S, Inesi G. Phospholipid orientation in sarcoplasmic membranes: spin-label ESR and proton MNR studies. Biochim Biophys Acta. 1972;282:174–9.

43. Lowry OH, Rosebrough NJ, Farr AL, Randall RJ. Protein measurement with the Folin phenol reagent. J Biol Chem. 1951;193:265–75.

44. Fabiato A. Computer programs for calculating total from specified free or free from specified total ionic concentrations in aqueous solutions containing multiple metals and ligands. Methods Enzymol. 1988;157:378–417.

45. Schwarzenbach G, Senn H, Komplexone AG, XXIX. Ein grosser Chelateffekt besonderer Art. Helv Chim Acta. 1957;40:1886–900.

46. Blinks JR, Wier WG, Hess P, Prendergast FG. Measurement of Ca2+ concentrations in living cells. Prog Biophys Mol Biol. 1982;40:1–114.

47. Lin TI, Morales MF. Application of a one-step procedure for measuring inorganic phosphate in the presence of proteins: the actomyosin ATPase system. Anal Biochem. 1977;77:10–7.

48. Martonosi A, Feretos R. Sarcoplasmic reticulum. I. The uptake of Ca++ by sarcoplasmic reticulum fragments. J Biol Chem. 1964;239:648–58.

49. Petretski JH, Wolosker H, de Meis L. Activation of Ca2+ uptake and inhibition of reversal of the sarcoplasmic reticulum Ca2+ pump by aromatic compounds. J Biol Chem. 1989;264:20339–43.

50. Wakabayashi S, Ogurusu T, Shigekawa M. Mechanism for 3,3',4',5-tetrachlorosalicylanilide-induced activation of sarcoplasmic reticulum ATPase. J Biol Chem. 1988;263:15304–12.

51. Martinez-Azorin F, Teruel JA, Fernandez-Belda F, Gomez-Fernandez JC. Effect of diethylstilbestrol and related compounds on the Ca(2+)-transporting ATPase of sarcoplasmic reticulum. J Biol Chem. 1992;267:11923–9.

52. Khan YM, Wictome M, East JM, Lee AG. Interactions of dihydroxybenzenes with the Ca(2+)-ATPase: separate binding sites for dihydroxybenzenes and sesquiterpene lactones. Biochemistry. 1995;34:14385–93.

53. Soler F, Plenge-Tellechea F, Fortea I, Fernandez-Belda F. Clomipramine and related structures as inhibitors of the skeletal sarcoplasmic reticulum Ca2+ pump. J Bioenerg Biomembr. 2000;32:133–42.

54. Logan-Smith MJ, Lockyer PJ, East JM, Lee AG. Curcumin, a molecule that inhibits the Ca2+−ATPase of sarcoplasmic reticulum but increases the rate of accumulation of Ca2+. J Biol Chem. 2001;276:46905–11.

55. Bhalla P, Agrawal D. Alterations in rat erythrocyte membrane due to hexachlorocyclohexane (technical) exposure. Hum Exp Toxicol. 1998;17:638–42.

56. Hara H, Kanazawa T. Selective inhibition by ionophore A23187 of the enzyme isomerization in the catalytic cycle of sarcoplasmic reticulum Ca2+−ATPase. J Biol Chem. 1986;261:16584–90.

57. Peter HW, Wolf HU. Kinetics of (Na+, K+)-ATPase of human erythrocyte membranes. I. Activation by Na+ and K+. Biochim Biophys Acta. 1972;290:300–9.

58. De Meis L, Tuena de Gomez Puyou M, Gomez Puyou A. Inhibition of mitochondrial F1 ATPase and sarcoplasmic reticulum ATPase by hydrophobic molecules. Eur J Biochem. 1988;171:343–9.

59. Lax A, Soler F, Fernandez-Belda F. Inhibition of sarcoplasmic reticulum Ca2+−ATPase by miconazole. Am J Physiol Cell Physiol. 2002;283:C85–92.

60. de Meis L, Hasselbach W, Machado RD. Characterization of calcium oxalate and calcium phosphate deposits in sarcoplasmic reticulum vesicles. J Cell Biol. 1974;62:505–9.

61. Ohmiya H, Kanazawa T. Inhibition by A23187 of conformational changes involved in the Ca(2+)-induced activation of sarcoplasmic reticulum Ca(2+)-ATPase. J Biochem. 1991;109:751–7.

62. Hara H, Ohmiya H, Kanazawa T. Selective inhibition by ionophore A23187 of the enzyme isomerization in the catalytic cycle of Na+, K+−ATPase. J Biol Chem. 1988;263:3183–7.

63. Wictome M, Michelangeli F, Lee AG, East JM. The inhibitors thapsigargin and 2,5-di(tert-butyl)-1,4-benzohydroquinone favour the E2 form of the Ca2+, Mg2+−ATPase. FEBS Lett. 1992;304:109–13.

64. Hua S, Xu C, Ma H, Inesi G. Interference with phosphoenzyme isomerization and inhibition of the sarco-endoplasmic reticulum Ca2+ ATPase by 1,3-dibromo-2,4,6-tris(methylisothiouronium) benzene. J Biol Chem. 2005;280:17579–83.

65. Berman M, Karlish S. Interaction of an aromatic dibromoisothiouronium derivate with the Ca2+-ATPase of skeletal muscle sarcoplasmic reticulum. Biochemistry. 2003;42:3356–66.

66. Smith WS, Broadbridge R, East JM, Lee AG. Sarcolipin uncouples hydrolysis of ATP from accumulation of Ca2+ by the Ca2+−ATPase of skeletal-muscle sarcoplasmic reticulum. Biochem J. 2002;361:277–86.

67. Reis M, Farage M, de Souza AC, de Meis L. Correlation between uncoupled ATP hydrolysis and heat production by the sarcoplasmic reticulum Ca2+−ATPase: coupling effect of fluoride. J Biol Chem. 2001;276:42793–800.

68. Wakabayashi S, Ogurusu T, Shigekawa M. Factors influencing calcium release from the ADP-sensitive phosphoenxyme intermediate of the sarcoplasmic reticulum ATPase. J Biol Chem. 1986;264:9762–9.

69. Masuda H, de Meis L. Effect of temperature on the Ca2+ transport ATPase of sarcoplasmic reticulum. J Biol Chem. 1977;252:8567–71.

70. Andersen JP, Jørgensen PL, Møller JV. Direct demonstration of structural changes in soluble, monomeric Ca2+−ATPase associated with Ca2+ release during the transport cycle. Proc Natl Acad Sci. 1985;82:4573–7.

71. Froehlich JP, Heller PF. Transient-state kinetics of the ADP-insensitive phosphoenzyme in sarcoplasmic reticulum: implications for transient-state calcium translocation. Biochemistry. 1985;24:126–36.

72. Covarrubias-Cervantes M, Champion D, Debeaufort F, Voilley A. Translational diffusion coefficients of volatile compounds in various aqueous solutions at low and subzero temperatures. J Agric Food Chem. 2005;53:6671–6.

# 7

# Linalool isomerase, a membrane-anchored enzyme in the anaerobic monoterpene degradation in *Thauera linaloolentis* 47Lol

Robert Marmulla[1], Barbara Šafarić[1], Stephanie Markert[2], Thomas Schweder[2] and Jens Harder[1]*

## Abstract

**Background:** *Thauera linaloolentis* 47Lol uses the tertiary monoterpene alcohol (*R,S*)-linalool as sole carbon and energy source under denitrifying conditions. The conversion of linalool to geraniol had been observed in carbon-excess cultures, suggesting the presence of a 3,1-hydroxyl-$\Delta^1$-$\Delta^2$-mutase (linalool isomerase) as responsible enzyme. To date, only a single enzyme catalyzing such a reaction is described: the linalool dehydratase/isomerase (Ldi) from *Castellaniella defragrans* 65Phen acting only on (*S*)-linalool.

**Results:** The linalool isomerase activity was located in the inner membrane. It was enriched by subcellular fractionation and sucrose gradient centrifugation. MALDI-ToF MS analysis of the enriched protein identified the corresponding gene named *lis* that codes for the protein in the strain with the highest similarity to the Ldi. Linalool isomerase is predicted to have four transmembrane helices at the N-terminal domain and a cytosolic domain. Enzyme activity required a reductant for activation. A specific activity of $3.42 \pm 0.28$ nkat mg * protein$^{-1}$ and a $k_M$ value of $455 \pm 124$ μM were determined for the thermodynamically favored isomerization of geraniol to both linalool isomers at optimal conditions of pH 8 and 35 °C.

**Conclusion:** The linalool isomerase from *T. linaloolentis* 47Lol represents a second member of the enzyme class 5.4.4.4, next to the linalool dehydratase/isomerase from *C. defragrans* 65Phen. Besides considerable amino acid sequence similarity both enzymes share common characteristics with respect to substrate affinity, pH and temperature optima, but differ in the dehydratase activity and the turnover of linalool isomers.

**Keywords:** Linalool, Geraniol, *Thauera*, Isomerase, Allyl alcohol, Monoterpene

## Background

Monoterpenes ($C_{10}H_{16}$), naturally occurring hydrocarbons, are the main constituents of essential oils and belong to the diverse group of terpenoids, of which more than 50000 structures are known to date [1, 2]. They are produced as secondary plant metabolites and serve diverse functions like signaling, attraction/repellence of pollinators and insects, thermotolerance and are involved in allelopathy [3, 4]. Atmospheric monoterpene emission from plants was estimated to be 127 Tg C yr$^{-1}$ [5], with half-lives of minutes to hours in the atmosphere due to their susceptibility to chemical and photooxidative reactions. Monoterpenes enter soils by precipitation from the atmosphere,

excretion from roots and by leaf fall [6–9]. The hydrophobic character of monoterpenes causes cell toxicity, mainly by accumulation into and destabilization of the cell membranes [10]. Below toxic concentrations, microorganisms can use monoterpenes as carbon and energy source for growth. Several bacteria have been described to transform monoterpenes in the presence of oxygen as a cosubstrate applying mono- and dioxygenases [11], but also other biotransformations are described [12]. Linalool, a tertiary monoterpene alcohol (Fig. 1), is the main component of essential oils in lavender and coriander. Its chemical structure prevents a direct oxidation of the hydroxyl group. Hence, another functionalization or isomerization is the initial biological degradation reaction. Linalool is oxidized to 8-hydroxylinalool under aerobic conditions [13, 14]. In the absence of oxygen, the primary alcohol geraniol is formed in linalool-grown cultures of *Thauera linaloolentis*

* Correspondence: jharder@mpi-bremen.de
[1]Department of Microbiology, Max Planck Institute for Marine Microbiology, Celsiusstr. 1, D-28359 Bremen, Germany
Full list of author information is available at the end of the article

**Fig. 1** Linalool isomerase catalyzes the reversible reaction from linalool to geraniol

47Lol (Fig. 1) [15]. This betaproteobacterium was isolated on linalool as sole carbon end energy source under denitrifying conditions [16]. A 3,1-hydroxyl-$\Delta^1$-$\Delta^2$-mutase was proposed as novel enzymatic function initializing the mineralization of linalool [15, 16].

A similar reaction was described for the bifunctional enzyme linalool dehydratase/isomerase from *Castellaniella defragrans* 65Phen. The enzyme catalyzes the reversible hydration of β-myrcene to (S)-linalool and its isomerization to geraniol. It is an oxygen-sensitive, periplasmatic protein of 43 kDa including a signal peptide for export [17, 18]. The tertiary alcohol 2-methyl-3-buten-2-ol (232-MB) may also be transformed by enzyme-catalyzed isomerization reactions. 232-MB is a metabolite in bacterial degradation of fuel oxygenates. Its mineralization may proceed via an initial isomerization to 3-methyl-2-buten-1-ol (prenol) in *Aquincola tertiaricarbonis* L108 and *Methylibium petroleiphilum* PM1 [19] as well as in *Pseudomonas putida* MB-1 [20]. The later was isolated on 232-MB as sole carbon source [20]. However, the corresponding enzymes were so far not characterized. For intramolecular hydroxyl-group transfer (EC 5.4.4.x), only seven different enzyme activities are described: (hydroxyamino) benzene mutase (EC 5.4.4.1), isochorismate synthase (EC 5.4.4.2), 3-(hydroxyamino) phenol mutase (EC 5.4.4.3), geraniol isomerase (EC 5.4.4.4), 9,12-octadecadienoate 8-hydroperoxide 8R-isomerase (EC 5.4.4.5), 9,12-octadecadienoate 8-hydroperoxide 8S-isomerase (EC 5.4.4.6) and hydroxyperoxy icosatetraenoate isomerase (EC 5.4.4.7).

We report the enrichment of the linalool isomerase activity in protein fractions of *Thauera linaloolentis* 47Lol and the kinetic properties of the enzyme. A corresponding gene was identified in the draft genome. We suggest to place the linalool isomerase of *Thauera linaloolentis* 47Lol as a new member in the enzyme family of intramolecular hydroxyl group transferases (EC 5.4.4.x) with the EC number 5.4.4.4 next to the geraniol isomerase function of the linalool dehydratase/isomerase from *Castellaniella defragrans* 65Phen.

## Results and discussion

### Identification of a candidate protein for linalool isomerase

The linalool dehydratase/isomerase (Ldi, NCBI:CBW30776) was used in similarity searches to identify a putative linalool isomerase protein in a draft genome of *T. linaloolentis* 47Lol. One protein showed a considerable similarity with an overall amino acid identity of 20 % (positives 33 %, E-value 3E-10, NCBI:ENO87364, Additional file 1: Figure S1). The corresponding gene is isolated from the adjacent genes (>150 bp) and encodes a protein of 644 amino acids with a calculated molecular weight of 71.8 kDa, an isoelectric point of 6.06 and a hydrophobicity of −0.115 (GRAVY) [21]. No signal peptide was predicted by the SignalP software. For the N-terminus, four transmembrane domains within the first 139 amino acids were predicted together with a localization in the inner membrane with the C-terminal protein fold in the cytoplasm. Conserved domains as described in Pfam were not present. The specific hydrophobicity values (GRAVY) were 0.94 and −0.406 for the N- and C-terminal parts of the protein (amino acids 1–139 and 140–644, respectively). The similarity to the Ldi was restricted to the C-terminal domain. Such a location at the cytoplasmic site of the inner membrane seems to be ideal for a catabolic enzyme acting on a hydrophobic substrate. This may maximize the contact with the substrate and reduces the intracellular concentration, but also produces geraniol for the next catabolic enzymes that likely depend on cytoplasmatic NAD$^+$ as electron acceptor. In contrast, the periplasmic location of Ldi is optimal for a defense enzyme. Myrcene is less toxic than the alcohols and can diffuse into the environment of the cell, thus keeping the damage at the inner membrane to a minimum.

### Enrichment of the linalool isomerase activity (Lis)

Lis activity was determined as geraniol isomerase activity and was detected in crude cell-free protein extracts also containing membrane fragments after application of high pressure cell disruption. Cell disintegration by osmosis or ultra sonification retained the activity in the membrane fraction. Due to the large cytoplasmatic domain of the candidate protein, we attempted the isolation as soluble enzyme from dialyzed crude extracts. The pH was increased to 9.5 to eventually increase the anionic character of the enzyme. The enzyme activity did not bind to a DEAE column. After ammonium sulfate addition (10 % v/v of a saturated solution), the enzyme activity was retained on a Phenyl-Sepharose column and eluted with pure water. A strong binding to hydrophobic columns has also been observed for Ldi [17] and other monoterpene-transforming enzymes [22]. A size determination of the eluted fraction on a size-exclusion column yielded a molecular weight of over

600 kDa, as native protein in phosphate buffer as well as in the presence of urea as denaturant. Visualization of the proteins on SDS-PAGE revealed the presence of many proteins, including abundant proteins between 60 and 70 kDa and around 33 kDa (data not shown).

The large size together with the predicted membrane association of the Lis candidate gene suggested an alternative purification approach, a subcellular fractionation including a sucrose-gradient centrifugation for the identification of membrane-associated proteins. First, the outer membrane and periplasmatic proteins were removed by spheroplast formation. After disintegration by high pressure release, the inner membrane (IM) fraction was separated from the cytosolic soluble protein (SP) fraction by ultracentrifugation (158 Svedberg). Total Lis activity was five times higher in the membrane fraction than in the soluble fraction (15.8 and 3.3 nkat, respectively) (Table 1 and Additional file 2: Figure S2). Both fractions were separated on sucrose gradients. The Lis activity of the membrane fraction (Fig. 2 and Table 2) was concentrated in two fractions with a sucrose density between 1.17 and 1.22 g mL$^{-1}$ (IM4 and IM5). For the soluble protein fraction (Fig. 3 and Table 3), Lis activity was also concentrated in the fractions with the aforementioned sucrose content (SP4 and SP5). Protein gels revealed an enrichment of proteins with a molecular size between 60 and 70 kDa and around 32 kDa for the active fractions from IM and SP. Lis activity was deposited from the active fractions (SP4 and SP5) by decrease of the sucrose concentration and ultracentrifugation. SDS-PAGE characterized the dissolved pellet as containing an approx. 60 kDa protein as most abundant protein (Fig. 4 and Table 4). This purification stage was used for enzyme kinetics.

Particles with a Svedberg constant of above 158 S were separated from the soluble fraction. High pressure cell disintegration produces a number of small membrane fragments and vesicles. Small vesicles (microsomes)

**Table 1** Enrichment of Lis activity from spheroplasts of *T. linaloolentis* 47Lol

| Sample | Protein [mg] | Activity [pkat] | Specific activity [pkat mg$^{-1}$] | Relative specific activity | Protein yield [%]* |
|--------|-----|------|-----|-----|-------|
| F1 | 71.9 | 7295 | 101 | | |
| F2 | 79.0 | 7800 | 99 | 1.0 | 100.0 |
| F3 | 67.6 | 3060 | 45 | 0.5 | |
| F4 | 51.8 | 24984 | 482 | 4.9 | 65.6 |
| F5 | 29.1 | 513 | 18 | 0.2 | 36.8 |
| F6 | 38.0 | 3283 | 86 | 0.9 | 73.3 |
| F7 | 21.4 | 15848 | 741 | 7.5 | 41.3 |

(F1) spheroplasts, (F2) lysed spheroplasts, (F3) outer membrane and periplasmatic proteins, (F4) crude extract from spheroplasts, (F5) unbroken spheroplast and cell debris, (F6) cytoplasmatic, soluble proteins, (F7) inner membrane fraction. *F4 and F5 are derived from F2. F6 and F7 are derived from F4

settle with a sedimentation coefficient between 100 and 10000 S [23]. We conclude from our observations that the Lis activity is located on a protein associated with the inner membrane. Only a minor fraction of the total enzyme activity, present as small membrane-protein aggregates, was enriched in the fraction that is traditionally considered to contain only cytosolic proteins.

The protein gels showed consistently protein(s) between 60 and 70 kDa in fractions with Lis activity. MALDI-ToF mass spectroscopy identified the protein band at approx. 60 kDa from several gels as NCBI:ENO87364, the protein that was predicted to be a candidate for the linalool isomerase. The discrepancy between the calculated molecular weight of 71.8 kDa and the observed weight of between 60 and 70 kDa is likely a result of the hydrophobic nature of the protein. Hydrophobic proteins, including membrane proteins tend to bind more SDS and show a faster migration in denaturing gel electrophoreses [24].

Further purification of the Lis activity was not successful, as a detergent-mediated release of the protein from the membrane was impaired by irreversible inhibitory effects on the enzyme activity. Detergents aid solubility of membrane proteins but also have an adverse effect on stability and functionality of proteins [25, 26]. Terpenoid synthases were shown to have a rather hydrophobic core shielding the catalytic center from surrounding bulk solvent [27]. Also the Lis may have a hydrophobic protein domain sensitive to detergents that change the conformation into a non-active state.

### Heterologous gene expression

Expression of the native *lis* gene in *E.coli* BL21(DE3) yielded an induced protein band around 60 kDa (Fig. 5). No inclusion bodies were observed. The expressed protein was located in the membrane fraction, but Lis activity was not observed in crude cell extracts or in cell-free protein fractions. Expression of a N-terminally truncated version of the linalool isomerase, representing the cytosolic part of the enzyme only, yielded a soluble protein that also did not show activity. The *lis* gene did not have rare codons. The folding in *E.coli* may have been incorrect.

### Characterization of linalool isomerase activity

The Lis activity was enriched without a chemical reductant in the buffer and enzymatically inactive. Activity was restored by addition of a reducing agent and under anoxic conditions in the enzyme assay. The activation could be initiated by cationic (ferrous iron), neutral (dithiothreitol) or anionic (dithionite) compounds. However, maximum activity was measured with 4 mM dithionite (268 pkat * mg protein$^{-1}$). A large excess of reductant caused a decrease in activity, e.g. 10 mM dithionite or 8 mM DTT were less suitable for activation (Table 5). Lis activity was

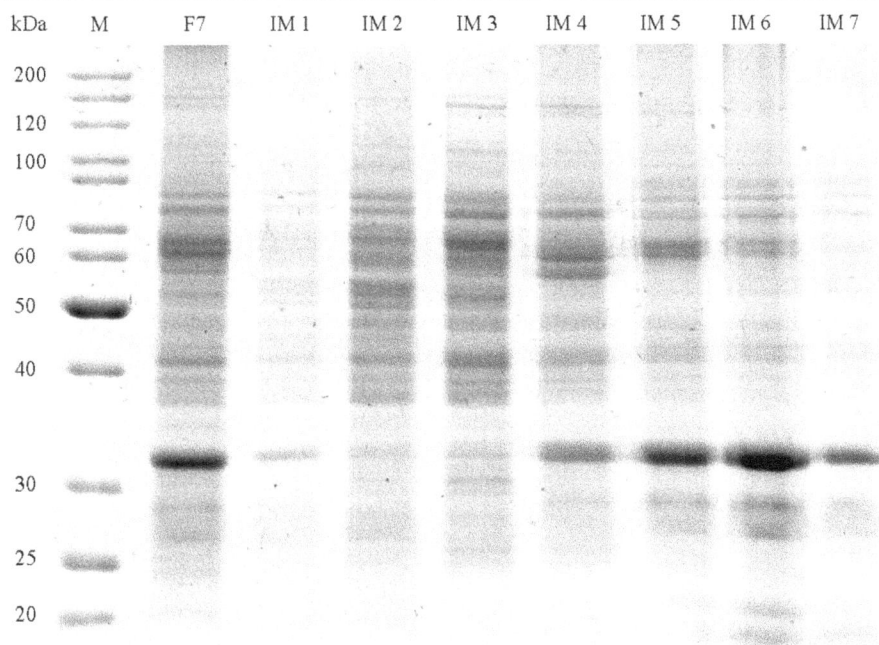

**Fig. 2** SDS-PAGE of sucrose gradient fractions after separation of the inner membrane (IM) fraction from spheroplast disintegration. (F7) inner membrane fraction from spheroplast disintegration, (IM 1) - (IM 7) fractions 1–7 (top-to-bottom) of sucrose gradient

determined in the thermodynamically favorable direction from geraniol to linalool [15] and was observed between pH 7 and 9.5, with an optimum around pH 8. The temperature optimum was 35 °C and the activation energy was $80.4 \pm 6.9$ kJ $mol^{-1}$. For comparison, the pH and temperature optimum of the Ldi enzyme were at pH 9 and 35 °C, respectively, and the activation energy was 68.6 kJ $mol^{-1}$ [17]. Geraniol formation from linalool was detected within 4 h of incubation (Fig. 6). Kinetic parameters were determined with the most enriched sample in duplicates for the geraniol isomerization to linalool. The enzyme followed Michaelis-Menten-kinetics with an

apparent $k_M$-value of $455 \pm 124$ μM and a $V_{max}$ of $3.42 \pm 0.28$ nkat $*$ mg protein$^{-1}$ (Fig. 7). A similar substrate affinity was determined for the linalool dehydratase/isomerase ($k_M$ 500 μM). Maximal velocity was higher for the Ldi ($V_{max}$ 410 nkat $*$ mg protein$^{-1}$) than for the Lis, however, the purification level for Lis was lower, and we do not know whether part of the enzyme is in an inactive state. Lis did not show a stereospecificity towards linalool isomers: both ($R$) and ($S$)-linalool were formed (Fig. 8). In contrast, the Ldi accepts only ($S$)-linalool as a substrate [17].

Earlier studies already showed the regiospecific formation of geraniol from linalool without nerol formation [15]. To confirm the regioselectivity, Lis activity was tested with nerol or citronellol and in combination with geraniol. No activity was measured with nerol or citronellol alone. Linalool isomerase activity dropped to approx. 50 % in the presence of nerol. Activity in the presence of citronellol and geraniol was barely detectable. Thus, the enzyme is regioselective and seems to bind nerol with a similar affinity as geraniol, whereas citronellol which lacks the C2-C3 double bond is stronger bound than geraniol.

UV-VIS spectroscopy in a range of 200–800 nm did not provide evidence for the presence of cofactors. Cofactor-independent enzymes are known for allylic rearrangements that are catalyzed by acid-base mechanisms [28]. The isomerization of geraniol to linalool requires a protonation of the hydroxyl group to leave as

**Table 2** Enrichment of Lis activity by sucrose gradient centrifugation from inner membrane pellets (F7) obtained from spheroplasts of *T. linaloolentis* 47Lol. 1 mL fractions from top to bottom are shown

| Sample | Protein [mg] | Activity [pkat] | Specific activity [pkat mg$^{-1}$] | Relative specific activity | Protein yield [%] |
|---|---|---|---|---|---|
| Applied sample (F7) | 6.74 | 5925 | 879 | 1 | 100.0 |
| IM 1 | 0.20 | 139 | 697 | 0.8 | 3.0 |
| IM 2 | 1.06 | 238 | 225 | 0.3 | 15.7 |
| IM 3 | 1.44 | 870 | 604 | 0.7 | 21.4 |
| IM 4 | 1.92 | 3085 | 1607 | 1.8 | 28.5 |
| IM 5 | 3.28 | 3513 | 1071 | 1.2 | 48.7 |
| IM 6 | 1.44 | 1050 | 729 | 0.8 | 21.4 |
| IM 7 | 0.02 | 0 | 0 | 0.0 | 0.3 |

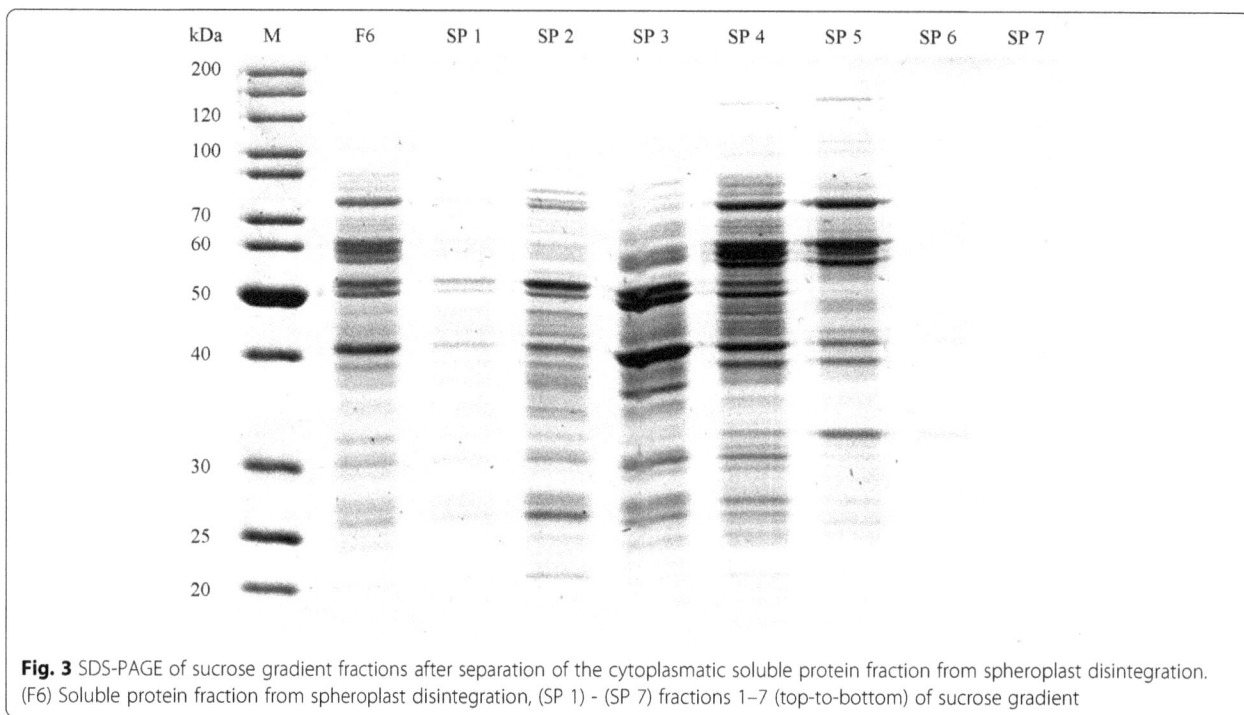

**Fig. 3** SDS-PAGE of sucrose gradient fractions after separation of the cytoplasmatic soluble protein fraction from spheroplast disintegration. (F6) Soluble protein fraction from spheroplast disintegration, (SP 1) - (SP 7) fractions 1–7 (top-to-bottom) of sucrose gradient

water, and a shift in electron density leading to a tertiary carbocation intermediate which may be attacked by water or a hydroxyl ion, resulting in the formation of linalool.

## Conclusion

We identified a linalool isomerase in *T. linaloolentis* 47Lol by partial protein purification. The gene encodes a two-domain protein with a N-terminal anchor in the inner membrane that is characterized by four transmembrane helices and a C-terminal cytosolic domain which showed considerable similarity to the linalool dehydratase/isomerase from *C. defragrans* 65Phen. The enzyme is active in the reduced state and sensitive towards

**Table 3** Enrichment of Lis activity by sucrose gradient centrifugation from cytoplasmatic soluble protein fraction (F6) obtained from spheroplasts of *T. linaloolentis* 47Lol. 1 mL fractions from top to bottom are shown

| Sample | Protein [mg] | Activity [pkat] | Specific activity [pkat mg$^{-1}$] | Relative specific activity | Protein yield [%] |
|---|---|---|---|---|---|
| Applied sample (F6) | 9.2 | 1863 | 202 | 1.0 | 100.0 |
| SP 1 | 0.6 | 14 | 24 | 0.1 | 6.1 |
| SP 2 | 1.5 | 16 | 11 | 0.1 | 16.7 |
| SP 3 | 3.1 | 165 | 54 | 0.3 | 33.2 |
| SP 4 | 2.7 | 273 | 100 | 0.5 | 29.5 |
| SP 5 | 1.2 | 141 | 114 | 0.6 | 13.3 |
| SP 6 | 0.3 | 13 | 47 | 0.2 | 3.0 |
| SP 7 | 0.3 | 0 | 0 | 0.0 | 2.8 |

detergents. It expands the enzyme class of intramolecular hydroxyl group transferases as a new member, linalool isomerase (5.4.4.4).

## Methods
### Bacterial strains, cultivation conditions and biomass harvest

*T. linaloolentis* 47Lol was cultivated under anaerobic, denitrifying conditions in artificial fresh water (AFW) medium. Medium was prepared as described by Foss et al. [29] with modifications: carbonate buffer was replaced by 10 mM $Na_2HPO_4/NaH_2PO_4$ and vitamins were omitted. The headspace contained only nitrogen gas. 1–2 mM (*R,S*)-linalool (>97 % purity; Sigma-Aldrich, Germany) were directly applied without carrier phase as sole carbon and energy source. Cultures were incubated at 28 °C under mild shaking (60 rpm). Alternatively, they were stirred. Bacterial biomass for protein purification was obtained from 2 L cultures grown on 2 mM linalool and 20 mM nitrate by centrifugation ($16000 \times g$, 25 min, 4 °C). If not used directly, biomass was frozen in liquid nitrogen and stored at –80 °C.

### Purification attempts by chromatography

First attempts on the purification of the linalool isomerase were performed by classical column chromatography using anion exchange and hydrophobic interaction columns (DEAE; binding buffer: 80 mM Tris-Cl, pH 9.5, elution buffer: 80 mM Tris-Cl, pH 9.5 with 1 M NaCl; Phenyl Sepharose; binding buffer: 80 mM Tris-Cl,

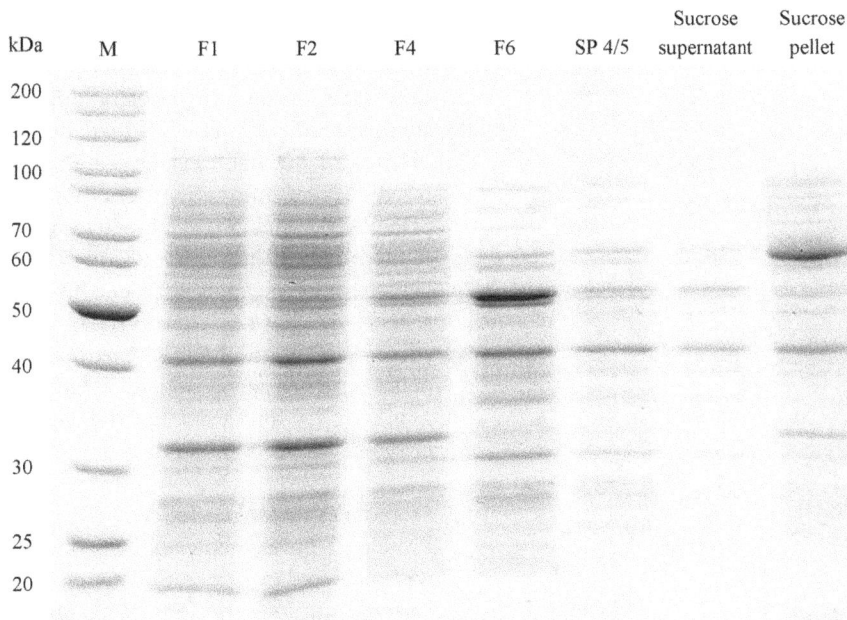

**Fig. 4** SDS-PAGE of Lis activity enrichment by spheroplast disintegration and sucrose gradient centrifugation. (F1) spheroplasts, (F2) spheroplasts after cell disintegration, (F4) crude extract after spheroplast disintegration, (F5) cytoplasmatic soluble protein fraction, (SP 4/5) fractions 4 and 5 from sucrose gradient centrifugation, (Sucrose supernatant) supernatant of second ultracentrifugation, (Sucrose pellet) protein pellet of second ultracentrifugation

pH 8.0 with 10 % v/v of a saturated ammonium sulfate solution, elution buffers: 80 mM Tris-Cl, pH 8.0 and water). Size-exclusion chromatography was performed on a HiLoad 16/60 Superdex200 column (GE healthcare, dimensions: $16 \times 600$ mm; 20 mM $KH_2PO_4/K_2HPO_4$, pH 8.0 with or without 6 M urea). Calibration was performed with thyroglobulin (670 kDa), bovine gamma-globulin (158 kDa), chicken ovalalbumin (44 kDa), equine myoglobin (17 kDa) and vitamin B12 (1.35 kDa).

Soluble protein extract was prepared by resuspending biomass in Tris-Cl buffer (80 mM, pH 9.5 for

DEAE and pH 8.0 for HIC) and fast thawing at room temperature. Cell disintegration was performed by mechanical sheering using a One-Shot cell disruptor (Constant Systems Ltd., Daventry, UK) at 1.7 GPa two times. The crude extract was clarified by ultracentrifugation ($150000 \times g$, 30 min, 4 °C).

All purification steps were performed on ice or at 5 °C. SDS-PAGE was used to characterize purification samples and protein concentrations were determined according to the method described by Bradford using bovine serum albumin as a calibration standard [30].

**Spheroplast preparation**

Spheroplasts were prepared according to the protocol for subcellular fractionation described by Koßmehl et al. [31]. Cells were washed with 1.5 M NaCl solution, collected by centrifugation ($14200 \times g$, 20 min, 4 °C) and resuspended in 20 % (w/v) sucrose, 30 mM Tris-Cl (pH 8.0) and 2 mM EDTA for osmotic shock treatment. The cell suspension was incubated for 20 min at 30 °C. A spheroplast enriched pellet was formed by centrifugation ($14200 \times g$, 20 min, 4 °C). The pellet was resuspended in 5 mL of ice-cold 40 mM Tris-Cl (pH 8.0) and disintegrated by a One Shot Cell Disruptor (Constants Systems Ltd., UK) in two passages at 1.7 GPa. After removal of larger cell debris and unbroken cells by centrifugation ($14200 \times g$, 20 min), the supernatant was further clarified by ultracentrifugation ($104000 \times g$, 1 h, 4 °C). The resulting pellet and supernatant corresponded

**Table 4** Enrichment of Lis activity from spheroplasts of *T. linaloolentis* 47Lol by sucrose gradient centrifugation

| Sample | Protein [mg] | Activity [pkat] | Specific activity [pkat mg$^{-1}$] | Relative specific activity | Protein yield [%] |
|---|---|---|---|---|---|
| F1 | 40.8 | 1408 | 35 | | |
| F2 | 111.7 | 3130 | 28 | 1 | 100.0 |
| F4 | 94.7 | 13572 | 143 | 5 | 84.8 |
| F6 | 58.8 | 3440 | 59 | 2 | 52.6 |
| SP 4/5 | 3.7 | 745 | 200 | 7 | 3.3 |
| Sucrose supernatant | - | - | - | - | - |
| Sucrose pellet | 1.9 | 323 | 171 | 6 | 1.7 |

(F1) spheroplasts, (F2) lysed spheroplasts, (F4) crude extract, (F6) cytoplasmatic, soluble protein fraction, (SP 4/5) fractions 4 and 5 from sucrose gradient, (Sucrose supernatant) supernatant after second ultracentrifugation, (Sucrose pellet) protein pellet after second ultracentrifugation

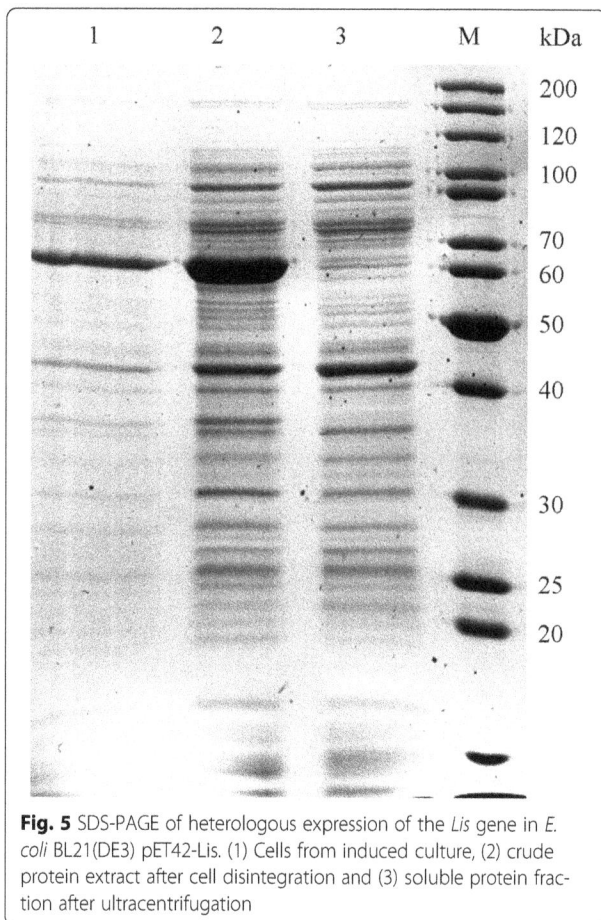

**Fig. 5** SDS-PAGE of heterologous expression of the *Lis* gene in *E. coli* BL21(DE3) pET42-Lis. (1) Cells from induced culture, (2) crude protein extract after cell disintegration and (3) soluble protein fraction after ultracentrifugation

to an inner membrane fraction (IM, F7, Additional file 1: Figure S1) and a soluble protein fraction (SP, F6, Additional file 1: Figure S1), respectively.

### Sucrose density gradient centrifugation

A linear sucrose gradient was created by overlaying a 20 % (w/v) over a 70 % (w/v) sucrose solution, 3 mL each. The tubes were incubated horizontally for 90 min to allow mixing. A 1 mL-protein sample was loaded

**Table 5** Influence of different reducing agents on Lis activity

| Reducing agent | Reduction potential [mV] | Concentration [mM] | Specific activity [pkat * mg protein$^{-1}$] |
|---|---|---|---|
| Dithionite | - 660 | 2 | 89 |
|  |  | 4 | 268 |
|  |  | 10 | 210 |
| Dithiothreitol | - 330 | 2 | 199 |
|  |  | 8 | 145 |
|  |  | 16 | 83 |
| Cysteine | - 220 | 5 | 23 |
|  |  | 10 | 7 |
| Ferrous iron | - 236 (Fe(OH)$_3$/Fe$^{2+}$) | 5 | 112 |

carefully on top of the gradient and centrifugation was performed in a L-70 ultracentrifuge (70.1Ti rotor, Beckmann Coulter) at 260000 × $g$, for 4 h at 4 °C, with slow acceleration and deceleration. The linearity of the sucrose gradient was confirmed by gravimetrical measurement. 1 mL fractions covering the gradient were analyzed for enzyme activity and for protein content on SDS-PAGE. The two most active fractions were pooled, diluted with 40 mM Tris-Cl (pH 8.0) to a final volume of 8 mL and a second ultracentrifugation step (260000 × $g$, 4 °C, 1.5 h ≡ 42 Svedberg) was performed to remove sucrose. This resulted in the formation of a protein pellet enriched in Lis activity, which was resuspended in 1 mL buffer and used for the characterization of the enzyme kinetic.

### Proteomics by MALDI-ToF MS

Protein samples, obtained from individual purifications, were analyzed by SDS-PAGE coupled with matrix-assisted laser desorption/ionization time of flight (MALDI-ToF) mass spectrometry (MS). Protein bands in gels were excised manually, and the Ettan Spot Handling Workstation (GE Healthcare) was used for trypsin digestion and embedding of the resulting peptide solutions in an α-cyano-4-hydroxycinnamic acid matrix for spotting onto MALDI targets. MALDI-ToF MS analysis was performed on an AB SCIEX TOF/TOF™ 5800 Analyzer (AB Sciex/MDS Analytical Technologies [32]. Spectra in a mass range from 900 to 3700 Da (focus 1700 Da) were recorded and analyzed by GPS Explorer™ Software Version 3.6 (build 332, Applied Biosystems) and the Mascot search engine version 2.4.0 (Matrix Science Ltd, London, UK) using the RAST draft genome as reference.

### Heterologous gene expression

The predicted *Lis* gene was isolated from genomic DNA of *T. linaloolentis* 47Lol by means of PCR, using the Phusion Polymerase according to the manufacturer manual (Life Technologies, Thermo Fisher Scientific, Waltham, USA) with the primer pair pLI_NdeI_FW (TCGTACA TATGATGAGCAATATGGAATCG) and pLI_BglII_RV (CGATAGATCTTCAGTGGCCCGGCTTG, annealing temperature 59 °C). Additionally, a N-terminal truncated version of the gene was constructed with the primer pair pLI_NdeI_FW-truncated-N (TCGTACATATGAT GCGCGGCGCCAAGC) and pLI_BglII_RV (annealing temperature 68 °C), which had an artificial start codon (ATG) and covered the amino acids 141–644 of the Lis. The genes were cloned into the pET42a overexpression plasmid and transformed into *E. coli* BL21(DE3). *E. coli* BL21(DE3) pET42-Lis or pET42-Lis-ΔN were grown in liquid LB medium at 37 °C and protein expression was induced by addition of 1 mM IPTG at an optical density (600 nm) of around 1. Cultures were further incubated at

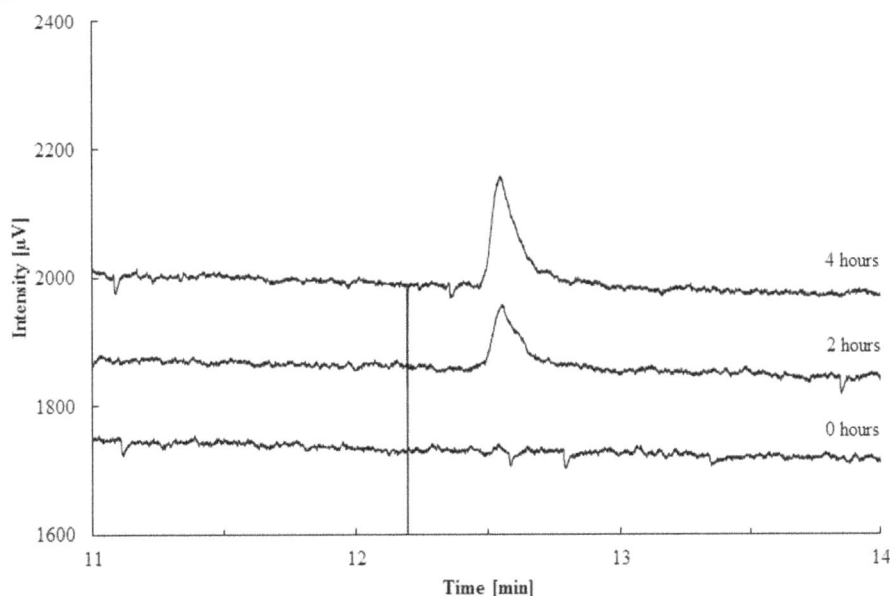

**Fig. 6** Formation of geraniol by linalool isomerase. The most enriched protein sample after sucrose gradient centrifugation and precipitation was used: 200 µL sample (1.04 mg mL$^{-1}$ protein), 5 mM dithionite, 1 µL (R,S)-linalool (≈30 mM). Incubation was performed at 28 °C and individual samples were extracted with 200 µL n-hexane after 0, 2 and 4 h (from bottom to top) and measured by gas chromatography (Shimadzu GC 14A, see methods). Retention time of geraniol was at 12.55 min

30 °C for 4–5 h. Biomass was harvested by centrifugation (16000 × g, 25 min, 4 °C). Protein extracts were prepared as described above.

### Geraniol isomerase activity

Lis activity was determined by end-point analysis for the thermodynamically favored reaction from geraniol to linlaool. Subfractions obtained during purification were dialyzed and adjusted to Tris-Cl buffer (40 mM, pH 8.0).

Individual assays were prepared in 4 mL glass vials in an anaerobic chamber with $N_2$ headspace, containing 300–500 µL of sample and 5 mM dithionite as reducing agent. Samples were incubated for 20 min prior addition of 200 µL organic phase (200 mM geraniol in 2,2,4,4,6,8,8-heptamethylnonane, HMN). Vials were air-tight closed with butyl rubber stoppers and incubated for 14–16 h at 28 °C under mild shaking. Product formation was determined by gas chromatography with flame ionization

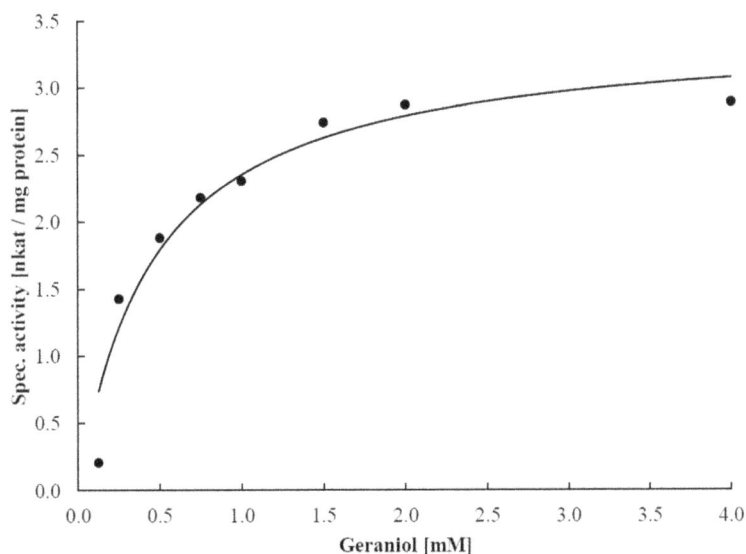

**Fig. 7** Michaelis-Menten plot of geraniol isomerization by linalool isomerase. An apparent $k_M$-value and $V_{max}$ value of 455 ± 124 µM and 3.42 ± 0.28 nkat * (mg protein)$^{-1}$ were determined for the isomerization of geraniol to linalool

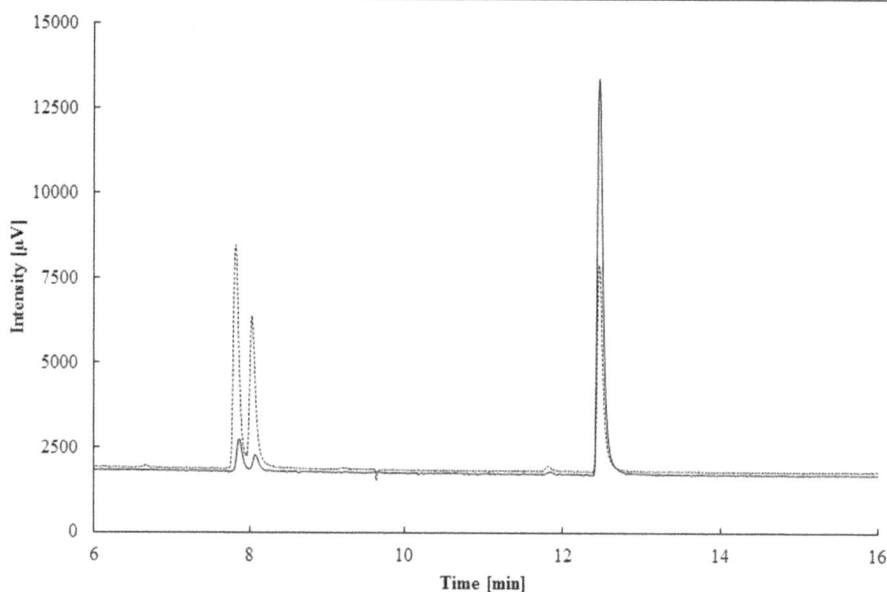

**Fig. 8** Gas chromatogram demonstration of the absence of stereoselectivity of the linalool isomerase. Soluble protein extract (1.4 mg protein; solid line) and inner membrane-enriched fraction (1.6 mg protein; dashed line) from spheroplast disintegration were incubated with geraniol, subsequently extracted with n-hexane and analyzed by gas chromatography. Retention times: (*R*)- and (*S*)-linalool 7.85 and 8.05 min, geraniol 12.45 min

detection (PerkinElmer Auto System XL, Überlingen, Germany). 1 μL of the HMN phase was injected onto an Optima-5 column (30 m × 0.32 mm, 0.25 μm film thickness; Macherey-Nagel, Germany) with hydrogen as carrier gas and the following temperature program: injection port 250 °C, detection port 350 °C, initial column temperature 40 °C for 2 min, increasing to 100 °C at a rate of 4 °C min$^{-1}$, keeping 100 °C for 0.1 min, followed by an increase to 320 °C at 45 °C min$^{-1}$ and hold for 3 min. The split ratio was set to 1:9.

The effect of detergents on Lis activity was tested for Triton X100 and Tween20 (0.5 %, 1 % w/v), CHAPS (0.1 % w/v) and n-octyl-α-D-glucoside (0.1 %, 0.5 %, 1 % w/v). Detergents were added to the soluble spheroplast fraction (protein concentration 0.1 mg mL$^{-1}$) and aforementioned enzyme assays were performed.

Reducing agents were tested with a dialyzed (Visking dialysis tubing 12–14 kDa cut-off, Serva) soluble protein extract (4 mg mL$^{-1}$ protein). The following reducing agents were added prior to the start of the assay: dithionite (2, 4 and 10 mM), dithiothreitol (2, 8 and 16 mM), cysteine (5 and 10 mM), or ferrous iron (5 mM).

Temperature dependency on Lis activity was determined between 12 and 50 °C. Samples (300 μL, 6.8 mg mL$^{-1}$ protein) were pre-incubated for 20 min at the individual temperatures prior to substrate addition. The assay was terminated after 8 h and analyzed by gas chromatography. Activation energy was calculated from the Arrhenius plot (y = −9664.3 x + 36.2; $R^2$ = 0.914).

The pH-optimum was determined by incubating crude cell lysate (20 mg mL$^{-1}$) in a pH-range from 7 to 9.5 in Tris-Cl (40 mM) applying the aforementioned enzyme assay.

Kinetic parameters ($k_M$ and $V_{max}$) were determined for the most enriched enzyme fraction in biological duplicates (68 and 80 μg mL$^{-1}$ protein; 10 to 20 μg total protein in final assay). Samples were incubated with geraniol concentration from 0.125 to 4 mM, directly applied without carrier phase. Both substrate and enzyme were prepared separately with 7 mM dithionite and pre-incubated. Reactions were started by injecting an equal volume (200 μL) of enzyme to the substrate solution. Samples were incubated at 28 °C for 90 min and terminated by addition of 100 mM NaOH (final concentration). 1 μL of sample was directly subjected to GC analysis. Kinetic parameters were calculated from primary data plotted in a Michaelis-Menthen-graph.

Substrate specificity of the Lis was tested with geraniol, nerol and citronellol. A 400 μL-sample (active fraction after sucrose gradient, SP 4/5) was incubated with 200 μL of 200 mM geraniol, nerol and citronellol in HMN as well as with 200 μL of geraniol-nerol and geraniol-citronellol mixtures in HMN (100 mM each). Assays were prepared as aforementioned.

Stereoselectivity was tested with soluble protein extract (1.4 mg protein) and inner membrane-enriched fraction (1.6 mg protein) from spheroplast disintegration. Samples were treated with 5 mM dithionite and incubated with 10 mM geraniol under anaerobic conditions at 28 °C for

Linalool isomerase, a membrane-anchored enzyme in the anaerobic monoterpene degradation...

69

14 h and subsequently extracted with 200 µL n-hexane. Monoterpene analysis was performed by gas chromatography with flame ionization detector (Shimadzu GC-14A, Shimadzu Corporation) on a Hydrodex-ß-6TBDM column (25 m × 0.25 mm, Macherey-Nagel, Germany) with the following temperature program: injection port 200 °C, detection port 250 °C, initial column temperature 60 °C for 1 min, increasing to 130 °C at a rate of 5 °C min$^{-1}$, keeping 130 °C for 0.5 min, followed by an increase to 230 °C at 20 °C min$^{-1}$ and hold for 4 min.

## Linalool isomerase activity
The forward reaction of the linalool isomerase - linalool to geraniol - was tested in a separate assay. 200 µL of enriched fraction after sucrose gradient centrifugation were treated with 5 mM dithionite and incubated with 1 µL (R,S)-linalool under anaerobic conditions at 28 °C for 0, 2 and 4 h. Samples were extracted with 200 µL n-hexane and analyzed by GC (Shimadzu GC-14A).

## UV-VIS spectrum for cofactors
The most enriched, active protein sample (0.95 mg mL$^{-1}$ protein) was analyzed by UV-VIS spectroscopy (Beckman DU-640 spectrophotometer) in the range of 200–800 nm to detect cofactors.

## Gene identification and bioinformatic analysis
A draft genome for *T. linaloolentis* 47Lol was obtained by merging data from two at NCBI public available draft genomes: ASM31020 (4.199 Mbp on 220 contigs, published 2012) and ASM62130 (4.214 Mbp on 46 contigs, published 2014). Contigs were automatically merged using Sequencher 4.6 with a minimum match percentage of 95 % and a minimum overlap of 50 bp. The resulting draft genome had 4.4 Mbp on 23 contigs and was uploaded to RAST for further analysis [33, 34]. Identification of a putative gene, coding for a linalool isomerase (*Lis*), was performed by homology search using the linalool dehydratase/isomerase sequence from *Castellaniella degragrans* 65Phen. The identified gene was analyzed by various bioinformatic tools: SignalP 4.1 for prediction of signal peptides [35], TMHMM, SOSUI and Philius for prediction of transmembrane helices [36–39] and the Pfam database to search for motifs and domain patterns [40].

## Additional files

Additional file 1: Figure S1. Alignment of the linalool dehydratase/isomerase (NCBI:CBW30776) from *C. defragrans* 65Phen and the linalool

isomerase from *T. linaloolentis* 47Lol (NCBI:ENO87364). Color indicates similarity. (PNG 106 kb)

Additional file 2: Figure S2. SDS-PAGE of different fractions during the inner membrane preparation from spheroplasts. (F1) spheroplasts, (F2) spheroplasts after cell disintegration, (F3) outer membrane and periplasmatic protein fraction, (F4) crude extract after spheroplast disintegration, (F5) unbroken spheroplast and cell debris, (F6) cytoplasmatic, soluble protein fraction, (F7) inner membrane fraction. (PNG 539 kb)

### Competing interests
The authors declare that they have no competing interests.

### Authors' contribution
RM and JH designed the study. RM and BS performed the purification and the experiments for enzyme characterization. SM and TS conducted mass-spectrometry based protein identifications and analyzed data. RM and JH drafted the manuscript. All authors read and approved the final manuscript.

### Acknowledgments
We thank Dirk Albrecht for MALDI-ToF measurements. This study was financed by the Max Planck Society.

### Author details
[1]Department of Microbiology, Max Planck Institute for Marine Microbiology, Celsiusstr. 1, D-28359 Bremen, Germany. [2]Institute for Pharmacy, Department of Pharmaceutical Biotechnology, University of Greifswald, Felix-Hausdorff-Str. 3, D-17487 Greifswald, Germany.

### References
1. Breitmaier E. Terpenes; flavors, fragrances, pharmaca, pheromones, vol. 1st edition. Weinheim, Wiley-VCH; 2006.
2. Ajikumar PK, Tyo K, Carlsen S, Mucha O, Phon TH, Stephanopoulos G. Terpenoids: opportunities for biosynthesis of natural product drugs using engineered microorganisms. Mol Pharm. 2008;5:167–90.
3. Dudareva N, Klempien A, Muhlemann JK, Kaplan I. Biosynthesis, function and metabolic engineering of plant volatile organic compounds. New Phytol. 2013; 198:16–32.
4. Sharkey TD, Wiberley AE, Donohue AR. Isoprene emission from plants: why and how. Ann Bot. 2008;101:5–18.
5. Günther A, Hewitt CN, Erickson D, Fall R, Geron C, Graedel T, Harley P, Klinger L, Lerdau M, McKay WA. A global model of natural volatile organic compound emission. J Geophys Res-Atmos. 1995;100:8873–92.
6. Atkinson R, Arey J. Atmospheric degradation of volatile organic compounds. Chem Rev. 2003;103:4605–38.
7. Kainulainen P, Holopainen JK. Concentrations of secondary compounds in Scots pine needles at different stages of decomposition. Soil Biol Biochem. 2002;34:37–42.
8. Wilt FM, Miller GC, Everett RL, Hackett M. Monoterpene concentrations in fresh, senescent, and decaying foliage of single-leaf pinyon (*Pinus monophylla* Torr. and Frem.: Pinaceae) from the western great basin. J Chem Ecol. 1993;19:185–94.
9. Ziemann PJ, Atkinson R. Kinetics, products, and mechanisms of secondary organic aerosol formation. Chem Soc Rev. 2012;41:6582–605.
10. Bakkali F, Averbeck S, Averbeck D, Waomar M. Biological effects of essential oils - a review. Food Chem Toxicol. 2008;46:446–75.
11. Schewe H, Mirata MA, Holtmann D, Schrader J. Biooxidation of monoterpenes with bacterial monooxygenases. Process Biochem. 2011;46:1885–99.
12. Marmulla R, Harder J. Microbial monoterpene transformations – a review. Front Microbiol. 2014;5.
13. Bell SG, Dale A, Rees NH, Wong LL. A cytochrome P450 class I electron transfer system from *Novosphingobium aromaticivorans*. Appl Microbiol Biotechnol. 2010;86:163–75.
14. Ullah AJH, Murray RI, Bhattacharyya PK, Wagner GC, Gunsalus IC. Protein components of a cytochrome P-450 linalool 8-methyl hydroxylase. J Biol Chem. 1990;265:1345–51.

15. Foss S, Harder J. Microbial transformation of a tertiary allylalcohol: Regioselective isomerisation of linalool to geraniol without nerol formation. FEMS Microbiol Lett. 1997;149:71–5.

16. Foss S, Harder J. *Thauera linaloolentis* sp. nov. and *Thauera terpenica* sp. nov., isolated on oxygen-containing monoterpenes (linalool, menthol, and eucalyptol) and nitrate. Syst Appl Microbiol. 1998;21:365–73.

17. Brodkorb D, Gottschall M, Marmulla R, Lüddeke F, Harder J. Linalool dehydratase-isomerase, a bifunctional enzyme in the anaerobic degradation of monoterpenes. J Biol Chem. 2010;285:30436–42.

18. Lüddeke F, Harder J. Enantiospecific (*S*)-(+)-linalool formation from beta-myrcene by linalool dehydratase-isomerase. Z Naturforsch C. 2011;66:409–12.

19. Schuster J, Schäfer F, Hübler N, Brandt A, Rosell M, Härtig C, Harms H, Müller RH, Rohwerder T. Bacterial degradation of *tert*-amyl alcohol proceeds via hemiterpene 2-methyl-3-buten-2-ol by employing the tertiary alcohol desaturase function of the Rieske nonheme mononuclear iron oxygenase MdpJ. J Bacteriol. 2012;194:972–81.

20. Malone VF, Chastain AJ, Ohlsson JT, Poneleit LS, Nemecek-Marshall M, Fall R. Characterization of a *Pseudomonas putida* allylic alcohol dehydrogenase induced by growth on 2-methyl-3-buten-2-ol. Appl Environ Microbiol. 1999; 65:2622–30.

21. Kyte J, Doolittle RF. A simple method for displaying the hydropathic character of a protein. J Mol Biol. 1982;157:105–32.

22. Lüddeke F, Wülfing A, Timke M, Germer F, Weber J, Dikfidan A, Rahnfeld T, Linder D, Meyerdierks A, Harder J. Geraniol and geranial dehydrogenases induced in anaerobic monoterpene degradation by *Castellaniella defragrans*. Appl Environ Microbiol. 2012;78:2128–36.

23. Kleinsmith LJ, Kish VM. Principles of cell and molecular biology, 2 edn. New York: HarperCollins; 1995.

24. Rath A, Glibowicka M, Nadeau VG, Chen G, Deber CM. Detergent binding explains anomalous SDS-PAGE migration of membrane proteins. Proc Natl Acad Sci. 2009;106:1760–5.

25. Gohon Y, Popot JL. Membrane protein-surfactant complexes. Curr Opin Colloid Interface Sci. 2003;8:15–22.

26. Seddon AM, Curnow P, Booth PJ. Membrane proteins, lipids and detergents: not just a soap opera. Biochim Biophys Acta. 2004;1666:105–17.

27. Christianson DW. Structural biology and chemistry of the terpenoid cyclases. Chem Rev. 2006;106:3412–42.

28. Schwab JM, Henderson BS. Enzyme-catalyzed allylic rearrangements. Chem Rev. 1990;90:1203–45.

29. Foss S, Heyen U, Harder J. *Alcaligenes defragrans* sp. nov., description of four strains isolated on alkenoic monoterpenes ((+)-menthene, alpha-pinene, 2-carene, and alpha-phellandrene) and nitrate. Syst Appl Microbiol. 1998;21:237–44.

30. Bradford MM. Rapid and sensitive method for quantitation of microgram quantities of protein utilizing principle of protein-dye binding. Anal Biochem. 1976;72:248–54.

31. Koßmehl S, Wöhlbrand L, Drüppel K, Feenders C, Blasius B, Rabus R. Subcellular protein localization (cell envelope) in *Phaeobacter inhibens* DSM 17395. Proteomics. 2013;13:2743–60.

32. Wolf C, Hochgräfe F, Kusch H, Albrecht D, Hecker M, Engelmann S. Proteomic analysis of antioxidant strategies of *Staphylococcus aureus*: diverse responses to different oxidants. Proteomics. 2008;8:3139–53.

33. Aziz RK, Bartels D, Best AA, DeJongh M, Disz T, Edwards RA, Formsma K, Gerdes S, Glass EM, Kubal M. The RAST server: rapid annotations using subsystems technology. BMC Genomics. 2008;9:75.

34. Overbeek R, Olson R, Pusch GD, Olsen GJ, Davis JJ, Disz T, Edwards RA, Gerdes S, Parrello B, Shukla M. The SEED and the Rapid Annotation of microbial genomes using Subsystems Technology (RAST). Nucleic Acids Res. 2014;42:D206–14.

35. Petersen TN, Brunak S, von Heijne G, Nielsen H. SignalP 4.0: discriminating signal peptides from transmembrane regions. Nat Methods. 2011;8:785–6.

36. Sonnhammer EL, von Heijne G, Krogh A. A hidden Markov model for predicting transmembrane helices in protein sequences. Proc Int Conf Intell Syst Mol Biol. 1998;6:175–82.

37. Hirokawa T, Boon-Chieng S, Mitaku S. SOSUI: classification and secondary structure prediction system for membrane proteins. Bioinformatics. 1998;14:378–9.

38. Imai K, Asakawa N, Tsuji T, Akazawa F, Ino A, Sonoyama M, Mitaku S. SOSUI-GramN: high performance prediction for sub-cellular localization of proteins in Gram-negative bacteria. Bioinformation. 2008;2:417–21.

39. Reynolds SM, Käll L, Riffle ME, Bilmes JA, Noble WS. Transmembrane topology and signal peptide prediction using dynamic bayesian networks. PLoS Comput Biol. 2008;4:e1000213.

40. Finn RD, Bateman A, Clements J, Coggill P, Eberhardt RY, Eddy SR, Heger A, Hetherington K, Holm L, Mistry J. Pfam: the protein families database. Nucleic Acids Res. 2014;42:D222–30.

# 8

# Tubulin is a molecular target of the Wnt-activating chemical probe

Yasunori Fukuda, Osamu Sano, Kenichi Kazetani, Koji Yamamoto, Hidehisa Iwata[*] and Junji Matsui[*]

## Abstract

**Background:** In drug discovery research, cell-based phenotypic screening is an essential method for obtaining potential drug candidates. Revealing the mechanism of action is a key step on the path to drug discovery. However, elucidating the target molecules of hit compounds from phenotypic screening campaigns remains a difficult and troublesome process. Simple and efficient methods for identifying the target molecules are essential.

**Results:** 2-Amino-4-(3,4-(methylenedioxy)benzylamino)-6-(3-methoxyphenyl)pyrimidine (AMBMP) was identified as a senescence inducer from a phenotypic screening campaign. The compound is widely used as a Wnt agonist, although its target molecules remain to be clarified. To identify its target proteins, we compared a series of cellular assay results for the compound with our pathway profiling database. The database comprises the activities of compounds from simple assays of cellular reporter genes and cellular proliferations. In this database, compounds were classified on the basis of statistical analysis of their activities, which corresponded to a mechanism of action by the representative compounds. In addition, the mechanisms of action of the compounds of interest could be predicted using the database. Based on our database analysis, the compound was anticipated to be a tubulin disruptor, which was subsequently confirmed by its inhibitory activity of tubulin polymerization.

**Conclusion:** These results demonstrate that tubulin is identified for the first time as a target molecule of the Wnt-activating small molecule and that this might have misled the conclusions of some previous studies. Moreover, the present study also emphasizes that our pathway profiling database is a simple and potent tool for revealing the mechanisms of action of hit compounds obtained from phenotypic screenings and off targets of chemical probes.

## Background

Drug candidate selection through small-molecule screening is a rational and widespread method in the current drug discovery cascade. Initially, drug discovery research involved cell-based phenotypic screening as a core approach to obtaining drug candidates [1]. However, since the completion of the Human Genome Project in 2003 and the finding that sequences include numerous potential target proteins for drug discovery, target-based drug screening has been pursued actively [2, 3]. In addition, target-based drug screening procedures were initially accelerated to increase the research and development productivity of drug discovery in pharmaceutical companies. However, the number of FDA-approved drugs screened from the target-based approach was much less than expected because a large number of drug candidates failed during drug development owing to safety issues and a lack of efficacy [4]. In contrast, recent analysis of all first-in-class new molecular entities showed that phenotypic screening approaches accounted for 37 % in comparison with 23 % from target-based approaches [1]. Accordingly, classical cellular phenotypic screenings, also called phenotypic drug discovery (PDD), are being reevaluated as complementary and efficient strategies for probing drug candidates.

Chemical probes are powerful tools for target validation of hit compounds from PDD. However, some well-known chemical probes have been used incorrectly and have resulted in misleading biological conclusions [5]. Therefore, target identification of these compounds is essential for PDD. To date, target identification methods that use chemical proteomics or activity-based proteomics have been developed, and they have uncovered many unique target proteins associated with bioactive compounds [6, 7]. Although they are certainly useful methods, they require mass spectrometry instrumentation and further

* Correspondence: hidehisa.iwata@takeda.com; junji.matsui@takeda.com
Pharmaceutical Research Division, Takeda Pharmaceutical Company Limited, 2-26-1, Muraokahigashi, Fujisawa, Kanagawa, Japan

chemical syntheses to add tags to compounds of interest without deteriorating their activities. To determine the target molecules of compounds without affinity tags, Petrone et al. developed the chemical biological descriptor "high-throughput screening finger-print (HTS-FP)" that employs accumulated HTS data [8]. On the other hand, Frederick et al. developed a screening platform that consists of a series of reporter gene assays to disclose the mechanisms of action (MOAs) of compounds and by conducting assays in a quantitative HTS format [9, 10]. To develop a much simpler target identification approach with tag-free compounds, we exploited a pathway profiling database using only tens of cellular assays representing cellular signaling cascades through evaluation of compounds at a single concentration.

Oncology has become one of the largest therapeutic areas in the pharmaceutical industry. Various kinds of molecular targets and cellular signals have been reported to inhibit cancer growth. Among them, cellular senescence is considered to be the most important cellular phenotype for permanently arresting the cell cycle [11]. To date, reports have shown that genetic mutations and cellular stressors such as oxidative stress enhance cellular senescence and that some small molecules induce cellular senescence [12, 13]. In particular, compounds that induce cellular senescence are expected to be potent drugs for suppressing cancer growth [14]. Here we conducted a phenotypic screening campaign based on high-content cellular imaging to probe small molecules that induce cellular senescence.

## Results

### Pathway profiling database classifies compounds according to their MOA

The pathway profiling database mainly comprises reporter gene assays using firefly luciferase that cover 13 different signaling pathways and cellular proliferation assays with 7 commercially available cell lines (Table 1). These types of cellular assays are widely used in cell biology research and are highly accessible because of their simple procedures and low cost. In addition, these assays are very robust and demonstrate high throughput, which enabled us to detect subtle signal changes in an HTS-compatible format. The assays were functionally validated using the dose-dependent response of a natural ligand or known inhibitors/activators.

Through the development of this database, we evaluated 1910 compounds from 3 commercial compound libraries that contained compounds with well-characterized MOAs and common experimentally used reference compounds. We evaluated these libraries at a single concentration of 3 µg/mL for the Natural Product Library and at 3 µM for the other libraries. After obtaining all data, the database was analyzed using hierarchical clustering of the activities using Ward's method in TIBCO Spotfire software (Fig. 1a).

**Table 1** Constituents of the pathway profiling database. The types of cellular signals for the reporter gene assays and cell lines of the proliferation assays are shown

| Cellular reporter gene assays | Cellular proliferation assays |
|---|---|
| cAMP response element (CRE) signal | HEK293T |
| Nuclear factor of activated T-cells (NFAT) signal | MRC5 (high density) |
| Nuclear factor kappa- light-chain-enhancer of activated B cells (NF-kB) signal | MRC5 (low density) |
| Serum response element (SRE) signal | A549 |
| Serum response factor (SRF) signal | PC3 |
| p53 signal | LNCaP |
| E2F signal | Jurkat |
| Activating transcription factor 6 signal | MDA-MB-231 |
| Hedgehog signal | |
| Hypoxia-inducible factor 1 (HIF1) signal | |
| Nuclear factor erythroid 2-related factor 2 (Nrf2) signal | |
| SMAD signal | |
| Wnt signal | |

As a result of the hierarchical clustering analysis, compounds that had similar activities in most assays were classified into the same cluster, enabling us to visually determine that they have similar molecular targets and signaling pathways.

Forskolin (Fig. 2), an adenylate cyclase activator [15], was included in each library, and all were grouped into one cluster (Fig. 1b). In the cluster, N-ethylcarboxamidoadenosine (NECA) (Fig. 2), an adenosine receptor agonist [16], was also included. This cluster was shown to gather compounds stimulating cAMP production via adenylate cyclase activation. This result indicates that the pathway profiling database classifies compounds according to their MOA. Similarly, phorbol 12-myristate 13-acetate (PMA) [17] and its structural analogs phorbol 12,13-dibutyrate [18], 13-O-acetylphorbol [19], and 12-deoxyphorbol 13-phenylacetate 20-acetate (dPPA) [20] (Fig. 2) were classified into the same cluster (Fig. 1c). In other words, the structural analogs that had the same effect on cellular signaling were categorized into one cluster, as expected.

Following these analyses, we investigated structurally diverse compounds that affect the same target proteins. We focused on the phosphodiesterase (PDE) inhibitors [21–24] (Fig. 2) contained in our database. To quantitatively compare differences in the structures and activities of each compound in our database, we employed Tanimoto structural similarity calculated by Daylight's fingerprints and Pearson's correlation coefficients (activity versus activity), respectively. The Tanimoto similarities ranged from 0.16 to 0.65, strongly indicating the broad structural diversity between the compounds in our database (Fig. 1d). In contrast,

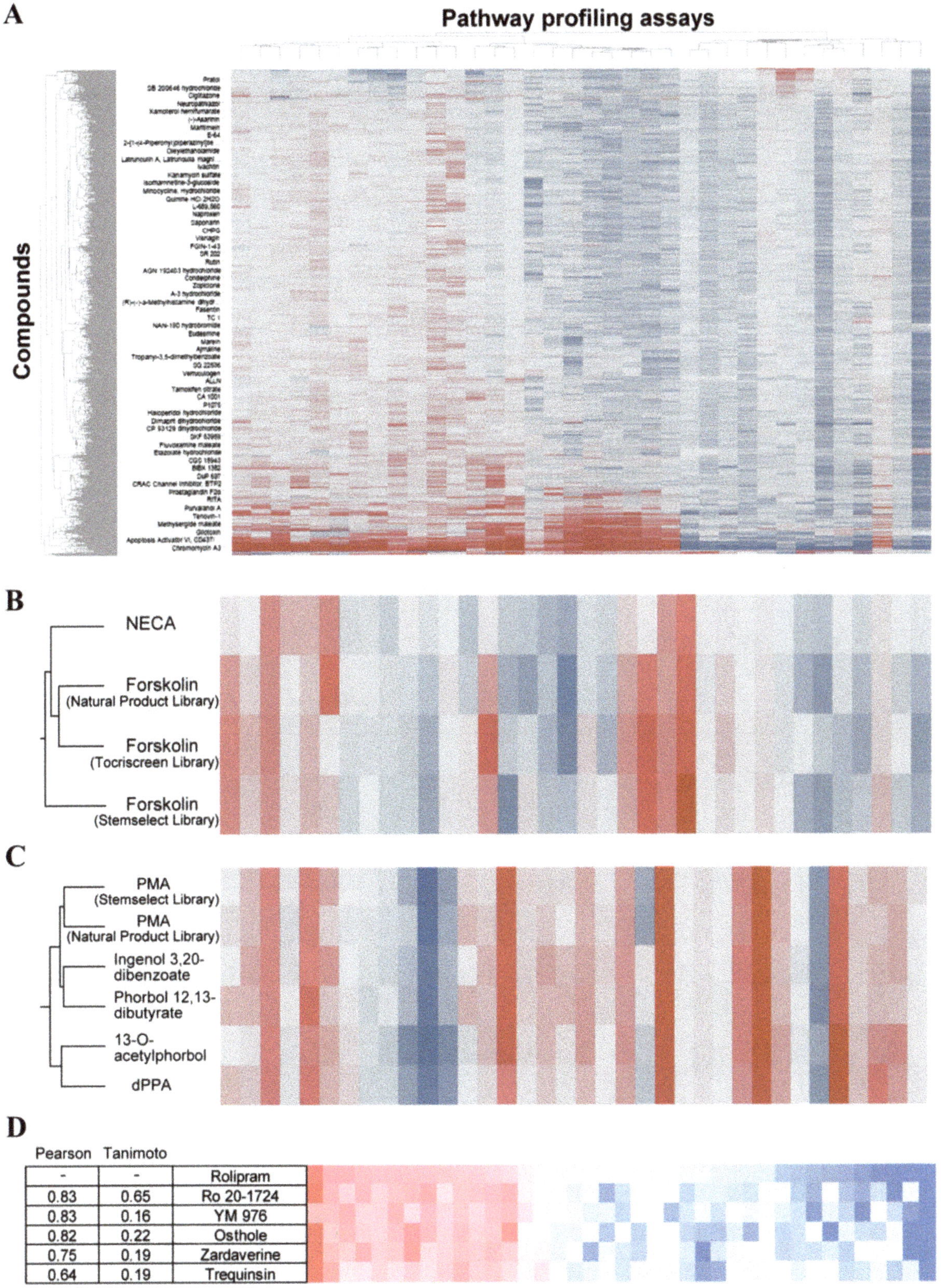

## A

### Pathway profiling assays

## B

## C

## D

| Pearson | Tanimoto | | | |
|---------|----------|--|--|--|
| - | - | Rolipram | | |
| 0.83 | 0.65 | Ro 20-1724 | | |
| 0.83 | 0.16 | YM 976 | | |
| 0.82 | 0.22 | Osthole | | |
| 0.75 | 0.19 | Zardaverine | | |
| 0.64 | 0.19 | Trequinsin | | |

**Fig. 1** (See legend on next page.)

(See figure on previous page.)
**Fig. 1** Analysis of the pathway profiling database. **a** The heat map was visualized with TIBCO Spotfire software for clustering analysis. This figure represents the entire heat map of the pathway profiling database. The activities of each assay are displayed as a gradient from minimum activities (blue) to maximum activities (red). For details of the assay lists, see Table 1. **b** This cluster contained forskolin derived from each commercial compound library and NECA, a potent adenosine receptor agonist. **c** PMA and its structural analogs were grouped in the cluster shown. **d** The Tanimoto structural similarities and Pearson's correlation coefficients (activity versus activity) were calculated for PDE4 inhibitors

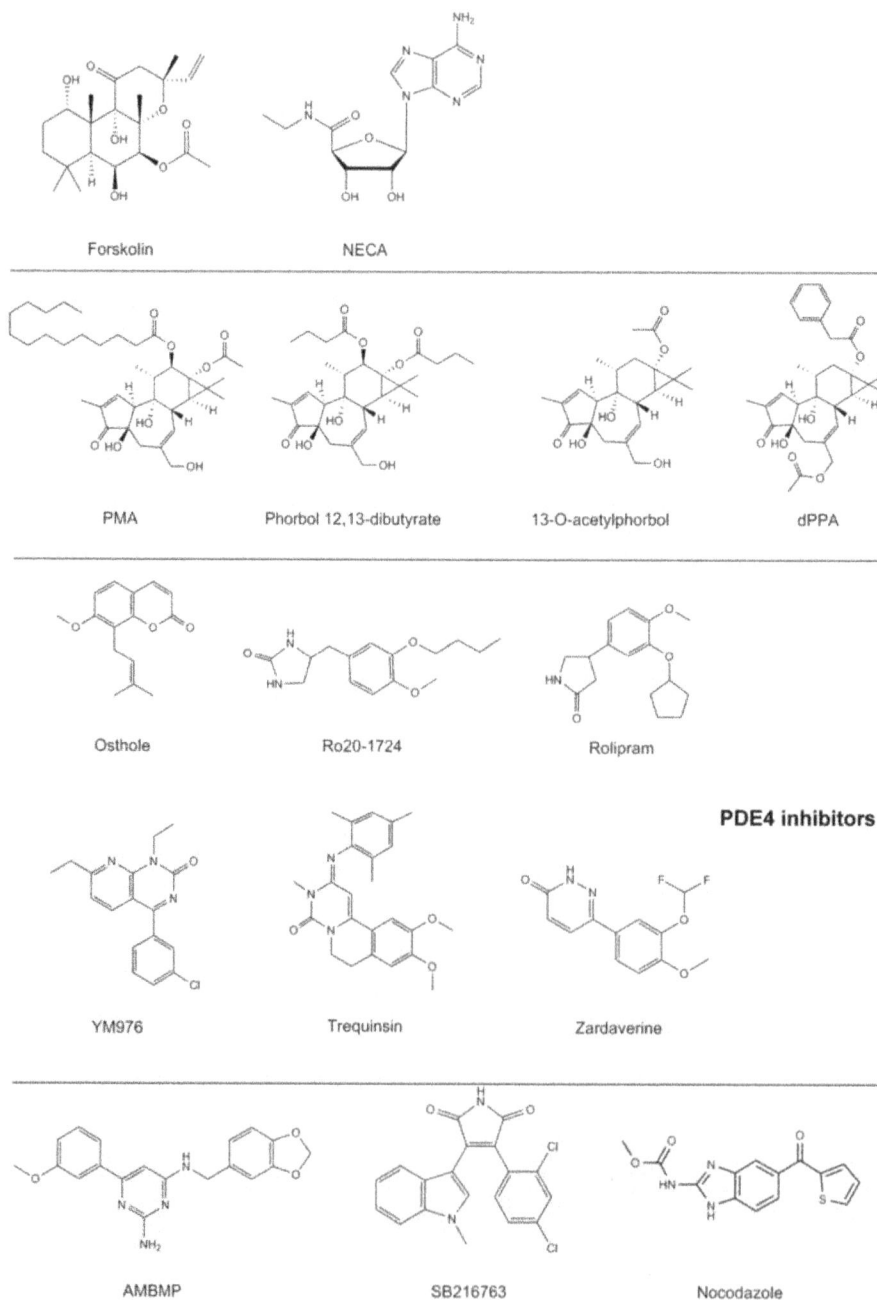

**Fig. 2** Chemical structures of the compounds discussed in this study

Pearson's correlation coefficients (activity versus activity) in our database ranged from 0.64 to 0.83 (Fig. 1d), showing their high bioactive similarities, despite their low structural similarities. These results indicate that our pathway profiling database based on the biological activities of compounds led to classifications corresponding to not only their structural similarities but also their MOAs.

## Wnt-activating small molecule is identified as a cellular senescence inducer

Triple-negative breast cancer has been a focus among the various cancer classes because of its lack of response to hormonal therapies, and new drugs with distinct MOAs are absolutely required to cure breast cancer patients [25]. Therefore, we employed MDA-MB-231 cells with triple-negative features to obtain cellular senescence inducers as anticancer agents [26]. In this strategy, we performed phenotypic screening on the basis of high-content cellular imaging, which is a very useful method to analyze altered cellular morphology. The cellular senescence morphology was reported to lead to a topologically enlarged appearance [11]. Sodium butyrate is a well-known senescence inducer [27], and we confirmed that it provoked the reported senescence phenotype in MDA-MB-231 cells and expanded cell shapes (Fig. 3a). In our study, this cellular morphology was defined as an indicator of cellular senescence.

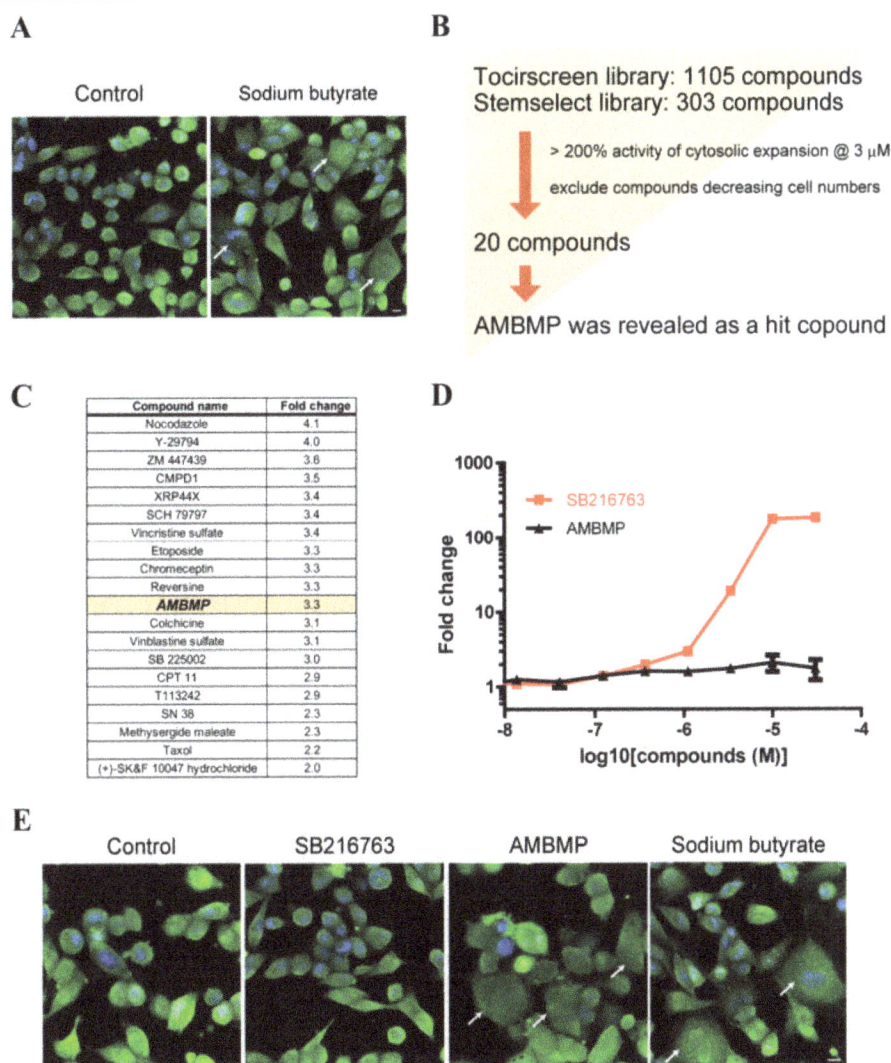

**Fig. 3** A cell-based assay for a screening campaign of cellular senescence morphology inducers by fluorescence microscopy. **a** MDA-MB231 cells were treated with 1 mM sodium butyrate. Hoechst 33342 was used as a nuclear marker (blue) and CMFDA was used to mark cytosols (green). Scale bar, 10 μm. **b** A compound selection scheme for the discovery of senescence inducers. AMBMP was obtained as a hit compound through the screening campaign. **c** Hit compound results from the screening campaign. Fold changes in the cellular area at 3 μM concentrations were calculated for the compounds with a custom-made image analysis algorithm. **d** Activity of the Wnt reporter gene assay with a potent GSK3β inhibitor, SB216763, and AMBMP is shown. The results are the mean of 3 replicate experiments (mean ± SD). **e** MDA-MB231 cells were treated with 3 μM SB216763, 3 μM AMBMP, and 1 mM sodium butyrate. Hoechst 33342 was used as a nuclear marker (blue) and CMFDA was used to mark cytosols (green). Scale bar, 10 μm

For high-content screening (HCS) of senescence inducers, we developed a cell-based assay to analyze cellular phenotypic changes in MDA-MB-231 cells. To determine the activities of compounds in this HCS, the cellular area, which plays a key role in the selection of senescence inducers, was calculated using a custom-made image analysis algorithm. We screened 1408 compounds in Tocriscreen (TOCRIS Bioscience) and StemSelect Small Molecule Regulators (Merck Millipore) at concentrations of 3 µM and obtained 20 compounds that induced a ≥2-fold enlargement of the cytosolic area (Fig. 3b). Of these 20 compounds identified as senescence inducers (Fig. 3c), molecular targets of 19 compounds have been clarified in past studies, but that of 2-amino-4-(3,4-(methylenedioxy)benzylamino)-6-(3-methoxyphenyl)pyrimidine (AMBMP) (Fig. 2) has not been revealed yet. Thus, we focused on AMBMP to elucidate its molecular target, which is described further in this report.

It is generally considered that Wnt signaling pathways play important roles during embryonic development [28]. AMBMP was first identified as a Wnt signal agonist through Wnt signal activator screening using a common reporter gene assay [29]. To date, the first report of AMBMP has been cited in 68 papers, and the compound itself and its 10 applications have been patented (SciFinder®). However, its binding proteins have not yet been identified. We initially measured the activity of AMBMP using a Wnt reporter gene assay, as reported previously by Liu et al. [29]. Unexpectedly, using the Wnt reporter assay, we detected a much lower efficacy of AMBMP than that of a widely known Wnt signal activator glycogen synthase kinase 3β (GSK3β) inhibitor (SB216763) [30] (Figs. 2 and 3d). In contrast, GSK3β inhibitors were not observed to induce the senescence morphology (Fig. 3e, Additional file 1: Figure S1). These results strongly suggest that Wnt signal activation is not directly related to its cellular senescence and that AMBMP has binding proteins responsible for inducing cellular senescence.

### Pathway profiling database identifies tubulin as a target protein of AMBMP

To identify an AMBMP target molecule, we compared the cellular assays with our pathway profiling database and calculated each Pearson's correlation coefficient (activity versus activity) between AMBMP and other all compounds in our database. As a result, 12 compounds demonstrated values above 0.8, which indicated high similarities (Fig. 4a). Moreover, 10 of the 12 compounds involved classical tubulin disruptors such as nocodazole (Fig. 2) and were thus known from previous reports to bind to tubulin [31–33]. Of these 10 compounds, only 2, KF 38789 and chromeceptin, had not been reported to induce tubulin depolymerization. The analyzed data allowed us to predict that AMBMP would directly interact with tubulin.

To test this hypothesis, we measured the tubulin disruption activity of AMBMP in a tubulin polymerization assay. Consequently, tubulin polymerization was detected by fluorescence enhancement following uptake of a fluorescent reporter molecule into the polymerized tubulin during polymerization [34].

We observed tubulin polymerization inhibition by AMBMP and nocodazole with $IC_{50}$ values of 0.33 µM and 0.34 µM, respectively (Fig. 4b, Additional file 1: Figure S2A). In this fluorescence-based polymerization assay, AMBMP was confirmed not to mediate the fluorescence interference through the observation of its UV–vis and fluorescence spectrum (Additional file 1: Figure S3). In addition, intrinsic fluorescence quenching was used to study the potential interaction between AMBMP and tubulin. The fluorescence intensity of tubulin was decreased gradually with increasing concentrations of AMBMP, confirming its binding to tubulin. (Fig. 4c). To determine the effects of these 2 compounds on the cellular microtubule network, we conducted a cell-based assay using cellular imaging techniques and fluorescent staining of tubulin. In the confocal image analysis, AMBMP and nocodazole were observed to clearly disrupt the intracellular microtubule network compared to control and SB216763-treated cells (Fig. 5A). Disturbance of the microtubule network by AMBMP and nocodazole was detected with $IC_{50}$ values of 0.34 µM and 1.7 µM, respectively (Additional file 1: Figure S2B). Furthermore, AMBMP as well as nocodazole was observed to inhibit cell proliferation and induce a cell cycle arrest in MDA-MB-231 cells (Additional file 1: Figure S4A, S4B). The effect of AMBMP on mitotic spindles was also observed with slightly shortening the spindle and astral microtubule at the low concentration of 30 nM and with significantly disrupting mitotic spindles at the higher concentrations of 0.3 and 3 µM (Fig. 5B), which was consistent with previous reports showing the effect of microtubule disruptors on mitotic spindles [35]. These results indicate that AMBMP had a strong inhibitory effect on tubulin polymerization, comparable to that of nocodazole. In addition, we had previously observed in our screening campaign that common tubulin disruptors induce cellular senescence (Fig. 3c) [36, 37].

### Discussion

In our study, the pathway profiling database based on the biological activities of compounds was confirmed to lead to classifications corresponding to both their structural similarities and their MOAs. Through operating the system, we will both maintain and obtain data at a lower cost and in a shorter period than the HTS-FP database and BioMAP™ (DiscoveRx), in which primary cells were utilized. However, our prediction method is limited to the range of target molecules of the reference compounds; however, to overcome this limitation, we will add various reference data for

**A**

| Pearson | Tanimoto | Compound | Biological activity | |
|---------|----------|----------|---------------------|---|
| - | - | Wnt Agonist | Wnt signal activator | |
| 0.93 | 0.24 | Nocodazole | Tubulin polymerization inhibitor | |
| 0.90 | 0.20 | Vincristine sulfate | Tubulin polymerization inhibitor | |
| 0.90 | 0.24 | Colchicine | Tubulin polymerization inhibitor | |
| 0.89 | 0.21 | D-64131 | Tubulin polymerization inhibitor | |
| 0.88 | 0.26 | 4'-Demethylpipodophyllotoxin | Tubulin polymerization inhibitor | |
| 0.87 | 0.21 | Vinorelbine | Tubulin polymerization inhibitor | |
| 0.85 | 0.16 | KF 38789 | Inhibitor of P-selectin-mediated cell adhesion | |
| 0.84 | 0.20 | Vinblastine sulfate | Tubulin polymerization inhibitor | |
| 0.84 | 0.22 | Chromeceptin | Insulin-like growth factor 2 signal inhibitor | |
| 0.83 | 0.26 | Podophyllotoxin | Tubulin polymerization inhibitor | |
| 0.82 | 0.12 | T113242 | Tubulin polymerization inhibitor | |
| 0.82 | 0.23 | JK184 | Hedgehog signal activator | |

**B**

**C**

Fig. 4 Target identification of AMBMP and its binding activity to tubulin. a The Tanimoto structural similarities and Pearson's correlation coefficients (activity versus activity) were calculated against AMBMP. b SB216763, AMBMP, and nocodazole were evaluated in a tubulin polymerization assay. The results are the mean of 3 replicate experiments (with SD not shown for graphical simplicity). c AMBMP induced intrinsic tryptophan fluorescence spectra changes of tubulin. The results are the mean of 3 replicate experiments (with SD not shown for graphical simplicity)

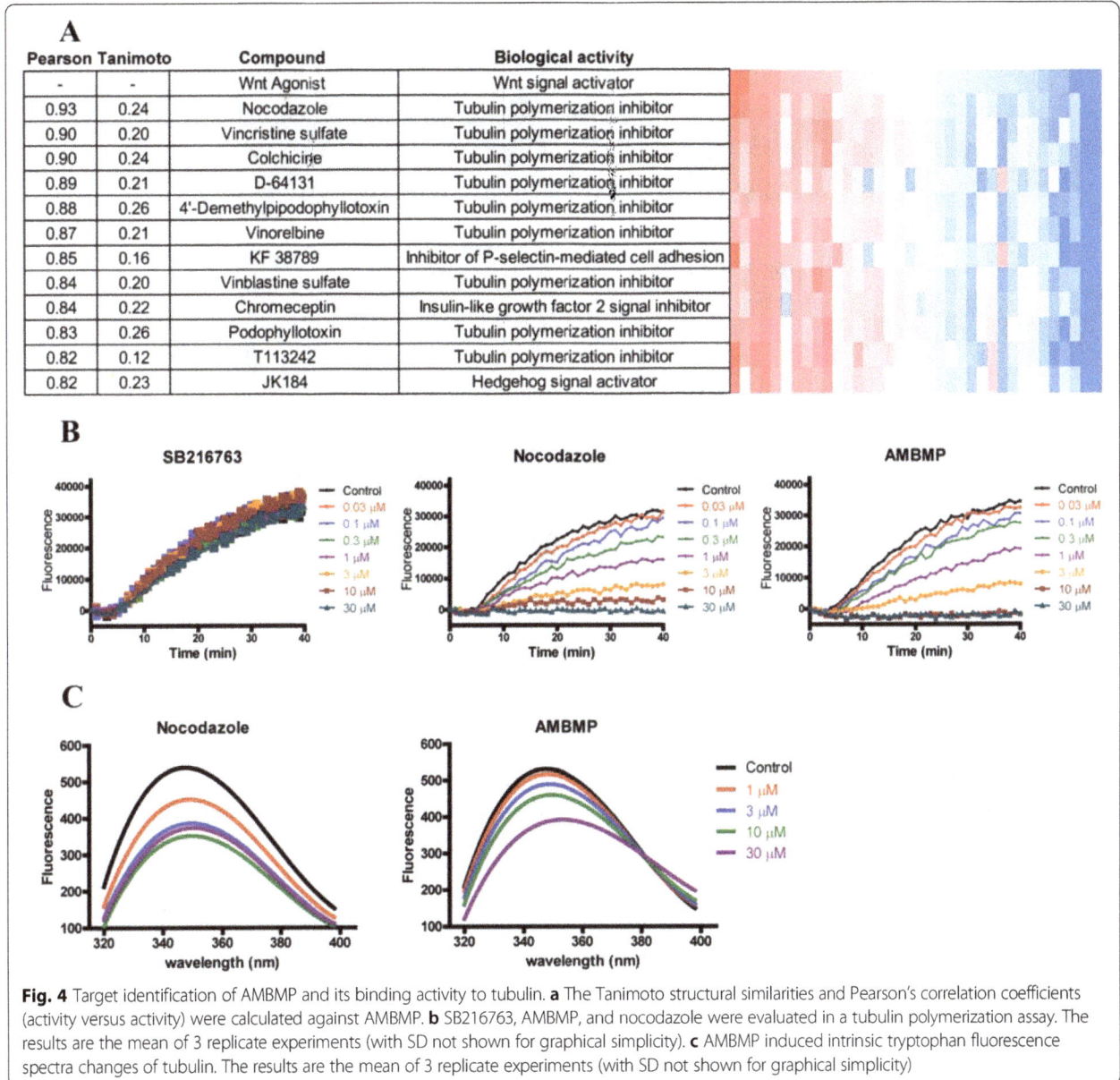

compounds that affect different types of target proteins other than those of the current compounds. In general, the accuracy of clustering analysis increases with a larger collection of datasets. Therefore, we will expand the cellular assays in the pathway profiling database to improve the accuracies of predicting both target molecules and cellular signaling properties. With these improvements in our system, we are attempting to perform target identification of other compounds, including our in-house compounds, with unknown targets.

In addition, we estimated the extent of cellular signaling pathways covered by our database through a computational approach. With Reactome Pathway Database [38], our pathway profiling database has the potential to detect cellular events involved in more than 200 canonical biological pathways. Moreover, 70 % of the tested compounds with well-characterized MOAs had detectable activity in at least one assay in our database. Consequently, our simple system is a promising and cost-effective tool for profiling phenotypes and for predicting molecular targets of hit compounds from PDD.

By applying our profiling system for target identification of AMBMP, we have revealed that AMBMP is a tubulin disrupting molecule for the first time since the compound was reported as a Wnt agonist. The Tanimoto similarities between AMBMP and tubulin disruptors ranged from 0.12 to 0.26 (Fig. 4a), which means that these compounds are apparently not structural analogs of AMBMP. Because of their low scores, the structural similarities did not lead us to hypothesize whether AMBMP could inhibit tubulin

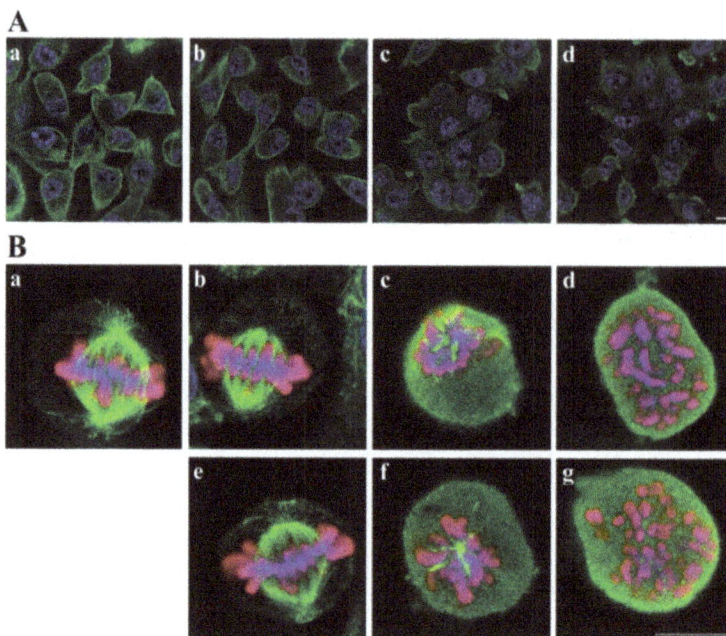

**Fig. 5** Effect of AMBMP on the cellular tubulin network and mitotic spindles. (A) The cellular tubulin network (green) was observed by fluorescence microscopy. Hoechst 33342 was used as a nuclear marker (blue). (a) control. (b) 3 μM SB216763. (c) 3 μM AMBMP. (d) 3 μM Nocodazole. Scale bar, 10 μm. (B) Control and compounds-treated MDA-MB-231 cells were stained with α-tubulin (green), phospho-histone H3 (red), and nuclei (blue). Phosphorylation at a highly conserved serine residue (Ser10) in the histone H3 is a key marker during the mitotic phase of the cell cycle. (a) control. (b) 30 nM nocodazole. (c) 0.3 μM nocodazole. (d) 3 μM nocodazole. (e) 30 nM AMBMP. (f) 0.3 μM AMBMP. (g) 3 μM AMBMP. Scale bar, 10 μm

polymerization. The achievement of AMBMP target identification supports the result that our pathway profiling database was extremely useful for predicting various pharmacological targets of compounds with unknown mechanisms. On the other hand, we consider that it is important to reveal the molecular mechanisms inhibiting tubulin polymerization by AMBMP. To address the issue, in future study, we will clarify its binding site on tubulin through a cocrystal structural analysis for AMBMP and tubulin complex.

Chemical probes are widely used to demonstrate target molecule proof-of-concept in drug discovery [39]. To this end, the selectivity of chemical probes against the intended targets is a key factor. If these chemical probes interact with unintentional molecules and induce cellular phenotypes through their off-target effects, then both time and money might be lost in the process of drug discovery research. Some past research that used AMBMP as a chemical probe for Wnt signal activation might have incorrectly generated misleading results due to inhibition of tubulin activity. Recently, the met proto-oncogene (c-MET) inhibitor tivantinib was confirmed to inhibit tubulin polymerization as well as AMBMP [40]. Through our study, KF 38789 and chromeceptin were also shown to have similar bioactive profiles to tubulin disruptors (Fig. 4b), generating the possibility that both compounds interact with tubulin. These compounds will be the subject of a future publication. In addition, a previous report revealed that the structural similarities of compounds do not provide sufficient information to speculate on their biological activities [41]. For an efficient drug discovery process, it is important to evaluate and profile chemical probes using various types of cellular assays, such as our pathway profiling database.

## Conclusion

Our pathway profiling database determined tubulin to be a target of AMBMP, which was unknown since the discovery of AMBMP, and our simple and efficient system proved to be a powerful method for predicting compound MOAs. AMBMP has been widely used as a chemical probe for Wnt signal activation, but the results for studies that used the compound might have been influenced by its modulation of tubulin activity and not Wnt signal activity. For proper utilization of chemical probes, it is potentially valuable to investigate their cellular profiles using multiple cellular assays, such our pathway profiling database, which provides beneficial information about representative cellular signaling processes. Moreover, in drug discovery, off-target interactions are strongly thought to lead to low efficacy and significant side effects in clinical trials; therefore, the development of target identification and prediction methods is now definitively required to determine not only on-target molecules but also off-target molecules. The system will certainly keep providing us with useful information for various stages of the drug discovery process through target prediction and drug safety research.

## Methods
### Chemical compounds
Tocriscreen (TOCRIS Bioscience), Natural Product Library (ENZO Life Sciences), and StemSelect Small Molecule Regulators (Merck Millipore) were all dissolved in DMSO (10 mM for Tocriscreen and StemSelect and 10 mg/mL for the Natural Product Library). AMBMP was obtained from Merck Millipore. Sodium butyrate, nocodazole, and SB216763 were sourced from Wako.

### Cell cultures
HEK293T, MRC5, A549, PC3, LNCaP, Jurkat, MDA-MB231, NIH-3 T3, and SW480 cells were purchased from ATCC. HEK293T, A549, MRC5, and MDA-MB-231 cells were cultured in DMEM containing 4.5 g/L glucose, 10 % fetal bovine serum (FBS), and penicillin/streptomycin. Jurkat, LNCaP, SW480, and PC3 cells were cultured in RPMI 1640 media containing 10 % FBS and penicillin/streptomycin. NIIH-3 T3 cells were cultured in DMEM containing 1.5 g/L glucose, 10 % FBS, and penicillin/

streptomycin. All cell culture reagents were purchased from Wako.

### Reporter gene assays in pathway profiling
We developed reporter gene assays using a firefly luciferase system purchased from Promega. Detailed assay conditions such as cell lines, cell densities, corresponding ligands, incubation time with compounds, and materials are shown (Table 2). All assays were performed in a 384-well plate format. Plasmids were constructed by inserting each response element sequence at a multi-cloning site upstream from firefly luciferase. Transient transfections of all plasmids were performed in corresponding cell lines with Fugene HD (Promega) according to the manufacturer's instructions. In each assay, we validated the assay condition with its ligand to perform a stable screening campaign (data not shown). In all assays, all compounds were diluted in complete media at a concentration of 3 μg/mL (Natural Product Library) and 3 μM (other libraries) and treated for the appropriate durations. After the addition of Steady-Glo (Promega) according to the manufacturer's instructions, luminescence signals were measured using a luminescence plate reader (EnVision; PerkinElmer). We typically obtained 2 parameters calculated from each assay: one was

**Table 2** Assay conditions such as cell lines, cell densities, corresponding ligands, incubation time with compounds, and materials for reporter gene assays in pathway profiling

| Cellular signal | Cell lines | Cell density (cells/well) | Ligand | Incubation time with compounds | Original materials or references |
|---|---|---|---|---|---|
| cAMP response element (CRE) signal | HEK293T | 5,000 | Forskolin (1 μM) | 5 h | pGL4.29 (Promega) |
| Nuclear factor of activated T-cells (NFAT) signal | HEK293T | 5,000 | Ionomycin (1 μM) PMA (10 ng/mL) | 5 h | pGL4.30 (Promega) |
| Nuclear factor kappa-light-chain-enhancer of activated B cells (NF-kB) signal | HEK293T | 10,000 | TNFα (3 ng/mL) | 20 h | pGL4.32 (Promega) |
| Serum response element (SRE) signal | HEK293T | 20,000 | FBS (15 %) PMA (30 ng/mL) | 20 h | pGL4.33 (Promega) |
| Serum response factor (SRF) signal | HEK293T | 20,000 | FBS (15 %) | 5 h | pGL4.34 (Promega) |
| p53 signal | HEK293T | 10,000 | Doxorubicin (1 μM) | 20 h | [43] |
| E2F signal | HEK293T | 20,000 | FBS (15 %) | 20 h | [44] |
| Activating transcription factor 6 signal | HEK293T | 5,000 | Thapsigargin (30 nM) | 5 h | [45] |
| Hedgehog signal | NIH3T3 | 7,500 | mouse sonic hedgehog | 20 h | [46] |
| Hypoxia-inducible factor 1 (HIF1) signal | HEK293T | 10,000 | hypoxia | 20 h | [47] |
| Nuclear factor erythroid 2-related factor 2 (Nrf2) signal | HEK293T | 5,000 | tert-butylhydroquinone (20 μM) | 20 h | pGL4.37 (Promega) |
| SMAD signal | HEK293T | 10,000 | TGFβ (0.2 ng/mL) | 20 h | [48] |
| Wnt signal | HEK293T | 10,000 | Wnt3a | 20 h | [49] |
| Wnt signal | SW480 | 10,000 | no ligand (constitutive active) | 20 h | [49] |
| IL17 signal | Jurkat | 5,000 | Ionomycin (400 nM) PMA (4 ng/mL) | 5 h | [50] |

the compound's inhibitory activity with ligand activation and the other was its agonistic activity without ligand activation.

### Cellular proliferation assays used in pathway profiling

Cell lines, cell densities, and incubation times with compounds are shown (Table 3). Cellular proliferation was detected with CellTiter-Glo (Promega). All assays were performed in a 384-well plate format. Luminescence signals were readout using a luminescence plate reader (EnVision; PerkinElmer). The proliferation assays with HEK293T cells and Jurkat cells were used as the counter-screen against reporter gene assays.

### Cell-based phenotypic assays for cellular senescence inducers

MDA-MB231 cells were seeded in a 384-well plate (3000 cells/well) for 20 h before the treatment of compounds. After seeding, the tested compounds were diluted in complete media and incubated with cells for 24 h, followed by cytosol and nuclear staining for 1 h with CellTracker Green CMFDA and Hoechst 33342 (Invitrogen), respectively. For cellular tubulin staining, tubulin tracker green was used according to the manufacturer's instructions (Invitrogen). Cellular images were recorded with an IN Cell Analyzer 6000 (GE Healthcare). After obtaining the images, the nuclear locations and cellular areas were stained with Hoechst 33342 and CMFDA, respectively, and quantitative signals from the images were calculated using a custom-made image analysis algorithm with IN Cell Developer Toolbox (GE Healthcare).

### Cluster analysis in the pathway profiling system

All compounds were utilized at a concentration of 3 μg/mL (Natural Product Library) or 3 μM (other libraries) in the pathway profiling assays. All calculated data, including percent inhibition and percent activation number, were first normalized in each assay using the Z-scoring method and then analyzed by hierarchical clustering analysis (Ward's method) with TIBCO Spotfire software (TIBCO).

**Table 3** Assay conditions such as cell lines, cell densities, and incubation times for cellular proliferation assays in pathway profiling

| Cell lines | Cell density (cells/well) | Incubation time with compounds |
|---|---|---|
| HEK293T | 5,000 | 20 h |
| Jurkat | 5,000 | 20 h |
| MRC5 | 1,000 | 72 h |
| MRC5 | 3,500 | 72 h |
| A549 | 1,000 | 72 h |
| PC3 | 1,000 | 72 h |
| LNCaP | 600 | 72 h |
| MDA-MB-231 | 500 | 72 h |

### Calculating Pearson's correlation coefficients

Pearson's correlation coefficients (Rp) were calculated using the following equation:

$$Rp = \frac{\sum_{i=1}^{N}(x_i - \bar{x})(y_i - \bar{y})}{\sqrt{\sum_{i=1}^{N}(x_i - \bar{x})^2}\sqrt{\sum_{i=1}^{N}(y_i - \bar{y})^2}}$$

where N equals 39 assay results and $x_i$ and $y_i$ are the activity values of each assay in our pathway profiling database for compounds A and B, respectively.

### Tubulin polymerization assay

Tubulin polymerization was performed using a tubulin polymerization assay kit (BK011P, Cytoskeleton). Compounds were evaluated according to the manufacturer's instructions.

### Tubulin binding assay with its intrinsic tryptophan fluorescence

4 μM of purified tubulin (Cytoskeleton) dissolved in general tubulin buffer (80 mM PIPES, pH 6.9, 2 mM $MgCl_2$, 0.5 mM EGTA) was pretreated with certain concentrations of compounds for 30 min. The intrinsic fluorescence spectra (320–400 nm) was measured with a fluorescence plate reader (EnVision; PerkinElmer) with the excitation wavelength 295 nm.

### Immunofluorescence microscopy

MDA-MB231 cells were incubated with compounds for 6 h and 24 h to observe the cellular microtubule network and the mitotic spindles respectively. Thereafter, the cells were fixed and permeabilized as described in the past report [42]. After blocking nonspecific binding with 1 % donkey serum/PBS, the cells were incubated with the mouse monoclonal anti-α-tubulin antibody (Cell Signaling Technology) (1:1000 dilution) followed by the Alexa-488 conjugated anti-mouse IgG antibody (Invitrogen) (1:500 dilution). To visualized nuclei, the cells were incubated with Hoechst33342. For staining phospho-Histone H3, the fixed cells were treated with the rabbit monoclonal anti-phospho-Histone H3 (Ser10) antibody (Cell Signaling Technology) (1:1000 dilution) followed by the Alexa-594 conjugated anti rabbit IgG antibody (Invitrogen) (1:500 dilution). Cellular images were obtained with SP8 confocal microscopy (Leica).

### Flow cytometric analysis

MDA-MB-231 cells were treated with compounds for 24 h, and fixed with ethanol. After fixation, cells were washed with PBS containing 2 % FCS, and, subsequently, treated with Guava Cell Cycle reagent (Merck Millipore) according to the manufacturer's instructions. The DNA contents were determined using a Guava easyCyte HT software (Merck Millipore).

## Abbreviations
FDA: The Food and Drug Administration; HTS: high-throughput screening.

## Acknowledgement
We thank K. Ishimaru for performing cell culture and cellular assays and S. Okubo for preparation of the recombinant plasmids. We also thank M. Oohori, I. Miyahisa and T. Kawamoto for discussions and editing. Finally, we thank N. Tarui for the encouragement to conduct this study.

## Authors' contributions
YF and OS conducted all experiments and data analysis. KK and KY performed statistical analyses for calculating the compounds' structural and bioactive similarities. YF, OS, KK, KY, HI, and JM conceived the study and participated in its design and drafting of the manuscript. All authors read and approved the final manuscript.

## Competing interests
The authors declare that they have no competing interests.

## References
1. Swinney DC, Anthony J. How were new medicines discovered? Nat Rev Drug Discov. 2011;10(7):507–19.
2. Imming P, Sinning C, Meyer A. Drugs, their targets and the nature and number of drug targets. Nat Rev Drug Discov. 2006;5(10):821–34.
3. Overington JP, Al-Lazikani B, Hopkins AL. How many drug targets are there? Nat Rev Drug Discov. 2006;5(12):993–6.
4. Arrowsmith J, Miller P. Trial Watch: Phase II and Phase III attrition rates 2011–2012. Nat Rev Drug Discov. 2013;12(8):569–9.
5. Arrowsmith CH, Audia JE, Austin C, Baell J, Bennett J, Blagg J, Bountra C, Brennan PE, Brown PJ, Bunnage ME. The promise and peril of chemical probes. Nat Chem Biol. 2015;11(8):536–41.
6. Kosaka T, Okuyama R, Sun W, Ogata T, Harada J, Araki K, Izumi M, Yoshida T, Okuno A, Fujiwara T. Identification of molecular target of AMP-activated protein kinase activator by affinity purification and mass spectrometry. Anal Chem. 2005;77(7):2050–5.
7. Rix U, Superti-Furga G. Target profiling of small molecules by chemical proteomics. Nat Chem Biol. 2009;5(9):616–24.
8. Petrone PM, Simms B, Nigsch F, Lounkine E, Kutchukian P, Cornett A, Deng Z, Davies JW, Jenkins JL, Glick M. Rethinking molecular similarity: comparing compounds on the basis of biological activity. ACS Chem Biol. 2012;7(8):1399–409.
9. Michael S, Auld D, Klumpp C, Jadhav A, Zheng W, Thorne N, Austin CP, Inglese J, Simeonov A. A robotic platform for quantitative high-throughput screening. ASSAY Drug Dev Techn. 2008;6(5):637–57.
10. King FJ, Selinger DW, Mapa FA, Janes J, Wu H, Smith TR, Wang Q-Y, Niyomrattanakitand P, Sipes DG, Brinker A. Pathway reporter assays reveal small molecule mechanisms of action. J Assoc Lab Autom. 2009;14(6):374–82.
11. Rodier F, Campisi J. Four faces of cellular senescence. J Cell Biol. 2011; 192(4):547–56.
12. Giorgio M, Trinei M, Migliaccio E, Pelicci PG. Hydrogen peroxide: a metabolic by-product or a common mediator of ageing signals? Nat Rev Mol Cell Biol. 2007;8(9):722–8.
13. Sayers CM, Papandreou I, Guttmann DM, Maas NL, Diehl JA, Witze ES, Koong AC, Koumenis C. Identification and characterization of a potent activator of p53-independent cellular senescence via a small-molecule screen for modifiers of the integrated stress response. Mol Pharmacol. 2013; 83(3):594–604.
14. Campisi J. Cellular senescence as a tumor-suppressor mechanism. Trends Cell Biol. 2001;11(11):S27–31.
15. McHugh Sutkowski E, Tang WJ, Broome CW, Robbins JD, Seamon KB. Regulation of forskolin interactions with type I, II, V, and VI adenylyl cyclases by Gs.alpha. Biochemistry. 1994;33(43):12852–9.
16. Cusack NJ, Hourani SM. 5'-N-ethylcarboxamidoadenosine: a potent inhibitor of human platelet aggregation. Br J Pharmacol. 1981;72(3):443–7.
17. Castagna M, Takai Y, Kaibuchi K, Sano K, Kikkawa U, Nishizuka Y. Direct activation of calcium-activated, phospholipid-dependent protein kinase by tumor-promoting phorbol esters. J Biol Chem. 1982;257(13):7847–51.
18. Middleton JP, Khan WA, Collinsworth G, Hannun YA, Medford RM. Heterogeneity of protein kinase C-mediated rapid regulation of Na/K-ATPase in kidney epithelial cells. J Biol Chem. 1993;268(21):15958–64.
19. Gustafson KR, Cardellina JH, McMahon JB, Gulakowski RJ, Ishitoya J, Szallasi Z, Lewin NE, Blumberg PM, Weislow OS. A nonpromoting phorbol from the Samoan medicinal plant Homalanthus nutans inhibits cell killing by HIV-1. J Med Chem. 1992;35(11):1978–86.
20. Ryves WJ, Evans AT, Olivier AR, Parker PJ, Evans FJ. Activation of the PKC-isotypes α, β1, γ, δ, and ε by phorbol esters of different biological activities. FEBS Lett. 1991;288(1–2):5–9.
21. Zhu WH, Majluf-Cruz A, Omburo GA. Cyclic AMP-specific phosphodiesterase inhibitor rolipram and RO-20-1724 promoted apoptosis in HL60 promyelocytic leukemic cells via cyclic AMP-independent mechanism. Life Sci. 1998;63(4):265–74.
22. Aoki M, Kobayashi M, Ishikawa J, Saita Y, Terai Y, Takayama K, Miyata K, Yamada T. A Novel Phosphodiesterase Type 4 Inhibitor, YM976 (4-(3-Chlorophenyl)-1,7-diethylpyrido[2,3-d]pyrimidin-2(1H)-one), with Little Emetogenic Activity. J Pharmacol Exp Ther. 2000;295(1):255–60.
23. Whitaker RM, Wills LP, Stallons LJ, Schnellmann RG. CGMP-selective phosphodiesterase inhibitors stimulate mitochondrial biogenesis and promote recovery from acute kidney injury. J Pharmacol Exp Ther. 2013; 347(3):626–34.
24. Schudt C, Winder S, Müller B, Ukena D. Zardaverine as a selective inhibitor of phosphodiesterase isozymes. Biochem Pharmacol. 1991;42(1):153–62.
25. Foulkes WD, Smith IE, Reis-Filho JS. Triple-negative breast cancer. N Engl J Med. 2010;363(20):1938–48.
26. Chavez KJ, Garimella SV, Lipkowitz S. Triple negative breast cancer cell lines: One tool in the search for better treatment of triple negative breast cancer. Breast Dis. 2010;32(1):35–48.
27. Terao Y, Nishida JI, Horiuchi S, Rong F, Ueoka Y, Matsuda T, Kato H, Furugen Y, Yoshida K, Kato K. Sodium butyrate induces growth arrest and senescence-like phenotypes in gynecologic cancer cells. Int J Cancer. 2001;94(2):257–67.
28. Clevers H, Nusse R. Wnt/β-Catenin signaling and disease. Cell. 2012;149(6): 1192–205.
29. Liu J, Wu X, Mitchell B, Kintner C, Ding S, Schultz PG. A small-molecule agonist of the Wnt signaling pathway. Angew Chem Int Ed. 2005;44(13): 1987–90.
30. Coghlan MP, Culbert AA, Cross DAE, Corcoran SL, Yates JW, Pearce NJ, Rausch OL, Murphy GJ, Carter PS, Roxbee Cox L. Selective small molecule inhibitors of glycogen synthase kinase-3 modulate glycogen metabolism and gene transcription. Chem Biol. 2000;7(10):793–803.
31. Cupido T, Rack PG, Firestone AJ, Hyman JM, Han K, Sinha S, Ocasio CA, Chen JK. The Imidazopyridine derivative JK184 reveals dual roles for microtubules in hedgehog signaling. Angew Chem Int Ed. 2009;48(13):2321–4.
32. Mahboobi S, Pongratz H, Hufsky H, Hockemeyer J, Frieser M, Lyssenko A, Paper DH, Bürgermeister J, Böhmer F-D, Fiebig H-H. Synthetic 2-Aroylindole derivatives as a New class of potent tubulin-inhibitory, antimitotic agents‖. J Med Chem. 2001;44(26):4535–53.
33. Dumontet C, Jordan MA. Microtubule-binding agents: a dynamic field of cancer therapeutics. Nat Rev Drug Discov. 2010;9(10):790–803.
34. Bonne D, Heuséle C, Simon C, Pantaloni D. 4',6-Diamidino-2-phenylindole, a fluorescent probe for tubulin and microtubules. J Biol Chem. 1985;260(5):2819–25.
35. Jordan MA, Thrower D, Wilson L. Effects of vinblastine, podophyllotoxin and nocodazole on mitotic spindles. Implications for the role of microtubule dynamics in mitosis. J Cell Sci. 1992;102(3):401–16.

36. Gewirtz DA, Holt SE, Elmore LW. Accelerated senescence: An emerging role in tumor cell response to chemotherapy and radiation. Biochem Pharmacol. 2008;76(8):947–57.

37. Tierno MB, Kitchens CA, Petrik B, Graham TH, Wipf P, Xu FL, Saunders WS, Raccor BS, Balachandran R, Day BW. Microtubule Binding and disruption and induction of premature senescence by disorazole C1. J Pharmacol Exp Ther. 2009;328(3):715–22.

38. Croft D, Mundo AF, Haw R, Milacic M, Weiser J, Wu G, Caudy M, Garapati P, Gillespie M, Kamdar MR. The reactome pathway knowledgebase. Nucleic Acids Res. 2014;42(Database issue):D472–7.

39. Bunnage ME, Chekler ELP, Jones LH. Target validation using chemical probes. Nat Chem Biol. 2013;9(4):195–9.

40. Katayama R, Aoyama A, Yamori T, Qi J, Oh-hara T, Song Y, Engelman JA, Fujita N. Cytotoxic activity of tivantinib (ARQ 197) is not due solely to c-MET inhibition. Cancer Res. 2013;73(10):3087–96.

41. Martin YC, Kofron JL, Traphagen LM. Do structurally similar molecules have similar biological activity? J Med Chem. 2002;45(19):4350–8.

42. Mohan R, Banerjee M, Ray A, Manna T, Wilson L, Owa T, Bhattacharyya B, Panda D. Antimitotic sulfonamides inhibit microtubule assembly dynamics and cancer cell proliferation†. Biochemistry. 2006;45(17):5440–9.

43. Wang Y, Debatin KM, Hug H. HIPK2 overexpression leads to stabilization of p53 protein and increased p53 transcriptional activity by decreasing Mdm2 protein levels. BMC Mol Biol. 2001;2:8–8.

44. Lukas J, Herzinger T, Hansen K, Moroni MC, Resnitzky D, Helin K, Reed SI, Bartek J. Cyclin E-induced S phase without activation of the pRb/E2F pathway. Genes Dev. 1997;11(11):1479–92.

45. Wang Y, Shen J, Arenzana N, Tirasophon W, Kaufman RJ, Prywes R. Activation of ATF6 and an ATF6 DNA binding site by the endoplasmic reticulum stress response. J Biol Chem. 2000;275(35):27013–20.

46. Kinzler KW, Vogelstein B. The GLI gene encodes a nuclear protein which binds specific sequences in the human genome. Mol Cell Biol. 1990;10(2):634–42.

47. Huang LE, Arany Z, Livingston DM, Bunn HF. Activation of hypoxia-inducible transcription factor depends primarily upon redox-sensitive stabilization of its α subunit. J Biol Chem. 1996;271(50):32253–9.

48. Piek E, Westermark U, Kastemar M, Heldin CH, van Zoelen EJ, Nistér M, Ten Dijke P. Expression of transforming-growth-factor (TGF)-β receptors and Smad proteins in glioblastoma cell lines with distinct responses to TGF-β1. Int J Cancer. 1999;80(5):756–63.

49. Molenaar M, van de Wetering M, Oosterwegel M, Peterson-Maduro J, Godsave S, Korinek V, Roose J, Destrée O, Clevers H. XTcf-3 Transcription factor mediates β-catenin-induced axis formation in xenopus embryos. Cell. 1996;86(3):391–9.

50. Yang XO, Pappu BP, Nurieva R, Akimzhanov A, Kang HS, Chung Y, Ma L, Shah B, Panopoulos AD, Schluns KS. T Helper 17 lineage differentiation is programmed by orphan nuclear receptors RORα and RORγ. Immunity. 2008; 28(1):29–39.

# 9

# Effect of mutations to amino acid A301 and F361 in thermostability and catalytic activity of the β-galactosidase from *Bacillus subtilis* VTCC-DVN-12-01

Thao Thi Nguyen, Hanh Van Vu, Nhung Thi Hong Nguyen, Tuyen Thi Do and Thanh Sy Le Nguyen[*]

## Abstract

**Background:** Beta-galactosidase (EC 3.2.1.23), a commercially important enzyme, catalyses the hydrolysis of β-1,3- and β-1,4-galactosyl bonds of polymer or oligosaccharidesas well as transglycosylation of β-galactopyranosides. Due to catalytic properties; β-galactosidase might be useful in the milk industry to hydrolyze lactose and produce prebiotic GOS. The purpose of this study is to characterize β-galactosidase mutants from *B. subtilis*.

**Results:** Using error prone rolling circle amplification (epRCA) to characterize some random mutants of the β-galactosidase (LacA) from *B. subtilis*VTCC-DVN-12-01, amino acid A301 and F361 has been demonstrated significantly effect on hydrolysis activity of LacA. Mutants A301V and F361Y had markedly reduced hydrolysis activity to 23.69 and 43.22 %, respectively. Mutants the site-saturation of A301 reduced catalysis efficiency of LacA to 20–50 %, while the substitution of F361 by difference amino acids (except tyrosine) lost all of enzymatic activity, indicating that A301 and F361 are important for the catalytic function. Interestingly, the mutant F361Y exhibited enhanced significantly thermostability of enzyme at 45–50 ℃. At 45 ℃, LacA-361Y retained over 93 % of its original activity for 48 h of incubation, whereas LacA-WT and LacA-301Vwere lost completely after 12 and 24 h of incubation, respectively. The half-life times of LacA-361Y and LacA-301 V were about 26.8 and 2.4 times higher, respectively, in comparison to the half-life time of LacA-WT. At temperature optimum 50 ℃, LacA-361Y shows more stable than LacA-WT and LacA-301 V, retaining 79.88 % of its original activities after 2 h of incubation, while the LacA-WT and LacA-301 V lost all essential activities. The half-life time of LacA-361Y was higher 12.7 and 9.39 times than that of LacA-WT and LacA-301 V, respectively. LacA-WT and mutant enzymes were stability at pH 5–9, retained over 90 % activity for 72 h of incubation at 30 ℃. However, LacA-WT showed a little bit more stability than LacA-301 V and LacA-361Y at pH 4.

**Conclusions:** Our findings demonstrated that the amino acids A301V and F361 play important role in hydrolysis activity of β -galactosidase from *B. subtilis*. Specially, amino acid F361 had noteworthy effect on both catalytic and thermostability of LacA enzyme, suggesting that F361 is responsible for functional requirement of the GH42 family.

**Keywords:** β-Galactosidase, Mutants, *Bacillus subtilis*, *Escherichia coli*, Error prone rolling circle amplification, Catalytic activity

* Correspondence: nslthanh@ibt.ac.vn
Institute of Biotechnology, Vietnam Academy of Science and Technology, 18 Hoang Quoc Viet Road, Distr. Caugiay, 10600 Hanoi, Vietnam

## Background

β-Galactosidase (β-D-galactoside galactohydrolase, E.C 3.2.1.23), a commercially important enzyme, catalyses the hydrolysis of β-D-galactoside linkage in polymers, oligosaccharides, or other secondary metabolic products [1]. Some β-galactosidases may have an activity of transferring one or more D-galactosyl units onto lactose [2]. Due to this property, β-galactosidase have two main applications containing the removal of lactose from milk products for lactose intolerant people and production of galactosylated products [3–5]. β-galactosidase is one of the most popular technologies to produce lactose reduced milk and related dairy products for consumption by lactose intolerant people. β-Galactosidase is widely used to improve sweetness, solubility, flavor and digestibility of dairy products [6, 7]. Besides, β-galactosidase shows a high transgalactosylation activity, so that they are used for the synthesis of prebiotic galacto-oligosaccharides [8], novel galactosides [9]. The β-galactosidase activity also contributes to glycoprotein degradation [10], the degradation of GM1 ganglioside and other glycolipids and glycoproteins with a terminal galactose moiety [11].

β-Galactosidases are widely distributed in nature and produced by microorganisms (yeasts, fungi, bacteria, and archaea), plants [12, 13], and animals [14, 15]. At present, based on their sequence similarity and reaction mechanisms, β-galactosidase are classified into four main glycoside hydrolase families (GHFs), GHF-1, GHF-2, GHF-35 and GHF-42 [16]. The catalytic residues of these enzyme group, which are located at the β-4 and β-7 of the triose phosphate isomerase (TIM) barrel fold, are a member of the 4/7-superfamily with a TIM container fold catalytic domain. In general, β-galactosidases of GHF-1 and GHF-2 are found in mesophiles and demonstrate lactase activity, while enzymes belonging to families GHF-35 and GHF-42 are usually found in thermophiles and preferentially degrade β-1,4-linkages between two galactose moieties. However, lactose hydrolysis activity of GHF-35 and GHF-42 are absent or weak [17, 18].

Nowadays, to improve the efficiency of using thermophilic β-galactosidases, different strategies have been used, including screening enzymes from different species of *Bacillus*, cloning and expressing in a heterologous host system [19–22], reconstruction of the enzyme by protein engineering to create enzymes with novel properties. Among them, protein engineering was an efficient way. Recently, efforts have been made to alter substrate specificity, stability and specific activity of β-galactosidases belong to GHF-42 [23–26].

In a previous study, we cloned a gene (*lacA*) coding for β-galactosidase of GHF-42 from *Bacillus subtilis* strain VTCC-DVN-12-01, expressed in *E. coli* and the β-galactosidase LacA was purified and characterized [20]. This β-galactosidase has a potential application in food industry. There was no report on random mutagenesis of the β-galactosidase using epRCA. For the first time, we used epRCA to characterize β-galactosidase mutants from *B. subtilis*.

## Results

### Library construction and prescreening for β-galactosidase activity of transformants

Plasmid pELacA was amplified by the rolling circle mechanism in the presence of manganese ions, which has been shown to reduce the fidelity of DNA polymerase and cause random mutagenesis during RCA. The results of electrophoresis on agarose gel indicated that the most of epRCA products had a size range of more than 10,000 bp, which was multimeric forms of two or more repeated sequences of pELacA (7500 bp, monomeric form). The epRCA products were directly transformed into *E. coli* JM109(DE3), resulting in colonies containing a randomly mutated plasmid library.

However, a few transformants (10–15 transformants) per 18 μg of epRCA resulted when the epRCA products directly transformed into *E. coli* JM109(DE3). Interestingly, the epRCA products was digested with a single-cut restriction enzyme *Mlu*I followed by self-ligation by treatment with T4 DNA ligase which dramatically increased the transformation efficiency. As a result, approximately 700 transformants were obtained from 72 ng of epRCA supplemented with 1.5 mM of manganese chloride, the transformation efficiency was 116 folds higher than that of direct transformation of the epRCA products without digestion and ligation.

Transformants containing putative mutants in the *lacA* gene were grown in LB medium in deep well microplates for the β-galactosidase production. The whole lysates were used as enzyme sources to determine the LacA activity in 96-well microplates using oNPG as a substrate. In total 10,000 transformants were screened hydrolysis activity of mutant library, the results have been shown that most transformants (75 %) were no significant change in hydrolysis activity in comparison with the wild-type. About 20 % of transformants were a complete loss of the activity and about 5 % of transformants strongly decreased in the hydrolysis activity. There were only 0.03 % of transformants with higher hydrolytic activity than wild-type.

### Mutation analysis

Among total 10,000 transformants obtained from the libraries, the transformants epRCA125, epRCA221, and epRCA887 (higher 1.5–2 times LacA activity than the wild-type by prescreening), epRCA259 and epRCA461 (≤50 % LacA activity than the wild-type) were randomly selected for DNA sequence analysis of the *lacA* gene.

The transformants epRCA259 and epRCA461 showed changes in amino acids, but epRCA125, epRCA221 and epRCA887 did not (Table 1). The mutation F361Y and A524T in LacA259 and mutation E62V, R77W, A191V, and A301V in LacA461 led to decrease in enzyme activity by nearly 50 % and 80 %, respectively.

However, to give a clearer explanation of effecting of the mutations of E62V, R77W, A191V, A301V, F361Y and A524T on decrease of enzyme activity, we made point mutations to confirm. The wild-type WT and mutants F361Y, A524T, E62V, R77W, A191V, A301V were cultivated to express the recombinant β-galactosidase in *E. coli* JM109(DE3). SDS polyacrylamide gel electrophoresis analysis showed the expression level of wild-type was similar to mutant enzymes (Fig. 1a). The expression level obtained 18,86–19,27 % of total protein of cells, that was estimated by using the Dolphin-1D software. This indicated that the mutations did not affect the expression of LacA variants in comparison to the wild-type.

The LacA-WT and mutants LacA-E62V, LacA-R77W, LacA-A191V, LacA-A301V, LacA-F361Y and LacA-A524T were purified to homogeneity using nickel chelate affinity chromatography and showed a single protein band corresponding to the molecular mass of 70 kDa of the wild type LacA (Fig. 1b). The purification factor and yield of variants were similar to that of LacA-WT. The purification factor and yield obtained 4.08–4.98 and 71–84.1 %, respectively (Table 2). The specific activities of mutants were determined with oNPG substrate and compared with that of the wild-type enzyme (Table 3). The specific activity of mutants LacA-A524T, LacA-E62V, LacA-R77W and LacA-A191V did not show any significant changes, whereas LacA-A301V and LacA-F361Y remained the relative specific activity 23.69 and 43.22 %, respectively, in comparison with LacA-WT. These results indicated mutations F361Y of epRCA259 and A301V of epRCA461 decreased in enzyme activity. This suggested that A301 and F361 could play a significant role maintaining activity of LacA.

### Site-saturation mutagenesis

Saturation mutagenesis at position 301 and 361 was carried out to give a deeper understanding of the role of A301 and F361 in hydrolysis activity of LacA. About 1,000–1,200

colonies of each mutant library were tested their hydrolysis activity in 96-well plates with oNPG substrate. The result of screening activity of colonies in mutants library of F361X showed that 94.4 % of colonies in this library completely lost hydrolysis activity, 5.3 % of colonies had hydrolysis activity similar to wild-type, and three colonies was a decrease in activity from wild-type of between 30 and 50 %. Interestingly, the analysis result of *lacA* gene from all three colonies decreased activity showed Phe361 was replaced with tyrosine. Whereas, sequencing of three *lacA* genes been randomly selected from colonies had activity similar to wild-type showed no change amino acid at this position. In addition, three *lacA* genes from colonies without hydrolysis activity were sequenced, and we found that Phe361 was replaced with valine, cysteine and leucine. Again, these results demonstrate role of Phe361 of LacA was important in substrate recognition. Any changes in amino acid residues of this position may disturb the hydrogen bond network to induce a loss or decrease in hydrolysis activity of LacA.

In contrast to F361 position, the result of screening activity of colonies in library of A301X indicated that 83.2 % selected colonies were decreased out of 1,000 (compare to the wild-type colonies the decreased activities ranged 20–40 %), 6.4 % of colonies had hydrolysis activity similar to wild-type, whereas 10.4 % had no activity. We determined the changes A301E and A301Y of LacA from some colonies decreased activity. This result demonstrated that any change in position A301 did not affect to hydrolysis activity of LacA as strong as that of F361 position. Amino acid A301 might not bind to substrate in hydrolysis but affect to structure of enzyme.

### pH and temperature dependency of mutant enzymes

The specific activity of LacA-361Y and LacA-301 V decreased dramatically to 43.09 and 23.7 % in comparison with the specific activity of the LacA-WT, respectively (Table 5). Both mutants remained the relative specific activity 44.8 % and 22.6 % in comparison with LacA-WT, when they were expressed using the initial vector pET22b + .

The optimum temperature and pH of LacA-WT, LacA-361Y and LacA-301 V were obtained at the same 50–55 °C and pH 6.5 (Fig. 2). However, mutants of

**Table 1** Mutations of selected epRCA transformants

| Transformant | RA (%) | Nucleotide substitution | Amino acid substitution |
|---|---|---|---|
| Wild-type | 100 | original | |
| LacA125 | 167 | No | |
| LacA221 | 150 | No | |
| LacA887 | 198 | ACG → ACA, CGG → CGA | T408, R624 |
| LacA259 | 52 | UUC → UAU, GCU → ACU | F361Y, A524T |
| LacA461 | 36 | GAG → GAA, CGG → UGG, GCG → GUG, GCG → GUG | E60, R77W, A191V, A301V |

Plasmids from selected colonies were isolated and sequenced to identify mutations in the *lacA* gene

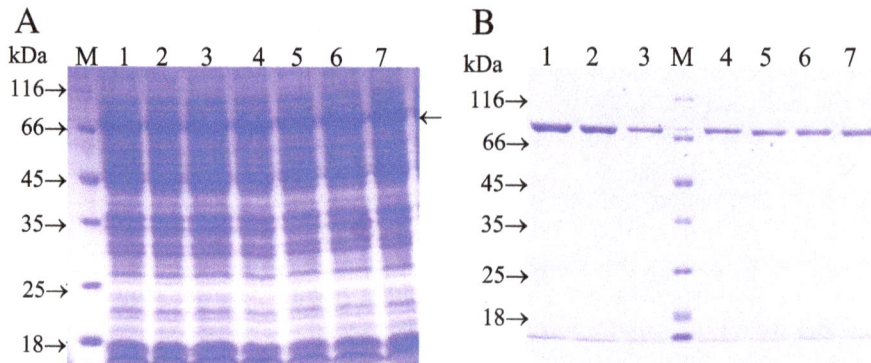

**Fig. 1** SDS-PAGE analysis of the level expression (A) and puried mutant proteins (B) of wild-type and mutants LacA. **a**. LacA-WT and mutants were expressed to homogeneity. Then total protein of cells were checked on SDS-PAGE to analysis the expression of LacA-WT and mutants. **b**. LacA-WT and mutants were puried to homogeneity using nickel chelate affinity chromatography. Homogeneous fractions were collected to analysis on SDS-PAGE. The protein samples were run on a 12.5 % SDS reducing gel and stained with Coomassie Briliant Blue R250. Lane 1–7: WT, E62V, R77W, A191V, A301V, F361Y and A524T, respectively; lane M: molecular marker

A301V and F361Y decreased the specific activity from $2.57 \pm 0.05$ U/mg (100 %) of the LacA-WT to $0.66 \pm 0.01$ U/mg (26 %) and $1.05 \pm 0.03$ U/mg (41 %), respectively, at 55 °C (Fig. 2a). At the optimum pH, the specific activity also decreased from $2.42 \pm 0.13$ (100 %) for LacA-WT to $0.54 \pm 0.001$ U/mg (22 %) and $1.1 \pm 0.008$ (45 %), respectively (Fig. 2b). Thus, mutant points A301V and F361Y were not effects on optimum temperature and pH of LacA.

The pH stability of LacA-WT, LacA-301 V and LacA-361Y remained most the same at pH 5.0 to pH 9.0. The hydrolytic activity of LacA-WT and mutant enzymes retained over 90 % at pH 5.0 to 9.0 after 72 h of incubation. However, LacA-WT showed a little more stability than LacA-361Y and LacA-301 V at pH 4 (Fig. 3). The

hydrolytic activity of Lac-WT retained 24 % for 72 h of incubation at 30 °C, whereas LacA-301 V and LacA-361Y did not have activity after 42 and 60 h, respectively.

In buffer, LacA-WT, LacA-301 V and LacA-361Y were stable in the temperature range of 30–40 °C with almost unchangeable activity after incubation for 72 h. The activity retained 95–100 % of their original activity. Interestingly, at 45–50 °C, LacA-361Y has been shown to be significantly more stable than LacA-WT and LacA-301 V (Fig. 4). LacA-361Y retained over 93 % of its original activity at 45 °C after 48 h of incubation, whereas LacA-WT and LacA-301 V lost completely after 12 and 24 h, respectively, of incubation (Fig. 4a). The half-life of LacA-361Y ($83.89 \pm 0.78$ h) and LacA-301 V ($7.6 \pm 0.19$ h) at 45 °C were about 26.8 and 2.4 times higher, respectively,

**Table 2** Purification steps of LacA-WT and mutants from JM109(DE3)

| Mutant LacA | Steps of purification | Total activity (U) | Total protein (mg) | Specific activity (U/mg) | Purification factor | Yield (%) |
|---|---|---|---|---|---|---|
| WT | Crude β-galactosidase | 5.13 | 9.04 | 0.57 | 1.0 | 100 |
| | Probond™resin | 4.12 | 1.66 | 2.48 | 4.53 | 80.49 |
| E62V | Crude β-galactosidase | 4.53 | 8.18 | 0.55 | 1.0 | 100 |
| | Probond™resin | 3.81 | 1.59 | 2.39 | 4.35 | 84.1 |
| R77W | Crude β-galactosidase | 5.02 | 9.25 | 0.54 | 1.0 | 100 |
| | Probond™resin | 3.56 | 1.39 | 2.56 | 4.74 | 71 |
| A191V | Crude β-galactosidase | 4.32 | 8.72 | 0.50 | 1.0 | 100 |
| | Probond™resin | 3.54 | 1.42 | 2.49 | 4.98 | 81.94 |
| A301V | Crude β-galactosidase | 1.15 | 8.66 | 0.13 | 1.0 | 100 |
| | Probond™resin | 0.89 | 1.51 | 0.59 | 4.54 | 77.1 |
| F361Y | Crude β-galactosidase | 2.06 | 9.11 | 0.24 | 1.0 | 100 |
| | Probond™resin | 1.72 | 1.55 | 1.11 | 4.63 | 83.5 |
| A524T | Crude β-galactosidase | 5.11 | 8.25 | 0.62 | 1.0 | 100 |
| | Probond™resin | 4.1 | 1.62 | 2.53 | 4.08 | 80.23 |

Wild-type and mutants of LacA were purified to homogeneityby affinity chromatography Ni$^{2+}$ ProBond™ resin. The relative activity of β-galactosidase was determined in 100 mM buffer Z (pH 7.0) at 55 °C with substrate oNPG (4 mg/ml)

**Table 3** Specific activities of the wild-type and mutant LacA enzymes using oNPG as substrate

| Transformant | Mutant LacA | Specific activity (U/mg) | Specific activity relative to wild-type (%) |
|---|---|---|---|
| WT | | 2.48 ± 0.06 | 100 |
| 259 | F361Y | 1.11 ± 0.075 | 44.42 ± 3.01 |
| | A524T | 2.53 ± 0.078 | 102.19 ± 3.16 |
| 461 | E62V | 2.39 ± 0.045 | 96.31 ± 1.8 |
| | R77W | 2.56 ± 0.068 | 103.39 ± 2.74 |
| | A191V | 2.49 ± 0.073 | 100.6 ± 2.95 |
| | A301V | 0.59 ± 0.014 | 23.69 ± 0.56 |

The relative activity of β-galactosidase was determined in 100 mM buffer Z (pH 7.0) at 55°C with substrate oNPG (4 mg/ml)

in comparison to the half-life of LacA-WT (Table 4). At temperature optimum 50 °C, LacA-361Y has also been shown more stable than LacA-WT and LacA-301 V. LacA-WT and LacA-301 V retained only 3.72 and 7.95 %, respectively, after 2 h of incubation, whereas LacA-361Y retained 79.88 % of its original activity (Fig. 4b). At 50 °C, LacA-WT and LacA-301 V hydrolysis activity were lost completely after 6 h of incubation, whereas LacA-361Y

activity was lost completely after 24 h. The half-life of LacA-361Y was higher 12.7 and 9.39 times than its of LacA-WT and LacA-301 V, respectively (Table 4). At 55–60 °C, all three enzymes have been shown to result in a complete loss of hydrolysis activity after 2 h of incubation.

In the presence of substrate oNPG, LacA-WT was a higher thermostability than that in buffer, and was the

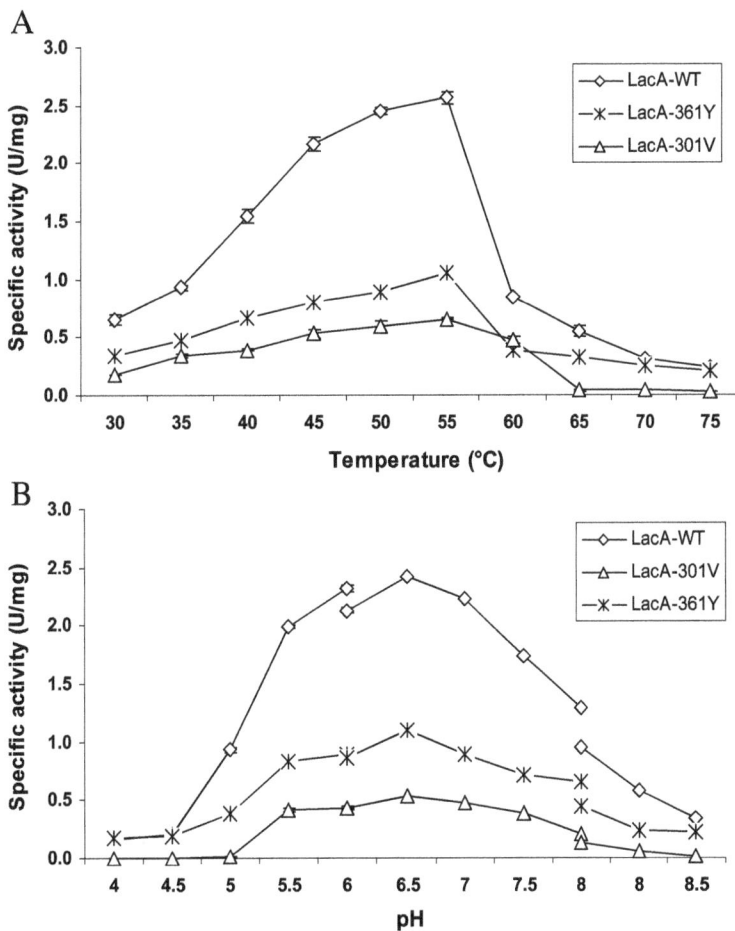

**Fig. 2** Temperature (**a**) and pH optimum (**b**) of Lac-WT, Lac-301 V and LacA-361Y. β-Galactosidase activity toward oNPG was determined in 100 mM buffer Z (pH 7.0). The pH was determined in different buffers by varying pH values from 4.0 to 9.0 at 55 °C using substrate oNPG (4 mg/ml). Na-acetate buffer pH 4–6, Na-phosphate buffer pH 6.0–8.0, Tris–HCl buffer pH 8.0–9.0

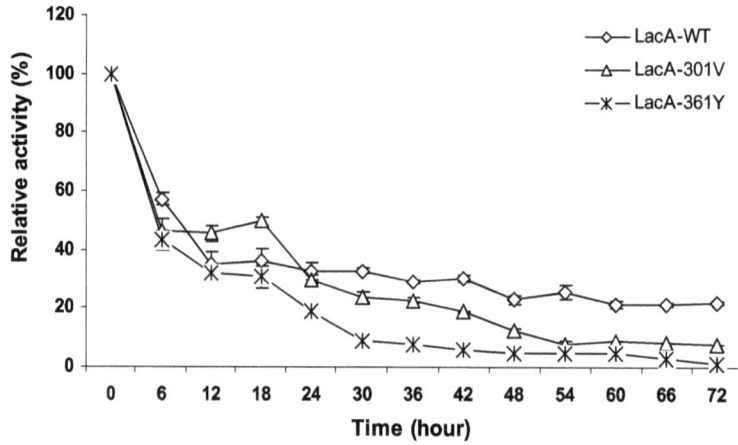

**Fig. 3** pH stability of Lac-WT, Lac-301 V and LacA-361Y. The purified LacA-WT, Lac-301 V and LacA-361Y were incubated in 100 mM Na-phosphate buffer pH 4. The relative activity of β-galactosidase toward oNPG (4 mg/ml) was determined in 100 mM buffer Z (pH 7.0) at 55 °C

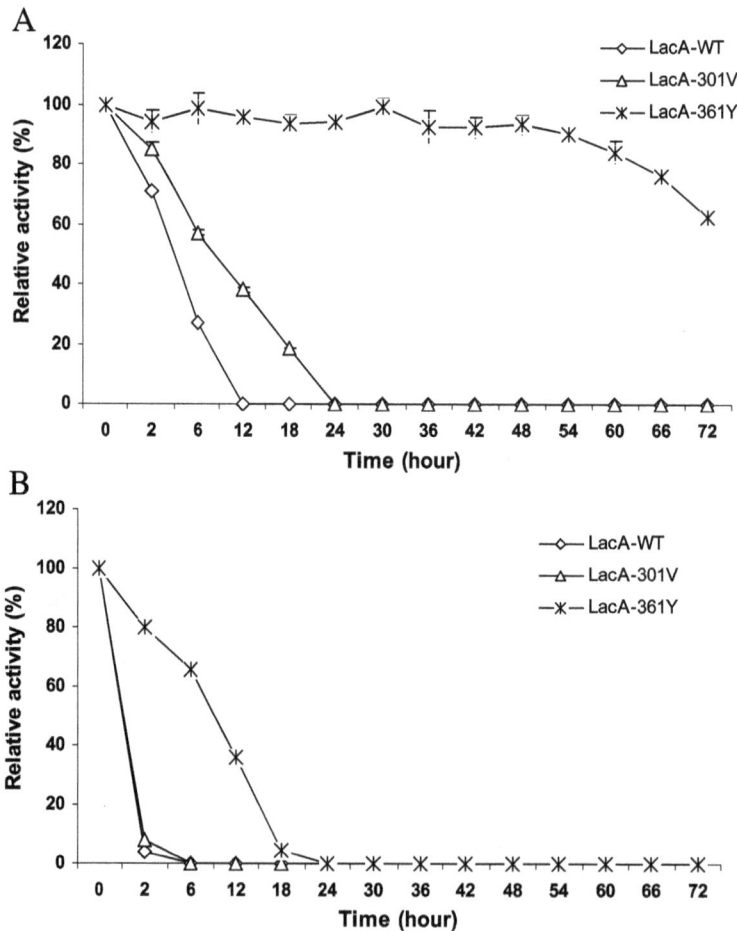

**Fig. 4** Thermostability of LacA-WT, Lac-301 V and LacA-361Y at 45 °C (**a**), and at 50 °C (**b**). The purified LacA-WT, Lac-301 V and LacA-361Y were incubated in 100 mM Na-phosphate buffer pH 7.0, and the relative activity of β-galactosidase toward oNPG (4 mg/ml) was determined in 100 mM buffer Z (pH 7.0) at 55 °C

**Table 4** The half-life of wild-type and mutants LacA at various initial oNPG concentrations (w/v) and temperatures

| Enzyme | Half-life (h) at various initial oNPG concentrations (w/v) and temperatures | | | | | | | |
|---|---|---|---|---|---|---|---|---|
| | 45 °C | | | | 50 °C | | | |
| | 0 % | 5 % | 10 % | 15 % | 0 % | 5 % | 10 % | 15 % |
| WT | 3.13 ± 0.14 | 29.36 ± 0.11 | 16.11 ± 0.15 | 11.81 ± 0.09 | 0.42 ± 0.03 | 3.75 ± 0.04 | 2.98 ± 0.06 | 1.79 ± 0.03 |
| A301V | 7.6 ± 0.19 | 6.52 ± 0.09 | 7.15 ± 0.12 | 6.79 ± 0.21 | 0.57 ± 0.03 | 0.81 ± 0.03 | 0.73 ± 0.008 | 0.49 ± 0.01 |
| F361Y | 83.89 ± 0.78 | 99 ± 0.33 | 90 ± 0.24 | 92.4 ± 0.52 | 5.4 ± 0.15 | 6.16 ± 0.23 | 7.1 ± 0.14 | 6.21 ± 0.08 |

The activity of β-galactosidase was determined in 100 mM buffer Z (pH 7.0) at 55 °C with substrate oNPG (4 mg/ml)

most stable in 5 % (w/v) of oNPG solution. At 45 °C and 50 °C, the half-life of LacA-WT in presenting of 5–15 % (w/v) oNPG substrate were more stable about 3.8–9.4 times and about 4.3–8.9 times, respectively, in comparison with enzyme incubating in the buffer (Table 4). Whereas, the thermostability of LacA-301 V and LacA-361Y in the presence of substrate were not much different compared to that of enzymes incubating in the buffer.

### Characterization of mutants

The enzyme kinetics toward the oNPG substrate were compared in Table 5. The mutant enzymes LacA-301 V and LacA-361Y had a decrease in Michaelis constant ($K_m$), the turnover rate ($K_{cat}$) and catalytic efficiency ($K_m/K_{cat}$). $K_m$ for LacA-301 V (5.57 mM) and LacA-361Y (8.02 mM) was lower than that in LacA-WT (9.28 mM) showed that the mutation of A301 to valine and F361 to tyrosine induced a decrease in the substrate affinity. The catalytic efficiency of LacA-301 V and LacA-361Y were 28.65 and 45.5 %, respectively, compared to wild-type enzyme. This result showed that A301 and F361 of LacA could be interacted with the substrate.

## Discussion

### Screening and identification of random mutagenesis

The random mutagenesis method epRCA is a rapid and simple method and epRCA products are directly transformed into E. coli JM109(DE3). Although, the transformation efficiency was significantly higher (116 times) when transfer the epRCA products digested with MluI and ligated into E. coli JM109(DE3). This result was in agreement with the report that the digestion of RCA products by a single-cut restriction enzyme and self-

ligation by T4 DNA ligase dramatically increased the E. coli XL1-blue transformation efficiency (approximately 5000 transformants per 50 ng of RCA products), whereas the direct transformation of these RCA products only resulted in a few transformants [27]. One possible explanation for this is smaller size of epRCA products after cutting by restriction enzyme MluI. Thus, the transformation efficiency of monomeric epRCA products is higher than the multimeric forms.

The result of site-mutagenesis demonstrated that A301 and F361 play an important role maintaining activity of LacA. Using SWISS-MODEL program (http://swissmodel .expasy.org/interactive), UniProtKB/Swiss-Prot to predict model of LacA showed the active site containing E159 and E323, and the substrate binding site containing R120, N158, W331, E371, K372, L373, and H374. Therefore, the amino acids of E62V, R77W, A191V, A301V, F361 and A524 were not the active site or substrate binding site of LacA (Fig. 5). However, when it using alignment of the putative amino acid sequences of β-galactosidase belonging to GH42 revealed the F361 of LacA corresponds to F347, an active site of a cold active β-galactosidase (bgaA, AF242542-1) from *Planococcus* sp. classified as a GH42. Surprisingly, the changing phenylalanine to tyrosine at position 347 of bgaA also caused a decrease in activity from wild-type of between 40 and 50 % on oNPG substrate [26]. This suggested that F361 might interact with the substrate in hydrolysis of LacA.

### Characterization of mutants

The optimum temperature and pH of LacA-WT, LacA-361Y and LacA-301 V were obtained at the same 50–55 °C and pH 6.5 (Fig. 2). These results were coincident with the optimum temperature and pH of the β-galactosidase from *B. subtilis* KL88 [27], *Arthrobacter* sp. 32c [28].

**Table 5** Kinetic parameters of the purified wild-type and mutant enzymes toward oNPG as substrate

| Enzyme | $V_{max}$ (IU/mg) | $K_m$ (mM) | $K_{cat}$ (s$^{-1}$) | $K_{cat}/K_m$ s$^{-1}$mM$^{-1}$ | $K_{cat}/K_m$ (%) |
|---|---|---|---|---|---|
| Wild-type LacA | 2.61 ± 0.12 | 9.28 ± 0.72 | 21.74 ± 0.99 | 2.35 ± 0.08 | 100 |
| LacA-301 V | 0.76 ± 0.02 | 5.57 ± 0.56 | 3.72 ± 0.09 | 0.67 ± 0.08 | 28.65 |
| LacA-361Y | 1.03 ± 0.03 | 8.02 ± 0.27 | 8.56 ± 0.24 | 1.07 ± 0.01 | 45.50 |

$K_m$ Michaelis constant, $K_{cat}$ turnover rate, $K_{cat}/K_m$ catalytic efficiency
The relative activity of β-galactosidase was determined in 100 mM buffer Z (pH 7.0) at 55 °C with oNPG as a substrate

```
Bacterial source        Partial sequence                                                      Accession
                                                                                              no
                        117-123155-161299-304328-334358-364369-376
B. subtilisG1           HGGRGDC--HISNEYG--ADLAMKV--AVNWHNV--VLYFQYR--SSEKLHGA                  EU585783
B. subtilis Bsn         HGGRGDC--HISNEYG--ADLAMKV--AVNWHNV--VLYFQYR--SSEKLHGA                  ADV94224-1
B. subtilis             HGGRGDC--HISNEYG--ADLAMKV--AVNWHNV--VLYFQYR--SSEKLHGA                  O07012
G. stearothemophilus    HGGRGEC--HVSNEYG--ADLAMKV--LVNWHKV--ILYFQYR--SFEKFHGA                  AEH26455-1
Planococcus sp.         FGSRHNS--HVSNEYG----------QQNWQPY--ILYFQLR--ACEKYHGA                   AF242542-1
```

**Fig. 5** Partial sequence alignment of amino acid sequences of LacA and β-galactosidase belonging to GHF-42 from different bacterial sources. Seven positions R120, N158, W331, E371, K372, L373 and H374 predicted in interaction with substrate were shown by underline. The black bars showed two mutagenesis positions 301 and 361

LacA-WT, LacA-301 V and LacA-361Y were the pH stability the same at pH 5.0 to pH 9.0. Other β-galactosidases from *B. coagulans* RCS3 and *B. megaterium* 2-37-4-1 were also reported to be stable at a neutral pH range: pH 6–9 [29, 30]. However, the interesting difference in the thermostability of mutant LacA-361Y. In the buffer, at 45–50 °C, LacA-361Y shown to be significantly more stable than LacA-WT and LacA-301 V. These results might explain that the hydroxyl group of the tyrosine interacted with the carboxyl group of a certain residue, therefore, it is intended in order to to increase the structure stability. In a recent report, Dong et al., [23] also found Ile42 of BgaB belong to GH42 from *Geobacillus stearothermophilus* affected to both catalysis and thermostability simultaneously. The replacement Ile42 with polar AA enhanced the thermostability but decreased the catalytic efficiency of BgaB [23].

In the presence of substrate oNPG, LacA-WT was a higher thermostability than that in buffer, whereas LacA-301 V and LacA-361Y were not. The higher thermostability of enzyme in the presence of substrate, which might be caused complexion to the substrate or with a remaining galactose, was expected by Warmerdam et al. [31]. This may explain why the thermostability of mutant LacA-301 V and LacA-361Y did not change significantly in presence of substrate because $K_m$ of mutant enzymes were lower than LacA-WT, and this result may decrease interaction of mutants LacA with substrate.

## Conclusions

Error prone rolling cycles amplification (epRCA) has been used in this study as a powerful tool to modify properties of the β-galactosidase from *B. subtilis* VTCC-DVN-12-01. These findings demonstrated the amino acids A301V and F361 play important role in hydrolysis activity of β-galactosidase from *B. subtilis*. Especially, amino acid F361 had significant effect on both catalytic and thermostability of LacA enzyme suggesting that F361 is responsible for functional requirement of the GH42 family. The finding could be applied to modify the other families of GH-42 β-galactosidase for the evolution properties of enzyme.

## Methods

### Chemicals and reagents

TempliPhi100DNA amplification kit was purchased from Roche (Basel, Switzerland). Ortho-nitrophenyl-β-D-galactopyranoside (oNPG), isopropyl thio-β-D-galactoside (IPTG), peptone, and yeast extract were provided from Bio Basic Inc. (Ontario, Canada). ProBon™nickel-chelating resin was supplied by Invitrogen Corp. (Carlsbad, CA, USA). The PCR reagents, restriction endonucleases, T4 DNA ligase and *Taq* polymerase, PCR primers(IDT, USA) were purchased from Fermentas (Thermo Fisher Scientific Inc., Waltham, USA).

### Bacterial strain and expression plasmid

The gene *lacA* (2061 bp, accession No EU585783) coding for β-galactosidase from *Bacillus subtilis* strain VTCC-DVN-12-01inserted into the expression vector pET22b(+) resulting in pELacA was described in a previous study [20]. The *Escherichia coli* strain JM109(DE3) [*end*A1, *rec*A1, *gyr*A96, *thi*, *hsd*R17 ($r_k^-$, $m_k^+$), *rel*A1, *sup*E44, λ-, Δ(*lac-pro*AB), [F′, *tra*D36, *pro*AB, *lac*I$^q$ZΔM15], IDE3] (Promega Corp, Madison, WI) and the plasmid pELacA were used for the expression of the wild-type and mutant β-galactosidase LacA and for screening of LacA mutants. *E. coli* cells were cultivated in Luria-Bertani (LB) medium containing1% (w/v) bacto tryptone, 0.5 % (w/v) yeast extract, 1 % (w/v) NaCl, pH 7–7.5 and 50 μg/ml of ampicillin. LB agar contained additionally 2 % (w/v) agar and 100 μg ampicillin/ml.

### Error-prone rolling circle amplification

The recombinant plasmid pELacA was used as a template for the epRCA reaction by using the TempliPhi 100 DNA amplification kit consisting of a sample buffer containing random hexamers that prime DNA synthesis nonspecifically, an enzyme mix containing Φ29 DNA polymerase and a reaction buffer containing deoxyribonucleotides. An amount of 25 pg of pELacA was mixed with 5 μl of sample buffer and heated at 95 °C for 3 min to denature the plasmid, then immediately cooled down to room temperature. The amplification was started by adding 5 μl of reaction buffer, 0.2 μl of enzyme mix and

MnCl$_2$ at the final concentration of 1.5 mM and incubated to 30 °C for 24 h. The mixture was heated at 65 °C for 10 min to inactivate the enzyme. The quantity of amplified DNA was estimated on 1 % (w/v) agarose gel electrophoresis and by measuring its absorbance at 260 nm with a spectrophotometer UV-2500 (LaboMed Inc., Culver City, CA, USA).

### Construction of mutant libraries
epRCA products were digested with *Mlu*I in a mixture contained 12 μl (600 ng) of ep-RCA product, 5 μl 10 × buffer R, 2 μl (20 U) *Mlu*I and 31 μl H$_2$O. After 6 h of incubation at 37 °C, the digested products were purified by using the MinElute Reaction Cleanup kit (Qiagen) and ligated with T4 ligase. The ligated epRCA transformed into *E. coli* JM109(DE3) by using standard electroporation method with a 0.1 cm electrode cuvette under the conditions at 1.8 kV, 200 Ω, and 25 F. Transformed cells were plated onto LB plates containing 50 μg/ml of ampicillin and incubated at 37 °C overnight. Colonies harboring putative mutation sites in *lacA* were prescreened for a higher β-galactosidase production.

### DNA manipulations
Plasmid DNA isolation was carried out by methods as previously described [20]. DNA sequencing was performed on ABI PRISM 3100 Avant Genetic Analyzer (Applied Biosystems Inc., Foster City, USA). Sequence alignments constructed and analyzed using the program MegAlign DNAStar. *E. coli* DH5α and JM109(DE3) cells were transformed using heat shock method that has been previously described [20].

### Screening β-galactosidase activity of mutants
To screen β-galactosidase activity of mutants, the individual transformants carrying the putative mutant *lacA* gene were randomly selected and grown in 300 μl LB medium containing 100 μg/ml ampicillin in 96-deep well plates at 37 °C overnight with agitation of 250 rpm. 25 μl of overnight culture was transferred from each well to second 96-deep well plates containing 300 μl LB medium containing ampicillin. The culture was cultivated at 37 °C with agitation of 300 rpm until an optical density (OD) at 600 nm of 0.6 to 0.8 was reached (for approximately 4 h), then 1 mM isopropyl-β-D-thiogalactopyranoside (IPTG) was added. The culture continuously incubated at 37 °C with agitation of 300 rpm for 16 h of induction. The cell cultures were used as the enzyme source to screen the activity.

The procedure for measuring β-galactosidase activity of colonies from site-saturation library was performed according to Griffith et al., [32] with oNPG substrate (4 mg/ml) in 0.1 M Na-phosphate buffer, pH 7. The absorbance of culture density at 600 nm, reaction mixture at 420 and 550 nm was measured in a microplate reader Elx800™ (BioTek Instruments Inc., Winooski, USA) and β-galactosidase activities were calculate in Miller units following equation: Miller Unit ($nM$/min/$OD_{cell}$) = [($OD_{420}$ − 1.75 × $OD_{550}$)] × $V_1$/($T$ × $V_2$ × $OD_{600}$). In that, $OD_{420}$ and $OD_{550}$ are read from the reaction mixture; $OD_{420}$−1.75 × $OD_{550}$, nmoles formed per milliliter; 1.75 × $OD_{550}$, light scattering at 420 nm; T, incubation time (min); $V_1$ (ml), total assay volume; $V_2$ (ml), volume of culture used in the assay; $OD_{600}$ reflects cell density in the washed cell suspension. All measurements were carried out in triplicate with the resulting values being the mean of the cumulative data obtained.

### Site-directed and saturation mutagenesis
Site-directed and saturation mutagenesis were performed by a one-step polymerase chain reaction (PCR) method, using plasmid pELacA as template and a pair of mutagenic primer. Randomization codon was performed with a pair of primer that introduced a codon NNK at selected positions. Whole pELacA plasmid was amplified in PCR mix containing 5 μl of 10× PCR buffer, 4 μl of 25 mM MgSO$_4$, 4 μl of 2.5 mM dNTP, 0.5 μl of 2.5 U/μl *Pfu* polymerase (Thermo), 1 μl of each primer (10 pmol), 1 μl of pELacA (50 ng), and 33.5 μl H$_2$O. The thermocycler conditions were performed as follow: 95 °C/4′; 18 cycles of 95 °C/30″, 54 °C/1′, 72 °C/8′; and 72 °C/10′. Then, 10 U of *Dpn*I restriction enzyme (Thermo) was added the reaction products and incubated for 2 h at 37 °C to digest pELacA template. The PCR products were purified using a PCR purification kit (Thermo). The resultant plasmid DNA was transformed into chemically competent JM109(DE3) cells. Transformants were selected on LB agar containing 100 μg/ml after incubation overnight at 37 °C.

### Enzyme expression
The transformants *E. coli* JM109(DE3)/pELacA harboring the wild type and mutant *lacA* gene were cultivated in 5 ml of LB medium containing 5 μl of 100 mg ampicillin/ml at 37 °C with agitation at 220 rpm. Five hundred μl of the overnight culture were transferred into 50 ml of LB medium containing 50 μl of 100 mg ampicillin/ml in a 250-ml Erlenmeyer flask. The culture was cultivated at 37 °C with agitation at 200 rpm until an optical density (OD) at 600 nm of 0.6–0.8 was reached (for approximately 4 h), then 50 μl of 100 mM IPTG was added. The culture was continuously incubated at 37 °C with agitation of 220 rpm for 6 h of induction. Cells were harvested by centrifugation at 5000 rpm for 10 min at 4 °C. Wet cells were used for enzyme purification.

## Enzyme purification

The recombinant LacA fused with a C-terminal 6 × histidine-tag was purified using affinity chromatography with Ni$^{2+}$-ProBond™ resin under native conditions. An amount of 500 mg wet cells from a 50-ml culture in LB medium was harvested by centrifugation at 4000 rpm and 4 °C for 10 min, washed with 8 ml of water and resuspended in 8 ml of 1× native purification buffer containing 50 mM NaH$_2$PO$_4$, 0.5 mM NaCl,100 mM imidazol, pH 8.0. To the mixture, lysozyme was added at a final concentration 0.5 mg/ml and incubated on ice bath for 30 min. The cell mixture was disintegrated by ultrasonic waves (3× 1 min with 1 min pause). The supernatant of the cell lysate was obtained by centrifugation at 13,000 rpm for 10 min and loaded on to a column containing 2 ml resin, which was equilibrated with native binding buffer and incubated for 45 min at room temperature with gentle hand shaking for several times. The column was washed with 3 times of 8 ml native wash buffer. The bound protein was eluted with 8 ml of 1× native purification buffer containing 250 mM imidazol. The enzyme solution was used for characterization.

## Electrophoresis analysis and protein concentration

The homogeneity and molecular mass of the β-galactosidase was determined by 12.5 % SDS polyacrylamide gel electrophoresis [33] with Biometra equipment (Göttingen, Germany). Proteins were visualized by staining with 0.1 % (w/v) Coomassie Brilliant Blue R-250. Protein concentrations were estimated by the method of Bradford with bovine serum albumin as standard [34].

## β-Galactosidase activity assay

To estimate the activity of the purified β-galactosidase, 1 μl (3.4 μg) purified enzyme solution was added to 74 μl 22 mM oNPG in 100 mM buffer Z (40 mM Na$_2$HPO$_4$.7H$_2$O, 60 mM NaH$_2$PO$_4$.H$_2$O, 10 mM KCl, 1 mM MgSO$_4$.7H$_2$O, 50 mM 2-Mercaptoethanol) pH 7, incubated at 50 °C for 10 min. Then the reaction was stopped by addition of 25 μl 1 M Na$_2$CO$_3$. The absorbance was read at 420 nm against a blank containing oNPG, buffer Z but without enzyme solution. The following equation was used to calculate units of β-galactosidase activity: $U/ml = [OD_{420} \times V_1]/[0.0045 \times 1000 \times T \times V_2]$ (ii). In the equation (ii), $OD_{420}/0.0045 \times 1000$: μmoles what formed per milliliter; T, incubation time (min); $V_1$ (ml), total assay volume; $V_2$ (ml), enzyme volume used in the assay.

## pH and temperature dependency of mutant enzymes

The temperature and pH optimum of the purified wild-type and mutant LacA, 3.4 μg enzyme for each reaction, were determined by measuring the activity as described above using 100 mM buffer Z (pH 7) at the temperature range of 30 to 70 °C and using different buffer pH 4.0–9.0 (100 mM potassium acetate buffer pH 4.0–6.0, 100 mM Na-phosphate buffer pH 6.0–8.0, or 100 mM Tris–HCl buffer pH 8.0–9.0) at 50 °C, respectively, for 5 min.

For the determination of temperature and pH stability, purified enzyme, 3.4 μg protein for each reaction, was incubated in 100 mM buffer Z at different temperatures from 30 to 60 °C for 0–72 h, and in 100 mM Na-phosphate buffer pH 4.0 and 9.0 at 30 °C for 0–72 h, respectively. The remaining activity was then determined. Half-life ($T_{1/2}$) of each mutant enzyme was determined as follows: $T_{1/2} = \ln2/k$, where $U_t$, $U_0$, and k are enzyme activity at t min, initial enzyme activity, and the apparent rate constant, respectively.

All data were averaged and experiments were performed three times. Specific activities were calculated from these averages. The entire assay experiments were than repeat two more times and three specific activity values were averaged. The errors (SD) were calculated by STDEV function of Excel.

## Characterization of mutants

The apparent kinetic parameters ($V_{max}$ and $K_m$) were determined against 0.1-6 mg/ml of oNPG as a substrate using Lineweaver-Burk plots. Activities were recorded at 55 °C and calculated on the basis of an extinction coefficient for o-nitrophenol of 4500 M$^{-1}$ cm$^{-1}$ at 420 nm.

## DNA and amino acid sequence analysis

Homologies of the DNA and amino acid sequences were determined with the program Megalign DNAStar.

### Abbreviations

Gene *lacA*, gene encoding β-galactosidase from *Bacillus subtilis* strain VTCC-DVN-12-01; Protein/Enzyme LacA, β-galactosidase from *Bacillus subtilis* strain VTCC-DVN-12-01

### Acknowledgements

This work was supported by the National Foundation for Science and Technology Development Vietnam (Nafosted), project 106.05-2010.04, 2011-2012.

### Funding

This work was supported by the National Foundation for Science and Technology Development Vietnam (Nafosted).

### Authors' contributions

TTN designed the experimental setup, assisted with data analysis and manuscript preparation. NTHN and DTT performed experiments of mutant libraries construction, expression, purification and characterization of the mutants. HVV initiated the project, read the final manuscript. NSLT read and approved the final manuscript. All authors read and approved the final manuscript.

### Competing interests

The authors declare that they have no competing interests.

**References**

1. Husain Q. β-Galactosidases and their potential applications: a review. Crit Rev Biotechnol. 2010;30(1):41–62.

2. Rabiu BA, Jay AJ, Gibson GR, Rastall RA. Synthesis and fermentation properties of novel galacto-oligosaccharides by beta-galactosidases from Bifidobacterium species. Appl Environ Microbiol. 2001;67(6):2526–30.

3. Haider T, Husain Q. β-Galactosidase by bioaffinity adsorption on immobilization of concanavalin A lay red calcium alginate-starch hybrid beads for the hyrolysis of lactose from whey/milk. Int Dairy J. 2009;19:172–7.

4. Heyman MB. Lactose intolerance in infants, children, and adolescents. Pediatrics. 2006;118(3):1279–86.

5. Neri DFM, Balcao VM, Carneiro-da-Cunha MG, Carvalho Jr LB, Teixeira JA. Immobilization of β-galactosidase from Kluyveromyces lactis onto a polysiloxane–polyvinyl alcohol magnetic (mPOS–PVA) composite for lactose hydrolysis. Catal Comm. 2008;4:2334–9.

6. Kim JW, Rajagopal SN. Isolation and characterization of β-galactosidase from Lactobacillus crispatus. Folia Microbiol (Praha). 2000;45:29–34.

7. Rhimi M, Boisson A, Dejob M, Boudebouze S, Maguin E, Haser R, Aghajari N. Efficient bioconversion of lactose in milk and whey: immobilization and biochemical characterization of a beta-galactosidase from the dairy Streptococcus thermophilus LMD9 strain. Res Microbiol. 2010;161(7):515–25.

8. Iqbal S, Nguyen TH, Nguyen TT, Maischberger T, Haltrich D. beta-Galactosidase from Lactobacillus plantarum WCFS1: biochemical characterization and formation of prebiotic galacto-oligosaccharides. Carbohydr Res. 2010;345(10):1408–16.

9. Irazoqui G, Giacomini C, Batista-Viera F, Brena BM, Cardelle-Cobas A, Corzo N, Jimeno ML. Characterization of galactosyl derivatives obtained by transgalactosylation of lactose and different polyols using immobilized beta-galactosidase from Aspergillus oryzae. J Agric Food Chem. 2009;57(23):11302–7.

10. Terra VS, Homer KA, Rao SG, Andrew PW, Yesilkaya H. Characterization of novel beta-galactosidase activity that contributes to glycoprotein degradation and virulence in Streptococcus pneumoniae. Infect Immun. 2010;78(1):348–57.

11. Samoylova TI, Martin DR, Morrison NE, Hwang M, Cochran AM, Samoylov AM, Baker HJ, Cox NR. Generation and characterization of recombinant feline beta-galactosidase for preclinical enzyme replacement therapy studies in GM1 gangliosidosis. Metab Brain Dis. 2008;23(2):161–73.

12. Biswas S, Kayastha AM, Seckler R. Purification and characterization of a thermostable beta-galactosidase from kidney beans (Phaseolus vulgaris L.) cv. PDR14. J Plant Physiol. 2003;160(4):327–37.

13. Kaneko S, Kobayashi H. Purification and characterization of extracellular beta-galactosidase secreted by supension cultured rice (Oryza sativa L.) cells. Biosci Biotechnol Biochem. 2003;67(3):627–30.

14. Ferreira AH, Terra WR, Ferreira C. Characterization of a beta-glycosidase highly active on disaccharides and of a beta-galactosidase from Tenebrio molitor midgut lumen. Insect Biochem Mol Biol. 2003;33(2):253–65.

15. Sopelsa AM, Severini MH, Da Silva CM, Tobo PR, Giugliani R, Coelho JC. Characterization of beta-galactosidase in leukocytes and fibroblasts of GM1 gangliosidosis heterozygotes compared to normal subjects. Clin Biochem. 2000;33(2):125–9.

16. Henrissat B, Davies G. Structural and sequence-based classification of glycoside hydrolases. Curr Opin Struct Biol. 1997;7:637–44.

17. Hinz SW, van den Boek LAM, Beldman G, Vincken J-P, Voragen AGJ. β-Galactosidase from Bifidobacterium adolescentis DSM20083 prefer β(1,4)-galactosides over lactose. Appl Microbiol Biotechnol. 2004;66:276–84.

18. Schwab C, Sorensen KI, Ganzle MG. Heterologous expression of glycoside hydrolase family 2 and 42 beta-galactosidases of lactic acid bacteria in Lactococcus lactis. Syst Appl Microbiol. 2010;33(6):300–7.

19. Ito Y, Sasaki T. Cloning and characterization of the gene encoding a novel beta-galactosidase from Bacillus circulans. Biosci Biotechnol Biochem. 1997; 61(8):1270–6.

20. Quyen DT, Nguyen TT, Nguyen SLT, Vu VH. Cloning, high-level expression and characterization of a β-galactosidase from Bacillus subtilis G1. Austr J Basic Appl Sci. 2011;7(5):193–9.

21. Trân LS, Szabó L, Fülöp L, Orosz L, Sík T, Holczinger A. Isolation of a beta-galactosidase-encoding gene from Bacillus licheniformis: purification and characterization of the recombinant enzyme expressed in Escherichia coli. Curr Microbiol. 1998;37(1):39–43.

22. Shaw GC, Kao HS, Chiou CY. Cloning, expression, and catabolite repression of a gene encoding beta-galactosidase of Bacillus megaterium ATCC 14581. J Bacteriol. 1998;180(17):4734–8.

23. Dong YN, Chen HQ, Sun YH, Zhang H, Chen W. A differentially conserved residue (Ile42) of GH42 β-galactosidase from Geobacillus stearothermophilus BgaB is involved in both catalysis and thermostability. J Dairy Sci. 2015;98(4):2268–76.

24. Dong YN, Liu XM, Chen HQ, Xia Y, Zhang HP, Zhang H, Chen W. Enhancement of the hydrolysis activity of β-galactosidase from Geobacillus stearothermophilus by saturation mutagenesis. J Dairy Sci. 2011;94(3):1176–84.

25. Placier G, Watzlawick H, Rabiller C, Mattes R. Evolved β-Galactosidase from Geobacillus stearothermophilus with improved transgalactosylation yield for galacto-oligosaccharide production. Appl Environ Microbiol. 2009;75(19):6312–21.

26. Shumway MV, Sheridan PP. Site-directed mutagenesis of a family 42 beta-galactosidase from an antarctic bacterium. Int J Biochem Mol Biol. 2012;3(2):209–18.

27. Torres MJ, Lee BH. Cloning and expression of β-galactosidase from psychotrophic Bacillus subtilis KL88 into Escherichia coli. Biotechnol Appl Biochem. 1995;17:123–8.

28. Hildebrandt P, Wanarska M, Kur J. A new cold-adapted beta-D-galactosidase from the Antarctic Arthrobacter sp 32c-gene cloning, overexpression, purification and properties. BMC Microbiol. 2009;9:151.

29. Batra N, Singh J, Banerjee UC, Patnaik PR, Sobti RC. Production and characterization of a thermostable beta-galactosidase from Bacillus coagulans RCS3. Biotechnol Appl Biochem. 2002;36(Pt 1):1–6.

30. Li L, Zhang M, Jiang Z, Tang L, Cong Q. Characterisation of a thermostable family 42 beta-galactosidase from Thermotoga maritima. Food Chem. 2009;112:844–50.

31. Warmerdam A, Boom RM, Janssen AE. β-galactosidase stability at high substrate concentrations. Springerplus. 2013;2:402.

32. Griffith KL, Wolf Jr RE. Measuring beta-galactosidase activity in bacteria: cell growth, permeabilization, and enzyme assays in 96-well arrays. Biochem Biophys Res Commun. 2002;290(1):397–402.

33. Laemmli UK. Clevage of structure proteins during the assembly of the head of bacteriophage T4. Nature. 1970;227:680–5.

34. Bradford MM. A rapid and sensitive method for the quantification of microgram quantities of protein utilizing the principle of protein-dye binding. Anal Biochem. 1976;72:248–54.

# 10

# Copper chelation and interleukin-6 proinflammatory cytokine effects on expression of different proteins involved in iron metabolism in HepG2 cell line

Luca Marco Di Bella[1,2], Roberto Alampi[1], Flavia Biundo[1], Giovanni Toscano[1] and Maria Rosa Felice[1*] ⓘ

**Abstract**

**Background:** In vertebrates, there is an intimate relationship between copper and iron homeostasis. Copper deficiency, which leads to a defect in ceruloplasmin enzymatic activity, has a strong effect on iron homeostasis resulting in cellular iron retention. Much is known about the mechanisms underlying cellular iron retention under "normal" conditions, however, less is known about the effect of copper deficiency during inflammation.

**Results:** We show that copper deficiency and the inflammatory cytokine interleukin-6 have different effects on the expression of proteins involved in iron and copper metabolism such as the soluble and glycosylphosphtidylinositol anchored forms of ceruloplasmin, hepcidin, ferroportin1, transferrin receptor1, divalent metal transporter1 and H-ferritin subunit. We demonstrate, using the human HepG2 cell line, that in addition to ceruloplasmin isoforms, copper deficiency affects other proteins, some posttranslationally and some at the transcriptional level. The addition of interleukin-6, moreover, has different effects on expression of ferroportin1 and ceruloplasmin, in which ferroportin1 is decreased while ceruloplasmin is increased. These effects are stronger when a copper chelating agent and IL-6 are used simultaneously.

**Conclusions:** These results suggest that copper chelation has effects not only on ceruloplasmin but also on other proteins involved in iron metabolism, sometimes at the mRNA level and, in inflammatory conditions, the functions of ferroportin and ceruloplasmin may be independent.

**Keywords:** Iron metabolism, Copper deficiency, Inflammation, Ceruloplasmin

## Background

Iron and copper are cofactors for numerous enzymes and are essential elements for all eukaryotes. They are, however, potentially dangerous because they can react with molecular oxygen generating reactive oxygen species that will damage DNA, lipids and proteins [1–3], and because they are both essential and dangerous their levels are strictly regulated. The copper-containing protein ceruloplasmin has an essential role in iron homeostasis. Its catalytic site has six copper atoms, four of which are involved in iron oxidation [4–6], converting $Fe^{2+}$ to $Fe^{3+}$ without generating reactive oxygen species.

In vertebrates, two forms of ceruloplasmin are expressed; the first is mainly produced by hepatocytes and is secreted into the circulation [7–9]. A second form, which is generated by alternative splicing, contains a glycosylphosphatidylinositol (GPI) moiety instead of the normal carboxyl terminal. The GPI anchors ceruloplasmin in the plasma membrane. GPI-Cp was found first in astrocytes where it represents the principal ferroxidase [10, 11]. GPI-Cp, however, is expressed by other cellular types such as leptomeningeal cells, Sertoli cells, and hepatocytes [12–15]. Another important ferroxidase is hephaestin, a transmembrane protein first detected in the small intestine [16, 17]. It mediates iron export from enterocytes to the bloodstream. Hephaestin and the two different forms of ceruloplasmin are suggested to interact with ferroportin, the only known protein involved in ferrous iron export

* Correspondence: mrfelice@unime.it
[1]Department of Chemical, Biological, Pharmaceutical, and Environmental Sciences, University of Messina, Viale F. Stagno D'Alcontres, 31, 98166 Messina, Italy
Full list of author information is available at the end of the article

from the cells [18–22]. $Fe^{3+}$ generated by ferroxidase activity, is loaded onto transferrin (Tf), the major iron-containing protein involved in plasma iron transport and distribution within organisms [23–25]. Diferric Tf binds to transferrin receptor 1 (TfR1) present on the plasma membrane of most cell types and in particular on developing red blood cells [26]. Once bound, the $Tf(Fe)_2$-TfR1 complex is internalized into an endosome where iron is released from Tf and is then exported to the cytoplasm by divalent metal transporter 1 protein (DMT1) [27–29]. The importance of ceruloplasmin in iron metabolism is demonstrated by the fact that decreases in active ceruloplasmin, as seen in Wilson or Menkes disease, is characterized by a strong accumulation of iron in liver, spleen, and brain [30–34]. Moreover, different studies have highlighted the importance of ceruloplasmin and iron metabolism in pathologies like Alzheimer and Parkinson diseases [35–38].

Systemic iron homeostasis is regulated by different stimuli and, in particular, inflammation can affect the concentration and accumulation of iron in the serum and in different organs [39]. Hepatocytes play a critical role in cellular iron as they are the major storage site for excess iron and are a central regulator of proteins (transferrin, ceruloplasmin and hepcidin) that play an important role in iron homeostasis. In particular, hepatocytes are the principal producers of the secreted peptide hormone, hepcidin. Hepcidin, by binding to the iron exporter ferroportin (Fpn1), induces its degradation resulting in reduced iron uptake from the diet and iron efflux from macrophages [40–42]. Hepcidin mRNA expression is increased by inflammatory cytokines [43, 44]. In particular, IL-6, a proinflammatory cytokine, induces the synthesis of hepcidin and it is responsible of a state of hypoferremia of inflammation [44, 45]. Pro-inflammatory cytokines can also regulate expression of other proteins involved in iron metabolism such as Fpn1, DMT1, TfR1 and ceruloplasmin [46–50].

Although studies have highlighted the effect exerted by copper deprivation or pro-inflammatory cytokines on expression of proteins involved in iron metabolism separately, it is not known if there is a synergistic effect of copper depletion and inflammation. The aim of this study was to analyse the effect of copper chelation and the pro-inflammatory cytokine interleukin-6 (IL-6) on the mRNA and protein levels of different proteins involved in iron metabolism using the human hepatocytoma cell line HepG2 as a model system.

## Methods

### Cell culture and treatment

The hepatocytoma cell line HepG2, kindly provided by prof. M.T. Sciortino (Department of Chemical, Biological, Pharmaceutical and Environmental Sciences, University of

Messina, Italy), was grown in Eagle's minimum essential medium (EMEM) (Lonza) supplemented with 10% Fetal Bovine Serum (Lonza), 1× non-essential amino acids (Lonza), 2 mM L-glutamine (Lonza), 100 μg/ml Streptomycin (Sigma), 100 U/ml Penicillin (Sigma), at 37 °C, and 5% $CO_2$. $4×10^5$ cells/ml were seeded in 6 well plates and incubated for 24 h in supplemented medium. Before treatment, cells were washed with PBS and incubated for an additional 16 h in serum-free, antibiotic-free medium, in the presence of 40 ng/ml of IL-6 (Cell Signaling Technology) [51] and/or 300 μM Bathophenanthroline disulfonate (BCS) (Sigma).

### RT-PCR analysis

Total RNA was extracted by EuroGold TriFast reagent (Euroclone) following the manufacturer's instructions. The concentration and purity of RNA was assayed at 260 nm and 280 nm by a DU 60 Beckman spectrophotometer. One μg of total RNA was retro-transcribed using oligo-dT (EuroClone) and PrimeScript MMLV-RT (Takara, Clontech) at 42 °C for 60 min followed by a denaturation step of 15 min at 70 °C. The primers used for PCR are listed in Table 1. The PCR reactions were run for 30 cycles in MyCycler instruments (BioRad) using EmeraldAmp Hot start DNA polymerase (Takara,

**Table 1** list of primers used in this study

| primer | Sequence 5' → 3' | Reference |
| --- | --- | --- |
| Fpn1AB Reverse | CATCCTCTCTGGCGGTTGTG | This study |
| Fpn1A Forward | TCCATAAGGCTTTGCCTTTCC | This study |
| Fpn1B Forward | GCATCTGGTTGGAGTTTCAAT | This study |
| GPI-Cp Reverse | GATTGGGTAGATCACATTCC | [90] |
| sCp Reverse | CCAATTTATTTCATTCAGCC | [90] |
| CP Forward | GTCTTTGACCTTATCCCTGG | This study |
| HAMP Forward | ATGGCACTGAGCTCCCAGAT | This study |
| HAMP Reverse | TTGCAGCACATCCCACACTTT | This study |
| β-actin Reverse | CACATCTGCTGGAAGGTGGA | This study |
| β-actin Forward | CATGAAGTGCGACGTTGACA | This study |
| qPCR Primers | | |
| TNF-α Forward | GCAGGTCTACTTTGGGATCATTG | A generous gift of prof. A. Mastino[a] |
| TNF-α Reverse | GCGTTTGGGAAGGTTGGA | A generous gift of prof. A. Mastino[a] |
| IL1B Forward | GCGAATGACAGAGGGTTTCTTAG | A generous gift of prof. A. Mastino[a] |
| IL1B Reverse | CACCTTCAGCTGCCCAGACT | A generous gift of prof. A. Mastino[a] |
| β-actin Forward | CATTCCAAATATGAGATGCGTTGT | This study |
| β-actin Reverse | TGTGGACTTGGGAGAGGACT | This study |

[a]Department of Chemical Biological Pharmaceutical and Environmental Sciences, University of Messina, Italy

Clontech). The PCR conditions adopted were: 98 °C for 10 s, 57 °C for 1 min, 72 °C for 30 s. The PCR amplicons were analyzed by 2.4% agarose gel electrophoresis, and images were acquired by KdS1D system (Kodak) and analyzed by ImageJ 1.47v software (http://imagej.nih.gov/ij). All the intensity values obtained for genes of interest were normalized with respect to β-actin.

Quantitative Real Time PCR, was performed on the same cDNA using StepOne Plus (Applied Biosystem, LifeTechnologies) and Sybr Premix Ex Taq II (Takara, Clontech) following manufacturer's instructions and primers listed in Table 1. The amplification was performed at 95 °C for 30 s (1 cycle), 95 °C for 5 s and 60 °C for 60 s (40 cycles). All samples were assayed in duplicate of three independent experiments and the results were normalized to the β-actin housekeeping gene using $\Delta\Delta C_T$ method [52, 53].

## Protein extraction and Western blot analysis

To analyze proteins, after specific incubations, cells were washed with PBS and then homogenized in specified lysis buffers. The specific buffer used differed depending on the protein being assayed. For immunodetection of DMT1, H-ferritin subunit, STAT3 and pSTAT3 the lysis buffer was composed of 25 mM MOPS pH 7.4 (Sigma), 150 mM NaCl (Applichem), 1% Triton X-100 (Sigma), and protease inhibitor cocktail (Sigma). The cells were homogenized by passage through a 28 gauge needle several times and left one hour at room temperature, before centrifugation at 15,400 × g for 30 min at 4 °C (Eppendorf 3417R). Total protein concentration of supernatant was assayed by BCA (Pierce) and an equal quantity of total proteins were analyzed by polyacrylamide gel electrophoresis using 16.5% Tris-Tricine SDS-PAGE for DMT1 and FTH1 [54], or 10% SDS-PAGE for STAT3 and pSTAT3, after denaturation at 95 °C for 10 min in the presence of 80 mM Dithiothreitol (DTT). For immunodetection of TfR1 and GPI-Cp cells were homogenized in a buffer composed of 25 mM MOPS pH 7.4 (Sigma), 75 mM NaCl (Applichem), and protease inhibitor cocktail (Sigma) by passage through a 28 gauge needle several times. The homogenate was centrifuged at 15,400 × g for 30 min at 4 °C and the pellet was incubated in extraction buffer composed of 25 mM MOPS pH 7.4, 150 mM NaCl (Sigma), 1% Triton X-100 (Sigma), and protease inhibitor cocktail (Sigma) for one hour before centrifugation as described above. Total protein concentration was assayed by BCA and equal quantities of total protein were separated by 10% Tris-Glycine SDS-PAGE after denaturation at 95 °C for 10 min in the presence of 80 mM DTT. The same membrane protein extraction protocol was used for Fpn1 with the exception that samples were incubated for 30 min at room temperature in the presence of 80 mM DTT before separation on 10% Tris-Glycine SDS-PAGE [55, 56].

After electrophoresis proteins were transferred to FluoroTransW PVDF Membrane (Pall Corporation) by Mini-Trans Blot (BioRad) and membranes were blocked with 5% skim milk (Applichem) before incubation overnight with primary antibodies that are listed in Table 2. After washing, the membranes were incubated with HRP-conjugated secondary antibodies and proteins were detected by etaC Westar ECL (Cyagen). The bands were analyzed by ImageJ 1.47v software (http://imagej.nih.gov/ij). The intensity values obtained for proteins of interest were normalized with respect to β-actin protein level.

## Cellular copper and iron concentration

Total cellular homogenates were obtained as described in section 2.3. For copper and iron mineralization, the homogenate was digested with 1:1 volume ratio of 65% $HNO_3$ overnight at 60 °C [57].

After a dilution of $HNO_3$ to 5%, equal aliquots of samples were used for copper and iron determination by a graphite furnace Perkin Elmer PinAACle 900H atomic absorption spectrophotometer, equipped with the autosampler AS900 and the Lumina cathode lamp (Perkin Elmer). Calibration was against a Cu or a Fe standard curve and the metal content was normalized to total cellular proteins concentration, determined by BCA assay kit, as described in section 2.3.

## Immunodetection and in-gel oxidase activity of secreted Ceruloplasmin

After specific treatments of cells, the medium was collected, concentrated, dialyzed by Centricon YM-50 (Millipore), and proteins were separated by 8% SDS-PAGE in non-denaturing condition for assay of oxidase

**Table 2** list of antibodies used in this study

| Target | Dilution | Host | Company |
|---|---|---|---|
| Fpn1 | 1:1,000 | Rabbit | Novus Biologicals |
| DMT1 | 1:1,000 | Mouse | Novus Biologicals |
| TfR1 | 1:5,000 | Mouse | Invitrogen |
| pSTAT3 | 1:1,000 | Rabbit | Cell Signalling |
| STAT3 | 1:1,000 | Rabbit | Cell Signalling |
| FTH1 | 1:1,000 | Rabbit | Cell Signalling |
| Human Ceruloplasmin | 1:5,000 | Goat | Sigma |
| β-Actin | 1:10,000 | Mouse | Sigma |
| Anti-goat HRP conjugated | 1:4,000 | Rabbit | Sigma |
| Anti-mouse HRP conjugated | 1:5,000 | Goat | Novex, ThermoFisher |
| Anti-rabbit HRP conjugated | 1:4,000 | Goat | Novex, ThermoFisher |

activity. To assay oxidase activity, gels were incubated in 0.1 M sodium acetate buffer pH 5.0 containing 0.5 µg/ml of o-dianisidine dihydrochloride (Sigma) [58]. Alternatively, samples were incubated under reducing conditions for Western blot analysis and immunodetection (as described above) or gels were stained with Coomassie Blue.

### Statistical analysis

The data were analyzed by GraphPad Prism 5.0. Values are expressed as the mean ± SEM. All assays were performed with samples obtained from six independent experiments. Statistical differences were determined by paired Student's $t$-test. Differences were considered significant at $p < 0.05$ level.

## Results

### Analysis of *Signal Transducer and Activator of Transcription 3* (STAT3) transcription factor

It is known that interleukin-6 is able to induce the phosphorylation and nuclear translocation of the transcription factor STAT3 [59, 60]. The level of pSTAT3 was analysed to verify if the concentration of IL-6 and the period of treatment adopted in the present study were able to evoke a response in the HepG2 cell line. A concentration of 40 ng/ml IL-6 was able to activate STAT3 and the presence of BCS did not affect STAT3 phosphorylation state in control cells or IL-6 treated cells (Fig. 1a).

### Effects of BCS and IL-6 on expression of secreted form of ceruloplasmin and determination of cellular copper concentration

The capacity of BCS to copper deprive cells was investigated by the analysis of the secreted form of ceruloplasmin (Cp), as copper deficiency is known to result in secretion of apoCp that is rapidly turned over [61]. Incubation of HepG2 cells with BCS results in the of loss Cp oxidase activity and immunodetectable Cp (Fig. 1b). IL-6 treatment is able to induce a strong signal of Cp protein with respect to control conditions, yet incubation with BCS results in the disappearance of Cp from the medium. A densitometric analysis comparison highlighted a strong correlation between soluble Cp oxidase activity and immunoreactivity (Fig. 1c). *Cp* mRNA levels were measured to determine if the decrease in Çp protein resulted from a decrease in *Cp* mRNA. Incubation with BCS resulted in a slight but statistically significant decrement of *Cp* mRNA compared to control conditions (Figs. 1d, and e). In contrast, treatment with IL-6 caused a threefold induction of *Cp* mRNA that was only slightly reduced by incubation with BCS, indicating that the absence of Cp in the media of cells treated with BCS was largely due to either a slower rate of protein secretion or degradation of the apo form of secreted Cp and was not the result of downregulation of gene expression.

To exclude a secondary effect exerted by BCS that is independent of copper deficiency, we have determined by atomic absorption copper intracellular concentration

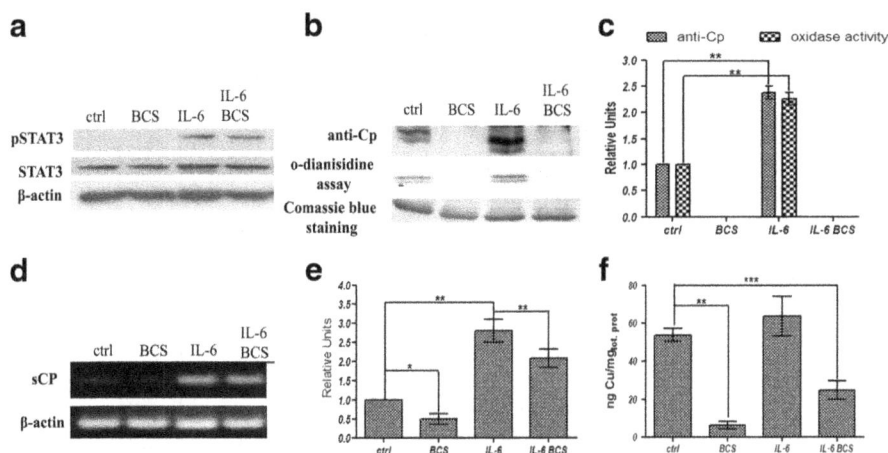

**Fig. 1** Western blot and RT-PCR analysis. HepG2 cells were treated for 16 h in serum-free medium with 300 µM BCS and/or 40 ng/ml of IL-6. **a** Western blot analysis of pSTAT3, STAT3, and β-actin proteins on total cell extracts as described in methods. **b** Western blot, Coomassie Blue staining of soluble Cp isoform, relative to denaturing SDS-PAGE, and in gel nondenaturing SDS-PAGE enzymatic activity of concentrated and dialyzed culture medium. Equal amounts of total proteins were loaded per lane. **c** relative densitometric analysis. **d** representative image of soluble Cp isoform RT-PCR product: after 16 h of treatment, RNA was isolated, reverse transcribed and subjected to PCR. The amplicons relative to soluble Cp isoform and β-actin were analysed by agarose gel electrophoresis and intensity of bands was determined by ImageJ 1.47v software (http://imagej.nih.gov/ij). The values of intensity relative to soluble Cp were normalized by using the β-actin housekeeping gene, (**e**) densitometric analysis of Cp RT-PCR results. **f** graph relative to intracellular copper concentration in HepG2 cells. Cells were extensively washed, lysed as described in methods and used for atomic absorption analysis. The copper content was normalized by cellular total protein concentration. All values are expressed as means ± SEM ($n = 6$). All indicated differences were statistically significant ($p < 0.05$). *$p \leq 0.05$; **$p \leq 0.01$; ***$p \leq 0.001$. Expression levels of control condition were normalized to one, and all values are expressed as relative units

and, in accord with the results reported above, the treatment of HepG2 cells with BCS induces a strong decrement of copper content and the cotreatment with IL-6 has only a slight positive effect (Fig. 1f).

### Effects of BCS and IL-6 on expression of GPI-anchored form of ceruloplasmin

In addition to the secreted form of Cp, hepatocytes express GPI-anchored Cp [15]. The effect of BCS alone or in combination with IL-6 on the GPI-anchored Cp was also investigated. At the transcriptional level (Fig. 2a and b), IL-6 induced a strong induction in *GPI-Cp* mRNA level compared to control cells. Treatment with BCS did not affect transcription in either control cells or in the IL-6 treated cells, indicating a behaviour very similar to that observed for expression of secreted Cp. The presence of BCS did not affect the amount of GPI-Cp present at the plasma membrane (Fig. 2c, and d). Unfortunately, the level of GPI-Cp was too low to assess enzymatic activity.

### Effect of BCS on HAMP and Fpn1 expression

Studies have shown a functional relationship between Cp and Fpn1 in which Cp is required to convert Fpn1-exported $Fe^{2+}$ to $Fe^{3+}$ for binding to Tf. Some studies have shown a physical relationship between GPI-Cp and Fpn1 [62, 63]. Based on these results we examined the effects of BCS and IL-6 on *Fpn1* and *HAMP*, the hepcidin gene. IL-6 was able to induce transcription of *HAMP* as previously reported [44], while the presence of BCS did not affect its expression level (Fig. 3a, and b). These results show the effect of BCS is specific for *Cp* expression but not for *HAMP* expression.

At the transcriptional level, expression of the two spliced variant forms of *Fpn1*, *1A* and *1B* [64, 65] (Fig. 3c, d, e, and f) were both decreased by BCS or IL-6. BCS had a similar effect on both isoforms while the negative effect of IL-6 is less evident in variant *1A* (50%) versus variant 1B (30%) (Fig. 3d, and f). Incubation of HepG2 cells with both IL-6 and BCS resulted in a small additive decrease, however, it did not reach statistical significance. A difference in the amount of Fpn1 protein was also observed when cells were treated with BCS or IL-6 (Fig. 3g, and h). Fpn1 levels were decreased 50 or 30% respectively. Further, incubation with BCS and IL-6 resulted in an additional protein decrement, which was statistically significant, indicating an additive effect of the two substrates. These experimental results highlight that the effect of this pro-inflammatory cytokine and copper chelation can negatively regulate Fpn1 expression.

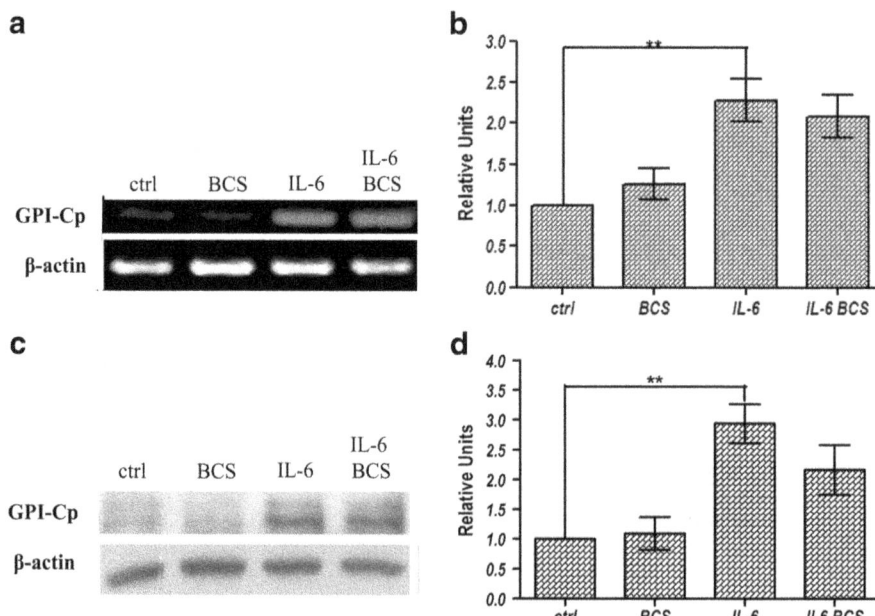

**Fig. 2** RT-PCR and Western blot analysis of GPI-Cp expression levels. HepG2 cells were treated for 16 h in serum-free medium with 300 μM BCS and/or 40 ng/ml IL-6. **a** and **b** after 16 h of treatment, RNA was isolated, reverse transcribed and subjected to PCR. The amplicons relative to *GPI-Cp* isoform and *β-actin* were analysed by agarose gel electrophoresis and intensity of bands was determined by ImageJ 1.47v software (http://imagej.nih.gov/ij). The values of intensity relative to GPI-Cp were normalized by using the β-actin housekeeping gene. **c** representative image of GPI-Cp isoform protein relative to membrane proteins extracts analysed by western blot. **d** densitometric analysis of GPI-Cp isoform protein. The values are normalized by β-actin protein level. All values are expressed as means ± SEM (*n* = 6). All indicated differences were statistically significant ($p < 0.05$). *$p \leq 0.05$; **$p \leq 0.01$; ***$p \leq 0.001$. Expression levels of control condition were normalized to one, and all values are expressed as relative units

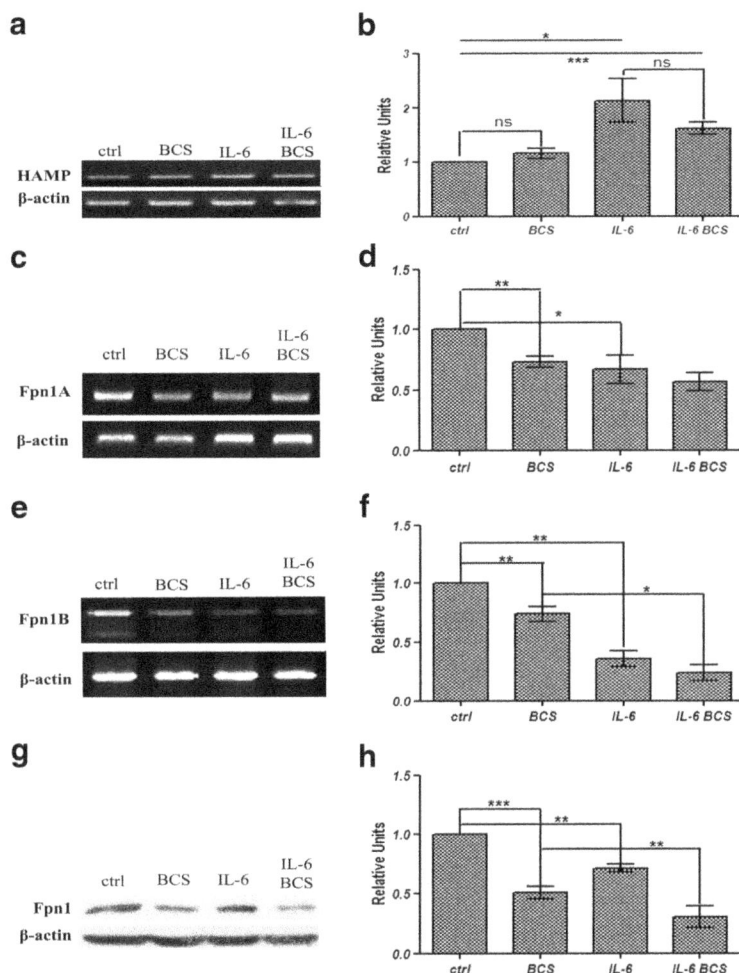

**Fig. 3** RT-PCR analysis of *HAMP* gene, and RT-PCR and western blot analysis of Fpn1 expression levels. HepG2 cells were treated for 16 h in serum-free medium with 300 μM BCS and/or 40 ng/ml IL-6. **a** after 16 h of treatment, RNA was isolated, reverse transcribed and subjected to PCR. The amplicons relative to *HAMP* and *β-actin* genes were analysed by agarose gel electrophoresis and intensity of bands was determined by ImageJ 1.47v software (http://imagej.nih.gov/ij). The values of intensity relative to HAMP were normalized by using the β-actin housekeeping gene. **b** densitometric analysis of *HAMP* gene RT-PCR results (**c**) representative image of *Fpn1A* isoform RT-PCR product analysed as described above and (**d**) densitometric analysis relative to *Fpn1A* RT-PCR results. The values were normalized by using the β-actin housekeeping gene. **e** representative image of *Fpn1B* isoform RT-PCR product analysed as described above and (**f**) densitometric analysis relative to *Fpn1B* RT-PCR results. The values were normalized by using the β-actin housekeeping gene. **g** representative image of Fpn1 protein immunoblot result relative to membrane proteins extracts. Equal amounts of proteins were loaded per lane. **h** densitometric analysis of Fpn1 protein. The values are normalized to β-actin protein level. All values are expressed as means ± SEM (n = 6). All indicated differences were statistically significant (p < 0.05). *p ≤ 0.05; **p ≤ 0.01; ***p ≤ 0.001. Expression levels of control condition were normalized to one, and all values are expressed as relative units

**Effects of BCS and IL-6 on TNF-α and IL-1β expression**

To test if the effect exerted by BCS alone or in combination with IL-6 was direct or indirect by production of other pro-inflammatory cytokines, the mRNA level of *TNF-alpha* and *IL-1B* was also assayed by qPCR. Unfortunately, the $C_T$ values relative to these two classes of mRNA were very low ($C_T$ 36–40) respect to *β-actin* mRNA level, and were not considered for further analysis.

**TfR1, and DMT1 expression**

Hepatocyte iron uptake through TfR1 and DMT1 is important in conditions of iron deficiency and it is also

important under culture conditions in which the amount of iron is limited. For these reasons, the expression levels of these two proteins were investigated under copper chelation and proinflammation. Treatment of HepG2 cells with BCS resulted in a 50% decrease in TfR1 protein levels (Fig. 4a and b) and IL-6 had almost the same effect. Incubation of cells with both BCS and IL-6 led to a further decrease of TfR1 indicating an additive effect. Given the functional relationship of TfR1 and DMT1 in TfR1-mediated iron uptake, the levels of DMT1 were also analysed. Incubation of HepG2 cells with BCS or IL-6 resulted in about a 50% decrement of

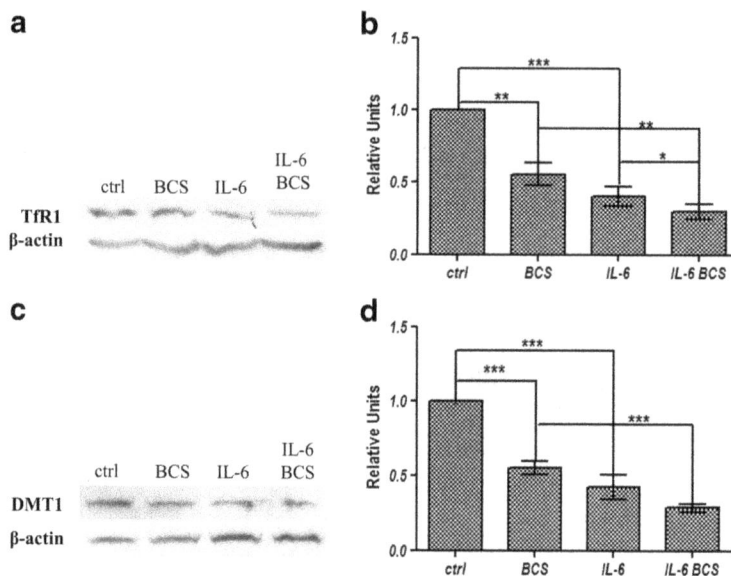

**Fig. 4** Western blot analysis of TfR1, and DMT1. HepG2 cells were treated for 16 h in serum-free medium with 300 μM BCS and/or 40 ng/ml IL-6. **a** representative image of TfR1 protein immunoblot relative to membrane proteins exstracts. Equal amounts of proteins were loaded per lane. **b** densitometric analysis of TfR1 protein. **c** representative image of DMT1 protein immunoblot relative to total cell extracts, after electrophoresis on 16.5% Tris-Tricine SDS-PAGE (**d**) densitometric analysis of DMT1 protein. The values are normalized to β-actin. All values are expressed as means ± SEM ($n = 6$). All indicated differences were statistically significant ($p < 0.05$). $*p \leq 0.05$; $**p \leq 0.01$; $***p \leq 0.001$. Expression levels of control condition were normalized to one, and all values are expressed as relative units

DMT1 protein, while treatment with both BCS and IL-6 led to an additional decrement, although it did not reach statistical significance (Fig. 4c, and d).

### FTH1 expression and cellular iron concentration

To test if the experimental conditions affected intracellular iron level, we examined ferritin heavy chain (FTH1) protein levels, an indicator of cytosolic iron. The treatment of cells with IL-6 did not affect FTH1 levels. In contrast, BCS treatment resulted in an increase in FTH1 levels suggesting an increase of cellular iron content (Fig. 5a, and b). Further, the addition of IL-6 together with BCS increased FTH1 protein levels suggesting the intracellular iron levels are greatly increased in copper chelation and proinflammatory conditions. To exclude secondary effects, the cellular iron concentration was determined and, as shown in Fig. 5c, the concentration of iron is coherent with ferritin protein amounts, indicating that BCS is able to induce an increase of intracellular iron concentration.

### Discussion

The copper-containing protein Cp has a key role in iron metabolism and its activity and level relies on appropriate copper acquisition. Accumulation of newly synthesized Cp is dependent on copper availability, as the stability of the apoprotein is severely reduced [61, 66, 67]. In Wilson Diseases caused by a mutation in *ATP7B* gene, a Golgi copper transporter, Cp is produced in the

apo-form that is secreted in the blood stream where it is rapidly degraded [68]. Decreased active Cp results in iron accumulation in liver and other organs due to a failure to export cellular iron [30–34]. Our study in the HepG2 cell line confirms that a deficiency of copper induces a strong reduction in the secreted form of Cp. Treatment of cells with IL-6 led to a strong induction in *Cp* mRNA and protein levels, consistent with previously published data [46, 47, 69]. The IL-6 induction of *Cp* mRNA, however, was not able to reverse the negative effect on protein secretion exerted by BCS. Of interest is that our results showed that the presence of BCS had minimal influence on stability of GPI-Cp present on the plasma membrane; unfortunately, we were not able to demonstrate a linear correlation between the amount of protein present and its enzymatic activity. These data are in accord with Mostad et al. [14], who demonstrated that copper deficiency has different effects on GPI-Cp protein level in different organs. Copper deficiency in the spleen induces a strong decrement of GPI-Cp protein levels, while only a slight reduction of the protein was found in liver. The different response of the two Cp isoforms to a copper deprivation state could be explained by different kinetic of secretion or degradation rates of the apoprotein dependent on tissue type.

Our results suggest that copper deficiency has an effect on other proteins involved in iron metabolism. It is known that cellular export of Fe(II) by Fpn1 requires Cp to oxidize $Fe^{2+}$ to $Fe^{3+}$. Studies using transfected C6

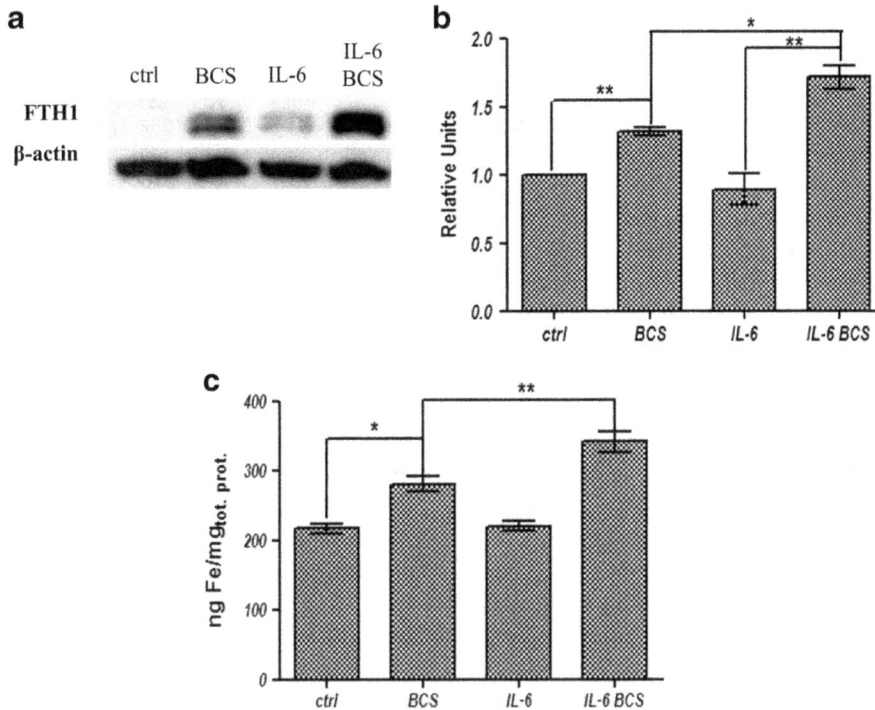

**Fig. 5** FTH1 expression levels and intracellular iron concentration. **a** representative image of FTH1 protein immunoblot relative to total cell extracts, after electrophoresis on 16.5% Tris-Tricine SDS-PAGE. **b** densitometric analysis of FTH1 protein. The values are normalized to β-actin. **c** graph relative to intracellular iron concentration in HepG2 cells. Cells were extensively washed, lysed as described in methods, and used for atomic absorption analysis. The iron content was normalized by cellular total protein concentration. All values are expressed as means ± SEM ($n = 6$). All indicated differences were statistically significant ($p < 0.05$). $*p \leq 0.05$; $**p \leq 0.01$; $***p \leq 0.001$. Expression levels of control condition were normalized to one, and all values are expressed as relative units

and HeLa cells showed that Cp activity is necessary for the stability of plasma membrane Fpn1 [63, 70, 71], and an interaction between the two proteins was also hypothesized [62]. The results reported in this study highlight that Fpn1 is only partially influenced by GPI-Cp protein amount; in fact, in conditions in which cells are treated with BCS, the decrement observed for Fpn1 protein is much more pronounced than that observed for GPI-Cp. This discrepancy can be explained considering the enzymatic activity rather than the protein amount. As mentioned, we do not know if the GPI-Cp protein present on the plasma membrane is also enzymatically active. A slight correlation is seen comparing Fpn1 and sCp protein amounts. The differences observed between our results and the reported published data could be explained with the use of different experimental models. In glioma cell lines the GPI-Cp is the isoform that is mostly highly expressed while in hepatocytes sCp is the most highly expressed isoform [11]. Different experimental models have reported some contrasting results such as animals fed a copper-deficient diet showed an increment of Fpn1 protein when whole liver was analysed. This apparent discrepancy could be due to a different response to the same stimuli between

the different cells present in this organ, e.g., Kupffer cells and hepatocytes [14, 72, 73].

To determine if copper deficiency could affect Fpn1 levels by inducing hepcidin we assayed *HAMP* mRNA levels. In our cells, *HAMP* mRNA levels were not affected by copper chelation. In contrast, copper chelation affected *Fpn1* transcripts including both *Fpn1A* and *Fpn1B* mRNA variants. To determine if the decrease observed was linked to a post-transcriptional regulation mechanism mediated by intracellular iron concentration, H-ferritin subunit protein was assayed as a measure of cellular iron content. Our results showed increased levels of H-ferritin suggesting an increase in cytosolic iron concentration. This result was confirmed by the determination of cellular iron concentration. Increased intracellular iron would be expected to increase *Fpn1* translation (IRP) and mRNA stability (mR485-3p), as increased Fpn1 activity is required to export cellular iron [74–76]. The finding that copper chelation leads to increased cellular iron retention and decreased *Fpn1* mRNA suggests a novel mechanism of Fpn1 regulation. The response of HepG2 cell line to BCS is indicative of a state in which the cells protect themselves from the accumulation of intracellular iron, probably because a not functional ceruloplasmin could cause a condition of iron overload. For this reason, it is possible that in the first period of treatment, ferroportin is downregulated causing an increase of cellular iron concentration. As consequence, TfR1-mediated iron uptake is also reduced. Some studies have reported that hepcidin activity can be dependent on copper availability [77]; in fact, it has an "ATCUN" (amino-terminal Cu-Ni)-binding motif in the N-terminal of the mature protein capable to bind copper and nickel, even if a recent study has questioning this possibility [77–79]. Tselepis et al. highlighted that the incapacity of hepcidin to bind copper, drastically reduce the capacity of hepcidin to induce ferroportin degradation [77]. Considering the results reported in this study and the possibility that hepcidin is not able to reduce ferroportin protein amount in condition of copper deficiency, a transcriptional downregulation of ferroportin can contrast a potential iron overload.

The apparent functional relationship between Fpn1 and Cp appears to break down in the face of inflammatory stimuli. Cp mRNA isoforms are strongly upregulated by IL-6, while FpnlA and FpnlB mRNAs seem to be downregulated. This effect is also seen on the protein level. The lower level of Fpn1 protein might be explained in part due to the post-translational hepcidin-mediated degradation mechanism [41], as hepcidin is upregulated in inflammation [44]. Our data confirm that in HepG2 cells treatment with IL-6 strongly induces HAMP gene expression. However, independent of post-translational regulation, our data show that IL-6 reduces Fpn1 mRNA. These results are consistent with published data, which demonstrated that IL-6 is able to downregulate Fpn1 levels in the HepG2 cell line [48] and upregulate the mRNA level of sCp [46, 47]. We demonstrate that the GPI-Cp isoform is also upregulated and the protein level of the two isoforms follow the same behaviour. The findings that IL-6 results in increased Cp levels but decreased Fpn1 indicates that the functions of these two proteins are not obligatorily linked together. As mentioned above, it is reported that treatment with IL-6 causes an increase in Cp mRNA level, probably in part by the transcription factor FOXO1 [47]. This protein is involved in cellular response to oxidative stress and upregulation of Cp can enter in the mechanism of correlation between oxidative stress and metal metabolism [80, 81]; in fact, CP ferroxidase activity is important in the loading of Fe(III) on transferrin, reducing the deleterious effect of Fe(II) oxidation and production of radical oxygen species [23, 82]. In this way, Cp enters in the circuit to limit NTBI (non-transferrin bound iron) in the serum with hepcidin that is strongly upregulated in IL-6 induced inflammation and, with its activity, limits the presence of iron in the plasma [43]. In addition to ferroxidase activity, Cp has other functions as Cu(I)

oxidation [83], NO-oxidase and $NO_2^-$ synthase [84], and superoxide dismutase [85]. Moreover, an interaction between Cp and myeloperoxidase (MPO) was also demonstrated and it is supposed that Cp inhibits prooxidant activity of MPO [86]; in fact, in systemic vasculitis, the interaction between Cp and MPO is prevented by auto-antibodies against MPO [87]. In vitro experiments have highlighted an interaction between Cp, MPO and lacto-ferrin (Lf). This ternary complex has different functions as reduce the activity of MPO, incorporate Fe(III) on Lf and protect Cp from proteolytic cleavage [88]. For these reasons, Cp can have a fundamental role in inflammation conditions and in autoimmunity diseases.

The dysfunction of cellular iron export resulting from copper chelation has an effect on TfR1-mediated iron delivery, resulting in decreased expression of TfR1 and DMT1. These results are consistent with published studies in which copper deficiency led to a decrement in TfR1 protein in the liver [73, 89].

## Conclusions

In summary, here we have demonstrated, using a hepatoma cell line, that IL-6 results in increased Cp levels and decreased Fpn1, indicating that the functions of these two proteins are not obligatorily linked together.

Moreover, we have demonstrated that copper chelation has effects not only on Cp but also on other proteins involved in iron metabolism, sometimes at the mRNA level.

### Acknowledgement
We are grateful for the critical reading, manuscript editing, and precious suggestions to prof. Jerry Kaplan and prof. Diane McVey Ward of the University of Utah.

### Funding
We thank, for their financial support, prof. A. Mastino (Grant sponsor: Italian Ministry for Education, University and Research, Research Project of National Interest; funding reference number: 2012SNMJRL_004), and prof. S. Cuzzocrea (Grant sponsor: Italian Ministry for Education, University and Research, Research Project of National Interest; funding reference number: 2012WBSSY4_003) of the Department of Chemical, Biological, Pharmaceutical, and Environmental Sciences, University of Messina.

### Authors' contributions
LMB performed the biological experiments, participated in the interpretation of data, and revised manuscript. RA, and FB, performed the biological experiments. GT performed the atomic absorption spectroscopy analysis. MRF designed the experiments, interpreted the data, and wrote the manuscript. All the authors read and approved the final manuscript.

### Competing interests
The authors declare that they have no competing interests.

## Author details

[1]Department of Chemical, Biological, Pharmaceutical, and Environmental Sciences, University of Messina, Viale F. Stagno D'Alcontres, 31, 98166 Messina, Italy. [2]Inter University National Group of Marine Sciences (CoNISMa), Piazzale Flaminio, 9, 00196 Rome, Italy.

## References

1. Aust SD, Morehouse LA, Thomas CE. Role of metals in oxygen radical reactions. Free Radic Biol Med. 1985;1:3–25.
2. Halliwell B. Oxidative DNAS damage: meaning and measurement. In: Halliwell B, Aruoma OI, editors. DNA and free radicals. Chichester: Ellis Horwood Ltd; 1993. p. 67–79.
3. Stadtman ER. Metal ion-catalyzed oxidation of proteins: biochemical mechanism and biological consequences. Free Radic Biol Med. 1990; 9(4):315–25.
4. Frieden E, Hsieh HS. Ceruloplasmin: the copper transport protein with essential oxidase activity. Adv Enzymol Relat Areas Mol Biol. 1976;44: 187–236.
5. Lindley PF, Card G, Zaitseva I, Zaitsev V, Reinhammar B, Selin-Lindgren E, et al. An X-ray structural study of human ceruloplasmin in relation to ferroxidase activity. J Biol Inorg Chem. 1997;2:454–63.
6. Bento I, Peixoto C, Zaitsev VN, Lindley PF. Ceruloplasmin revisited: structural and functional roles of various metal cation-binding sites. Acta Crystallogr D Biol Crystallogr. 2007;63(2):240–8.
7. Yang FM, Friedrichs WE, Cupples RL, Bonifacio MJ, Sanford JA, Horton WA, et al. Human ceruloplasmin. Tissue specific expression of transcripts produced by alternative splicing. J Biol Chem. 1990;265:10780–5.
8. Healy J, Tipton K. Ceruloplasmin and what it might do. J Neural Transm. 2007;114:777–81.
9. Ortel TL, Takahashi N, Putnam FW. Structural model of human ceruloplasmin based on internal triplication, hydrophilic/hydrophobic character, and secondary structure of domains. Proc Natl Acad Sci U S A. 1984;81(15):4761–5.
10. Patel BN, David S. A novel glycosylphosphatidylinositol-anchored form of ceruloplasmin is expressed by mammalian astrocytes. J Biol Chem. 1997;272:20185–90.
11. Patel BN, Dunn RJ, David S. Alternative RNA splicing generates a glycosylphosphatidylinositol-anchored form of ceruloplasmin in mammalian brain. J Biol Chem. 2000;275(6):4305–10.
12. Fortna RR, Watson HA, Nyquist SE. Glycosyl phosphatidylinositol-anchored ceruloplasmin is expressed by rat Sertoli cells and is concentrated in detergent-insoluble membrane fractions. Biol Reprod. 1999;61:1042–9.
13. Mittal B, Doroudchi MM, Jeong SY, Patel BN, David S. Expression of a membrane-bound form of the ferroxidase ceruloplasmin by leptomeningeal cells. Glia. 2003;41(4):337–46.
14. Mostad EJ, Prohaska JR. Glycosylphosphatidylinositol-linked ceruloplasmin is expressed in multiple rodent organs and is lower following dietary copper deficiency. Exp Biol Med. 2011;236(3):298–308.
15. Marques L, Auriac A, Willemetz A, Banha J, Silva B, Canonne-Hergaux F, et al. Immune cells and hepatocytes express glycosylphosphatidylinositol-anchored ceruloplasmin at their cell surface. Blood Cells Mol Dis. 2012;48(2):110–20.
16. Vulpe CD, Kuo YM, Murphy TL, Cowley L, Askwith C, Libina N, et al. Hephaestin, a ceruloplasmin homologue implicated in intestinal iron transport, is defective in the sla mouse. Nat Genet. 1999;21:195–9.
17. Frazer DM, Vulpe CD, McKie AT, Wilkins SJ, Trinder D, Cleghorn GJ, et al. Cloning and gastrointestinal expression of rat hephaestin: relationship to other iron transport proteins. Am J Physiol Gastrointest Liver Physiol. 2001;281:G931–9.
18. Ganz T. Cellular iron: ferroportin is the only way out. Cell Metab. 2005;1:155–7.
19. Donovan A, Brownlie A, Zhou Y, Shepard J, Pratt SJ, Moynihan J, et al. Positional cloning of zebrafish ferroportin1 identifies a conserved vertebrate iron exporter. Nature. 2000;403:776–81.
20. Han O, Kim EY. Colocalization of ferroportin-1 with hephaestin on the basolateral membrane of human intestinal absorptive cells. J Cell Biochem. 2007;101:1000–10.
21. Yeh KY, Yeh M, Glass J. Interactions between ferroportin and hephaestin in rat enterocytes are reduced after iron ingestion. Gastroenterology. 2011;141:292–9.
22. Yeh KY, Yeh M, Mims L, Glass J. Iron feeding induces ferroportin 1 and hephaestin migration and interaction in rat duodenal epithelium. Am J Physiol Gastrointest Liver Physiol. 2009;296:G55–65.
23. Roeser HP, Lee GR, Nacht S, Cartwright GE. The role of ceruloplasmin in iron metabolism. J Clin Invest. 1970;49:2408–17.
24. Jandl JH, Inman JK, Simmons RL, Allen DW. Transfer of iron from serum iron-binding protein to human reticulocytes. J Clin Invest. 1959;38:161–85.
25. Morgan EH, Appleton TC. Autoradiographic localization of 125-I-labelled transferrin in rabbit reticulocytes. Nature. 1969;223:1371–2.
26. Van Renswoude J, Bridges KR, Harford JB, Klausner RD. Receptor-mediated endocytosis of transferrin and the uptake of Fe in K562 cells: identification of a nonlysosomal acidic compartment. Proc Natl Acad Sci U S A. 1982;79: 6186–90.
27. Dautry-Varsat A, Ciechanover A, Lodish HF. pH and the recycling of transferrin during receptor-mediated endocytosis. Proc Natl Acad Sci U S A. 1983;80:2258–62.
28. Conner SD, Schmid SL. Differential requirements for AP-2 in clathrin-mediated endocytosis. J Cell Biol. 2003;162:773–9.
29. Frazer DM, Anderson GJ. Iron imports. I. Intestinal iron absorption and its regulation. Am J Physiol Gastrointest Liver Physiol. 2005;289:G631–5.
30. Yoshida K, Furihata K, Takeda S, Nakamura A, Yamamoto K, Morita H, et al. A mutation in the ceruloplasmin gene is associated with systemic hemosiderosis in humans. Nat Genet. 1995;9(3):267–72.
31. Harris ZL, Durley AP, Man TK, Gitlin JD. Targeted gene disruption reveals an essential role for ceruloplasmin in cellular iron efflux. Proc Natl Acad Sci U S A. 1999;96:10812–7.
32. Panagiotakaki E, Tzetis M, Manolaki N, Loudianos G, Papatheodorou A, Manesis E, et al. Genotype-phenotype correlations for a wide spectrum of mutations in the Wilson disease gene (ATP7B). Am J Med Genet A. 2004; 131(3):168–73.
33. Hayashi H, Yano M, Fujita Y, Wakusawa S. Compound overload of copper and iron in patients with Wilson's disease. Med Mol Morphol. 2006;39(3):121–6.
34. Kodama H, Fujisawa C. Copper metabolism and inherited copper transporter disorders: molecular mechanisms, screening, and treatment. Metallomics. 2009;1:42–52.
35. Torsdottir G, Kristinsson J, Sveinbjornsdottir S, Snaedal J, Jóhannesson T. Copper, ceruloplasmin, superoxide dismutase and iron parameters in Parkinson's disease. Pharmacol Toxicol. 1999;85(5):239–43.
36. Jin L, Wang J, Zhao L, Jin H, Fei G, Zhang Y, Zeng M, Zhong C. Decreased serum ceruloplasmin levels characteristically aggravate nigral iron deposition in Parkinson's disease. Brain. 2011;134(1):50–8.
37. Kristinsson J, Snaedal J, Tórsd tir G, Jóhannesson T. Ceruloplasmin and iron in Alzheimer's disease and Parkinson's disease: a synopsis of recent studies. Neuropsychiatr Dis Treat. 2012;8:515–21.
38. Ayton S, Lei P, Duce JA, Wong BX, Sedjahtera A, Adlard PA, et al. Ceruloplasmin dysfunction and therapeutic potential for Parkinson's disease. Ann Neurol. 2013;73(4):554–9.
39. Weiss G. Modification of iron regulation by the inflammatory response. Best Pract Res Clin Haematol. 2005;18:183–201.
40. Ganz T. Hepcidin, a key regulator of iron metabolism and mediator of anemia of inflammation. Blood. 2003;102(3):783–8.
41. Nemeth E, Tuttle MS, Powelson J, Vaughn MB, Donovan A, Ward DM, Ganz T, Kaplan J. Hepcidin regulates cellular iron efflux by binding to ferroportin and inducing its internalization. Science. 2004;306:2090–3.
42. Nemeth E, Ganz T. The role of hepcidin in iron metabolism. Acta Haematol. 2009;122(2–3):78–86.
43. Nemeth E, Valore EV, Territo M, Schiller G, Lichtenstein A, Ganz T. Hepcidin, a putative mediator of anemia of inflammation, is a type II acute-phase protein. Blood. 2003;101(7):2461–3.
44. Nemeth E, Rivera S, Gabayan V, Keller C, Taudorf S, Pedersen BK, Ganz T. IL-6 mediates hypoferremia of inflammation by inducing the synthesis of the iron regulatory hormone hepcidin. J Clin Invest. 2004;113(9):1271–6.
45. Ganz T, Nemeth E. Iron sequestration and anemia of inflammation. Semin Hematol. 2009;46(4):387–93.
46. Conley L, Geurs TL, Levin LA. Transcriptional regulation of ceruloplasmin by an IL-6 response element pathway. Brain Res Mol Brain Res. 2005;139:235–41.
47. Sidhu A, Miller PJ, Hollenbach AD. FOXO1 stimulates ceruloplasmin promoter activity in human hepatoma cells treated with IL-6. Biochem Biophys Res Commun. 2011;404:963–7.

48. Naz N, Malik IA, Sheikh N, Ahmad S, Khan S, Blaschke M, et al. Ferroportin-1 is a 'nuclear'-negative acute-phase protein in rat liver: a comparison with other iron-transport proteins. Lab Invest. 2012;92(6):842–56.

49. Ahmad S, Sultan S, Naz N, Ahmad G, Alwahsh SM, Cameron S, et al. Regulation of iron uptake in primary culture rat hepatocytes: the role of acute-phase cytokines. Shock. 2014;41(4):337–45.

50. Tacchini L, Gammella E, De Ponti C, Recalcati S, Cairo G. Role of HIF-1 and NF-kappaB transcription factors in the modulation of transferrin receptor by inflammatory and anti-inflammatory signals. J Biol Chem. 2008;283(30): 20674–86.

51. Fein E, Merle U, Ehehalt R, Herrman T, Kulaksiz H. Regulation of hepcidin in HepG2 and RINm5F cells. Peptides. 2007;28(5):951–7.

52. Livak KJ. ABI Prism 7700 Sequence detection System User Bulletin #2 Relative quantification of gene expression; 1997 & 2001. http://docs. appliedbiosystems.com/pebiodocs/04303859.pdf.

53. Livak KJ, Schmittgen TD. Analysis of relative gene expression data using real-time quantitative PCR and the 2^[−delta delta C(T)] Method. Methods. 2001;25(4):402–8.

54. Schägger H, von Jagow G. Tricine–sodium dodecyl sulfate polyacrylamide gel electrophoresis for the separation of proteins in the range from 1–100 kDalton. Anal Biochem. 1987;166:368–79.

55. Delaby C, Pilard N, Gonçalves AS, Beaumont C, Canonne-Hergaux F. Presence of the iron exporter ferroportin at the plasma membrane of macrophages is enhanced by iron loading and down-regulated by hepcidin. Blood. 2005;106(12):3979–84.

56. Delaby C, Pilard N, Puy H, Canonne-Hergaux F. Sequential regulation of ferroportin expression after erythrophagocytosis in murine macrophages: early mRNA induction by haem, followed by iron-dependent protein expression. Biochem J. 2008;411:123–31.

57. Vazquez MC, Martinez P, Alvarez AR, Gonzalez M, Zanlungo S. Increased copper levels in in vitro and in vivo models of Niemann-Pick C disease. Biometals. 2012;25:777–86.

58. di Patti Bonaccorsi MC, Bellenchi GC, Bielli P, Calabrese L. Release of highly active Fet3 from membranes of the yeast Pichia pastoris by limited proteolysis. Arch Biochem Biophys. 1999;372:295–9.

59. Heinrich PC, Behrmann I, Haan S, Hermanns HM, Muller-Newen G, Schaper F. Principles of interleukin (IL)-6-type cytokine signalling and its regulation. Biochem J. 2003;374(1):1–20.

60. Levy DE, Darnell Jr JE. Stats: transcriptional control and biological impact. Nat Rev Mol Cell Biol. 2002;3(9):651–62.

61. Gitlin JD, Schroeder JJ, Lee-Ambrose LM, Cousins RJ. Mechanisms of caeruloplasmin biosynthesis in normal and copper-deficient rats. Biochem J. 1992;282:835–9.

62. Jeong SY, David S. Glycosylphosphatidylinositol-anchored ceruloplasminis required for iron efflux from cells in the central nervous system. J Biol Chem. 2003;278(29):27144–8.

63. Bonaccorsi di Patti MC, Maio N, Rizzo G, De Francesco G, Persichini T, Colasanti M, et al. Dominant mutant of ceruloplasmin impair the copper loading machinery in aceruloplasminemia. J Biol Chem. 2009;284(7):4545–54.

64. Cianetti L, Segnalini P, Calzolari A, Morsilli O, Felicetti F, Ramoni C, et al. Expression of alternative transcripts of ferroportin-1 during human erythroid differentiation. Haematologica. 2005;90:1595–606.

65. Zhang DL, Hughes RM, Ollivierre-Wilson H, Ghosh MC, Rouault TA. A Ferroportin transcript that lack an iron responsive element enables duodenal and erythroid precursor cells to evade translational repression. Cell Metab. 2009;9(5):461–73.

66. Holtzman NA, Gaumnitz BM. Identification of an apoceruloplasmin-like substance in the plasma of copper-deficient rats. J Biol Chem. 1970;245(9): 2350–3.

67. Broderius M, Mostad E, Wendroth K, Prohaska JR. Levels of plasma ceruloplasmin protein are markedly lower following dietary copper deficiency in rodents. Comp Biochem Physiol C Toxicol Pharmacol. 2010;151(4):473–9.

68. Huster D, Finegold MJ, Morgan CT, Burkhead JL, Nixon R, Vanderwerf SM, et al. Consequences of copper accumulation in the livers of Atp7b−/− (Wilson disease gene) knockout mice. Am J Pathol. 2006;168(2):423–34.

69. McCarthy RC, Kosman DJ. Activation of C6 glioblastoma cell ceruloplasmin expression by neighboring human brain endothelia-derived interleukins in an in vitro blood-brain barrier model system. Cell Commun Signal. 2014;12:65.

70. De Domenico I, Ward DM, Di Patti Bonaccorsi MC, Jeong SY, David S, Musci G, et al. Ferroxidase activity is required for the stability of cell surface ferroportin in cells expressing GPI-ceruloplasmin. EMBO J. 2007;26(12):2823–31.

71. Kono S, Yoshida K, Tomosugi N, Terada T, Hamaya Y, Kanaoka S, et al. Biological effects of mutant ceruloplasmin on hepcidin-mediated internalization of ferroportin. Biochim Biophys Acta. 2010;1802(11):968–75.

72. Jenkitkasemwong S, Broderius M, Nam H, Prohaska JR, Knutson MD. Anemic copper-deficient rats, but not mice, display low hepcidin expression and high ferroportin levels. J Nutr. 2010;140(4):723–30.

73. Broderius M, Mostad E, Prohaska JR. Suppressed hepcidin expression correlates with hypotransferrinemia in copper-deficient rat pups but not dams. Genes Nutr. 2012;7(3):405–14.

74. Muckenthaler MU, Galy B, Hentze MW. Systemic iron homeostasis and the iron-responsive element/iron-regulatory protein (IRE/IRP) regulatory network. Annu Rev Nutr. 2008;28:197–213.

75. Lymboussaki A, Pignatti E, Montosi G, Garuti C, Haile DJ, Pietrangelo A. The role of the iron responsive element in the control of ferroportin1/IREG1/ MTP1 gene expression. J Hepatol. 2003;39:710–5.

76. Sangokoya C, Doss JF, Chi JT. Iron-responsive miR-485-3p regulates cellular iron homeostasis by targeting ferroportin. PLoS Genet. 2013;9(4):1–11.

77. Tselepis C, Ford SJ, McKie AT, Vogel W, Zoller H, Simpson RJ, Castro JD, Iqbal TH, Ward DG. Characterization of the transition-metal binding properties of hepcidin. Biochem J. 2010;427:289–96.

78. Melino S, Garlando L, Patamia M, Paci M, Petruzzelli R. A metal-binding site is present in the amino terminal region of the bioactive iron regulator hepcidin-25. J Pept Res. 2005;66:65–71.

79. Kulprachakarn K, Chen YL, Kong X, Arno MC, Hider RC, Srichairatanakool S, et al. Copper(II) binding properties of hepcidin. J Biol Inorg Chem. 2016;21:329–38.

80. Walter PL, Steinbrenner H, Barthel A, Klotz LO. Stimulation of selenoprotein P promoter activity in hepatoma cells by FoxO1a transcription factor. Biochem Biophys Res Commun. 2008;365:316–21.

81. de Candia P, Blekhman R, Chabot AE, Oshlack A, Gilad Y. A combination of genomic approaches reveals the role of FOXO1a in regulating an oxidative stress response pathway. PLoS ONE. 2008;3(2):e1670.

82. Bielli P, Calabrese L. Structure to function relationships in ceruloplasmin: a moonlighting protein. Cell Mol Life Sci. 2002;59:1413–27.

83. Stoj C, Kosman DJ. Cuprous oxidase activity of yeast Fet3p and human ceruloplasmin: implication for function. FEBS Lett. 2003;554:422–6.

84. Shiva S, Wang X, Ringwood LA, Xu X, Yuditskaya S, et al. Ceruloplasmin is a NO oxidase and nitrite synthase that determines endocrine NO homeostasis. Nat Chem Biol. 2006;2:486–93.

85. Vasil'ev VB, Kachurin AM, Soroka NV. Dismutation of superoxide radicals by ceruloplasmin - details of the mechanism. Biokhimiia. 1988;53:2051–8.

86. Segelmark M, Persson B, Hellmark T, Wieslander J. Binding and inhibition of myeloperoxidase (MPO): a major function of ceruloplasmin? Clin Exp Immunol. 1997;108:167–74.

87. Griffin SV, Chapman PT, Lianos EA, Lockwood CM. The inhibition of myeloperoxidase by ceruloplasmin can be reversed by anti-myeloperoxidase antibodies. Kidney Int. 1999;55:917–25.

88. Samygina VR, Sokolov AV, Bourenkov G, Petoukhov MV, Pulina MO, Zakharova ET, et al. Ceruloplasmin: macromolecular assemblies with iron-containing acute phase proteins. PLoS ONE. 2013;8(7):e67145.

89. Chung J, Prohaska JR, Wessling-Resnick M. Ferroportin-1 is not upregulated in copper-deficient mice. J Nutr. 2004;134(3):517–21.

90. Banha J, Marques L, Oliveira R, Martis Mde F, Paixao E, Pereira D, et al. Ceruloplasmin expression by human peripheral blood lymphocytes: a new link between immunity and iron metabolism. Free Radic Biol Med. 2008;44(3):483–92.

# 11

# Systematic substitutions at BLIP position 50 result in changes in binding specificity for class A β-lactamases

Carolyn J. Adamski[1,2] and Timothy Palzkill[1,2]* ⓘ

## Abstract

**Background:** The production of β-lactamases by bacteria is the most common mechanism of resistance to the widely prescribed β-lactam antibiotics. β-lactamase inhibitory protein (BLIP) competitively inhibits class A β-lactamases via two binding loops that occlude the active site. It has been shown that BLIP Tyr50 is a specificity determinant in that substitutions at this position result in large differential changes in the relative affinity of BLIP for class A β-lactamases.

**Results:** In this study, the effect of systematic substitutions at BLIP position 50 on binding to class A β-lactamases was examined to further explore the role of BLIP Tyr50 in modulating specificity. The results indicate the sequence requirements at position 50 are widely different depending on the target β-lactamase. Stringent sequence requirements were observed at Tyr50 for binding *Bacillus anthracis* Bla1 while moderate requirements for binding TEM-1 and relaxed requirements for binding KPC-2 β-lactamase were seen. These findings cannot be easily rationalized based on the β-lactamase residues in direct contact with BLIP Tyr50 since they are identical for Bla1 and KPC-2 suggesting that differences in the BLIP-β-lactamase interface outside the local environment of Tyr50 influence the effect of substitutions.

**Conclusions:** Results from this study and previous studies suggest that substitutions at BLIP Tyr50 may induce changes at the interface outside its local environment and point to the complexity of predicting the impact of substitutions at a protein-protein interaction interface.

**Keywords:** β-lactamase, Binding specificity, Systematic substitutions, Beatmusic

## Background

Interactions between proteins play an essential role in nearly every cellular process. Each protein in a cell is estimated to interact with approximately five other proteins, forming a complex interaction network [1]. A better understanding of protein-protein interactions, in particular what regulates rates of formation and dissociation and the molecular basis of specificity, would have applications ranging across fields from protein engineering to drug design [2]. Numerous protein-protein interactions have been studied and provide details about the roles of shape complementarity, long- and short-range interactions and solvent in binding [3–10]. However, even with this large accumulation of data, prediction programs often have limited success, largely because of challenges posed by cooperativity between residues, flexibility, and rearrangement at the large, multifaceted interface upon binding [5, 11]. Some success has been shown for predicting changes upon mutation to alanine; however, predicting the effects of mutations to the other 19 amino acids often falls short because residues other than alanine lose interactions and also have the ability to form new ones. It is important to understand how mutations to all possible amino acids modify protein-protein interactions for protein engineering and because mutations other than alanine are frequently seen in nature.

Various protein-protein complexes such as BLIP and β-lactamases have developed into model systems to examine the basic principles underlying protein-protein interactions [6, 7, 12–15]. Studies of a variety of protein complexes indicate that only a subset of residues at the interface contributes substantially to binding affinity and these residues are termed "hot spots" [12, 13, 16–19]. Specificity determinants, i.e., residues where energetic contributions vary depending on the binding partners,

---

\* Correspondence: timothyp@bcm.edu
[1]Department of Biochemistry and Molecular Biology, Baylor College of Medicine, Houston, TX, USA
[2]Department of Pharmacology, Baylor College of Medicine, Houston, TX, USA

were found among the hot spot residues in the BLIP-β-lactamase interaction [12]. Specificity determinants are of particular interest because of their ability to exhibit a significantly different effect on binding affinity to different protein binding partners when mutated. For example, when BLIP Tyr50 was mutated to alanine it exhibited a 50-fold increase in binding affinity for one binding partner (TEM-1) and a 65-fold decrease in binding affinity for another (Sme-1) (Table 1) [12].

The current study focuses on further examination of the specificity determinant BLIP Tyr50 in the interaction of BLIP with various class A β-lactamases. BLIP is a 17.5 kDa protein produced from the soil bacterium *Streptomyces clavuligerus* that inhibits class A β-lactamases with varying affinities (subnanomolar to micromolar) (Table 1) [20–22]. Class A β-lactamases hydrolyze the commonly prescribed β-lactam antibiotics rendering them inactive [23]. The production of β-lactamases is the most common mechanism of bacterial resistance in Gram-negative bacteria [23]. BLIP inhibits class A β-lactamases by docking its predominantly polar, concave surface onto the enzyme, burying approximately 2,600 Å$^2$ of surface area [22]. BLIP competitively inhibits class A β-lactamases via two binding loops that occlude the active site of the enzymes (Fig. 1a & b) [22]. The tertiary structures of class A β-lactamases are homologous but the sequences vary in identity from 30-70% (Fig. 2) (Table 2) [23].

BLIP positions Tyr50, Glu73, Lys74 and Tyr143 were previously identified as specificity determinants in that substitutions at these positions result in large changes in the relative affinity of BLIP for various class A β-lactamases [13]. BLIP Tyr50 resides on the 46-53 loop that contains two hotspots for binding – Asp49 and Tyr53 [13]. A concerted rearrangement occurs at both interfaces upon binding of BLIP and TEM-1 β-lactamase; of particular interest, TEM-1 Tyr105 rearranges upon complex formation to relieve a steric clash with BLIP Tyr50 (Fig. 1c) [16, 22, 24]. In addition, residue 105, which is tryptophan in KPC-2 β-lactamase, is in a similar position as Tyr105 of TEM-1 and also undergoes a rearrangement in the BLIP-KPC-2 complex (Fig. 1d) [25]. This rearrangement of β-lactamase position 105 may be a contributing factor to changes in binding affinity upon mutation of BLIP Tyr50. BLIP Tyr50 forms van der Waals contacts with the β3 strand of TEM-1 and KPC-2, and also interacts directly with positions 107, 129 and 216 on the β-lactamase interface (Fig. 2) [16, 22]. A structural alignment of the

positions on the β-lactamase interface is shown in Fig. 2 with TEM-1 and KPC-2 from the apo form and Bla1 from a BLIP-II bound form as no apo structure of Bla1 is available [26]. β-lactamase positions 107, 129 and 216 have the same sequence and similar structure for Bla1 and KPC-2 while TEM-1 differs at positions 129 and 216 (Fig. 2). As discussed above and seen in Fig. 2, the Tyr105 and Trp105 residues of TEM-1 and KPC-2 are in a similar position in apo forms of the enzyme. The Tyr105 residue of Bla1 is in an altered position in the structural alignment, however, this is likely due to the structure originating from the BLIP-II-Bla1 complex (Fig. 2) [26].

In this study, the effect of systematic substitutions at BLIP Tyr50 is examined using kinetic analysis to determine how specificity can be modulated for binding TEM-1, KPC-2 and Bla1 β-lactamases. These experiments were also performed computationally to assess the current success rate of an available protein binding prediction program. A deeper understanding of the interactions of BLIP with β-lactamases offers an opportunity to explore how specificity can be introduced into proteins rationally, by design.

## Results

### Determination of inhibition constants for β-lactamases

BLIP Y50 was substituted to all 19 amino acids to investigate the role of this residue in modulating specificity. The mutant proteins were purified (with the exception of BLIP Y50I which could not be purified due to low expression and yield) and assayed with TEM-1, KPC-2 and Bla1 β-lactamases to determine the inhibition constants (Table 3, Fig. 3). It was previously reported that the BLIP Y50A substitution alters the binding specificity for β-lactamases; however, the extent to which other substitutions at position 50 alter binding specificity is unknown. Overall, the results of this study support the hypothesis that BLIP Y50 makes important contributions to the binding specificity of BLIP for class A β-lactamases.

BLIP is a potent inhibitor of each of the β-lactamases studied with K$_i$ values of 0.5 nM for TEM-1, 1.5 nM for KPC-2 and 2.5 nM for Bla2 (Table 3). The effect of substitutions at BLIP Y50 on the binding affinity for the enzymes, however, is widely different. Most BLIP Y50 substitutions retain tight binding for KPC-2 while many substitutions reduce binding to TEM-1 and the majority of substitutions are detrimental for binding Bla1 (Table 3). This is apparent from the finding that only 3 substitutions result in a greater than 10-fold loss in affinity for KPC-2 while 10 substitutions reduce binding by >10-fold for TEM-1 and 15 result in a >10-fold loss in affinity for Bla1 (Table 3). The changes in binding constants for the substitutions were normalized by calculating the changes in free energy of the complex using the following equation: $\Delta\Delta G = -RT \ln (K_i^{WT} / K_i^{MUT})$

**Table 1** Binding Constants of BLIP and BLIPY50A for Class A β-lactamases

| BLIP | Ki (nM) TEM-1 | Ki (nM) SHV-1 | Ki (nM) Sme-1 |
|------|---------------|---------------|---------------|
| WT | 0.5[a] | 1130[a] | 2.4[a] |
| Y50A | 0.011[a] | 34[a] | 32[a] |

[a]values from Zhang Z, et al. [13]

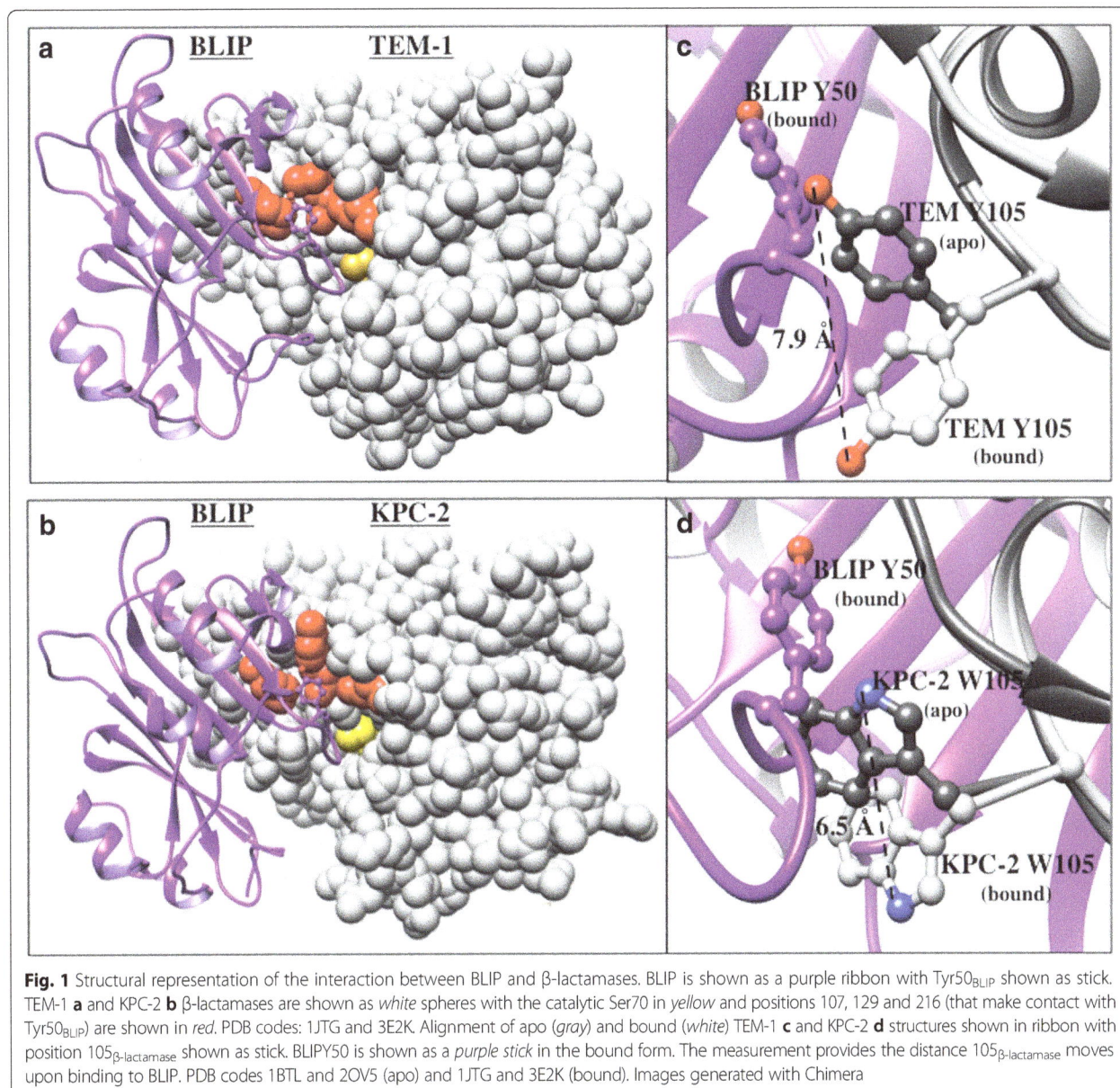

**Fig. 1** Structural representation of the interaction between BLIP and β-lactamases. BLIP is shown as a purple ribbon with Tyr50$_{BLIP}$ shown as stick. TEM-1 **a** and KPC-2 **b** β-lactamases are shown as *white* spheres with the catalytic Ser70 in *yellow* and positions 107, 129 and 216 (that make contact with Tyr50$_{BLIP}$) are shown in *red*. PDB codes: 1JTG and 3E2K. Alignment of apo (*gray*) and bound (*white*) TEM-1 **c** and KPC-2 **d** structures shown in ribbon with position 105$_{β-lactamase}$ shown as stick. BLIPY50 is shown as a *purple stick* in the bound form. The measurement provides the distance 105$_{β-lactamase}$ moves upon binding to BLIP. PDB codes 1BTL and 2OV5 (apo) and 1JTG and 3E2K (bound). Images generated with Chimera

(Table 4). A negative $\Delta\Delta G$ value is indicative of an increase in binding affinity as compared to wild type while a positive value corresponds to a decrease in binding affinity. Using these values, the wide difference in tolerance to BLIP Y50 substitutions for binding β-lactamases is clear in that the average effect of substitutions ($\Delta\Delta G$) on binding KPC-2 was 0.7 kcal/mol while that for binding TEM-1 was 1.3 kcal/mol and that for binding Bla1 was 2.3 kcal/mol (Table 4 and Fig. 4). These results indicate that the sequence requirements at BLIP position 50 are significantly less stringent for binding KPC-2 compared to the requirements for binding TEM-1 and Bla1.

Examination of the substitution results in Tables 3 and 4 reveals some common sequence requirements at BLIP position 50 for binding all three β-lactamases. For

example, cysteine, phenylalanine and lysine substitutions showed a greater than 10-fold decrease in binding affinity for all three enzymes (Table 3). Cysteine may decrease binding affinity because of the potential of being oxidized, which would disrupt binding. Phenylalanine has a similar van der Waals volume as the wild-type tyrosine residue but does not have the hydrogen bonding capacity and could potentially interrupt the organization of structural waters at the interface because of its strong hydrophobic properties. The decrease in binding affinity when lysine is substituted at BLIP position 50 is likely due to introduction of an unpaired charge in the interface. Although lysine was the only charged residue to globally decrease binding affinity by 10-fold, the general finding is that charged

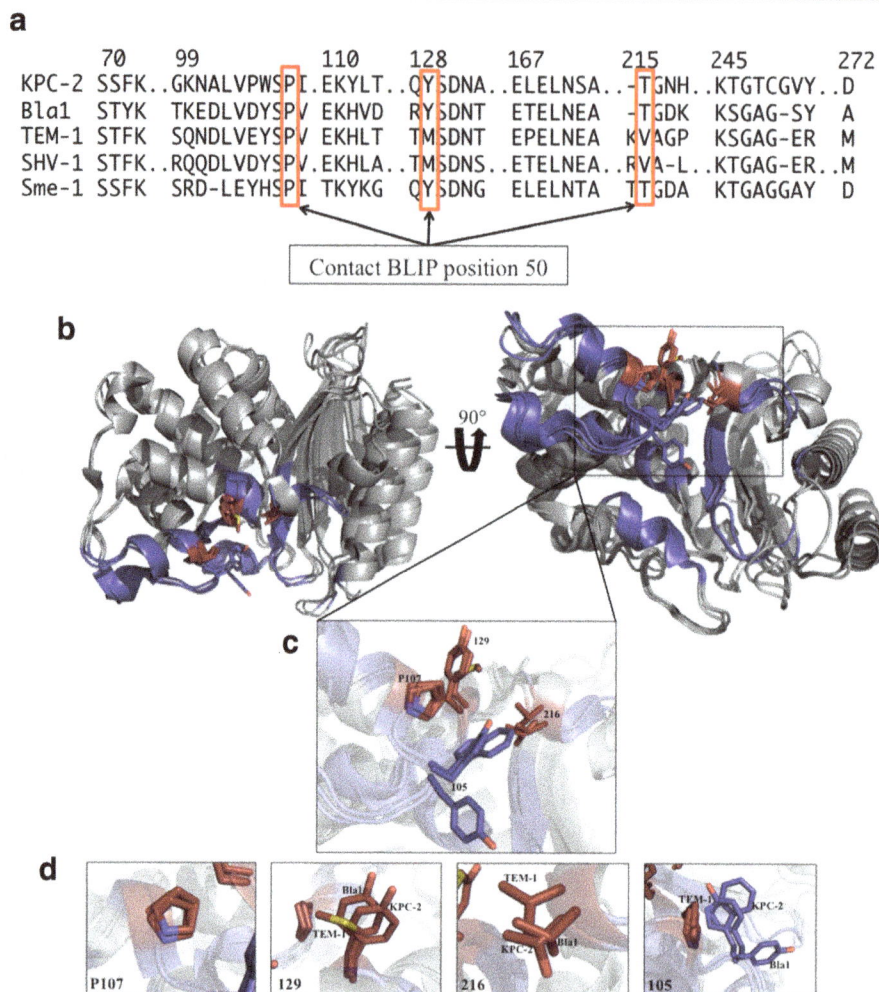

**Fig. 2** Alignment of class A β-lactamase residues at the BLIP interface. **a** The alignment is based on the structure of class A β-lactamase residues found at the BLIP interface as defined by the TEM-1/BLIP complex X-ray structure. The positions that contact BLIP position 50 are boxed in *red*. PDB codes used for the structural alignment are as follows: 2OV5 for KPC-2, 3QHY for Bla1, 1BTL for TEM-1, 1SHV for SHV-1 and 1DY6 for SME-1. Structural alignment performed in Chimera [41]. **b** β-lactamase structures are shown as *grey ribbon* and were aligned using MacPyMOL. Interface residues are shown in navy *blue* with β-lactamase position 105 shown as *blue sticks*. Residues that make direct contact with Tyr50$_{BLIP}$ (107, 129 and 216) are shown as *red sticks*. A global structural alignment of TEM-1, Bla1 and KPC-2 β-lactamases is shown in two orientations. **c** A close-up view of an alignment of the β-lactamase residues that make contact with Tyr50$_{BLIP}$. β-lactamase position 105 is also shown as stick model as it has been shown to make structural rearrangements upon binding to BLIP [14, 16, 35]. **d** A close up alignment of β-lactamase positions 107, 129, 216 and 105 are shown with changes in orientation made for ease of viewing the structural alignment. Residues are labeled with their corresponding β-lactamase. PDB codes used for generation of images were as follows: 1BTL for TEM-1, 3QHY for Bla1 and 2OV5 for KPC-2. Images generated in MacPyMOL

**Table 2** Sequence Identity Comparison of Class A β-lactamases

|  | Percent sequence identity | |
|---|---|---|
|  | Total protein | Interface residues |
| KPC-2 & TEM-1 | 38 | 39 |
| KPC-2 & Bla1 | 38 | 52 |
| TEM-1 & Bla1 | 38 | 61 |

residues at position 50 result in a decrease in binding affinity for all β-lactamases tested, although the effect is less pronounced for binding KPC-2 (Table 3). This is supported by the fact that BLIP containing arginine, glutamate or aspartate at position 50 exhibited decreased affinity for TEM-1 and Bla1 by greater than 100-fold and also exhibited decreased affinity for KPC-2 (Table 3). A proline at position 50 is also generally disruptive in that it decreased the binding affinity of BLIP for all three β-lactamases, possibly by altering the conformation or flexibility of the Y50 loop, which includes two hot spot residues for binding [27].

**Table 3** Inhibition Constants of BLIPY50 mutants for Class A β-lactamases

|           | BLIP Mutant | Ki (nM) KPC-2 | Ki (nM) TEM-1 | Ki (nM) Bla1 |
|-----------|-------------|---------------|---------------|--------------|
| Nonpolar  | WT          | 1.5 ± 0.2     | 0.5 ± 0.1     | 2.5 ± 0.3    |
|           | Y50A        | 1.9 ± 0.3     | 0.010 ± 0.001 | 1.2 ± 0.3    |
|           | Y50F        | 15 ± 2        | 10.3 ± 0.8    | 90 ± 14      |
|           | Y50L        | 0.7 ± 0.2     | 0.17 ± 0.02   | 160 ± 39     |
|           | Y50M        | 0.5 ± 0.1     | 0.11 ± 0.01   | 34 ± 9       |
|           | Y50V        | 10 ± 1        | 3.5 ± 0.6     | 200 ± 11     |
|           | Y50W        | 1.1 ± 0.2     | 11 ± 1        | 90 ± 11      |
| Polar     | Y50C        | 42 ± 6        | 60 ± 10       | 360 ± 5      |
|           | Y50N        | 3.8 ± 0.8     | 7.8 ± 0.5     | 180 ± 43     |
|           | Y50Q        | 1.0 ± 0.5     | 0.24 ± 0.02   | 1.3 ± 0.3    |
|           | Y50S        | 5 ± 1         | 1.1 ± 0.1     | 3110 ± 5     |
|           | Y50T        | 3.5 ± 0.4     | 2.5 ± 0.2     | 290 ± 53     |
| Small     | Y50G        | 0.72 ± 0.02   | 0.40 ± 0.06   | 0.58 ± 0.07  |
|           | Y50P        | 5 ± 1         | 40 ± 3        | 220 ± 28     |
| Charged   | Y50H        | 1.4 ± 0.1     | 5.0 ± 0.1     | 230 ± 2      |
|           | Y50R        | 11 ± 1        | 69 ± 2        | >1500        |
|           | Y50K        | 260 ± 10      | 195.3 ± 0.2   | 180 ± 47     |
|           | Y50D        | 3.2 ± 0.3     | 160 ± 23      | >4500        |
|           | Y50E        | 5.5 ± 0.1     | 76 ± 6        | 331 ± 6      |

BLIP Y50I could not be purified

Another common trend for binding all three β-lactamases is that substitutions of BLIP Y50 by the small amino acids alanine and glycine either does not affect or improves affinity (Table 3). For example, BLIP Y50A retains affinity for KPC-2 and Bla1 and exhibits 50-fold tighter binding of TEM-1 while Y50G shows a small increase in affinity for all three enzymes. As noted above and shown in Figs. 1 and 2, Tyr105 in TEM-1 and the equivalent Trp105 in KPC-2 change position in the BLIP-β-lactamase complexes compared to the apo-enzymes in order to avoid a steric clash with BLIP Y50. It is possible that substitution of Y50 with alanine or glycine avoids the clash and allows β-lactamase residue 105 to retain its apo-position in the complex, which may result in improved affinity.

Finally, polar residue substitutions at BLIP Y50 have quite disparate effects on binding the β-lactamases. For example, serine and threonine substitutions have relatively small effects on binding KPC-2 and TEM-1 but result in greatly decreased binding to Bla1 (Table 3). In addition, glutamine at BLIP position 50 does not affect binding to any of the β-lactamases while an asparagine substitution results in decreased affinity for all three β-lactamases, including a greater than 10-fold decrease for binding TEM-1 and Bla1 (Table 3). This result cannot easily be explained because asparagine has similar properties to glutamine, which had little to no effect on binding any of the β-lactamases.

## Impact of BLIP Y50 on binding specificity

Because the purpose of this study was to examine the role of BLIP position 50 as a specificity determinant, substitutions that have differential effects on binding are of interest. As indicated above and is apparent in Tables 3 and 4, many substitutions at BLIP Y50 have differential effects on β-lactamase binding, the most clear example being the numerous substitutions that retain or modestly impact binding to KPC-2 while greatly decreasing binding to Bla1 (Y50-L,M,W,N,S,T,H), and the subset of substitutions that retain binding to KPC-1 and TEM-1 while losing affinity for Bla1 (Y50-L,M,S,T). In contrast, there are no BLIP Y50 substitutions that retain affinity for Bla1 while losing affinity for KPC-2 or TEM-1 (Table 3). Thus, the large differences in stringency of sequence requirements at position 50 results in BLIP variants that bind KPC-2 but not TEM-1 and Bla1 as well as those that bind KPC-2 and TEM-1 but not Bla1. However, substitutions at BLIP Y50 do not produce a variant that binds Bla1 but not TEM-1 or KPC-2.

It was next of interest to examine a possible structural basis for the observed differences in sequence requirements for BLIP Y50 substitutions for binding Bla1 versus KPC-2 and TEM-1. The side chain of BLIP Y50 is in direct contact with β-lactamase residues 107, 129 and 216 in the crystal structures of the BLIP-TEM-1 and BLIP-KPC-2 complexes (Fig. 2). These β-lactamase contact residues are identical between KPC-2 and Bla1 (P107-Y129-T216) while TEM-1 differs at 2 of the 3 positions (P107-M129-V216) (Fig. 2). Based on these sequences, it would be expected that substitutions at BLIP Y50 would have similar effects on binding KPC-2 and Bla1. The results indicate this is clearly not the case. Therefore, a simple comparison of the β-lactamase contact residues for BLIP Y50 does not explain the observed differences in effects of substitutions on binding the β-lactamases. Although KPC-2 and Bla1 have the same amino acids at positions 107, 129 and 216, the overall sequence identity of all β-lactamase residues at the interface is higher between Bla1 and TEM-1 compared to KPC-2 (Table 2). There are a total of 10 positions (71, 102, 106, 109, 112, 133, 172, 246, 248 and 249) on the β-lactamase interface where TEM-1 and Bla1 have the same sequence and the sequence of KPC-2 differs (Fig. 2). KPC-2 is much better at accommodating changes at BLIP Y50 than both Bla1 and TEM-1 and has the most sequence differences at the interface. Therefore, more widespread differences in the entire interface may influence the effect of substitutions at BLIP Y50. This may be due to changes at the interface induced by mutation of BLIP Y50 that propagate outside of its local environment. In fact, previous studies have shown that BLIP Tyr50 is energetically coupled to both positions Tyr143 and Glu73,

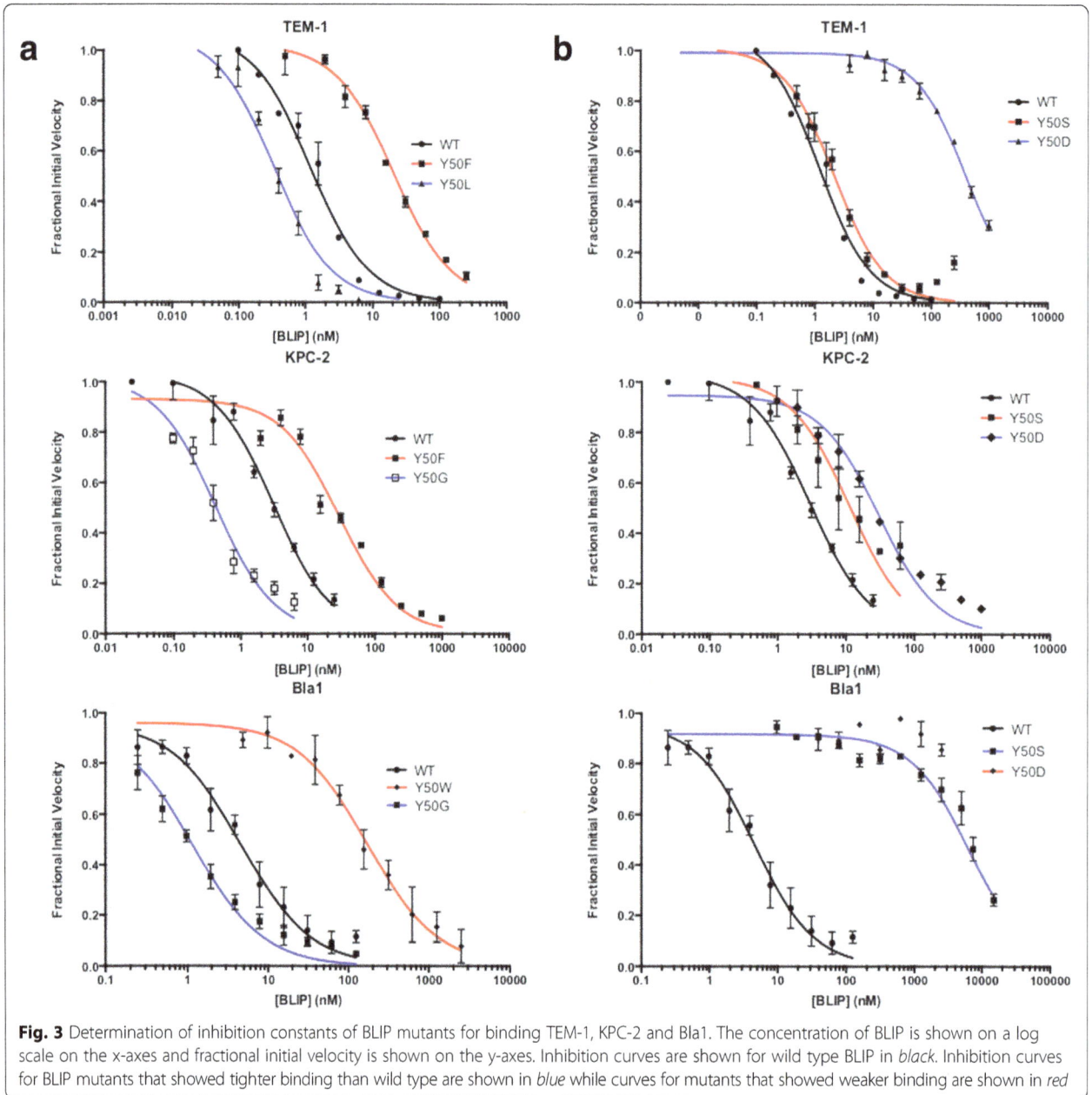

**Fig. 3** Determination of inhibition constants of BLIP mutants for binding TEM-1, KPC-2 and Bla1. The concentration of BLIP is shown on a log scale on the x-axes and fractional initial velocity is shown on the y-axes. Inhibition curves are shown for wild type BLIP in *black*. Inhibition curves for BLIP mutants that showed tighter binding than wild type are shown in *blue* while curves for mutants that showed weaker binding are shown in *red*

which are not in direct contact with Tyr50 [13]. The hypothesis that changes at position 50 are influenced by other sites in the interface and vice versa is consistent with previous observations of structural plasticity and cooperativity of the BLIP interface upon mutation [12, 13, 15, 16, 27]. For example, it has been shown that the BLIP W150A mutation induces a greater than 4 Å shift in residue Asp49, demonstrating both structural flexibility of the loop containing Tyr50 and long distance coupling at the BLIP interface [28].

An interesting question is whether there are also more stringent sequence requirements at other BLIP positions in the interface for binding Bla1 versus

TEM-1 and KPC-2. A previous alanine-scanning mutagenesis study for 23 BLIP residues that contact β-lactamase in the bound complex evaluated binding to TEM-1 and Bla1 (KPC-2 was not evaluated) [27]. The results of this study suggest that the sequence requirements are not generally more stringent for BLIP binding Bla1 in that the average ΔΔG effect for the 23 alanine substitutions was less detrimental for binding Bla1 (avg ΔΔG = 0.4) than for TEM-1 (avg ΔΔG = 1.0) [27]. Therefore, the stringent sequence requirements observed here for Bla1 binding are unique to position 50 and not a general property of all interface positions for binding Bla1.

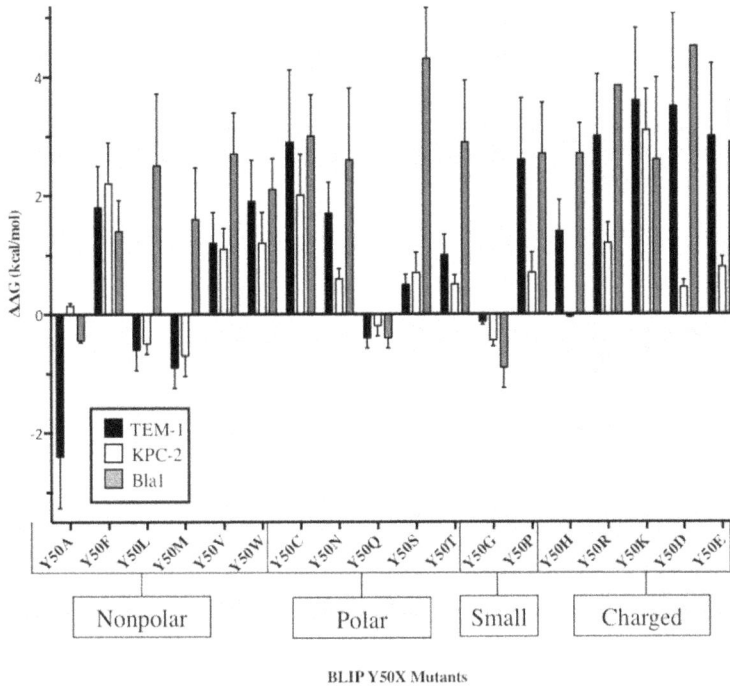

**Fig. 4** Comparison of ΔΔG values of BLIP mutants for binding TEM-1, KPC-2 and Bla1 β-lactamases. BLIP Y50X mutants are shown on the x-axis and the calculated change in free energy is shown on the y-axis. Values for TEM-1 are shown in *black*, KPC-2 in *white* and Bla1 in *gray*

**Table 4** ΔΔG values for Class A β-lactamases and BLIP Y50 mutants

|  | BLIP Mutant | ΔΔG(kcal/mol) KPC-2 | ΔΔG(kcal/mol) TEM-1 | ΔΔG(kcal/mol) Bla1 |
|---|---|---|---|---|
| Nonpolar | Y50A | 0.14 ± 0.03 | -2.4 ± 0.5 | -0.44 ± 0.07 |
|  | Y50F | 2.2 ± 0.4 | 1.8 ± 0.4 | 1.4 ± 0.3 |
|  | Y50L | -0.5 ± 0.1 | -0.6 ± 0.2 | 2.5 ± 0.7 |
|  | Y50M | -0.7 ± 0.2 | -0.9 ± 0.2 | 1.6 ± 0.5 |
|  | Y50V | 1.1 ± 0.2 | 1.2 ± 0.3 | 2.7 ± 0.4 |
|  | Y50W | 1.2 ± 0.3 | 1.9 ± 0.4 | 2.1 ± 0.4 |
| Polar | Y50C | 2.0 ± 0.4 | 2.9 ± 0.7 | 3.0 ± 0.4 |
|  | Y50N | 0.6 ± 0.1 | 1.7 ± 0.3 | 2.6 ± 0.7 |
|  | Y50Q | -0.2 ± 0.1 | -0.4 ± 0.1 | -0.4 ± 0.1 |
|  | Y50S | 0.7 ± 0.2 | 0.5 ± 0.1 | 4.3 ± 0.5 |
|  | Y50T | 0.51 ± 0.09 | 1.0 ± 0.2 | 2.9 ± 0.6 |
| Small | Y50G | -0.44 ± 0.06 | -0.13 ± 0.03 | -0.9 ± 0.2 |
|  | Y50P | 0.7 ± 0.2 | 2.6 ± 0.6 | 2.7 ± 0.5 |
| Charged | Y50H | 0.042 ± 0.006 | 1.4 ± 0.3 | 2.7 ± 0.3 |
|  | Y50R | 1.2 ± 0.2 | 3.0 ± 0.6 | >3.9 |
|  | Y50K | 3.1 ± 0.4 | 3.6 ± 0.7 | 2.6 ± 0.7 |
|  | Y50D | 0.46 ± 0.07 | 3.5 ± 0.9 | >4.5 |
|  | Y50E | 0.8 ± 0.1 | 3.0 ± 0.7 | 2.9 ± 0.4 |

BLIP Y50I could not be purified

## Comparison of experimental and predicted ΔΔG values

It is appealing to use computational methods to guide engineering of binding specificity in protein-protein interactions. Therefore, we were interested to examine how computational methods compared to our experimental results. The same mutagenesis experiment was performed computationally on BLIP position 50 with the BeAtMuSiC server, which computes theoretical ΔΔG values based on a set of statistical potentials derived from known protein structures [11]. The TEM-1-BLIP (PDB code: 1JTG) and KPC-2-BLIP (PDB code: 2OV5) complexes were submitted for analysis. The BLIP-Bla1 interaction was not analyzed because there is currently no crystal structure available of this complex.

The ΔΔG values generated by the BeAtMuSiC server and the experimentally determined ΔΔG values are plotted in Fig. 5. The average predicted ΔΔG value for the TEM-1/BLIP interaction was 2.0 kcal/mol while the experimentally determined average ΔΔG value was 1.3 kcal/mol. For the KPC-2-BLIP interaction, the average predicted ΔΔG was 2.2 kcal/mol and the experimentally determined average ΔΔG was 0.7 kcal/mol. Therefore, the BeAtMuSiC server was more accurate at predicting ΔΔG values for the BLIP-TEM-1 interaction than the BLIP-KPC-2 interaction.

## Discussion

Specificity determinants are often identified through alanine scanning of interface residues [16–18, 29, 30].

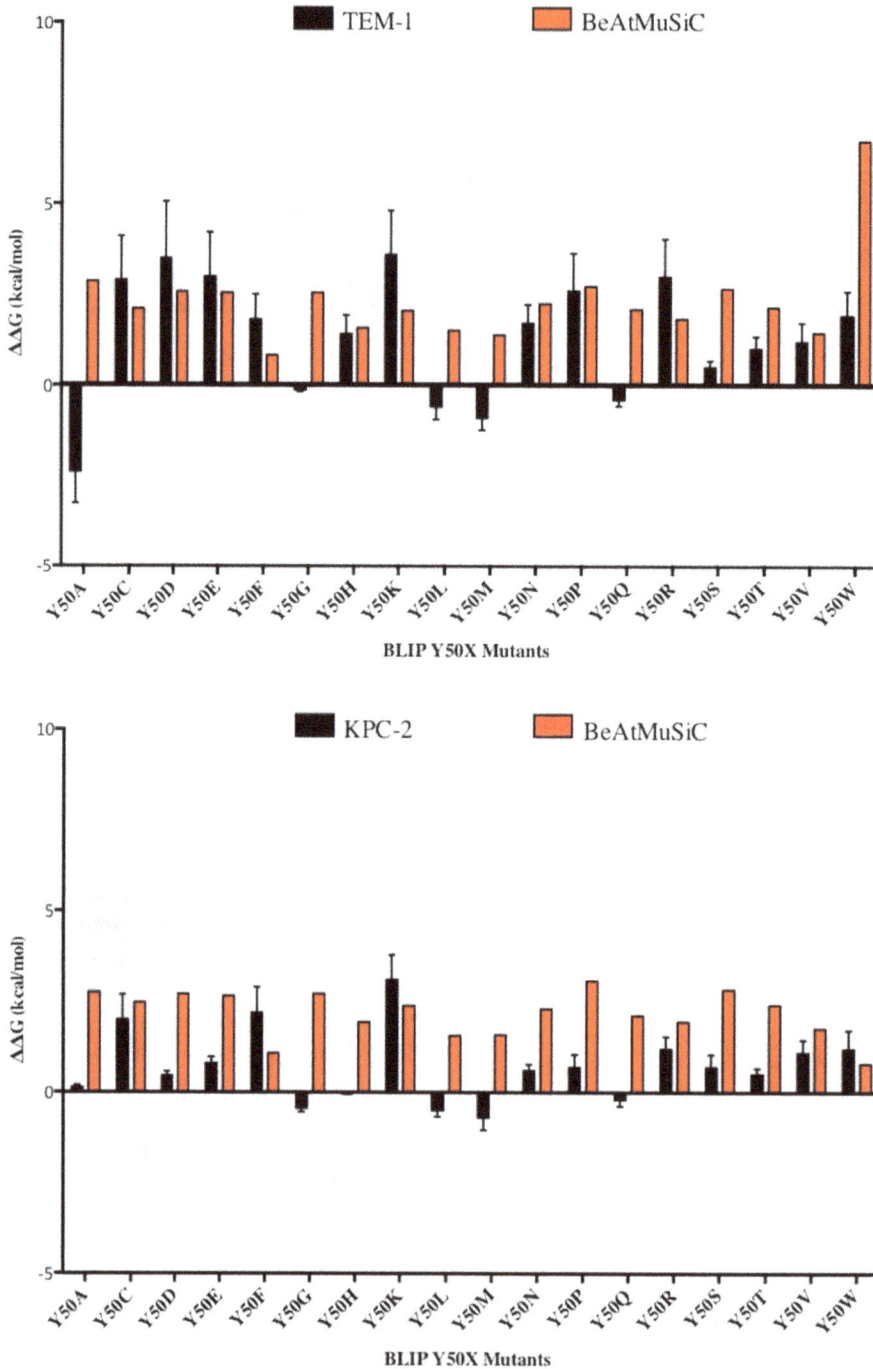

**Fig. 5** Comparison of experimental and predicted ΔΔG values for BLIP Y50 variants. A comparison of the experimental and predicted ΔΔG values of the BLIP Y50X mutants for binding TEM-1 are shown on the top panel. A comparison of the experimental and predicted ΔΔG values of the BLIP Y50X mutants for binding KPC-2 are shown on the bottom panel. The BeAtMuSiC server was used to generate predicted ΔΔG values. The predicted values are shown in *red* and the experimental values are shown in *black*

Whether mutations to other amino acids would also identify these residues as specificity determinants is unknown. Here, we present data supporting the role of BLIP Y50 as a specificity determinant and furthermore, provide evidence that this position can be targeted to

engineer binding specificity of BLIP for a range of class A β-lactamases.

The BeAtMuSiC server did not predict any negative changes in free energy meaning that the energy of the wild-type complex is predicted to be more stable than any of the mutants. However, the BLIP Y50A substitution was shown experimentally to bind TEM-1 β-lactamase 50-fold tighter than wild type BLIP. It is known that some residues rearrange upon binding of BLIP and class A β-lactamases and this could influence the accuracy of the prediction programs [27]. These rearrangements could play into the differences seen (about 1.4 kcal/mol) between our experimental $\Delta\Delta G$ values and the predicted values. Furthermore, BLIP Y50 is located on a loop that has two hot spots for binding; therefore, even small alterations in the placement of this loop induced by Y50 substitutions could result in large changes in binding affinity. In addition, as described above, the effect of substitutions at BLIP Y50 may be influenced by positions outside of the direct contact residues through coupled interactions and therefore predictions of effects of substitutions poses a significant challenge for computational prediction programs. However, it may be these same properties that provide BLIP with its unique ability to bind structurally homologous proteins with a wide-range of affinities.

Because BLIP binds homologous β-lactamase structures that only differ by small changes in sequence at the interface, the BLIP-β-lactamase system is useful for examining how sequence dictates binding affinity. However, an alignment of the β-lactamase sequences in direct contact with the BLIP Y50 residue (TEM-1 residues 107, 129 and 216) suggests that KPC-2 and Bla1 would exhibit the same changes in binding affinity upon mutation of BLIP Y50 because they have the same amino acids in similar conformations at these positions; however, this was not the case (Fig. 2). Although KPC-2 and Bla1 have the same amino acids at positions 107, 129 and 216 that directly interact with Y50, the overall sequence identity of all β-lactamase residues at the interface is higher between Bla1 and TEM-1 compared to KPC-2 (Table 2). KPC-2 was much better at accommodating changes at BLIP Y50 than both Bla1 and TEM-1 and had the most sequence differences at the interface. This suggests that simply comparing sequence identity of positions that make direct interactions with BLIP Y50 (or any BLIP residue) is not sufficient to predict changes in binding affinity upon mutation. Therefore, more widespread differences in the entire interface may influence the effect of substitutions at BLIP Y50. This may be due to changes at the interface induced by mutation of BLIP Y50 that propagate outside of its local environment due to structural plasticity and coupled interactions.

## Conclusions

Properties such as structural plasticity and cooperativity between residues are important for mediating protein interactions and critical for allosteric regulation in various cell processes [31–34]. Understanding how these properties contribute to binding specificity would greatly improve current protein binding prediction programs. This is an active area of investigation in G-protein coupled receptors, the human growth hormone receptor and other proteins [31–34]. Numerous studies such as these have established that the dynamic nature of proteins is critical to binding and proper functioning; however, this dynamic nature is challenging to predict and structurally understand, as flexible proteins are inherently difficult to model and crystallize. Here, we demonstrate the complexity of predicting the impact of substitutions using the well-studied BLIP-β-lactamase protein-protein interaction model. Furthermore, we have shown that surveying sequence homology and the structural interface of a complex are not sufficient in predicting the impact of mutations.

Currently, protein prediction programs are unable to reliably predict changes in binding affinity upon mutation at the protein interface. Systematic studies such as these could improve the current state by providing experimental data to be incorporated into these programs. Lastly, there is a pressing need for new detection methods for β-lactamases, which are a widespread source of resistance to β-lactam antibiotics. Identification of specificity determinants in BLIP could be useful in the development of BLIP-based diagnostic reagents that can discriminate between class A β-lactamases and inform treatment options for clinicians.

## Methods
### Construction of BLIP Y50 mutants
BLIP position 50 was mutated to all 19 amino acids using the Quickchange method (Stratagene) and *Pfu* polymerase (Stratagene) on the pGR32 plasmid with an N-terminal His-tag as previously described [35]. DNA sequencing was used to confirm the mutations and that no extraneous mutations occurred elsewhere on the BLIP gene in each of the mutants (Lonestar Labs).

### Protein purification
N-terminal His-tagged BLIP mutants were purified using the TALON Metal Affinity Resin (Clontech) [35]. Despite multiple attempts, the BLIP Y50I mutant could not be purified due to poor expression and yield. The TEM-1 and Bla1 proteins were purified as previously described using a zinc chelating column and elution with a pH gradient [36]. KPC-2 was purified as previously described using a HiTrap SP column and elution with an NaCl gradient [37]. The BLIP mutants and the various β-lactamases were each concentrated and injected onto a Superdex 75 gel filtration size exclusion column as a final purification step. Fractions with greater than 90% purity as determined by SDS-PAGE were combined, concentrated and used in the inhibition assay. The

protein concentrations for the β-lactamases and BLIP mutants were determined by a Bradford assay where they were compared with a curve that was calibrated by quantitative amino acid analysis specific to each protein. The concentrations for all proteins were confirmed by measuring absorbance at 280 nm and using the extinction coefficient as determined by the ExPASy ProtParam tool [38]. Kinetic parameters ($k_{cat}$ and $K_m$) were determined to confirm activity for each β-lactamase using the chromogenic β-lactam substrate, nitrocefin (data not shown).

### β-lactamase inhibition assay

Inhibition constants for BLIP mutants binding to the β-lactamases were determined as previously described [35]. Increasing concentrations of BLIP were incubated with a constant concentration of β-lactamase (1 nM) for 1 h at room temperature in 50 mM sodium phosphate buffer pH 7.0. The chromogenic substrate, nitrocefin, was then added at the $K_m$ concentration for the β-lactamases and the initial velocity was measured at 482 nm in 20 s intervals. The experiments were performed in at least duplicate. The $K_i^{app}$ for each BLIP mutant was determined by fitting the initial velocities to the Morrison tight-binding equation [39]:

$$E_{free} = [E_0] - \frac{[E_0] + [I_0] + K_i^{app} - \sqrt{\left([E_0] + [I_0] + K_i^{app}\right)^2 - (4[E_0][I_0])}}{2} \quad (1)$$

where $E_{free}$ is the concentration of free enzyme determined by residual activity of the β-lactamase by comparison with the initial velocity of nitrocefin hydrolysis by the uninhibited β-lactamase, $[E_0]$ is the total enzyme concentration and $[I_0]$ is the total BLIP concentration. The errors reported were calculated based on the fit of the curve. The $K_i$ values were calculated from the $K_i^{app}$ values as previously described using eq. 2 [40]:

$$Ki = K_i^{app}/(1 + ([S]/K_M)) \quad (2)$$

### ΔΔG calculations

ΔΔG was calculated using the following equation:

$$\Delta\Delta G = -RT \ln\left(K_i^{WT}/K_i^{MUT}\right) \quad (3)$$

Using this equation, a decrease in $K_i$ upon mutation would result in a negative ΔΔG value while an increase in $K_i$ would be reported as a positive change in free energy. Error for ΔΔG values was calculated using the following equation:

$$\Delta\Delta G \text{ error} = \Delta\Delta G \sqrt{\frac{SEM^2}{K_i^{WT}} + \frac{SEM^2}{K_i^{MUT}}} \quad (4)$$

Where 'SEM' represents standard error of the mean, 'WT' represents wild-type BLIP and 'MUT' represents the mutant protein.

### Computational prediction of the effect of BLIP mutations on binding affinity

The BeAtMuSiC online server was used to predict changes in binding affinity of the BLIP mutants on complex formation with TEM-1 and KPC-2 β-lactamases [11]. The BeAtMuSiC server relies on a set of statistical potentials derived from known protein structures and predicts the changes in binding affinity by the combined effect of the mutation on the overall stability of the complex and the interface [11]. PDB codes 1JTG (BLIP/TEM-1) and 3E2K (BLIP/KPC-2) (chains A and B) were submitted for analysis.

### Abbreviation
BLIP: β-lactamase inhibitory protein

### Acknowledgement
None declared.

### Funding
This work was supported by NIH Grant no. AI32956 to T.P. This work was also supported by a training fellowship from the Houston Area Molecular Biophysics Training Program (NIGMS Grant No. T32 GM008280) of the Gulf Coast Consortia to C.J.A.

### Authors' contributions
CJA and TP designed experiments. CJA performed experiments. CJA and TP prepared the manuscript. Both author read and approved the final manuscript.

### Competing interests
The authors declare that they have no competing interests.

### References
1.  Kuzu G, Keskin O, Gursoy A, Nussinov R. Constructing structural networks of signaling pathways on the proteome scale. Curr Opin Struc Biol. sciencedirect; 2012;22:367–377
2.  Jubb H, Higueruelo A, Winter A, Blundell T. Structural biology and drug discovery for protein–protein interactions. Trends Pharmacol Sci. 2012;33: 241–248.
3.  Shoichet BK, Baase WA, Kuroki R, Matthews BW. A relationship between protein stability and protein function. Proc Nat Acad Sci. 1995;92(2):452–6.
4.  Selzer T, Albeck S, Schreiber G. Rational design of faster associating and tighter binding protein complexes. Nat. Struct. Biol. 2000;7:537–41.
5.  Kim DE, Chivian D, Baker D. Protein structure prediction and analysis using the Robetta server. Nucleic Acids Res. 2004;32(Web Server issue):31.
6.  Reichmann D, Rahat O, Albeck S, Meged R, Dym O, Schreiber G. The modular architecture of protein-protein binding interfaces. Proc Nat Acad Sci. 2005; 102(1):57–62.
7.  Reichmann D, Phillip Y, Carmi A, Schreiber G. On the contribution of water-mediated interactions to protein-complex stability. Biochemistry. 2008;47:1051–60.

8. Harel M, Spaar A, Schreiber G. Fruitful and futile encounters along the association reaction between proteins. Biophysical J. 2009;96(10):4237–48.

9. Jr R, Poelwijk F, Raman A, Gosal W, Ranganathan R. The spatial architecture of protein function and adaptation. Nature. nature; 2012;491:138–142.

10. Baskaran K, Duarte JM, Biyani N, Bliven S, Capitani G. A PDB-wide, evolution-based assessment of protein-protein interfaces. BMC Struct Biol. 2014;14(1):22.

11. Dehouck Y, Kwasigroch JM, Rooman M, Gilis D. BeAtMuSiC: Prediction of changes in protein-protein binding affinity on mutations. Nucleic Acids Res. 2013;41(Web Server issue):9.

12. Zhang Z, Palzkill T. Determinants of binding affinity and specificity for the interaction of TEM-1 and SME-1 beta-lactamase with beta-lactamase inhibitory protein. J. Biol. Chem. 2003;278:45706–12.

13. Zhang Z, Palzkill T. Dissecting the Protein-Protein Interface between β-Lactamase Inhibitory Protein and Class A β-Lactamases. J Biol Chem. highwire; 2004;279:42860–42866.

14. Reynolds KA, Thomson JM, Corbett KD, Bethel CR, Berger JM, Kirsch JF, et al. Structural and computational characterization of the SHV-1 beta-lactamase-beta-lactamase inhibitor protein interface. J. Biol. Chem. 2006;281:26745–53.

15. Gretes M, Lim DC, de Castro L, Jensen SE, Kang SG, Lee KJ, Strynadka NC. Insights into positive and negative requirements for protein-protein interactions by crystallographic analysis of the beta-lactamase inhibitory proteins BLIP, BLIP-I, and BLP. J Mol Biol. 2009;389(2):289–305.

16. Reichmann D, Cohen M, Abramovich R, Dym O, Lim D, Strynadka NC, Schreiber G. Binding hot spots in the TEM1-BLIP interface in light of its modular architecture. J Mol Biol. 2007;365(3):663–79.

17. Keskin O, Ma B, Nussinov R. Hot regions in protein–protein interactions: the organization and contribution of structurally conserved hot spot residues. J Mol Biol. 2005;345(5):1281–94.

18. Moreira I, Fernandes P, Ramos M. Hot spots—A review of the protein–protein interface determinant amino-acid residues. Proteins Struct Funct Bioinform. Wiley. 2007;68:803–812.

19. Kozakov D, Hall DR, Chuang G-YY, Cencic R, Brenke R, Grove LE, Beglov D, Pelletier J, Whitty A, Vajda S. Structural conservation of druggable hot spots in protein-protein interfaces. Proc Nat Acad Sci. 2011;108(33):13528–33.

20. Doran JL, Leskiw BK, Aippersbach S, Jensen SE. Isolation and characterization of a beta-lactamase-inhibitory protein from Streptomyces clavuligerus and cloning and analysis of the corresponding gene. J Bacteriology. 1990;172(9):4909–18.

21. Strynadka NC, Jensen SE, Johns K, Blanchard H, Page M, Matagne A, et al. Structural and kinetic characterization of a beta-lactamase-inhibitor protein. Nature. 1994;368:657–60.

22. Strynadka NC, Jensen SE, Alzari PM, James MN. A potent new mode of beta-lactamase inhibition revealed by the 1.7 A X-ray crystallographic structure of the TEM-1-BLIP complex. Nat. Struct. Biol. 1996;3:290–7.

23. Bush K, Fisher JF. Epidemiological expansion, structural studies, and clinical challenges of new β-lactamases from Gram-negative bacteria. Ann Rev Microbiol. 2011;65:455–78.

24. Fishovitz J, Hermoso JA, Chang M, Mobashery S. Penicillin-binding protein 2a of methicillin-resistant Staphylococcus aureus. IUBMB Life. 2014;66:572–7.

25. Hanes MS, Jude KM, Berger JM, Bonomo RA, Handel TM. Structural and biochemical characterization of the interaction between KPC-2 beta-lactamase and beta-lactamase inhibitor protein. Biochemistry. 2009;48(39):9185–93.

26. Brown NG, Chow DC, Sankaran B, Zwart P, Prasad BV, Palzkill T. Analysis of the binding forces driving the tight interactions between beta-lactamase inhibitory protein-II (BLIP-II) and class A beta-lactamases. J Biol Chem. 2011;286(37):32723–35.

27. Yuan J, Huang W, Chow DC, Palzkill T. Fine mapping of the sequence requirements for binding of beta-lactamase inhibitory protein (BLIP) to TEM-1 beta-lactamase using a genetic screen for BLIP function. J. Mol. Biol. 2009;389:401–12.

28. Wang J, Palzkill T, Chow DC. Structural insight into the kinetics and DeltaCp of interactions between TEM-1 beta-lactamase and beta-lactamase inhibitory protein (BLIP). J. Biol. Chem. 2009;284:595–609.

29. Bogan AA, Thorn KS. Anatomy of hot spots in protein interfaces. J Mol Biol. 1998;280(1):1–9.

30. DeLano WL. Unraveling hot spots in binding interfaces: progress and challenges. Curr Opin Struct Biol. 2002;12(1):14–20.

31. Christopoulos A. Advances in G protein-coupled receptor allostery: from function to structure. Mol Pharm. 2014;86(5):463–78.

32. Comps-Agrar L, Maurel D, Rondard P, Pin J-PP, Trinquet E, Prézeau L. Cell-surface protein-protein interaction analysis with time-resolved FRET and snap-tag technologies: application to G protein-coupled receptor oligomerization. Meth Mol Biol (Clifton, NJ). 2010;756:201–14.

33. Pál G, Ultsch MH, Clark KP, Currell B, Kossiakoff AA, Sidhu SS. Intramolecular cooperativity in a protein binding site assessed by combinatorial shotgun scanning mutagenesis. J Mol Biol. 2005;347(3):489–94.

34. Xue L, Rovira X, Scholler P, Zhao H, Liu J, Pin J-PP, Rondard P. Major ligand-induced rearrangement of the heptahelical domain interface in a GPCR dimer. Nature Chem Biol. 2015;11(2):134–40.

35. Petrosino J, Rudgers G, Gilbert H, Palzkill T. Contributions of aspartate 49 and phenylalanine 142 residues of a tight binding inhibitory protein of beta-lactamases. J. Biol. Chem. 1999;274:2394–400.

36. Marciano DC, Brown NG, Palzkill T. Analysis of the plasticity of location of positive charge within the active site of the TEM-1 β-lactamase. Protein Sci. 2009;18:2080–9.

37. Brown NG, Chow DC, Ruprecht KE, Palzkill T. Identification of the β-lactamase inhibitor protein-II (BLIP-II) interface residues essential for binding affinity and specificity for class A β-lactamases. J Biol Chem. 2013;288: 17156–66.

38. Gasteiger EHC, Gattiker A, Duvaud S, Wilkins MR, Appel RD, Bairoch A.Protein Identification and Analysis Tools on the ExPASy Server. The Proteomics Protocols Handbook. Walker JM (Ed). Totowa: Humana Press; 2005:571-607.

39. Murphy DJ. Determination of accurate KI values for tight-binding enzyme inhibitors: an in silico study of experimental error and assay design. Anal Biochem. 2004;327(1):61–7.

40. Brown NG, Palzkill T. Identification and characterization of beta-lactamase inhibitor protein-II (BLIP-II) interactions with beta-lactamases using phage display. PEDS. 2010;23(6):469–78.

41. Pettersen EF, Goddard TD, Huang CC, Couch GS, Greenblatt DM, Meng EC, Ferrin TE. UCSF chimera - a visualization system for exploratory research and analysis. J Comput Chem. 2004;25:1605–12.

# 12

# Binding of smoothelin-like 1 to tropomyosin and calmodulin is mutually exclusive and regulated by phosphorylation

Annegret Ulke-Lemée[1], David Hao Sun[1], Hiroaki Ishida[2], Hans J. Vogel[2] and Justin A. MacDonald[1*]

## Abstract

**Background:** The smoothelin-like 1 protein (SMTNL1) can associate with tropomyosin (Tpm) and calmodulin (CaM), two proteins essential to the smooth muscle contractile process. SMTNL1 is phosphorylated at Ser301 by protein kinase A during calcium desensitization in smooth muscle, yet the effect of SMTNL1 phosphorylation on Tpm- and CaM-binding has yet to be investigated.

**Results:** Using pull down studies with Tpm-Sepharose and CaM-Sepharose, we examined the interplay between Tpm binding, CaM binding, phosphorylation of SMTNL1 and calcium concentration. Phosphorylation greatly enhanced the ability of SMTNL1 to associate with Tpm in vitro; surface plasmon resonance yielded a 10-fold enhancement in $K_D$ value with phosphorylation. The effect on CaM binding is more complex and varies with the availability of calcium.

**Conclusions:** Combining both CaM and Tpm with SMTNL1 shows that the binding to both is mutually exclusive.

**Keywords:** SMTNL1, CHASM, Smooth muscle, Thin filament, Protein binding

## Background

Smoothelin-like 1 protein (SMTNL1), also termed calponin homology-associated with smooth muscle (CHASM) protein, was originally identified as a protein phosphorylated in response to cyclic nucleotide-dependent protein kinase (i.e., protein kinase A (PKA) and protein kinase G (PKG)) activation during calcium desensitization of gastrointestinal smooth muscle [1]. Additional studies have revealed a role for SMTNL1 in vascular smooth muscle contractile activity, as well as cardiovascular and skeletal muscle adaptation to exercise, development and pregnancy [2–6].

The ubiquitous calmodulin (CaM) is the primary cellular signal transducer that provides spatial and temporal responses to changes in intracellular calcium levels [7]. CaM provides signaling by association with specific binding sites in target proteins, both in its calcium-saturated (Ca-CaM) or its calcium-free (apo-CaM) form [8, 9]. We have previously reported that SMTNL1 possesses two CaM-binding sites: a classic CaM-binding domain (CBD1, amino acids 310–325) in the center of the protein as well as an IQ-motif that serves as an apo-CaM-binding site (CDB2, amino acids 439–457) located in the carboxy-terminal calponin homology domain [10–12]. SMTNL1 was established as a *bona fide* CaM-binding protein within aortic A7r5 smooth muscle cells using the proximity ligation assay. In this regard, CBD1 is thought to provide the majority of CaM-binding in situ; however, CBD2 may contribute cooperatively to binding [12].

A second binding partner of SMTNL1 has also been identified in smooth muscle. Tropomyosin (Tpm), a muscle protein that aids in the stabilization of actin filaments, is incorporated as super-helical polymers along smooth muscle thin filaments and contribute to the regulation of actin-myosin cross bridge cycling during smooth muscle contraction [13–15]. In smooth muscle, two Tpm isoforms (Tpm1.4 and Tpm2.1; previously identified as Tmsm-α/Tm6 and Tmsm-β/Tm1β, respectively [16]) are predominantly expressed [17]. SMTNL1 was shown to associate with smooth muscle Tpm α/β dimers with a binding surface comprising of its N-terminal intrinsically-disordered region and the C-terminal calponin homology domain [18].

* Correspondence: jmacdo@ucalgary.ca
[1]Department of Biochemistry & Molecular Biology, University of Calgary, Cumming School of Medicine, 3280 Hospital Drive NW, Calgary, AB T2N 4Z6, Canada
Full list of author information is available at the end of the article

Previous in vitro studies have defined SMTNL1 as a Tpm- and CaM-binding protein [10–12, 18, 19]. Moreover, the phosphorylation of SMTNL1 was identified during calcium desensitization of smooth muscle [1, 5]. However, the effect of phosphorylation on the ability of SMTNL1 to associate with Tpm and/or CaM has not been investigated. Intriguingly, the binding domains for CaM (CBD1 and CBD2) are localized in close vicinity to the Tpm-binding surface on the SMTNL1 protein. In addition, the Ser301 phosphorylation site targeted by cyclic nucleotide-dependent protein kinases (i.e., PKA and PKG) is located near CBD1 and within the Tpm-binding region [11, 18]. This led us to inquire about the interplay of SMTNL1 phosphorylation and calcium levels on the multi-functionality of Tpm- and CaM-binding. Herein, we demonstrate that phosphorylation of Ser301 greatly enhanced the ability of SMTNL1 to associate with smooth muscle Tpm in vitro. The effect of Ser301 phosphorylation on CaM binding was more complex and varied with the availability of calcium. Combining both CaM and Tpm revealed that binding with SMTNL1 was mutually exclusive.

## Methods

### Materials

PreScission Protease, glutathione-Sepharose and CNBr-activated Sepharose were from GE Healthcare (Piscataway, NJ). The Phos-tag acrylamide reagent was from NARD Chemicals (Kobe City, Japan). Smooth muscle Tpm was a purified α/β heterodimer of chicken gizzard Tpm1.4$_{sm(a.a.b.d)}$ (equivalent to Tm6 or Tmsm-α) and Tpm2.1$_{sm(a.b.a.d)}$ (equivalent to TM-1, βTm or Tmsm-β) [16] and was provided by Dr. Michael Walsh (University of Calgary). CaM was provided by Dr. Hans Vogel (University of Calgary). PKA was purified from bovine heart using the method described in [20].

### Expression and Purification of SMTNL1

A clone comprising the Tpm-binding region of SMTNL1 (SMTNL1-TMB, amino acids 195–459; NP_077192) was described previously [18, 19]. SMTNL1-TMB was expressed as a glutathione S-transferase (GST)-fusion protein in E. coli and purified with glutathione-Sepharose. The fusion protein was cleaved with PreScission Protease, and GST was removed by an additional pass over glutathione-Sepharose along with a final cleanup using MonoQ anion exchange chromatography. In some cases, SMTNL1-TMB was phosphorylated with PKA at a molar ratio of 500:1 (SMTNL1: PKA) in a buffer consisting of 2 mM MgCl$_2$, 0.2 mM ATP, 25 mM HEPES and 150 mM NaCl. The phosphorylated protein was buffer-exchanged and concentrated using a 3 kDa cutoff centrifugation filter unit (EMD Millipore, Billerica, MA).

### Tpm-Sepharose binding assay

Purified smooth muscle Tpm or Tris-HCl (as control) was covalently bound to CNBr-activated Sepharose following the manufacturers' instructions and as described in [19]. SMTNL1-TMB (200 µg) was incubated with 40 µL of Tpm-Sepharose or Control-Sepharose in binding buffer (20 mM Tris-HCl, pH 7.5 with 0.1% (v/v) β-mercaptoethanol) for 2 h at 4 °C and then washed extensively with the same buffer supplemented with 150 mM NaCl. Bound protein was eluted with SDS-PAGE loading buffer, separated by gel electrophoresis, and detected by Coomassie stain.

### Calmodulin-Sepharose binding assay

Calmodulin (CaM)-Sepharose pull-down experiments were completed as described previously [11]. Briefly, SMTNL1-TMB (200 µg) was incubated with 40 µL of CaM-Sepharose in binding buffer (20 mM HEPES, pH 7.0 in the presence of 5 mM CaCl$_2$ (Ca-CaM) or 1 mM EDTA (apo-CaM)). After incubation for 1 h at 4 °C, the CaM-Sepharose was washed extensively with the same buffer supplemented with 150 mM NaCl. Bound SMTNL1-TMB protein was eluted with SDS-PAGE loading buffer, separated by gel electrophoresis, and detected by Coomassie stain. In some experiments, SMTNL1-TMB was premixed with Tpm in different molar ratios, allowed to form complexes for 1 h at 4 °C and then added to the CaM-Sepharose.

### Surface Plasmon Resonance (SPR)

The binding between Tpm and SMTNL1-TMB with or without phosphorylation was evaluated by SPR using a BIAcore X100 instrument (GE Healthcare). Purified Tpm was immobilized via amine-coupling onto a CM5 sensor chip (GE Healthcare). The running buffer contained 20 mM HEPES pH 7.5, 100 mM KCl, 1 mM DTT, and 0.005% (v/v) Tween-20. Five concentrations of SMTNL1-TMB from 0.12 to 10 µM were prepared by serial dilution and were injected at a flow rate of 30 µL/min with a contact time of 1 min at 25 °C. The chip was regenerated by injecting 1 M NaCl for 4 min followed by glycine-HCl (10 mM, pH 3.0) for 1 min. Each experiment was repeated three times ($n = 3$) to obtain a standard error (SE). The BIAevaluation software 2.0 (GE Healthcare) was used to analyze the SPR sensorgrams and to obtain the dissociation constants (Kd).

### Data analysis

Values are presented as the mean ± S.E.M., with $n$ indicating the number of independent experiments. Data were analyzed with two-tailed Student's $t$ test, or for comparison of multiple groups, with one-way analysis of variance (ANOVA) and Tukey's post hoc test. Differences were considered to be statistically significant for $p < 0.05$. All statistical analyses were performed using the GraphPad Prism 6.0 program.

## Results

### Phosphorylation of SMTNL1 by PKA enhances Tpm-binding potential

Full-length SMTNL1 protein exhibits reduced stability that complicates conclusions from in vitro binding experiments. A truncated form of SMTNL1 (i.e., SMTNL1-TMB) was used since this clone is stable under the conditions and times used; furthermore, SMTNL1-TMB contains all known functional elements of SMTNL1 (e.g., CBD1, CBD2, calponin-homology (CH) domain, Tpm-binding regions and Ser301 phosphorylation site; Fig. 1a). SMTNL1-TMB could be effectively phosphorylated with PKA; the phosphorylation stoichiometry was assessed with Phos-tag SDS-PAGE and judged to be complete (i.e., ~ 1 mol phosphate/mol SMTNL1) after 16 h incubation at 4 °C (Fig. 1b). The binding of SMTNL1-TMB to Tpm was examined in pull-down assays with purified smooth muscle Tpm (mixture of Tpm1.4 and Tpm2.1 isoforms) immobilized to Sepharose resin (Tpm-Sepharose). SMTNL1-TMB was recovered from Tpm-Sepharose while no interaction with Control-Sepharose was observed (Fig. 1c). Some Tpm could be released from the Sepharose column with application of SDS elution buffer and boiling. This was anticipated given that smooth muscle Tpm exists as coiled-coil heterodimers [21], and it was unlikely that all Tpm molecules could be covalently coupled to the CNBr-Sepharose resin. Approximately 2-fold more phosphorylated SMTNL1-TMB could be recovered from pull-down assays with Tpm-Sepharose (Fig. 1d) when compared with unphosphorylated SMTNL1-TMB. Binding constants for a kinetic series, 0–10 μM phosphorylated SMTNL1-TMB provided a $K_D$ of $3.2 \times 10^{-7}$M with biophysical analyses by SPR (Additional file 1: Figure S1), a response that was enhanced 10-fold when compared to unphosphorylated SMTNL1-TMB ($3.0 \times 10^{-6}$M). The SPR assays provided a similar $K_D$ for unphosphorylated SMTNL1 and Tpm as previously defined by isothermal titration calorimetry (ITC) assays [19]. We previously reported that the SMTNL1 binding surface for Tpm included a central portion of the intrinsically-disordered region, located proximal to the Ser301 phosphorylation site, as well as a surface of the calponin homology (CH) domain located at the C-terminus [18]. So, it is not unexpected that phosphorylation at Ser301 could alter the electrostatic properties and/or the intramolecular conformation of the Tpm-binding surface on SMTNL1.

### Phosphorylation of SMTNL1-TMB influences binding to CaM-Sepharose

The Ser301 phosphorylation site of SMTNL1 is located in close proximity to the Ca-CaM binding domain (CBD1, amino acids 310–325 [11]). It was expected that the phosphorylation of SMTNL1-TMB by PKA would impact the association with Ca-CaM and not with apo-

CaM, since the binding of the latter was linked to an IQ-domain at the C-terminus (CBD2, amino acids 439–457 [10]). Indeed, we observed reductions in Ca-CaM-Sepharose binding for phosphorylated SMTNL1-TMB, with recovery of approximately 30% less bound material (Fig. 2a). Surprisingly, the phosphorylation of SMTNL1-TMB protein also impacted upon binding to apo-CaM-Sepharose (Fig. 2b). Although a general reduction in SMTNL1-TMB bound to CaM was observed in the apo condition (i.e., ~60% less SMTNL1-TMB recovered on the resin), the recovery of phosphorylated SMTNL1-TMB from the resin was further reduced to approximately 20% of the unphosphorylated material (Fig. 2b). We have previously demonstrated that the binding of SMTNL1-TMB to CaM was influenced by calcium, with SMTNL1 binding to CaM-Sepharose decreased by approximately 50% in the apo-calcium condition [11].

### Phosphorylation and calcium influence SMTNL1-TMB binding to CaM-Sepharose in the presence of Tpm

Equimolar amounts of SMTNL1-TMB and Tpm proteins were pre-incubated and then applied to CaM-Sepharose so that competition between CaM and Tpm for binding to SMTNL1-TMB could be assessed. Binding of SMTNL1-TMB to apo-CaM-Sepharose was reduced by approximately 60% by the addition of Tpm to SMTNL1-TMB prior to capture with the resin (Fig. 2b). These data indicate that Tpm could interfere with the apo-CaM binding properties of SMTNL1-TMB, likely due to the proximity of the two binding sites within the calponin homology (CH) domain. Furthermore, we can conclude that SMTNL1-TMB interacts with Tpm more strongly than with apo-CaM. This was not unexpected as the interaction of SMTNL1 with apo-CaM is weak (Kd ~$10^{-6}$ M [10]). The results of the binding assays were not affected by the order of addition of proteins since pre-incubating SMTNL1-TMB with CaM-Sepharose followed by competition with Tpm showed the same reduction in binding (data not shown). The subsequent addition of Tpm to the mixture of phosphorylated SMTNL1 and apo-CaM-Sepharose, shown in Fig. 2b, further decreased binding potential to barely detectably levels. When investigating interactions with Ca-CaM, the binding of unphosphorylated SMTNL1-TMB to Ca-CaM-Sepharose was not significantly influenced by the addition of equimolar amounts of Tpm (Fig. 2a). However, the addition of Tpm did significantly reduce the recovery of phosphorylated SMTNL-TMB on the Ca-CaM-Sepharose resin.

### SMTNL1 phosphorylation diminishes Ca-CaM-binding potential under conditions of high Tpm content

The intracellular Tpm concentration is predicted to be significantly higher than that of SMTNL1, as judged

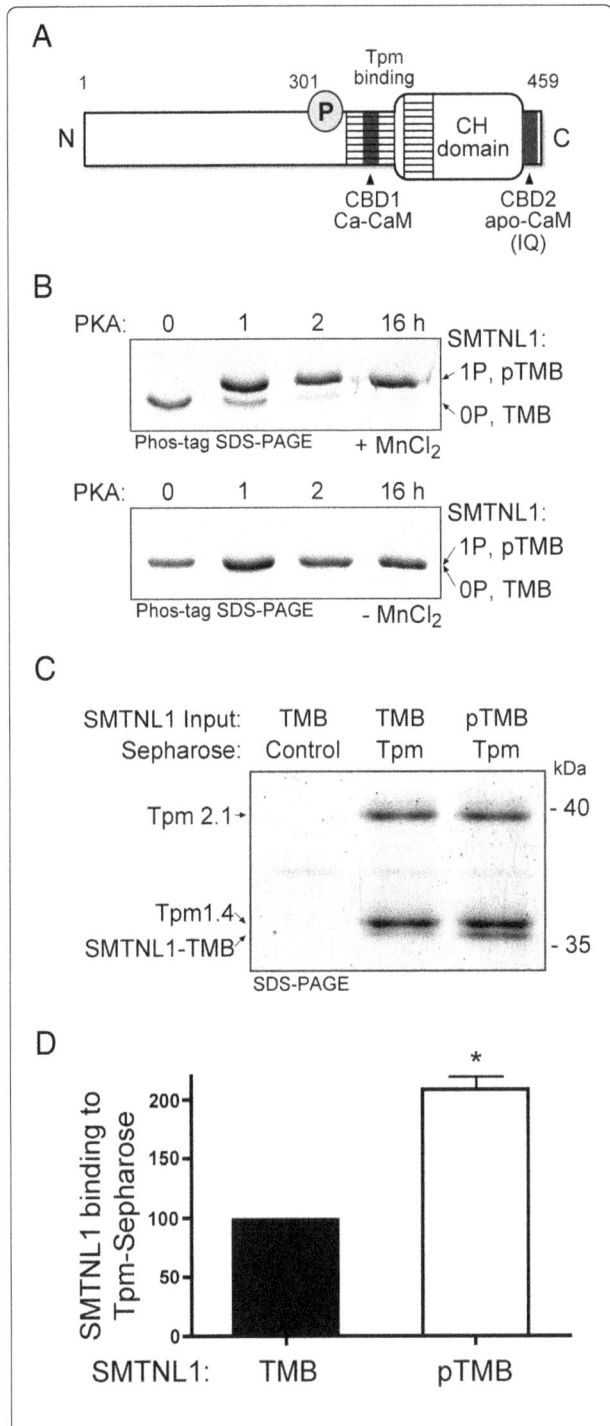

Fig. 1 The phosphorylation of SMTNL1 by PKA alters its tropomyosin-binding potential. **a** The SMTNL1 protein contains: a C-terminal calponin homology (CH) domain, Ser301 PKA-phosphorylation site, Ca-calmodulin (CaM)-binding domain (CBD1), apo-CaM-binding domain (CBD2), and tropomyosin (Tpm)-binding domain (indicated as the hatched area). **b** Purified recombinant SMTNL1-TMB was incubated with PKA as described in the Methods section. Samples were withdrawn at the indicated times and subjected to Phos-tag SDS-PAGE; two discrete bands representing unphosphorylated SMTNL1-TMB (0P, TMB) or SMTNL1-TMB phosphorylated at Ser301 (1P, pTMB) were detected by Coomassie stain. Samples subjected to Phos-tag SDS-PAGE in the absence of $MnCl_2$ confirmed the shift in band migration to be a result of phosphorylation. **c** Unphosphorylated or phosphorylated SMTNL1-TMB (200 μg) was incubated with 40 μL of Tpm-Sepharose (α/β-heterodimer: Tpm1.4/Tpm2.1). Bound SMTNL1-TMB was eluted with boiling 0.1% SDS solution and detected with Coomassie staining of SDS-PAGE gels. **d** The SMTNL1-TMB bands were quantified by densitometry, and binding to Tpm-Sepharose was expressed as percentage of the SMTNL1-TMB binding found for the unphosphorylated state. All experiments are $n = 3–5$ and were analyzed by Student's t-test. *- Significantly different from unphosphorylated SMTNL1-TMB, $p < 0.05$

from the relative staining intensities of 2D-SDS-PAGE gels [1]. So, the incubation of SMTNL1-TMB with different molar ratios of Tpm prior to the addition of CaM-Sepharose was used to reveal additional information about the character of the protein complex formation. Increasing the amounts of Tpm resulted in a notable decrease in the binding of SMTNL1-TMB to apo-CaM-Sepharose (Fig. 3a). Moreover, there was minimal influence of increased Tpm on the binding of phosphorylated SMTNL1-TMB in the apo-CaM condition (Fig. 3b). The small inhibitory effect of increasing Tpm concentration on the amount of SMTNL1-TMB retained on Ca-CaM Sepharose suggests a greater affinity of SMTNL1-TMB for Ca-CaM over Tpm (Fig. 3c). However, distinct effects were observed for Ca-CaM binding of phosphorylated SMTNL1-TMB under conditions of high Tpm (Fig. 3d). In this case, the phosphorylation of SMTNL1-TMB suppressed Ca-CaM binding potential in favour of Tpm.

## Discussion

Phosphorylation events are known to regulate binding of CaM to target proteins. Several examples of phosphorylation within or near CaM-binding domains (CBDs) have been reported in the literature, including but not limited to smooth muscle myosin light chain kinase [22], MARKS protein [23], plasma membrane $Ca^{2+}$-ATPase [24], endothelial nitric oxide synthase (eNOS) [25], and the $Ca^{2+}$-dependent $K^+$-channel [26]. CaM binds to its targets by wrapping around a short amphipathic helix within the CBD [8, 9], thus the phosphorylation of residues within a CBD can disrupt CaM interactions even if the Ser/Thr residues subject to phosphorylation are

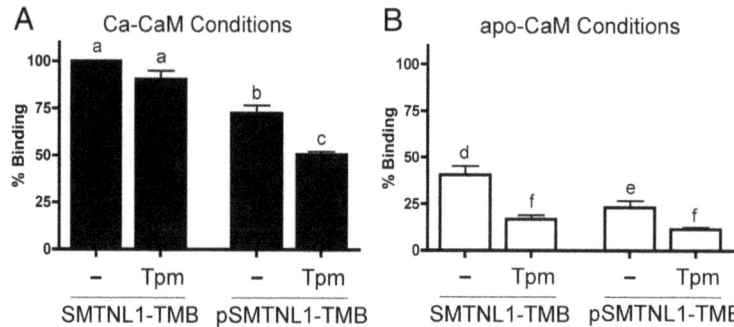

**Fig. 2** Interdependency of SMTNL1 phosphorylation, calcium and tropomyosin on calmodulin-binding. SMTNL1-TMB (200 µg, 7 nmol) was incubated with an equimolar amount of Tpm. The mixture was then incubated with CaM-Sepharose (40 µL; ligand density of ~10–14 µmol/mL) in the presence (**a**; Ca-CaM, 5 mM CaCl$_2$) or absence (**b**; apo-CaM, 1 mM EDTA) of calcium. Some experiments were completed with SMTNL1-TMB that had been previously phosphorylated with PKA. After washing, the retention of SMTNL1-TMB or phosphorylated SMTNL1-TMB was analyzed. The band densities were quantified and binding to CaM-Sepharose expressed as percentage of SMTNL1-TMB recovered under maximal binding conditions (i.e., Ca-CaM in the absence of Tpm). All experiments are $n = 3$–5 and were analyzed by one-way ANOVA with Tukey's *post hoc* analysis. Different letters indicate significant differences among groups (a,b,c: comparison among all Ca-CaM conditions; d,e,f: comparison among all apo-CaM conditions; $p < 0.05$)

not necessarily required for CaM binding [22]. Herein, we describe a novel incidence where phosphorylation regulates CaM binding to a target protein. In this case, the PKA-dependent phosphorylation of SMTNL1 at Ser301 modulates both Ca- and apo-CaM binding. Ser301 lies well upstream of the core hydrophobic residues within CBD1 required for Ca-CaM-binding (i.e., 20 amino acids from Phe321 which is the critical contributor [11]). It is likely that the distance between Ser301 and the hydrophobic patch of CBD1 is too distant for phosphorylation to directly block Ca-CaM docking. This agrees with our observation that the rate of PKA-dependent phosphorylation at Ser301 was not influenced when apo- or Ca-CaM was included in the kinase assay (unpublished results). We have previously demonstrated that intramolecular interactions between the *C*-terminal IQ motif (localization of the apo-CaM binding site, CBD2) and the intrinsically disordered region (localization of the phosphorylation site) influence apo-CaM binding [11, 12]. Furthermore, a portion of the *N*-terminal intrinsically disordered region forms intramolecular contacts with the globular *C*-terminal calponin homology (CH) domain [18]. Thus, it is conceivable that phosphorylation within the *N*-terminal region has far-reaching effects in the *C*-terminal domain and can attenuate both Ca- and apo-CaM binding potentials.

The phosphorylation of SMTNL1-TMB enhanced the Tpm-binding potential in both pull down studies and by confirming SPR assays. Inconsequential amounts of non-specific Tpm-binding were found with CaM-Sepharose, and its binding was not influenced by the presence of SMTNL1-TMB, irrespective of phosphorylation or calcium. Thus, we conclude that a heterotrimeric complex of SMTNL1-TMB, CaM and Tpm was not formed under any of the binding conditions employed. Since no

heterotrimer was generated, we suggest that Tpm and Ca-CaM compete for a similar binding site on SMTNL1-TMB. This conclusion is further supported by the fact that the binding surfaces for CaM and Tpm overlap on SMTNL1. Indeed, only very weak binding of Tpm is observed with the isolated CH domain (containing: CBD2; apo-CaM binding) in the absence of the intrinsically-disordered upstream sequence (containing: CBD1, Ca-CaM binding; and Ser301, phosphorylation consensus sequence) [18].

Our findings place SMTNL1 in a position to influence both Ca$^{2+}$/CaM-regulated contractile events and Ca$^{2+}$ desensitization pathways in smooth muscle. Depending on intracellular calcium concentration and cyclic nucleotide-dependent kinase activation, SMTNL1 is predicted to cycle between Tpm-bound and Ca-CaM-bound complexes. The impact of SMTNL1 on smooth muscle contractile processes may be via its binding partners and associations with the contractile filament. In its dephosphorylated state SMTNL1 has weaker affinity for Tpm and would be more likely to associate with CaM. With an increase in intracellular calcium due to a contractile signal, Ca-CaM could bind to SMTNL1 and initiate the dissociation of the complex from the thin filament. Unphosphorylated SMTNL1 can inhibit myosin light chain phosphatase (MLCP) activity in vitro [27], so the release of SMTNL1 could attenuate MLCP-dependent dephosphorylation of myosin regulatory light chain (LC20). Conversely, SMTNL1 phosphorylation during calcium desensitization of smooth muscle could provide enhanced binding to Tpm and restrict binding of CaM in the apo-state, securing phosphorylated SMTNL1 with Tpm on the thin filament. The weak binding modes of SMTNL1 with apo-CaM might not possess enough stability to occur in vivo [12]. While the physiological significance of

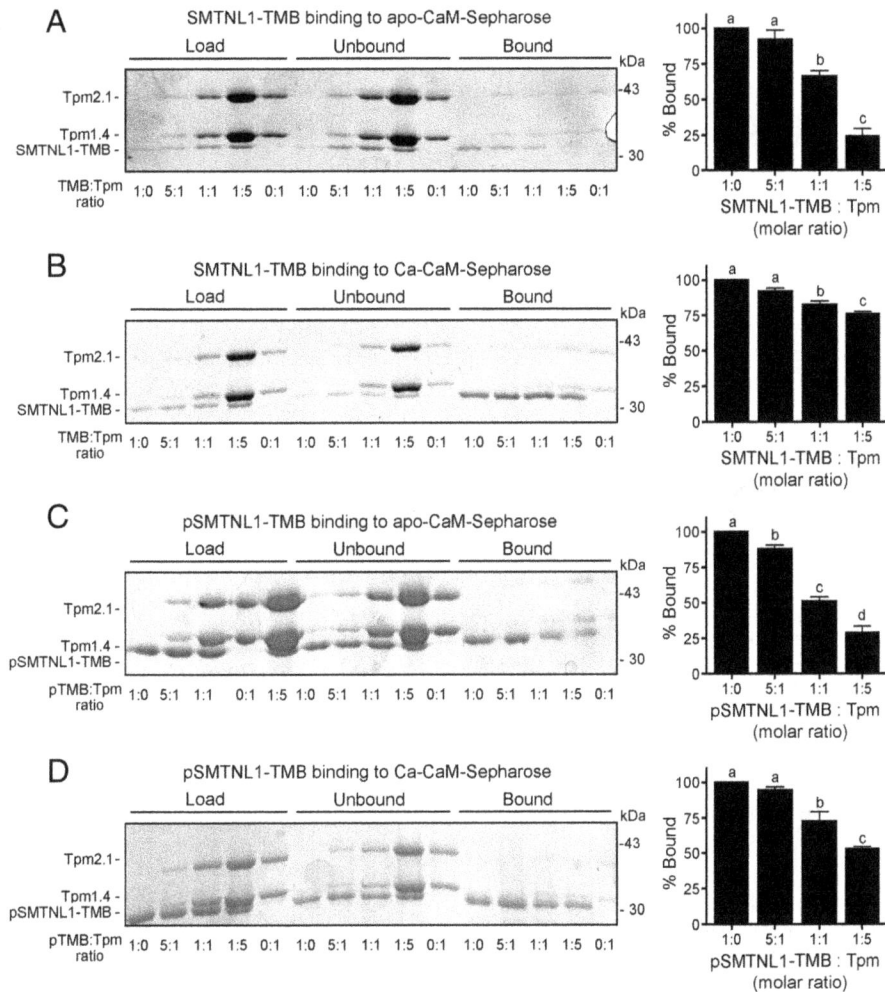

**Fig. 3** SMTNL1 phosphorylation precludes Ca-CaM-binding in favour of Tpm-binding under conditions of high Tpm content. SMTNL1-TMB or phosphorylated SMTNL1-TMB (concentrations as described in Fig. 2) was pre-incubated with the indicated molar ratio of purified α/β Tpm dimer. The mixtures were then added to CaM-Sepharose in the absence (**a**, **c**; apo-CaM, 1 mM EDTA), or presence (**b**, **d**; Ca-CaM, 5 mM CaCl$_2$) of calcium. After extensive washing, bound SMTNL1-TMB was eluted and detected with Coomassie staining of SDS-PAGE gels. The band density was quantified and binding expressed as percentage of maximum binding within each condition. All experiments are $n = 3$ and were analyzed by one-way ANOVA with Tukey's *post hoc* analysis. Different letters indicate significant differences among groups ($p < 0.05$)

SMTNL1 association with the contractile filament in situ has not been investigated, we speculate that the protein may act in analogous manners to caldesmon or calponin on actin-myosin cross bridge cycling. Calponin interacts with many cytoskeleton and related proteins (including tropomyosin) and functions as an inhibitor of actin-activated myosin ATPase [28]. The binding of Ca-CaM dissociates calponin from the actin filament to facilitate smooth muscle contraction [29]. Calponin can also be phosphorylated in response to cyclic nucleotide signals to cause a reduced affinity for the thin filament [30]. Caldesmon is another actin filament-associated regulatory protein; it also binds to Tpm and regulates cross-bridge cycling in a Ca-CaM dependent manner [31]. A detailed physiological analysis of the CaM- and Tpm-interactions of SMTNL1

and their functional implications are required to provide a clear description of role of SMTNL1 in smooth muscle contractility. Further investigation will be needed to verify the role of SMTNL1 in fine-tuning thin filament dynamics.

## Conclusions

Herein, we provide evidence that calcium availability increases CaM association with SMTNL1, and CaM and tropomyosin binding to SMTNL1 are mutually exclusive events. Moreover, SMTNL1 binding to CaM is reduced by phosphorylation of Ser301 while binding to tropomyosin is enhanced. We propose a model whereby SMTNL1 can exist in distinct complexes with either Tpm or CaM, depending on availability of calcium and activity of PKA. This switching mechanism could aid in the fine-tuning of smooth muscle contraction.

## Abbreviations

apo-CaM: Calcium-free calmodulin; Ca-CaM: Calcium-saturated calmodulin; CaM: Calmodulin; CBD1 and CBD2: Calmodulin binding domain 1 and 2, respectively; PKA: Protein kinase A; PKG: Protein kinase G; SMTNL1: Smoothelin-like 1; SMTNL1-TMB: Tropomyosin-binding region of SMTNL1; Tpm: Tropomyosin

## Acknowledgements

Not applicable.

## Funding

This work was supported by the Canadian Institutes of Health Research (MOP-97931). JAM held an Alberta Innovates – Health Solutions (AIHS) Senior Scholar Award and was recipient of a Canada Research Chair (Tier 2) in Smooth Muscle Pathophysiology. AUL was a Heart & Stroke Foundation of Canada Postdoctoral Fellow.

## Authors' contributions

AUL, HI and DHS designed, performed and analyzed the majority of experiments. AUL and JAM wrote the manuscript and drafted figures. JAM and HJV conceived and coordinated the study. All authors reviewed the results and approved the final version of the manuscript.

## Competing interests

The authors declare that they have no competing interests.

## Author details

[1]Department of Biochemistry & Molecular Biology, University of Calgary, Cumming School of Medicine, 3280 Hospital Drive NW, Calgary, AB T2N 4Z6, Canada.
[2]Biochemistry Research Group, Department of Biological Sciences, University of Calgary, 2500 University Drive NW, Calgary, AB T2N 1 N4, Canada.

## References

1. Borman MA, MacDonald JA, Haystead TA. Modulation of smooth muscle contractility by CHASM, a novel member of the smoothelin family of proteins. FEBS Lett. 2004;573:207–13.
2. Lontay B, Bodoor K, Sipos A, Weitzel DH, Loiselle D, et al. Pregnancy and smoothelin-like protein 1 (SMTNL1) deletion promote the switching of skeletal muscle to a glycolytic phenotype in human and mice. J Biol Chem. 2015;290:17985–98.
3. Bodoor K, Lontay B, Safi R, Weitzel DH, Loiselle D, et al. Smoothelin-like 1 protein is a bifunctional regulator of the progesterone receptor during pregnancy. J Biol Chem. 2011;286:31839–51.
4. Lontay B, Bodoor K, Weitzel DH, Loiselle D, Fortner C, et al. Smoothelin-like 1 protein regulates myosin phosphatase-targeting subunit 1 expression during sexual development and pregnancy. J Biol Chem. 2010;285:29357–66.
5. Wooldridge AA, Fortner CN, Lontay B, Akimoto T, Neppl RL, et al. Deletion of the protein kinase A/protein kinase G target SMTNL1 promotes an exercise-adapted phenotype in vascular smooth muscle. J Biol Chem. 2008;283:11850–9.
6. Turner SR, MacDonald JA. Novel contributions of the smoothelin-like 1 protein in vascular smooth muscle contraction and its potential involvement in myogenic tone. Microcirculation. 2014;21:249–58.
7. Chin D, Means AR. Calmodulin: a prototypical calcium sensor. Trends Cell Biol. 2000;10:322–8.
8. Villarroel A, Taglialatela M, Bernardo-Seisdedos G, Alaimo A, Agirre J, et al. The ever changing moods of calmodulin: how structural plasticity entails transductional adaptability. J Mol Biol. 2014;426:2717–35.

9. Ishida H, Vogel HJ. Protein-peptide interaction studies demonstrate the versatility of calmodulin target protein binding. Protein Pept Lett. 2006;13:455–65.
10. Ishida H, Borman MA, Ostrander J, Vogel HJ, MacDonald JA. Solution structure of the calponin homology (CH) domain from the smoothelin-like 1 protein: a unique apocalmodulin-binding mode and the possible role of the C-terminal type-2 CH-domain in smooth muscle relaxation. J Biol Chem. 2008;283:20569–78.
11. Ulke-Lemée A, Ishida H, Chappellaz M, Vogel HJ, MacDonald JA. Two domains of the smoothelin-like 1 protein bind apo- and calcium-calmodulin independently. Biochim Biophys Acta. 1844;2014:1580–90.
12. Ulke-Lemée A, Turner SR, MacDonald JA. In situ analysis of smoothelin-like 1 and calmodulin interactions in smooth muscle cells by proximity ligation. J Cell Biochem. 2015;116:2667–75.
13. Gunning PW, O'Neill G, Hardeman E. Tropomyosin-based regulation of the actin cytoskeleton in time and space. Physiol Rev. 2008;88:1–35.
14. Wang CL, Coluccio LM. New insights into the regulation of the actin cytoskeleton by tropomyosin. Int Rev Cell Mol Biol. 2010;281:91–128.
15. Gunning PW, Hardeman EC, Lappalainen P, Mulvihill DP. Tropomyosin - master regulator of actin filament function in the cytoskeleton. J Cell Sci. 2015;128:2965–74.
16. Geeves MA, Hitchcock-DeGregori SE, Gunning PW. A systematic nomenclature for mammalian tropomyosin isoforms. J Muscle Res Cell Motil. 2015;36:147–53.
17. Rao JN, Rivera-Santiago R, Li XE, Lehman W, Dominguez R. Structural analysis of smooth muscle tropomyosin alpha and beta isoforms. J Biol Chem. 2012;287:3165–74.
18. MacDonald JA, Ishida H, Butler EI, Ulke-Lemée A, Chappellaz M, et al. Intrinsically disordered N-terminus of calponin homology-associated smooth muscle protein (CHASM) interacts with the calponin homology domain to enable tropomyosin binding. Biochemistry. 2012;51:2694–705.
19. Ulke-Lemée A, Ishida H, Borman MA, Valderrama A, Vogel HJ, MacDonald JA. Tropomyosin-binding properties of the CHASM protein are dependent upon its calponin homology domain. FEBS Lett. 2010;584:3311–6.
20. Rannels SR, Beasley A, Corbin JD. Regulatory subunits of bovine heart and rabbit skeletal muscle cAMP-dependent protein kinase isozymes. Methods Enzymol. 1983;99:55–62.
21. Jancso A, Graceffa P. Smooth muscle tropomyosin coiled-coil dimers. Subunit composition, assembly, and end-to-end interaction. J Biol Chem. 1991;266:5891–7.
22. Lukas TJ, Burgess WH, Prendergast FG, Lau W, Watterson DM. Calmodulin binding domains: characterization of a phosphorylation and calmodulin binding site from myosin light chain kinase. Biochemistry. 1986;25:1458–64.
23. Graff JM, Young TN, Johnson JD, Blackshear PJ. Phosphorylation-regulated calmodulin binding to a prominent cellular substrate for protein kinase C. J Biol Chem. 1989;264:21818–23.
24. Hofmann F, Anagli J, Carafoli E, Vorherr T. Phosphorylation of the calmodulin binding domain of the plasma membrane Ca2+ pump by protein kinase C reduces its interaction with calmodulin and with its pump receptor site. J Biol Chem. 1994;269:24298–303.
25. Fleming I, Fisslthaler B, Dimmeler S, Kemp BE, Busse R. Phosphorylation of Thr(495) regulates Ca(2+)/calmodulin-dependent endothelial nitric oxide synthase activity. Circ Res. 2001;88:E68–75.
26. Wong R, Schlichter LC. PKA reduces the rat and human KCa3.1 current, CaM binding, and Ca2+ signaling, which requires Ser332/334 in the CaM-binding C terminus. J Neurosci. 2014;34:13371–83.
27. Borman MA, Freed TA, Haystead TA, MacDonald JA. The role of the calponin homology domain of smoothelin-like 1 (SMTNL1) in myosin phosphatase inhibition and smooth muscle contraction. Mol Cell Biochem. 2009;327:93–100.
28. Liu R, Jin JP. Calponin isoforms CNN1, CNN2 and CNN3: Regulators for actin cytoskeleton functions in smooth muscle and non-muscle cells. Gene. 2016;585:143–53.
29. Naka M, Kureishi Y, Muroga Y, Takahashi K, Ito M, Tanaka T. Modulation of smooth muscle calponin by protein kinase C and calmodulin. Biochem Biophys Res Commun. 1990;171:933–7.
30. Winder SJ, Walsh MP. Smooth muscle calponin. Inhibition of actomyosin MgATPase and regulation by phosphorylation. J Biol Chem. 1990;265:10148–55.
31. Marston SB, Redwood CS. The essential role of tropomyosin in cooperative regulation of smooth muscle thin filament activity by caldesmon. J Biol Chem. 1993;268:12317–20.

# *CrMAPK3* regulates the expression of iron-deficiency-responsive genes in *Chlamydomonas reinhardtii*

Xiaowen Fei[1†], Junmei Yu[2†], Yajun Li[2] and Xiaodong Deng[2*]

## Abstract

**Background:** Under iron-deficient conditions, *Chlamydomonas* exhibits high affinity for iron absorption. Nevertheless, the response, transmission, and regulation of downstream gene expression in algae cells have not to be investigated. Considering that the MAPK pathway is essential for abiotic stress responses, we determined whether this pathway is involved in iron deficiency signal transduction in *Chlamydomonas*.

**Results:** *Arabidopsis* MAPK gene sequences were used as entry data to search for homologous genes in *Chlamydomonas reinhardtii* genome database to investigate the functions of mitogen-activated protein kinase (MAPK) gene family in *C. reinhardtii* under iron-free conditions. Results revealed 16 *C. reinhardtii* MAPK genes labeled *CrMAPK2–CrMAPK17* with TXY conserved domains and low homology to MAPK in yeast, *Arabidopsis*, and humans. The expression levels of these genes were then analyzed through qRT-PCR and exposure to high salt (150 mM NaCl), low nitrogen, or iron-free conditions. The expression levels of these genes were also subjected to adverse stress conditions. The mRNA levels of *CrMAPK2*, *CrMAPK3*, *CrMAPK4*, *CrMAPK5*, *CrMAPK6*, *CrMAPK8*, *CrMAPK9*, and *CrMAPK11* were remarkably upregulated under iron-deficient stress. The increase in *CrMAPK3* expression was 43-fold greater than that in the control. An RNA interference vector was constructed and transformed into *C. reinhardtii* 2A38, an algal strain with an exogenous *FOX1:ARS* chimeric gene, to silence *CrMAPK3*. After this gene was silenced, the mRNA levels and ARS activities of *FOX1:ARS* chimeric gene and endogenous *CrFOX1* were decreased. The mRNA levels of iron-responsive genes, such as *CrNRAMP2*, *CrATX1*, *CrFTR1*, and *CrFEA1*, were also remarkably reduced.

**Conclusion:** *CrMAPK3* regulates the expression of iron-deficiency-responsive genes in *C. reinhardtii*.

**Keywords:** *Chlamydomonas reinhardtii*, Mitogen-activated protein kinases, Iron deficiency, Real-time PCR

## Background

*Chlamydomonas reinhardtii* (Volvocales, Chlorophyta) is a single-celled eukaryotic and flagellated green alga, whose three genetic systems located in the nucleus, chloroplast, and mitochondria can be used for transformation. This alga is regarded as a "photosynthetic yeast" because of its easy culturing process, rapid growth, short life cycle, and high photosynthetic efficiency. With its three genome sequences, this model organism is highly useful for cell and molecular biology research [1].

In phosphorylation cascades, mitogen-activated protein kinases (MAPKs) are eukaryotic signal proteins involved in extracellular signal amplification and intracellular signal transduction in yeasts, animals, and plants [2–4]. Combined with other signal molecules, MAPKs transfer external stimuli via successive phosphorylation reactions: MAPKKKs → MAPKKs → MAPKs. Progressively and continuously enlarged signals, such as environmental stress factors, including high salinity, high osmotic pressure, and low temperature, reach the nucleus and regulate downstream gene expression [5, 6]. In eukaryotic cells, phosphorylation cascades are composed of MAPKs, MAPKKs, and MAPKKKs. *Homo sapiens* possesses 15 MAPKs, 7 MAPKKs, and 16 MAPKKKs, while *Arabidopsis* contains 20 MAPKs, 10 MAPKKs, and 80 MAPKKKs. Few MAPK cascades have been described because of

* Correspondence: xiaodong9deng@hotmail.com
†Equal contributors
2Institute of Tropical Bioscience and Biotechnology, Chinese Academy of Tropical Agricultural Science, Key Laboratory of Tropical Crop Biotechnology, Ministry of Agriculture, Haikou 571101, China
Full list of author information is available at the end of the article

the complexity of genetic networks and pleiotropic and interaction effects. MAPK genes have been identified in plants, such as *Arabidopsis*, rice, corn, wheat, and barley [7–13]. MAPKs function through stress-response pathways [14, 15].

Iron is an essential trace element for most living organisms. A precise iron regulation system is necessary to maintain the dynamic equilibrium of iron [16] because iron overload and deficiency cause metabolic disorders. Following nitrogen and phosphate deficiencies, iron deficiency restricts plant growth and yield and consequently induces crop chlorosis and yields low productivity. In humans, insufficient iron concentrations trigger iron deficiency anemia or iron deficiency syndrome. Iron has also been considered a growth-limiting factor in some tumor cells. Therefore, iron chelators are clinically used for cancer suppression.

Under iron-deficient conditions, *Chlamydomonas* exhibits high affinity for iron absorption that slightly differs from iron absorption in plants. Environmental ferric iron is reduced to ferrous iron via FRE1 (homology of *Arabidopsis* FRO2 [17]) on the plasma membrane and then putatively transferred to FOX1 by FEA1 [18]. Afterward, FOX1 oxidizes ferrous iron to ferric iron, which is then transported to the cytoplasm by FTR1 on the plasma membrane [19–21]. The expression of the genes encoding these proteins is significantly increased under iron-deficient conditions, and this phenomenon indicates that iron deficiency signals in these genes are regulated. Nevertheless, the response, transmission, and regulation of downstream

gene expression in algal cells have yet to be investigated. Considering that the MAPK pathway is essential for non-biological stress responses, we determined whether this pathway is involved in iron deficiency signal transduction in *Chlamydomonas*. In this study, *Arabidopsis* MAPKs were used to search for the corresponding genes in the *Chlamydomonas* genome database (https://phytozome.jgi.doe.gov/pz/portal.html #), and 16 homologous genes, namely, CrMAPK2–CrMAPK17, were obtained. The mRNA expression level variation of these genes exposed to different stressors, such as −Fe, −N, and osmotic shock (150 mM NaCl), was also detected. Among these genes, *CrMAPK3* is specifically functionally analyzed by RNA silencing.

## Results
### Bioinformatics Analysis of MAPK Genes in Chlamydomonas

Sixteen homologous genes (Table 1), which are localized in chromosomes 1, 2, 3, 8, 12, 13, 16, and 17, were identified by searching the *Chlamydomonas* genome database with Blast. The predicted open reading frames of these genes were 1062–5301 bp in length, and their protein products contained 353–1766 amino acids with molecular weights of 39.8–178.76 kD and isoelectric points of 5.68–9.5. Fourteen of the MAPKs located in the cytosome were predicted by Euk-mPLoc2.0 except *CrMAPK6* and *CrMAPK14*, which exist in the nucleus. Using PROSITE predictions, we verified that the 16 *CrMAPKs* were mitogen-activated protein kinases. Multi-sequence alignment of the MAPK-specific

**Table 1** List of the 16 MAPK genes identified in C. reinhardtii and their sequence characteristics

| Name | Locus Name | ORF (bp) | Amino Acids | kD | pI | Chromosomal localization | Sub cellular location |
|------|-----------|----------|-------------|------|------|--------------------------|----------------------|
| CrMAPK2 | Cre08.g385050 | 2223 | 740 | 79.0 | 8.61 | chr8:4906426..4913322 F | Cytoplasm |
| CrMAPK3 | Cre12.g509000 | 1062 | 353 | 39.8 | 8.76 | chr12:2119590..2122035 R | Cytoplasm |
| CrMAPK4 | Cre17.g745447 | 2298 | 765 | 81.2 | 7.20 | chr17:6855222..6862749 F | Cytoplasm |
| CrMAPK5 | Cre13.g607300 | 1302 | 433 | 48.5 | 8.88 | chr13:5094976..5099387 F | Cytoplasm |
| CrMAPK6 | Cre12.g508900 | 1128 | 375 | 42.5 | 7.66 | chr12:2129460..2133325 R | Nuclear |
| CrMAPK7 | Cre16.g661100 | 1758 | 585 | 63.3 | 9.7 | chr16:2519399..2524020 F | Cytoplasm |
| CrMAPK8 | Cre01.g010000 | 1170 | 389 | 44.0 | 5.68 | chr1:1838122..1842287 F | Cytoplasm |
| CrMAPK9 | Cre12.g538300 | 1923 | 640 | 68.4 | 9.1 | chr12:6473633..6479228 F | Cytoplasm |
| CrMAPK10 | Cre01.g052800 | 3692 | 1163 | 117 | 9.91 | chr1:7342585..7351700 R | Cytoplasm |
| CrMAPK11 | Cre01.g052850 | 5199 | 1732 | 168 | 9.5 | chr1:7356808..7367075 F | Cytoplasm |
| CrMAPK12 | Cre03.g200200 | 5301 | 1766 | 178 | 9.36 | chr3:8184080..8192783 R | Cytoplasm |
| CrMAPK13 | Cre10.g432250 | 4752 | 1583 | 160 | 8.31 | chr10:1948371..1956866 R | Cytoplasm |
| CrMAPK14 | Cre03.g169500 | 3321 | 1106 | 114 | 9.49 | chr3:3709775..3716106 R | Nuclear |
| CrMAPK15 | Cre02.g111014 | 4212 | 1403 | 142 | 9.15 | chr2:5820048..5827789 F | Cytoplasm |
| CrMAPK16 | Cre17.g709500 | 3237 | 1078 | 107 | 7.93 | chr17:1790415..1798683 R | Cytoplasm |
| CrMAPK17 | Cre17.g709750 | 4956 | 1651 | 165 | 8.56 | chr17:1821974..1831590 F | Cytoplasm |

F and R represent the forward and reverse directions on the chromosome, respectively. In total, 16 CrMAPK proteins were obtained by BLASTP search using the C. reinhardtii V5.5 proteome database and MAPK proteins from Arabidopsis thaliana as queries. The 16 CrMAPK genes were named based on their name annotated in JGI database. The molecular weights and pIs of the 16 CrMAPK proteins were predicted using ExPASy. The CrMAPK sub-cellular locations were predicted using the Euk-mPLoc2.0 program

TXY motifs in the CrMAPK proteins revealed that the T(D/E/T/S/P)Y activation loop motifs were conserved in the serine-threonine kinase (S-Tkc) domain in the 16 CrMAPKs (Fig. 1). CrMAPK3, CrMAPK6, and CrMAPK8 contain TEY; CrMAPK2, CrMAPK4, CrMAPK5, CrMAPK7, CrMAPK9, CrMAPK10, and CrMAPK13 comprise TDY; CrMAPK11, CrMAPK12, CrMAPK15, and CrMAPK16 possess TSY; CrMAPK17 is composed of TPY; and CrMAPK14 consists of TTY. Chlamydomonas MAPKs were divided into two groups by using MEGA6. Group I contained CrMAPK2 to CrMAPK10, whereas Group II comprised CrMAPK11 to CrMAPK17 (Fig. 2a). All of the MAPK genes contained 6 to 10 exons. The gene length ranged from 2.8 kb to 10.5 kb. Among these genes, CrMAPK3 is the shortest and CrMAPK11 is the longest (Fig. 2b). In addition to the S-Tkc-conserved region, 8 other protein domains/motifs, such as Syn N(Syntaxin N-terminal domain), RIO (RIO-like kinase), CUE (domain that may be involved in binding ubiquitin-conjugating enzymes), and Tyr-kc (tyrosine kinase catalytic domain) motifs (Fig. 2c), are present in CrMAPKs. The annotated CrMAPK1(Cre13.g582650) in JGI database is a small protein with 149 aa. After the alignment was compared with the other CrMAPKs(CrMAPK2-CrMAPK17), the results revealed that the conserved domain (T(D/E/T/S/P)YXTRWYRAPEL(V)) in the MAPK family could not be found in CrMAPK1. As such, CrMAPK1 is not included in Table 1.

## Analysis of mRNA levels of MAPK gene under – Fe, –N, and 150 mM NaCl stress conditions

The RNA extracted from the samples of Chlamydomonas cultivated under – Fe, –N, and 150 mM NaCl conditions was used for quantitative analysis, and the results are shown in Fig. 3. CrMAPK2–CrMAPK17 expression levels were affected by iron deficiency, nitrogen deficiency, and high salt concentration. Compared with the expression of the gene in the TAP medium, the mRNA expression levels of CrMAPK2, CrMAPK3, CrMAPK4, CrMAPK5, CrMAPK6, CrMAPK8, CrMAPK9, and CrMAPK11 were increased by iron deficiency to various degrees. CrMAPK3, CrMAPK5, and CrMAPK11 respectively increased by 43-, 5-, and 40-fold after cultivation for 48 h. However, iron deficiency decreased the mRNA levels of CrMAPK7, CrMAPK12, CrMAPK13, CrMAPK15, CrMAPK16, and CrMAPK17 after cultivation for 48 h. In nitrogen deficiency, the mRNA expression levels of CrMAPK6 and CrMAPK14 were significantly increased, whereas the expression levels of most MAPKs, such as those of CrMAPK2, CrMAPK3, CrMAPK5, CrMAPK5, CrMAPK8, CrMAPK9, CrMAPK10, CrMAPK11, CrMAPK12, CrMAPK13, CrMAPK15, CrMAPK16, and CrMAPK17, of C. reinhardtii were inhibited, and the mRNA expression levels of these genes were significantly decreased. The mRNA expression levels of all MAPK genes were also inhibited under high salt (150 mM NaCl) condition, and the mRNA expression of CrMAPK17 was reduced by 10E10.

## CrMAPK3 positively regulates the expression of CrFOX1 gene

The C. reinhardtii 2A38 strain was prepared by using the integrated FOX1 promoter:ARS box into the chromosome of the CC425 strain. Under iron-deficient conditions, CrFOX1 promoted the ARS reporter expression and appeared deep blue when this gene was mixed with XSO4 substrate or yellow when this gene was mixed with p-nitrophenylsulfate. A total of 133 colonies were obtained after Maa7IR/CrMAPK3IR was transformed into C. reinhardtii 2A38 and then transferred onto –

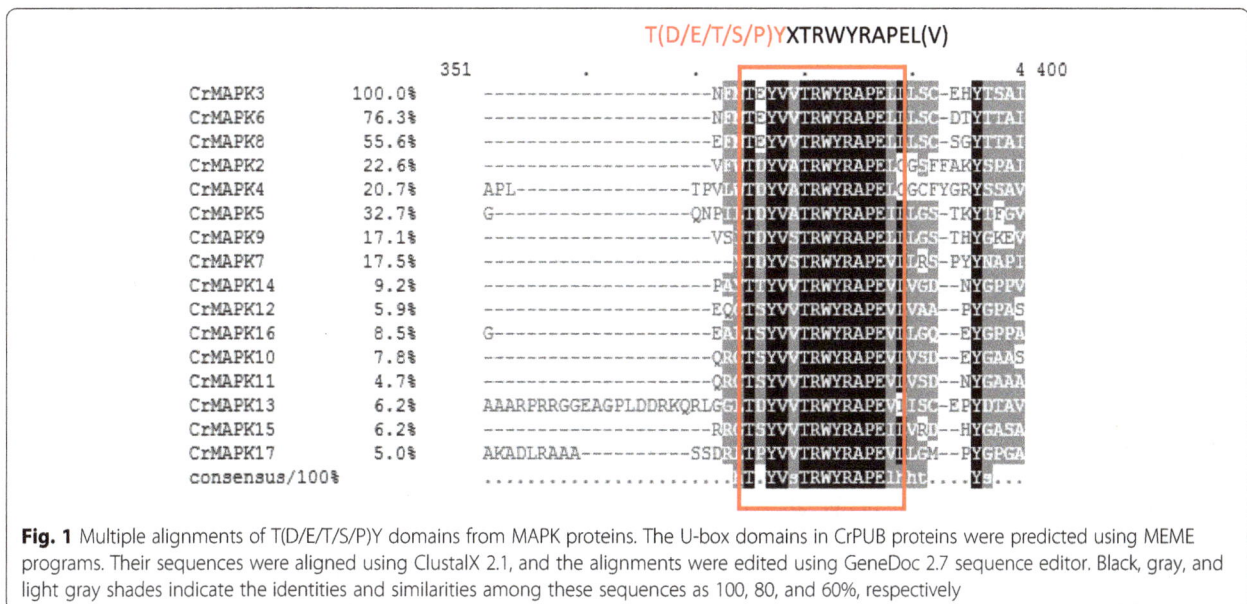

**Fig. 1** Multiple alignments of T(D/E/T/S/P)Y domains from MAPK proteins. The U-box domains in CrPUB proteins were predicted using MEME programs. Their sequences were aligned using ClustalX 2.1, and the alignments were edited using GeneDoc 2.7 sequence editor. Black, gray, and light gray shades indicate the identities and similarities among these sequences as 100, 80, and 60%, respectively

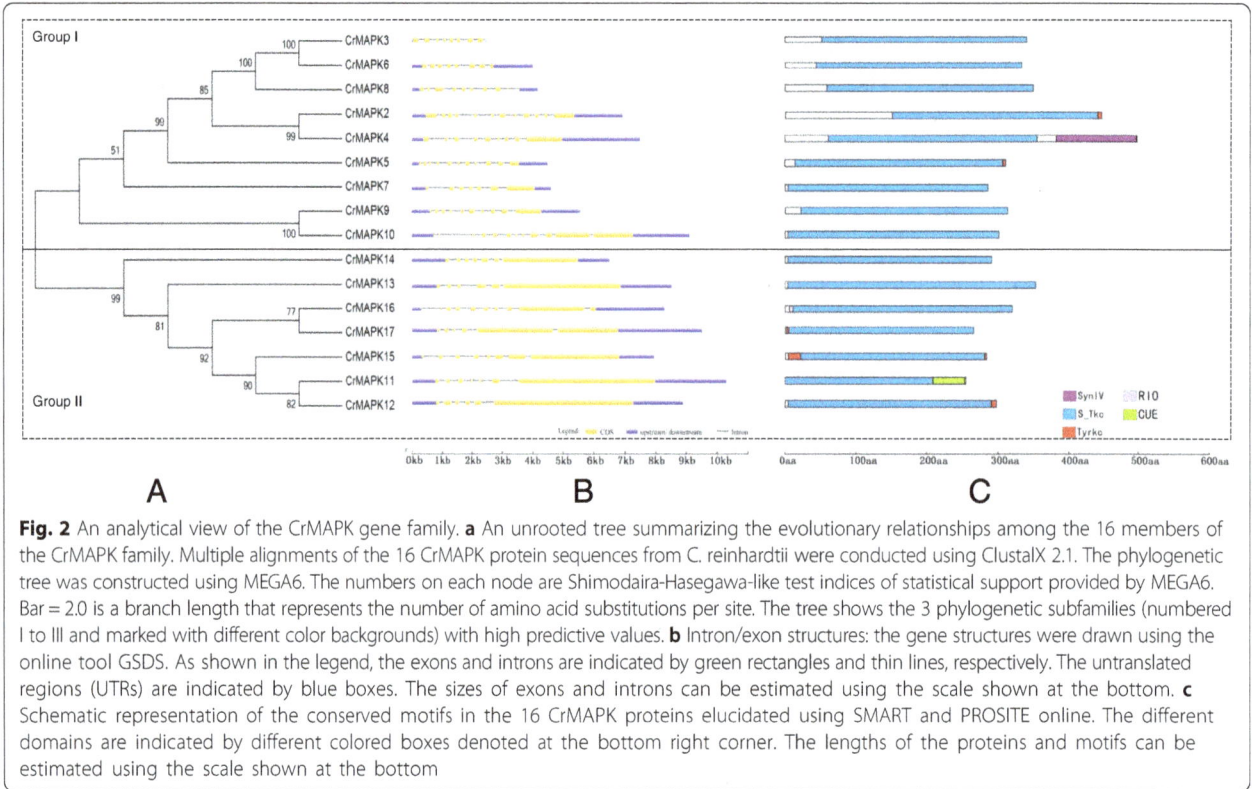

**Fig. 2** An analytical view of the CrMAPK gene family. **a** An unrooted tree summarizing the evolutionary relationships among the 16 members of the CrMAPK family. Multiple alignments of the 16 CrMAPK protein sequences from C. reinhardtii were conducted using ClustalX 2.1. The phylogenetic tree was constructed using MEGA6. The numbers on each node are Shimodaira-Hasegawa-like test indices of statistical support provided by MEGA6. Bar = 2.0 is a branch length that represents the number of amino acid substitutions per site. The tree shows the 3 phylogenetic subfamilies (numbered I to III and marked with different color backgrounds) with high predictive values. **b** Intron/exon structures: the gene structures were drawn using the online tool GSDS. As shown in the legend, the exons and introns are indicated by green rectangles and thin lines, respectively. The untranslated regions (UTRs) are indicated by blue boxes. The sizes of exons and introns can be estimated using the scale shown at the bottom. **c** Schematic representation of the conserved motifs in the 16 CrMAPK proteins elucidated using SMART and PROSITE online. The different domains are indicated by different colored boxes denoted at the bottom right corner. The lengths of the proteins and motifs can be estimated using the scale shown at the bottom

Fe plates with XSO4 to determine the ARS activities. Only 44 colonies and the control sample of 2A38 appeared blue, whereas the 99 other colonies were colorless or light blue. Furthermore, 74.4% of chromogenic reactions indicated that *CrMAPK3* silencing affected the FOX1 promoter function in −Fe. The transformants of RNAi11, RNAi37, and RNAi62 appeared colorless in both + Fe (16 uM) and −Fe except the non-transgenic *C. reinhardtii* 2A38 control, which appeared deep blue under −Fe conditions (Fig. 4a). These results were further confirmed by the ARS activity in transgenic strains. The ARS activities of RNAi11, RNAi37, and RNAi62 respectively decreased by 82, 85, and 83% compared with those of the *C. reinhardtii* 2A38 control (Fig. 4b). The mRNA of the ARS2 of the transgenic stains decreased by more than 97% in −Fe (Fig. 4c). The mRNA levels of the target *CrMAPK3* of the transgenic strains RNAi11, RNAi37, and RNAi62 decreased remarkably by 97, 97, and 98%, respectively (Fig. 4d). These data implied that *CrMAPK3* silencing positively regulated *FOX1:ARS* expression. *CrMAPK3* silencing also decreased the gene expression of endogenous *FOX1* in *Chlamydomonas*. The mRNA levels of *CrFOX1* of the RNAi11 strain decreased by 63, 54, and 71% when this strain was cultured for 12, 24, and 48 h under the −Fe condition, respectively. *CrMAPK3* silencing also repressed the iron-induced upregulation of *CrFOX1* gene expression. Therefore, *CrMAPK3* positively regulated the endogenous expression of *CrFOX1* (Fig. 4e).

### CrMAPK3 positively regulates the expression levels of iron uptake-associated genes

*Chlamydomonas* exhibits an iron uptake pattern similar to Type I plants. During iron deficiency, *Chlamydomonas* cells undergo the affinity iron absorption mechanism by inducing the expression of *CrFOX1* [19], *CrNRAMP2* [18], *CrATX1* [22], *CrFTR1* [20], and *CrFEA1* to enhance iron absorption [18]. *CrFEA1* is located in the cell walls and responsible for the transport of reduced $Fe^{2+}$ via FRE1 (homolog of *Arabidopsis* FRO2) to *CrFOX1*, which is found in the plasma membrane (homolog of yeast FET3 [23]), and reoxidizes $Fe^{2+}$ to $Fe^{3+}$. *CrFOX1* is then transported inside the cells through the plasma membrane protein *CrFTR1*, and ATX1 of yeast transports $Cu^{2+}$ to the cytoplasm. Thus far, direct evidence supporting iron transmission has yet to be obtained, but studies have shown that ATX1 is an iron-deficiency-inducible protein. The *NRAMP* gene family is located on the vacuole membrane, and it shuttles $Fe^{2+}$ between vacuole membranes to maintain the iron concentration in the cytoplasm. The mRNA levels of iron absorption-related genes, such as *CrNRAMP2*, *CrATX1*, *CrFTR1*, and *CrFEA1*, in the *CrMAPK3* RNAi transgenic strain RNAi11 are shown in Fig. 5. The mRNA levels of the genes decreased after the strains were cultivated for 48 h under −Fe. The mRNA level of *CrNRAMP2* was decreased by 86% compared with the control after cultivation for 48 h in −Fe. Similarly, the mRNA levels of *CrATX1*, *CrFTR1*, and *CrFEA1* were

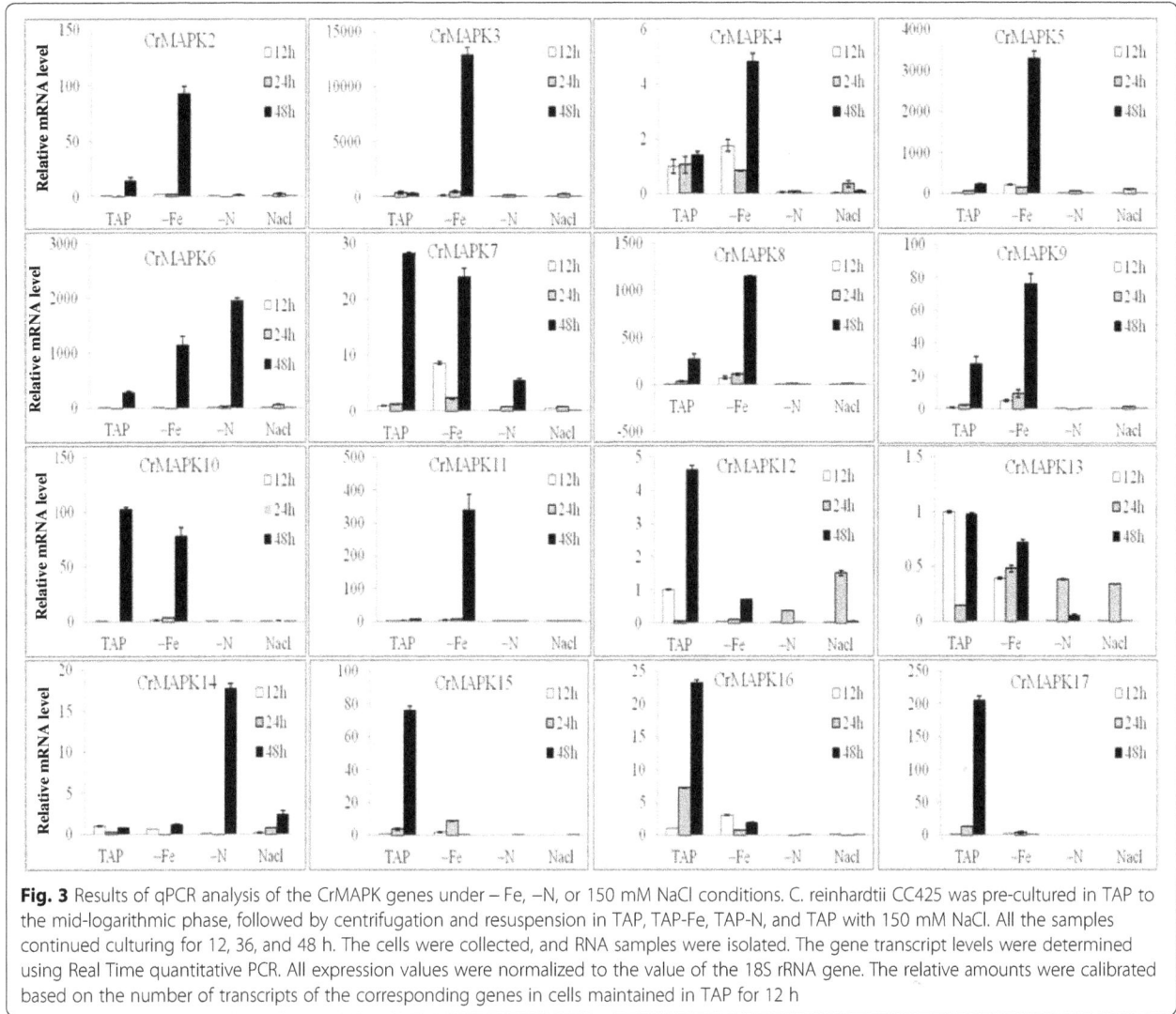

**Fig. 3** Results of qPCR analysis of the CrMAPK genes under − Fe, −N, or 150 mM NaCl conditions. C. reinhardtii CC425 was pre-cultured in TAP to the mid-logarithmic phase, followed by centrifugation and resuspension in TAP, TAP-Fe, TAP-N, and TAP with 150 mM NaCl. All the samples continued culturing for 12, 36, and 48 h. The cells were collected, and RNA samples were isolated. The gene transcript levels were determined using Real Time quantitative PCR. All expression values were normalized to the value of the 18S rRNA gene. The relative amounts were calibrated based on the number of transcripts of the corresponding genes in cells maintained in TAP for 12 h

decreased by 96, 96, and approximately 53%, respectively. These results indicated the association of *CrMAPK3* of *Chlamydomonas* with the iron metabolism-related genes. In the *CrMAPK3*-silenced strain RNAi11, the mRNA levels of the genes, including *CrNRAMP2*, *CrATX1*, *CrFTR1*, and *CrFEA1*, were also decreased when the mRNA level of *CrMAPK3* was decreased. Thus, *CrMAPK3* might positively regulate the expression of iron-uptake-associated genes, such as *CrNRAMP2*, *CrATX1*, *CrFTR1*, and *CrFEA1*.

## Discussion

MAPKs are widely distributed in eukaryotic organisms, such as yeast, humans, and plants, and are involved in phosphorylation signaling cascades in extracellular amplification and intracellular transduction [23]. The MAPK pathway is responsive to biological and non-biological stress stimuli, hormones, or growth factors and to cell division and apoptosis. Moreover, the MAPK pathway comprises MAPKKK, MAPKK, and MAPK and amplifies signals via

subsequent phosphorylation by using protein kinases and by migrating to the nucleus; thus, the extracellular stimuli of membrane receptors are connected to the molecular effectors of the cytoplasm and the nucleus [24, 25]. A few MAPKs, including 20 in *Arabidopsis*, 17 in rice, 19 in corn, 21 in aspen (*Populus*), 17 in tobacco, 16 in tomato, and 26 in apple, have been identified [26–28]. Proteins encoded by MAPKs in different species contain various domains. In *Chlamydomonas*, 3 of TEY, 7 of TDY, 4 of TSY, 1 of TPY, and 1 of TTY exist. In *Arabidopsis*, 8 of TDY and 12 of TEY are present. These diversities of types and kinase domains demonstrate that MAPKs participate in many metabolic activities. Through cluster analysis, we found that TDY and TEY of *Chlamydomonas* kinases are highly homologous to those of *Arabidopsis* kinases possibly because only TDY and TEY domains are found in *Arabidopsis*. Other domains are highly similar to human kinases.

Organisms need iron for respiration, DNA synthesis, and enzyme reactions. Transport systems have been developed

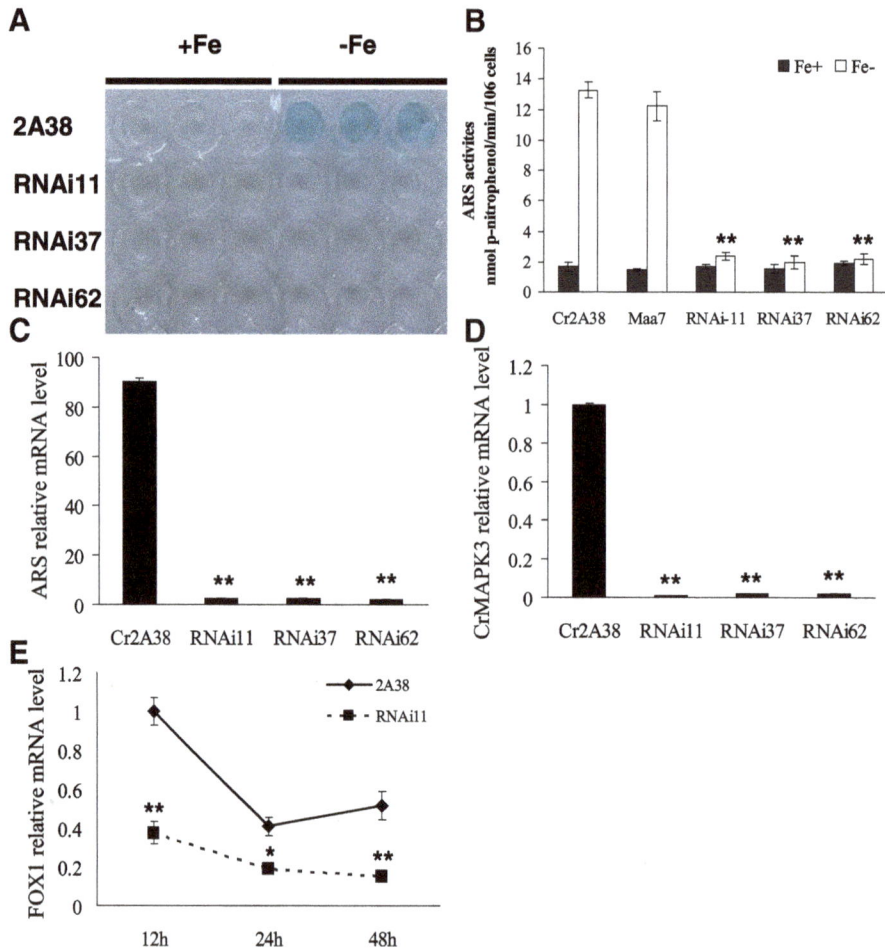

**Fig. 4** Analysis of CrMAPK3 RNAi transgenic algal strains. Of the 133 CrMAPK3 RNAi transformants, 99 were colorless or light blue. Among them, the ARS activities of transgenic strains RNAi11, RNAi37, and RNAi62 were significantly decreased under – Fe (**a, b**). Moreover, the mRNA levels of ARS were significantly decreased (**c**). The mRNA level of target gene CrMAPK3 was decreased by 97–98% compared with the control (**d**), indicating that CrMAPK3 in transgenic strains of RNAi11, RNAi37, and RNAi62 has been effectively silenced. The mRNA levels of endogenous CrFOX1 were reduced by 63, 54, and 71%, respectively, at 12, 24, and 48 h post-incubation in –Fe, indicating that CrMAPK3 positively regulates the expression of CrFOX1 gene (**e**). The data are shown as the means (±SD, $n = 3$). Significance is indicated as $*P < 0.05$, $**P < 0.01$

for iron absorption because iron balance is vital. Iron regulation, especially iron absorption and transportation, has been extensively investigated, but iron signal response systems have been rarely explored. Iron deficiency in humans causes iron deficiency anemia and adolescent iron deficiency 1 syndrome. Iron is an important element required by the body; excessive or scarce amounts of iron likely cause metabolic disorders; therefore, organisms should have a sophisticated control system to regulate the dynamic balance of iron elements [16].

Iron deficiency is the third-most important limiting factor of plant growth and yield in agriculture. Photosynthetic plants reduce their chlorophyll synthesis and photosynthesis rate under iron-deficient conditions.

In humans, iron deficiency causes anemia. Conversely, excess iron increases the risk of liver disease, heart attack, and hypothyroidism. Iron is also a limiting factor in the

growth of some tumor cells, and iron chelators are used clinically to inhibit tumor cell growth. Furthermore, studies on iron MAPK signal cascades have focused on human cancers. Iron deficiency inhibits the mitosis of lung carcinoma cells, melanoma cells, and dysembryoplastic neuroepithelial tumor cells and thus induces cell apoptosis [29–31]. Therefore, iron chelators, desferrioxamine (DFO), and Dp44mT are used to treat these cancers clinically [32, 33]. Iron deficiency signals are also transduced through the activation of JNA and P38 by ASK1 (MAPKKK) to regulate the suspension of the mitotic activity and apoptosis of cancer cells [34].

Plant MAPK gene responses to various stresses have also been detected. In our study, gene expression analysis revealed that 16 MAPK genes in *Chlamydomonas* were involved in response to stress. During iron deficiency, 8 MAPK genes, including *CrMAPK3*, were upregulated.

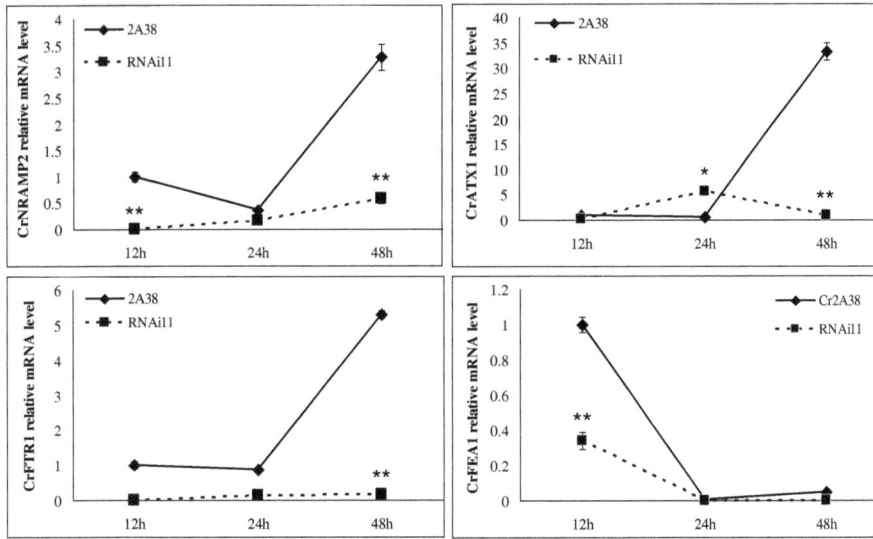

**Fig. 5** mRNA expressions of iron absorption-related genes in CrMAPK3 RNAi transgenic strain RNAi11 in –Fe. In the CrMAPK3-silenced strain RNAi11, the mRNA levels of genes, including CrNRAMP2, CrATX1, CrFTR1, and CrFEA1, were decreased when the mRNA level of CrMAPK3 was decreased and after they were incubated with –Fe after 48 h. This finding indicated that CrMAPK3 may positively regulate the expression of iron uptake-associated genes, such as CrNRAMP2, CrATX1, CrFTR1, and CrFEA1. The data are shown as means ± SD ($n = 3$). Significance is indicated as *$P < 0.05$, **$P < 0.01$

Therefore, *CrMAPK3* possibly responded to iron regulation. These findings were further verified by silencing *CrMAPK3*, and our results demonstrated that the mRNA levels of FOX1-ARS, the enzyme activities of ARS, and the endogenous mRNA level of *CrFOX1* decreased. Therefore, *CrMAPK3* positively regulated *CrFOX1* expression. The mRNA levels of –Fe-inducing genes, including *CrNRAMP2*, *CrATX1*, *CrFTR1*, and *CrFEA1*, and the expression of *CrMAPK3* were reduced. These findings confirmed that *CrMAPK3* positively regulated the expression of iron-absorption genes. However, the exact proteins upstream and downstream of *CrMAPK3* should be identified to reveal the MAPK pathway of iron deficiency response in *Chlamydomonas*.

## Methods
### Algal strains and culture conditions
*C. reinhardtii* CC425 (mt) was purchased from the *Chlamydomonas* Genetics Center at Duke University. *C. reinhardtii* 2A38 is a transgenic strain with an integrated *Fox1 promoter:ARS* chimeric gene in *C. reinhardtii* CC425 genome. Under iron-deficient conditions, the *CrFOX1* promoter in 2A38 strain stimulated the *ARS* gene expression and appeared blue when the XSO4 substrate was added. Liquid cultures were grown in the TAP medium at 26 °C with agitation at 220 rpm under 110 $\mu mol \cdot m^{-2} s^{-1}$ of continuous light for 3 days and then to the TAP, TAP-Fe, TAP-N, or TAP + 150 mM NaCl media for various time periods (12, 24, and 36 h). Total RNA was extracted to prepare cDNA for gene cloning and real-time PCR assay.

All *Chlamydomonas* strains were cultured in the TAP or deficiency medium of TAP with Hunter's trace element mix.

### Bioinformatics analysis of MAPK gene family of Chlamydomonas
*Chlamydomonas* MAPK homologous genes were retrieved from *Chlamydomonas* database (https://phytozome.jgi.doe.gov/pz/portal.html) by using the BLAST of *Arabidopsis* MAPK. Multiple sequence alignments were generated using ClustalX 2.1 and MEGA6. The following parameters were predicted: molecular weights and isoelectric points of proteins in Expasy (http://web.expasy.org/compute_pi/); protein structures in SMART; and conserved protein motifs in PROSITE (http://prosite.expasy.org/) and MEME (http://meme.nbcr.net/meme/). The structures of *CrMAPK* genes were generated online by using the Gene Structure Display Server (GSDS) (http://gsds.cbi.pku.edu.cn/), and the homologous chromosome segments were detected using a synteny plot in Plaza (http://bioinformatics.psb.ugent.be/plaza/versions/pico-plaza/synteny/index). The *CrMAPK* genes were subjected to BLAST analysis in Plaza, and their duplication patterns were detected using a synteny plot. The subcellular localization of *Chlamydomonas* MAPKs was performed using Euk-mPLoc2.0 (http://www.csbio.sjtu.edu.cn/bioinf/euk-multi-2/).

### Statistical analyses
Data were presented as mean ± S.D. One-way ANOVA followed by Duncan's post-test was performed to examine

significant differences between means. In all cases, comparisons showing $P < 0.05$ were considered significant.

## mRNA abundance detection

Three independent cell populations exposed to different stress conditions in various periods were collected. Total RNA was extracted and real-time PCR was performed as described previously [35]. The primers used in this study are listed in Tables 2 and 3. All the products were 300 +/- 50 nt The amplification rate of each transcript (Ct) was calculated using the PCR Base Line Subtracted method performed in the iCycler software at a constant fluorescence

**Table 2** Primer sequences for amplifying the Chlamydomonas MAPK genes

| Primer Name | Primer Sequences |
| --- | --- |
| CrMAPK2-F | GAGGCAAACCGATACACGAT |
| CrMAPK2-R | CGAGTACTTGGCGAAGAAGG |
| CrMAPK3-F | CGTCCGCAAAAGACAGTGTA |
| CrMAPK3-R | GGAGCACCTGGTAGACGAAG |
| CrMAPK4-F | GCACAGCCTCATAGGAAAGG |
| CrMAPK4-R | CCAATCACTGTGTGCAGGTC |
| CrMAPK5-F | GGAGGTGCACAAGCAGTACA |
| CrMAPK5-R | GTGCTACCCAGCAGGATCTC |
| CrMAPK6-F | CCAGCTCAAGCTCATCATCA |
| CrMAPK6-R | CATCTCGTCAAAGTCCAGCA |
| CrMAPK7-F | AAGACCGGGACAAGTTCCTT |
| CrMAPK7-R | ATGTACAGCTCCGCCATGAT |
| CrMAPK8-F | AGAGATTTGAAGCCCAGCAA |
| CrMAPK8-R | ACCTTGGTGATGAGGGACAG |
| CrMAPK9-F | GCTGTGCGACTTTGGCTTTGC |
| CrMAPK9-R | TACTTGCGGTCCAGCGTCTCCG |
| CrMAPK10-F | CGTTGTGGCGGTGAAAGG |
| CrMAPK10-R | GAACCCGAAGTCGCACAGC |
| CrMAPK11-F | ATCAAGCCCGCCAACATCC |
| CrMAPK11-R | GTTGTCAGATACCAGCACCTCC |
| CrMAPK12-F | GATGCCTTCAAATCTAAAACCG |
| CrMAPK12-R | AAGTCGCACAGCCGCACCAC |
| CrMAPK13-F | TCGGGCACGCCGCTGTTC |
| CrMAPK13-R | TAGCGTGCGACCTGGTGCGG |
| CrMAPK14-F | GGGCGTATGGGTCGGTAT |
| CrMAPK14-R | GCAGGTTGACGATGTTCAC |
| CrMAPK15-F | CACGATTGATGCGATAGAGGAG |
| CrMAPK15-R | AAGGTCGGGAACGGTGGA |
| CrMAPK16-F | TGGCAGTTAGCCTCCGTCATT |
| CrMAPK16-R | CGCAAAGCCAAAGTCGCA |
| CrMAPK17-F | GAGGTTACCGTGCTCAATGGC |
| CrMAPK17-R | GGGTAGCGGTCCTCCAACA |

**Table 3** Primer sequences for amplifying Chlamydomonas iron responsive genes

| Gene Name | Locus Name | Primer Name | Primer Sequences |
| --- | --- | --- | --- |
| CrARS2 | Cre16.g671350 | ARS2-F | ATGGGTGCCCTCGCGGTGTTC |
| | | ARS2-R | GTAGCGGATGTACTTGTGCAG |
| CrFOX1 | Cre09.g393150 | FOX1-F | GACGTGGAGGCCCAGAAG |
| | | FOX1-R | CGCGACGAAGTAGGTGTTG |
| CrFTR1 | Cre03.g192050 | FTR1-F | TCTTTCGGGAGACCATTGAG |
| | | FTR1-R | GAAGCATAGCAAAGCCAAGG |
| CrFEA1 | Cre12.g546550 | FEA1-F | CTCAAGTACCACCTGCACGA |
| | | FEA1-R | ACATAGCTCTTGCCGAGGAA |
| CrATX1 | Cre09.g392467 | ATX1-F | AGCTCGTGTCCTCGTAAAGC |
| | | ATX1-R | CTGCAACAGGTTCCGTGTAA |
| CrNRAMP2 | Cre07.g321951 | NRAMP2-F | CTGTCGCAGGTGATCCTGT |
| | | NRAMP2-R | TTTGCACCACCAGGTTAATG |

level. Cts were determined with three repeats. Relative fold differences were calculated on the basis of the relative quantification analytical method ($2^{-\Delta\Delta CT}$) using 18 s rRNA amplification as internal standard [36].

## Construction of CrMAPK3 RNA interference vector and transformation of Chlamydomonas

Using *Chlamydomonas* cDNA as a template, we amplified the fragments through PCR with forward primer *CrMAPK3*-F: CGTCCGCAAAAGACAGTGTA and reverse primer *CrMAPK3*-R: CTTCGTCTACCAGGTGC TCC. We then inserted the amplified fragments into pMD18-T vector to generate *CrMAPK3*-18 T, which was further digested with HindIII and BamHI and ligated into the intermediate vector T282 to produce *CrMAPK3*-T282. *CrMAPK3*-T282-*CrMAPK3* with inverted repeat sequence of *CrMAPK3* (*CrMAPK3IR*) was developed by digesting *CrMAPK3*-18 T and *CrMAPK3*-T282. *CrMAP K3IR* was inserted into EcoRI-digested *pMaa7/XIR* to produce *Maa7IR/CrMAPK3IR*. *Maa7IR/CrMAPK3IR* was then transformed into *C. reinhardtii* 2A38 by applying the glass bead procedure [37].

## ARS (arylsulfatase) activity detection

ARS activity was determined as described by Davies and Grossman [38]. XSO4 (10 mM) was added to plates with −Fe TAP solid medium and scribed before clones were inoculated. After 1 day, the transformants that expressed ARS activity were identified using blue halos around their colonies. The cells were initially collected by centrifugation to quantify the ARS activity. The supernatant was mixed with 0.1 M glycine–NaOH at pH 9.0, 10 mM imidazole, and 4.5 mM *p*-nitrophenyl sulfate. The reaction mixture was incubated at 27 °C for 30 min. The reaction was terminated by adding 0.25 M NaOH, and its absorbance at 410 nm was determined. A standard curve of *p*-nitrophenol (Sigma Chemical Co.) was obtained using 0.2 M NaOH.

## Conclusions

Silencing *CrMAPK3* decreased the mRNA levels and ARS activities of FOX1:ARS chimeric gene and endogenous *CrFOX1*. The mRNA levels of iron-responsive genes, such as *CrNRAMP2*, *CrATX1*, *CrFTR1*, and *CrFEA1*, were also remarkably reduced. Therefore, *CrMAPK3* regulated the expression of iron-deficiency-responsive genes in *C. reinhardtii*.

### Acknowledgments
Not applicable.

### Funding
This study was supported by the National Natural Science Foundation of China (31160050, 31360051), the Funds of Hainan Engineering and Technological Research (GCZX2011006, GCZX2012004, GCZX2013004), and the Key Projects of Hainan Province (ZDYF2016021).

### Authors' contributions
XW Fei and XD Deng designed experiments. XW Fei and JM Yu performed experiments. JM Yu and YJ Li analyzed data. XW Fei and XD Deng wrote the manuscript. All authors read and approved the final manuscript.

### Competing interest
The authors declare that they have no competing interests.

### Author details
[1]School of Science, Hainan Medical College, Haikou 571101, China. [2]Institute of Tropical Bioscience and Biotechnology, Chinese Academy of Tropical Agricultural Science, Key Laboratory of Tropical Crop Biotechnology, Ministry of Agriculture, Haikou 571101, China.

### References
1. Fei XW, Deng XD. A novel Fe deficiency responsive element (FeRE) regulates the expression of atx1 in *Chlamydomonas reinhardtii*. Plant Cell Physiol. 2007;48:1496–503.
2. Mohanta KT, Arora PK. Identification of new members of the MAPK gene family in plants shows diverse conserved domains and novel activation loop variants. BMC Genomics. 2015;6:16–58.
3. Besteiro MAG, Ulm R. Phosphorylation and stabilization of arabidopsis MAP kinase phosphatase 1 in response to UV-B stress. J Biol Chem. 2013;288:480–6.
4. Caunt CJ, Keyse SM. Dual-specificity MAP kinase phosphatases (MKPs)Shaping the outcome of MAP kinase signalling. FEBS J. 2013;280:489–504.
5. Rodriguez MC, Petersen M, Mundy J. Mitogen-activated proteinkinase signaling in plants. Annu Rev Plant Biol. 2010;61:621–49.
6. Nakagami H, Pitzschke A, Hirt H. Emerging MAP kinase pathways in plant stress signalling[J]. Trends Plant Sci. 2005;10:1360–85.
7. Mizoguchi T, Hayashida N, Yamaguchi-Shinozaki K. ATMPKs: A gene family of plant MAP kinases in Arabidopsis thaliana. FEBS Lett. 1993;336:440–4.
8. Lieberherr D, Thao NP, Nakashima A. A sphing olipid elicitor-inducible mitogen-activated protein kinase is regulated by the small GTPase OsRac1 and heterotrimeric G-protein in rice. Plant Physiol. 2005;138:1644–52.
9. Lalle M, Visconti S, Marra M. ZmMPK6, a novel maize MAP kinase that interacts with 14-3-3 proteins. Plant Mol Biol. 2005;59:713–22.
10. Takezawa D. Elicitor- and A23187-induced expression of WCK-1, a gene encoding mitogen-activated protein kinase in wheat. Plant Mol Biol. 1999; 40:921–33.
11. Knetsch M, Wang M, Snaar-Jagalska BE. Abscisic acid induces mitogen-activated protein kinase activation in barley aleuron protoplasts. Plant Cell. 1996;8:1061–7.
12. Huttly AK, Phillips AL. Gibberellin-regulated expression in oataleurone cells of two kinases that show homology to MAP kinase and a ribosomal protein kinase. Plant Mol Biol. 1995;27:1043–52.
13. Wilson C, Anglmayer R, Vicente O. Molecular cloning, functional expression in Escherichia coli, and characterization of multiple mitogen-activated protein kinases from tobacco. Eur J Biochem. 1995;233:249–57.
14. Jonak C, Ökrész L, Bögre L. Complexity, cross talk and integration of plant MAP kinase signaling. Curr Opin Plant Biol. 2002;5:415–24.
15. Singh R, Lee MO, Lee JE. Rice mitogen-activated protein kinase interactome analysis using the yeast two-hybrid system. Plant Physiol. 2012;160:477–87.
16. Deng XD, Eriksson M. Two iron-responsive promoter elements control expression of FOX1 in *Chlamydomonas reinhardtii*. Eukaryot Cell. 2007;6:2163–7.
17. Robinson NJ, Procter CM, Connolly EL, Guerinot ML. A ferric-chelate reductase for iron uptake from soils. Nature. 1999;397:694–7.
18. Allen MD, del Campo JA, Kropat J, Merchant SS. FEA1, FEA2, and FRE1, encoding two homologous secreted proteins and a candidate ferrireductase, are expressed coordinately with FOX1 and FTR1 in iron-deficient *Chlamydomonas reinhardtii*. Eukaryot Cell. 2007;6:1841–52.
19. Chen JC, Hsieh SI, Kropat J, Merchant SS. A ferroxidase encoded by FOX1 contributes to iron assimilation under conditions of poor iron nutrition in *Chlamydomonas*. Eukaryot Cell. 2008;7:541–5.
20. La Fontaine S, Quinn JM, Nakamoto SS, Page MD, Göhre V, Moseley JL, Kropat J, Merchant SS. Copper-dependent iron assimilation pathway in the model photosynthetic eukaryote *Chlamydomonas reinhardtii*. Eukaryot Cell. 2002;1:736–57.
21. Terzulli A, Kosman DJ. Analysis of the high-affinity iron uptake system at the *Chlamydomonas reinhardtii* plasma membrane. Eukaryot Cell. 2010;9:815–26.
22. Pufahl RA, Singer CP, Peariso KL, Lin SJ, Schmidt PJ, Fahrni CJ, Culotta VC, Penner-Hahn JE, O'Halloran TV. Metal ion chaperone function of the soluble Cu (I) receptor Atx1. Science. 1997;278:853–6.
23. Liang W, Yang B, Yu BJ, Zhou Z, Li C, Jia M. Identification and analysis of MKK and MPK gene families in canola (Brassica napus L.). BMC Genomics. 2013;14:392.
24. Chen L, Hu W, Tan S, Wang M, Ma Z, Zhou S. Genome-wide identification and analysis of MAPK and MAPKK gene families in *Brachypodium distachyon*. PLoS One. 2012;7:e46744.
25. Janitza P, Ullrich KK, Quint M. Toward a comprehensive phylogenetic reconstruction of the evolutionary history of mitogen-activated protein kinases in the plant kingdom. Front Plant Sci. 2012;3:1–11.
26. Zhang S, Xu R, Luo X, Jiang Z, Shu H. Genome-wide identification and expression analysis of MAPK and MAPKK gene family in *Malus domestica*. Gene. 2013;531:377–87.
27. Nicole MC, Hamel LP, Morency MJ. MAP-ping genomic organization and organ-specific expression profiles of poplar MAP kinases and MAP kinase kinases. BMC Genom. 2006;7:223.
28. Kong F, Wang J, Cheng L. Genome-wide analysis of the mitogen-activated protein kinase gene family in *Solanum lycopersicum*. Gene. 2012;499:108–20.
29. Le NTV, Richardson DR. The role of iron in cell cycle progression and the proliferation of neoplastic cells. Biochim Biophys Acta. 2002;1603:31–46.
30. Brodie C, Siriwardana G, Lucas J, Schleicher R, Terada N, Szepesi A, Gelfand E, Seligman P. Neuroblastoma sensitivity to growth inhibition by deferrioxamine: evidence for a block in G1 phase of the cell cycle. Cancer Res. 1993;53:3968–75.
31. Hileti D, Panayiotidis P, Hoffbrand AV. Iron chelators induce apoptosis in proliferating cells. Br J Haematol. 1995;89:181–7.
32. Buss JL, Torti FM, Torti SV. The role of iron chelation in cancer therapy. Cur Med Chem. 2003;10:1021–34.
33. Kalinowski DS, Richardson DR. The evolution of iron chelators for the treatment of iron overload disease and cancer. Pharmacol Rev. 2005;57:547–83.
34. Yu Y, Richardson D. Cellular iron depletion stimulates the JNK and p38 MAPK signaling transduction pathways, dissociation of ASK1-thioredoxin, and activation of ASK1. J Biol Chem. 2011;286:15413–27.
35. Luo Q, Li Y, Wang W, Fei X, Deng X. Genome-wide survey and expression analysis of Chlamydomonas reinhardtii U-box E3 ubiquitin ligases (CrPUBs) reveal a functional lipid metabolism module. PLoS One. 2015;10:e0122600.
36. Livak KJ, Schmittgen TD. Analysis of relative gene expression data using real-time quantitative PCR and the 2(-Delta Delta C(T)) Method. Methods. 2001;25:402–8.
37. Kindle KL. High frequency nuclear transformation of *Chlamydomonas reinhardtii*. Proc Natl Acad Sci U S A. 1990;87:1228–32.
38. Davies JP, Grossman AR. Sequences controlling transcription of the *Chlamydomonas reinhardtii* beta 2-tubulin gene after deflagellation and during the cell cycle. Mol Cell Biol. 1994;14:5165–74.

# 14

# Glycyl-alanyl-histidine protects PC12 cells against hydrogen peroxide toxicity

Hideki Shimura[1,3]* iD, Ryota Tanaka[2], Yoshiaki Shimada[1], Kazuo Yamashiro[2], Nobutaka Hattori[2] and Takao Urabe[1]

## Abstract

**Background:** Peptides with cytoprotective functions, including antioxidants and anti-infectives, could be useful therapeutics. Carnosine, β-alanine-histidine, is a dipeptide with anti-oxidant properties. Tripeptides of Ala-His-Lys, Pro-His-His, or Tyr-His-Tyr are also of interest in this respect.

**Results:** We synthesized several histidine-containing peptides including glycine or alanine, and tested their cytoprotective effects on hydrogen peroxide toxicity for PC12 cells. Of all these peptides (Gly-His-His, Ala-His-His, Ala-His-Ala, Ala-Ala-His, Ala-Gly-His, Gly-Ala-His (GAH), Ala-His-Gly, His-Ala-Gly, His-His-His, Gly-His-Ala, and Gly-Gly-His), GAH was found to have the strongest cytoprotective activity. GAH decreased lactate dehydrogenase (LDH) leakage, apoptosis, morphological changes, and nuclear membrane permeability changes against hydrogen peroxide toxicity in PC12 cells. The cytoprotective activity of GAH was superior to that of carnosine against hydrogen peroxide toxicity in PC12 cells. GAH also protected PC12 cells against damage caused by actinomycin D and staurosporine. Additionally, it was found that GAH also protected SH-SY5Y and Jurkat cells from damage caused by hydrogen peroxide, as assessed by LDH leakage.

**Conclusion:** Thus, a novel tripeptide, GAH, has been identified as having broad cytoprotective effects against hydrogen peroxide-induced cell damage.

## Backgrounds

More than 7000 peptides have been identified as playing crucial roles in human physiology, including those acting as hormones, neurotransmitters, growth factors, ion channel ligands, or having anti- microbial activity [1–4]. Peptides are selective and efficacious signaling molecules that bind to specific cell surface receptors where they induce intracellular effects. Peptides represent an excellent starting point for the design of novel therapeutics. Even small peptides, such as dipeptides and tripeptides, may also have potent functions [5–8]. Some have cytoprotective functions and have been used in clinical trials for human disease [9]. Carnosine is a well-characterized antioxidant dipeptide composed of β-alanine and histidine. It has cytoprotective activity against various stresses as determined in both in vitro and in vivo models [10]. The imidazole ring of histidine is reported to have an important role in antioxidant cell protection [11]. Carnosine is a more effective singlet oxygen scavenger than L-histidine, although both compounds have been shown to protect against oxidative DNA damage and against liposome oxidation induced experimentally in vitro [12].

Histidine is a scavenger of hydroxyl radicals [13], and may interact chemically with toxic oxygen species through at least two distinct mechanisms: (1) by interfering with the redox reactions involving metal ions that produce the hydroxyl radical, and (2) by direct interactions of the histidine imidazole ring with singlet oxygen [14]. The imidazole ring of L-histidine has been shown to be responsible for the antioxidant activity of several biologically important dipeptides, including carnosine (β-alanyl-L-histidine), anserine (β-alanyl-3-methyl-L-histidine), and homocarnosine (l-aminobutyryl-L-histidine) [12]. Ala-His-Lys, Pro-His-His, and Tyr-His-Tyr were also reported to have antioxidant properties [15–17].

We hypothesized that histidine-containing tripeptides might also have antioxidant activity. In the present study, we synthesized and determined the antioxidant activities of tripeptides containing histidine and the small amino acids alanine and glycine.

* Correspondence: miurashimura@yahoo.co.jp
[1]Department of Neurology, Juntendo University Urayasu Hospital, 2-1-1 Tomioka, Urayasu, Chiba, Japan
[3]Institute for Environment and Gender Specific Medicine, Juntendo University School of Medicine, Chiba, Japan
Full list of author information is available at the end of the article

## Methods

### Peptides

The histidine-containing tripeptides Gly-His-His (GHH), Ala-His-His (AHH), Ala-His-Ala (AHA), Ala-Ala-His (AAH), Ala-Gly-His (AGH), Gly-Ala-His (GAH), Ala-His-Gly (AHG), His-Ala-Gly (HAG), His-His-His (HHH), Gly-His-Ala (GHA), and Gly-Gly-His (GGH) were synthesized by and purchased from Biogate (Gifu, Japan). Carnosine was purchased from Sigma-Aldrich (St. Louis, MO, USA). GAH at 1 µg/µl = 3530 µM.

### Cell culture

PC12 (CRL-1721), SH-SY5Y (CRL-2266) and Jurkat cells (CRL-1990) were purchased from ATCC. PC12, Jurkat, and SH-SY5Y cells were grown at 37 °C (5% $CO_2$ atmosphere) in RPMI 1640 medium supplemented with 10% heat-inactivated horse serum, 5% heat-inactivated fetal bovine serum, 2 mM L-glutamine, 1 mM sodium pyruvate, 100 U/mL of penicillin, and 100 mg/mL of streptomycin. Cell culture medium was changed three times per week, and when confluent, cells were split 1:6. For the experiments reported here, subconfluent cells were treated with different concentrations of 100–10,000 µM hydrogen peroxide, 10 µM staurosporine, or 500 µg/mL of actinomycin D.

### Lactate dehydrogenase assay

To assess cytotoxicity, lactate dehydrogenase (LDH) activity was measured using LDH cytotoxicity detection kits (Takara, Otsu, Shiga Japan). PC-12 cells were seeded into a 96-well plate at $2 \times 10^6$ cells/mL with assay medium, for a period of 18 h at 37 °C in a 5% $CO_2$ humidified incubator. The culture medium was then removed and replaced with serum-free medium and 1% bovine serum albumin (BSA) was added. The plates were treated with 100–5000 µM hydrogen peroxide for 1–24 h. After incubation, the samples were centrifuged for 10 min at 250 g. One hundred µL/well of supernatant was removed, without disturbing the cell pellet, and transferred into corresponding wells of a new 96-well plate. Solution C (100 µL, the reaction mixture) was added to each well and incubated for 30 min at room temperature. The 96-well plate was protected from light during this time. The absorbance of the samples was measured at 490 nm using the ARVO SX 1420 Multilabel Counter (PerkinElmer Wallac Inc., Turku, Finland).

### Flow cytometry analysis

Cells were washed twice with buffer (140 mM NaCl, 10 mM HEPES, 2.5 mM $CaCl_2$, pH 7.4), resuspended in 1 mL of the same buffer, and incubated on ice for 30 min with 5 µL of propidium iodide (50 µg/mL $H_2O$ stock solution) added to each sample. They were then analyzed by flow cytometry (Becton Dickinson, San Jose, CA, USA) [18].

### Terminal deoxynucleotidyl transferase dUTP nick end labeling (TUNEL) assay

The TUNEL assay of PC12 cells was conducted using the Click-iT TUNEL Alexa Fluor 594 Imaging Assay Kit (Molecular Probes™, Eugene, OR, USA). DNase I was used to generate strand breaks in the DNA to provide a positive TUNEL reaction control. The number of Alexa Fluor 594-positive cells was counted using BZ-II Analyzer software (Keyence, Japan). Nuclear was stained with 4′, 6-diamidino-2-phenylindole (DAPI).

### Dead cell images

PC12 cells were incubated for 30 min at 37 °C with 4 µM of the EthD-1 (LIVE/DEAD viability/cytotoxicity kit reagent (Thermo Fisher Scientific Inc., Waltham, MA, USA). EthD-1 enters cells with damaged membranes and undergoes a 40-fold enhancement of fluorescence upon binding to nucleic acids, thereby producing a bright red fluorescence in dead cells (excitation/emission, ~495 nm/~635 nm, respectively). The number of Alexa Fluor 594-positive cells was counted using BZ-II Analyzer software (Keyence).

### Cell viability analysis

Cell growth was assessed using the Cell Counting Kit-8 (CCK-8; Dojindo, Japan) assay. The cells ($2 \times 10^4$) were plated in 100 µL of media and added to 100 µL of hydrogen peroxide with or without GAH or carnosine in each well of a 96-well flat-bottomed microtiter plate. Assays were done in triplicate cultures and incubated at 37 °C in an incubator with 5% $CO_2$. Ten µL of the CCK8 solution was added to each well after 24 h of treatment, and the cells were cultured for another 2 h at 37 °C. The absorbance was measured using a microplate reader (Nanoquant Plate™; Tecan, Männedorf, Switzerland), at 450 nm with 600 nm used as the reference wavelength. The cell viability was expressed as a percentage of absorbance in cells with indicated treatments to that of the control cells.

### Statistical analysis

All values are expressed as mean ± SEM. One-way analysis of variance and post hoc Fisher's protected least significant difference tests were used to determine the significance of differences between the groups. $P$ values $< 0.05$ indicated significant difference.

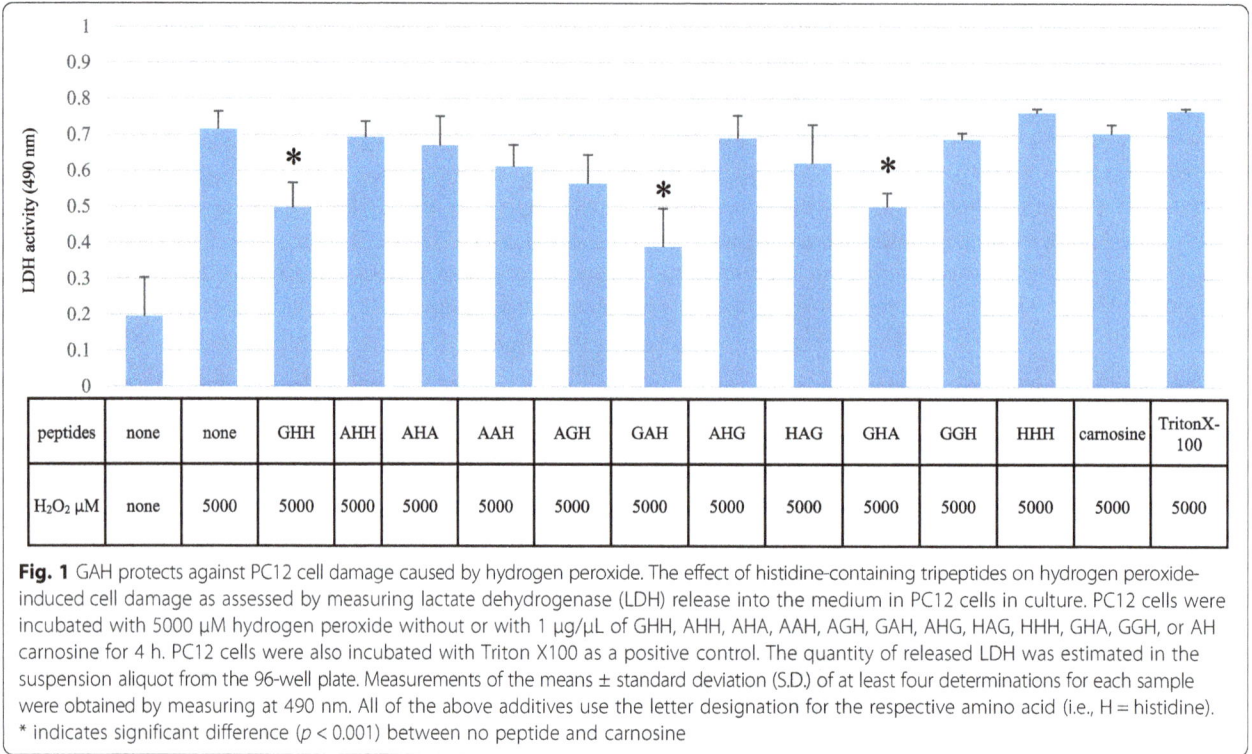

**Fig. 1** GAH protects against PC12 cell damage caused by hydrogen peroxide. The effect of histidine-containing tripeptides on hydrogen peroxide-induced cell damage as assessed by measuring lactate dehydrogenase (LDH) release into the medium in PC12 cells in culture. PC12 cells were incubated with 5000 μM hydrogen peroxide without or with 1 μg/μL of GHH, AHH, AHA, AAH, AGH, GAH, AHG, HAG, HHH, GHA, GGH, or AH carnosine for 4 h. PC12 cells were also incubated with Triton X100 as a positive control. The quantity of released LDH was estimated in the suspension aliquot from the 96-well plate. Measurements of the means ± standard deviation (S.D.) of at least four determinations for each sample were obtained by measuring at 490 nm. All of the above additives use the letter designation for the respective amino acid (i.e., H = histidine). * indicates significant difference ($p < 0.001$) between no peptide and carnosine

## Results

We incubated PC12 cell with 5000 μM hydrogen peroxide with 1 μg/μL of GHH, AHH, AHA, AAH, AGH, GAH, AHG, HAG, HHH, GHA, GGH, or carnosine for 4 h. GAH, GHH and GHA decreased LDH leakage compared with either no peptide or with carnosine ($p < 0.01$) (Fig. 1). GAH had the strongest hydrogen peroxide-induced cell death-inhibiting effect of all the tested peptides, and most effectively decreased cell damage in each experiment.

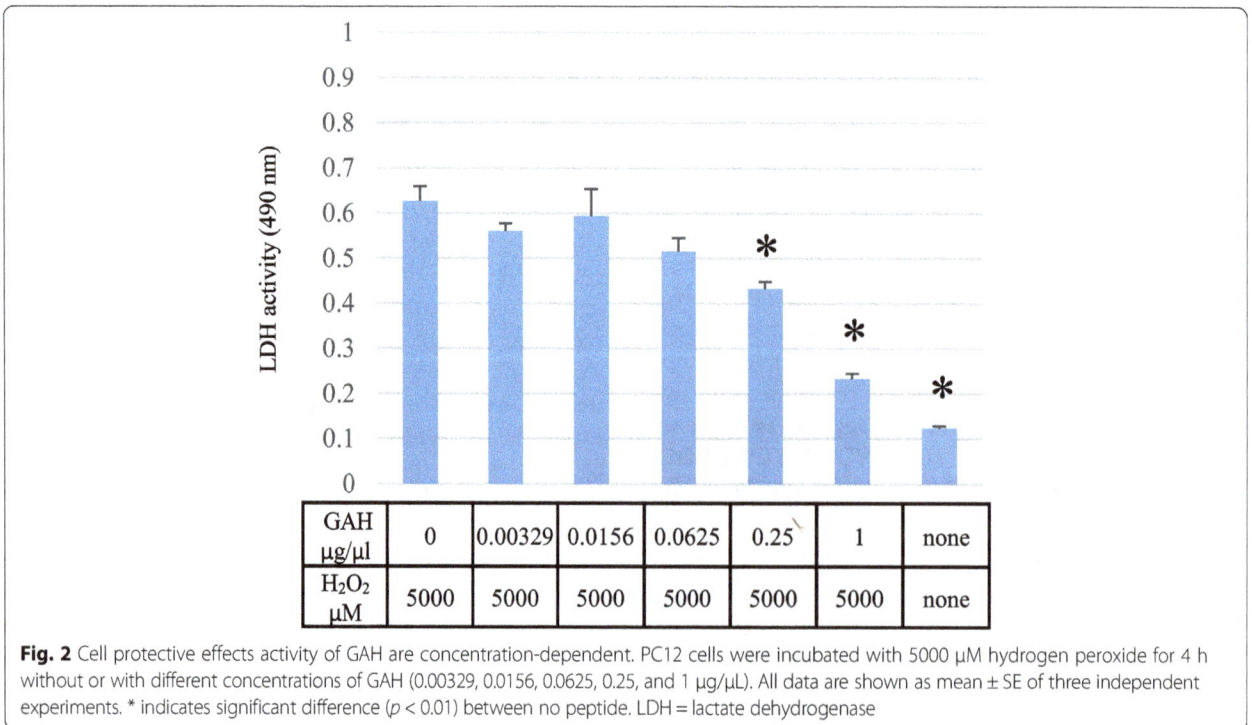

**Fig. 2** Cell protective effects activity of GAH are concentration-dependent. PC12 cells were incubated with 5000 μM hydrogen peroxide for 4 h without or with different concentrations of GAH (0.00329, 0.0156, 0.0625, 0.25, and 1 μg/μL). All data are shown as mean ± SE of three independent experiments. * indicates significant difference ($p < 0.01$) between no peptide. LDH = lactate dehydrogenase

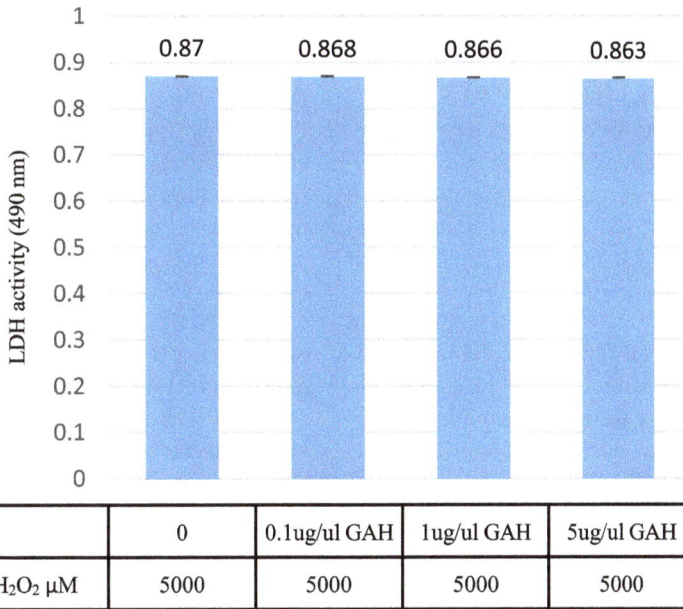

**Fig. 3** GAH does not directly inhibit LDH enzyme activity. We added 0.1 μg/μL, 1 μg/μL, or 5 μg/μL of GAH to the media of PC12 cells incubated with 5000 μM hydrogen peroxide for 4 h and subsequently measured the LDH activity. All data are shown as mean ± SE

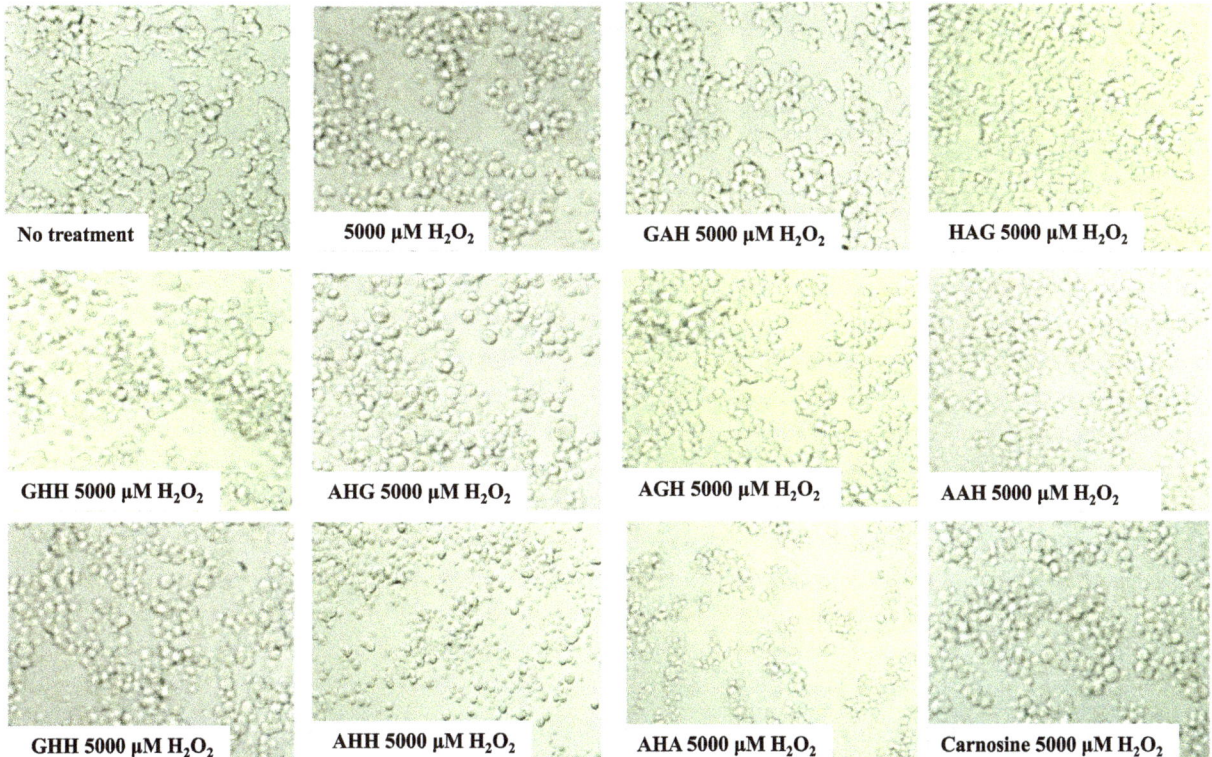

**Fig. 4** Micrographs representative of shape changes of PC12 cells. PC12 cells were incubated with 5000 μM hydrogen peroxide for 4 h with or without 1 μg/μL of GHH, AHH, AHA, AAH, AGH, GAH, AHG, HAG, HHH, GHA, GGH, or AH carnosine for 4 h

We confirmed GAH cytoprotective activity in at least 20 independent experiments. We then analyzed the cytoprotective functions of GAH, compared to the other tripeptides used in this study.

To determine whether the cell protective activity of GAH was concentration-dependent, we incubated the cells with 5000 μM of hydrogen peroxide for 4 h without or with different concentrations of GAH (0.00329, 0.0156, 0.0625, 0.25, and 1 μg/μL). GAH at 0.25 μg/μL decreased the amount of LDH release ($p < 0.01$), and 1 μg/μL GAH was even more effective in this assay ($p < 0.01$) (Fig. 2). Thus, the highest concentration of GAH tested here had the best cytoprotective activity. We then determined whether GAH directly inhibited LDH enzyme activity. PC12 cells were incubated with 5000 μM hydrogen peroxide and culture media was collected and assayed. Different concentrations of GAH were added to the media and the LDH activity was determined. GAH at 0.1 μg/μL, 1 μg/μL, and 5 μg/μL did not inhibit enzyme activity (Fig. 3). These results indicate that GAH did not directly block LDH activity, but most likely prevented its leakage into the media, caused by loss of membrane integrity.

Cell morphology assessments showed that hydrogen peroxide treatment decreased the number of adherent PC12 cells and increased the number of nonadherent cells, resulting in a change from a flat-shaped to a round-shaped cell. The addition of GAH together with hydrogen peroxide increased the number of adherent cells and maintained the flat shape of the cell compared to the effect of hydrogen peroxide alone or after coculture with HAG, GHH, AHG, AGH, AAH, GHH, AHH, AHA, or carnosine (Fig. 4). We next examined the mechanism of cell protection by GAH using the ethidium bromide homodimer 1 from the LIVE/DEAD® Viability/Cytotoxicity Kit (Molecular Probes™) to stain dead cells. PC12 cells were incubated with ethidium homodimer 1 after treatment with 5000 μM hydrogen peroxide for 4 h in the presence or absence of GAH. The number of dead cells per 100 cells was calculated. GAH and carnosine significantly protected against hydrogen peroxide-induced cell death ($p < 0.01$), GAH more strongly than carnosine ($p < 0.01$) (Fig. 5).

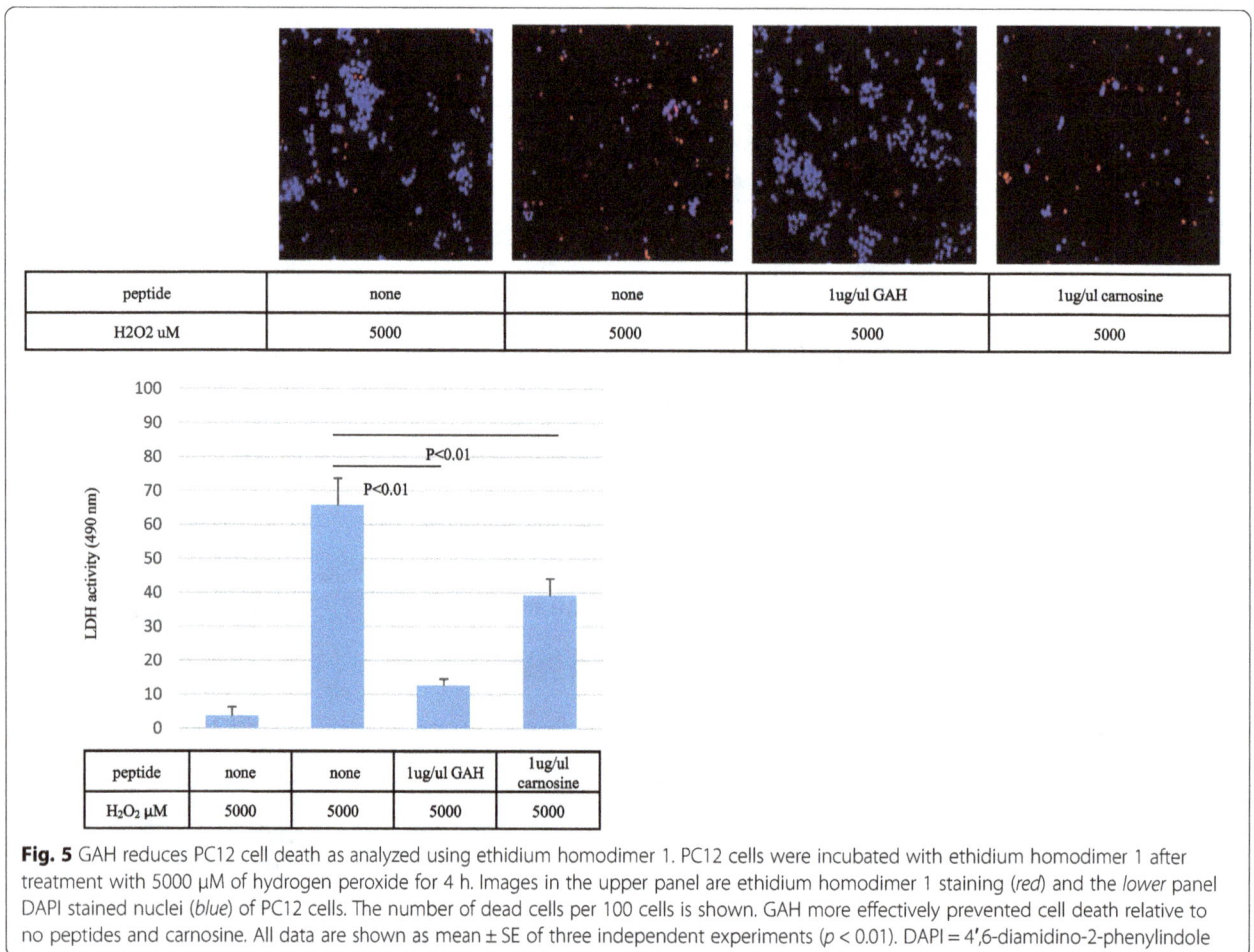

| peptide | none | none | 1ug/ul GAH | 1ug/ul carnosine |
|---|---|---|---|---|
| H2O2 uM | 5000 | 5000 | 5000 | 5000 |

| peptide | none | none | 1ug/ul GAH | 1ug/ul carnosine |
|---|---|---|---|---|
| H$_2$O$_2$ μM | 5000 | 5000 | 5000 | 5000 |

**Fig. 5** GAH reduces PC12 cell death as analyzed using ethidium homodimer 1. PC12 cells were incubated with ethidium homodimer 1 after treatment with 5000 μM of hydrogen peroxide for 4 h. Images in the upper panel are ethidium homodimer 1 staining (*red*) and the *lower* panel DAPI stained nuclei (*blue*) of PC12 cells. The number of dead cells per 100 cells is shown. GAH more effectively prevented cell death relative to no peptides and carnosine. All data are shown as mean ± SE of three independent experiments ($p < 0.01$). DAPI = 4',6-diamidino-2-phenylindole

Flow cytometric analyses were performed on the cells after propidium iodide (PI) staining to evaluate the effects of GAH on PC12 cells treated with 500 μM of hydrogen peroxide for 4 h. As shown in Fig. 6, the percentage of PI-positive PC12 cells was $28.13 \pm 4.88\%$ without hydrogen peroxide, and $60.95 \pm 14.82\%$ with 500 μM of hydrogen peroxide, after 4 h. GAH decreased the percentage of PI-positive PC12 cells treated with 500 μM hydrogen peroxide by 11.13% (from $60.95 \pm 14.82\%$ to $49.04 \pm 10.49\%$; $p = 0.0219$). In contrast, carnosine did not significantly affect this parameter (reduced from $60.95 \pm 14.82\%$ to $59.16 \pm 14.45\%$; $p = 0.748$). Thus, GAH exerted a cell protective effect superior to that of carnosine.

The percentage of TUNEL-positive PC12 cells was $18.4 \pm 0.87\%$ without hydrogen peroxide, and $73.1 \pm 1.17\%$ after treatment with 500 μM hydrogen peroxide for 24 h. GAH decreased the percentage of TUNEL-positive hydrogen peroxide-treated PC12 cells by 60.9% (from $73.1 \pm 1.17\%$ to $12.21 \pm 0.078\%$; $p < 0.01$) (Fig. 7). Carnosine also decreased the percentage of TUNEL-positive PC12 cells (by 42%, from $73.1 \pm 1.17\%$ to $31 \pm$ 3.75%; $p < 0.01$). GAH thus prevented apoptosis better than carnosine. Whether the viability of PC12 cells exposed to 100 μM hydrogen peroxide was protected by GAH was investigated using the CCK-8 assay. The results showed that cell viability (as measured at OD 450 nm) was significantly increased in the GAH ($0.625 \pm 0.023$)- and carnosine ($0.711 \pm 0.026$)-treated groups relative to controls ($0.144 \pm 0.012$, p < 0.01) (Fig. 8). Thus, in this assay, there was no difference between GAH and carnosine.

We also examined the protective effects of GAH against actinomycin D and staurosporine-induced apoptotic cell death of PC12 cells [19, 20]. Cells were incubated with 500 μg/mL of actinomycin D or 10 μM of staurosporine for 12 h, and viability again measured by the LDH leakage assay. GAH did prevent cell death induced by actinomycin D ($p < 0.001$) and staurosporine ($p < 0.001$)) similar to its effects on hydrogen peroxide toxicity (Fig. 9).

We also examined the effect of GAH on Jurkat cells (human lymphocyte cell type) and SH-SY5Y cells (human neuroblastoma cells) in culture. Jurkat cells were

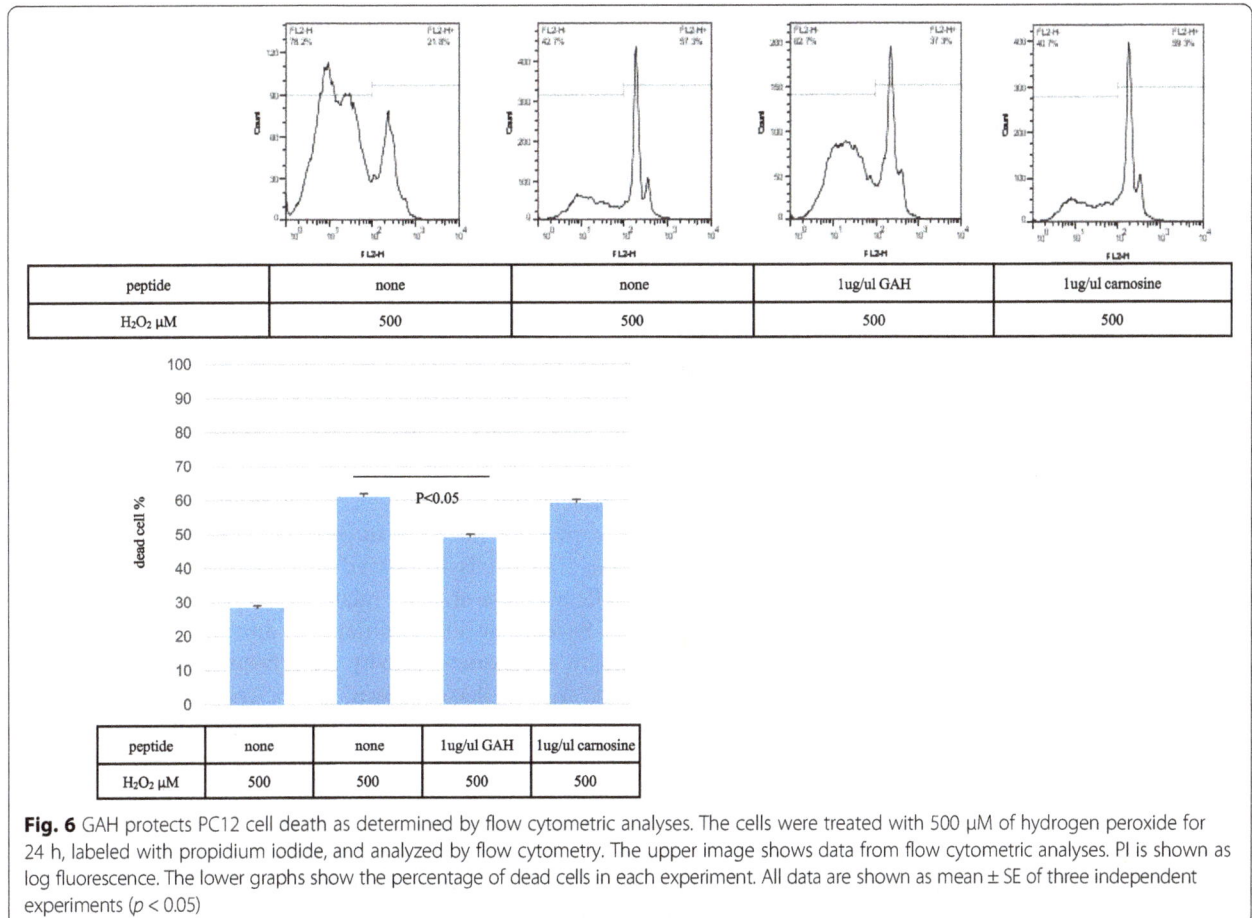

| peptide | none | none | 1ug/ul GAH | 1ug/ul carnosine |
|---------|------|------|------------|------------------|
| $H_2O_2$ μM | 500 | 500 | 500 | 500 |

| peptide | none | none | 1ug/ul GAH | 1ug/ul carnosine |
|---------|------|------|------------|------------------|
| $H_2O_2$ μM | 500 | 500 | 500 | 500 |

**Fig. 6** GAH protects PC12 cell death as determined by flow cytometric analyses. The cells were treated with 500 μM of hydrogen peroxide for 24 h, labeled with propidium iodide, and analyzed by flow cytometry. The upper image shows data from flow cytometric analyses. PI is shown as log fluorescence. The lower graphs show the percentage of dead cells in each experiment. All data are shown as mean ± SE of three independent experiments ($p < 0.05$)

| peptide | Dnase | none | none | 1ug/ul GAH | 1ug/ul carnosine |
|---------|-------|------|------|------------|------------------|
| H2O2 uM | none | none | 500 | 500 | 500 |

| peptide | Dnase | none | none | 1ug/ul GAH | 1ug/ul carnosine |
|---------|-------|------|------|------------|------------------|
| H2O2 uM | none | none | 500 | 500 | 500 |

**Fig. 7** GAH decreases the number of TUNEL-positive PC12 cells. PC12 cells were exposed to 500 μM hydrogen peroxide for 24 h. The TUNEL assay was performed using the Click-iT TUNEL Alexa Fluor 594 Imaging Assay Kit. The TUNEL signal is shown in *red* (*top* of figure, *bottom* panels) and Hoechst 33,342 nuclear staining is shown in *blue* (*top* of figure, *top* panels). The *upper* panels are representative images of TUNEL staining. All data are shown as mean ± SE of three independent experiments ($p < 0.01$)

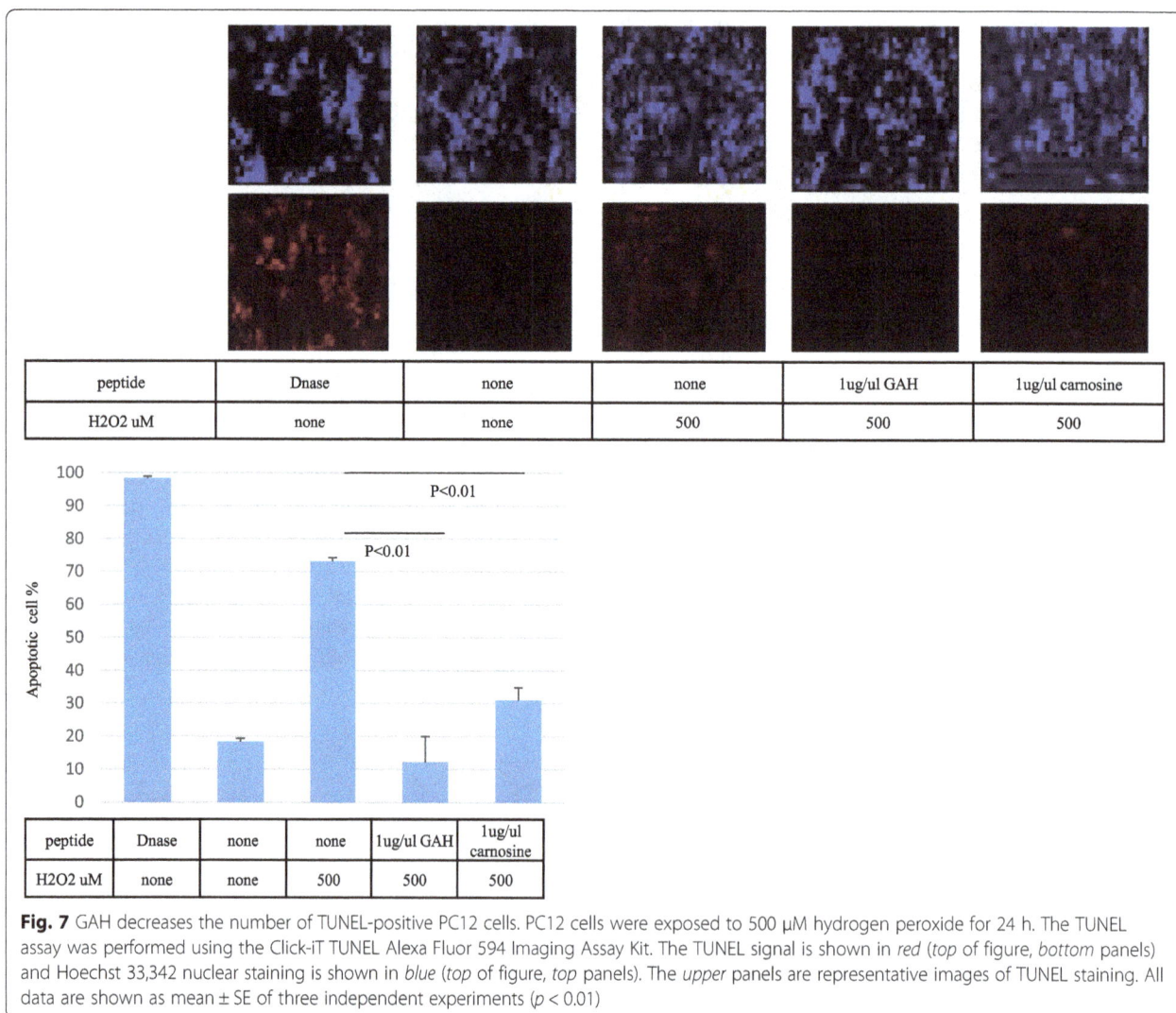

incubated with 10,000 μM of hydrogen peroxide for 4 h. GAH at 1 μg/μL decreased the amount of LDH release from Jurkat cells ($p < 0.001$) (Fig. 10a). We also incubated SH-SY5Y cells with 1000 μM of hydrogen peroxide for 24 h in the presence or absence of GAH. At 1 μg/μL, GAH also decreased the amount of LDH released from SH-SY5Y cells ($p = 0.0039$) (Fig. 10b). These findings indicate that GAH prevents Jurkat and SH-SY5Y cell membrane damage caused by hydrogen peroxide.

## Discussion

We screened newly synthesized histidine-containing tripeptides for their radical scavenging activity. GHH, AHH, AHA, AAH, AGH, GAH, AHG, HAG, HHH, GHA, and GGH were screened for their ability to decrease LDH leakage from PC12 cells treated with hydrogen peroxide. Of these peptides, GAH at 1 μg/μL had

the strongest protective effect against cell damage as assessed by LDH leakage, by ethidium bromide staining, cell morphology, TUNEL assays, CCK-8 assays, and PI assays. GAH also protected Jurkat cells and SH-SY5Y cells; it may therefore have a protective effect against many different types of cells. GAH was not effective for SH-SY5Y cells compared to PC12 cells suggesting that effectiveness of GAH might depends on cell type.

Several histidine-containing dipeptides or tripeptides with antioxidant activity have been identified. Hartman et al. [21] have shown that carnosine is an efficient singlet-oxygen scavenger, quenching singlet oxygen more effectively than histidine. They also reported that carnosine, anserine, and histidine protect phages against gamma-irradiation, which gives rise to oxidative DNA damage. Tsuge et al. reported the isolation of a potent antioxidative peptide, Ala-His-Lys, from an egg white albumin hydrolysate [15]. Chen et al. reported that Pro-

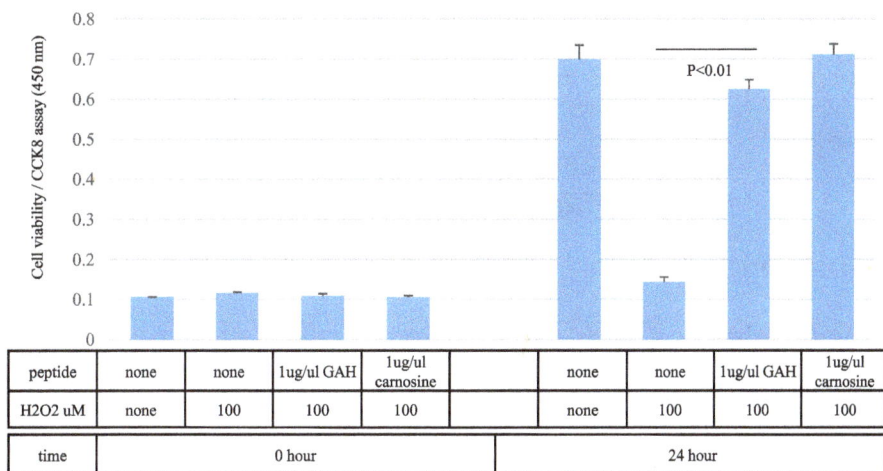

**Fig. 8** GAH protects PC12 cells as measured by the CCK-8 assay. PC12 cells were exposed to 100 μM hydrogen peroxide. Cell viability was measured using the CCK-8 assay at the start and after 24 h incubation. All data are shown as mean ± SE of three independent experiments ($p < 0.01$)

His-His was the most active antioxidant among the 28 synthetic peptides that were structurally related to Leu-Leu-Pro-His-His [16]. Saito et al. reported that Tyr-His-Tyr had a strong synergistic effect with phenolic antioxidants [17]. As reported here, GAH attenuated cell damage by hydrogen peroxide, suggesting that it might be an efficient singlet oxygen scavenger, similar to carnosine, Pro-His-His, and Tyr-His-Tyr.

GAH protected cells not only against hydrogen peroxide damage, but also prevented apoptosis induced by staurosporine and actinomycin D. Staurosporine is a broad-spectrum inhibitor of protein kinases, and has been widely used for the induction of apoptosis in diverse cellular models [22, 23]. Staurosporine preferentially activates the mitochondrial apoptotic pathway, relying on caspase activation to cause cell death. Actinomycin D, on the other hand, is a widely-used intercalating transcription inhibitor. [24]. Protection by GAH against hydrogen peroxide, staurosporine, and actinomycin D suggests that it might not only be a radical scavenger but could also protect by other mechanisms. The protective effects of carnosine include actions on glycolytic enzymes, metabolic regulatory activities, redox biology, protein glycation, glyoxalase activity, apoptosis, gene expression, and cancer cell metastasis [25].

In this study, we did not address the mechanism of how GAH attenuated cell death. Further studies will be needed to clarify the mechanism of GAH cytoprotection.

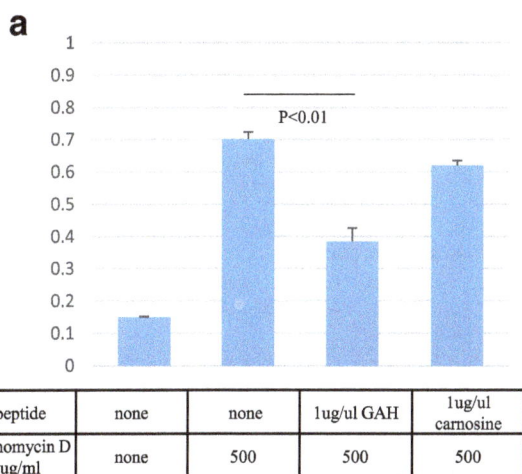

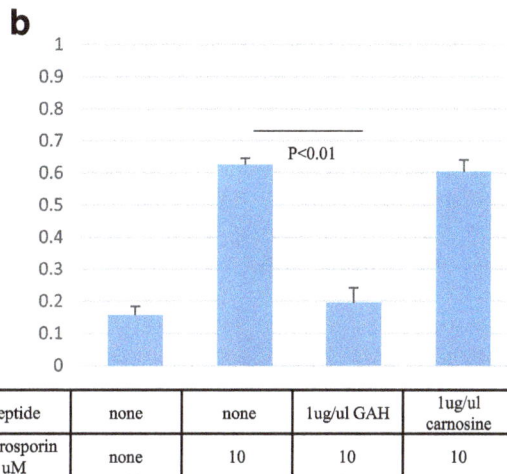

**Fig. 9** GAH protects against PC12 cell damage caused by actinomycin D and staurosporine. PC12 cells were exposed to 500 μg/mL of actinomycin D (**a**) or 10 μM of staurosporine (**b**) for 12 h. Cell viability was measured by the LDH leakage assay. All data are shown as mean ± SE of three independent experiments ($p < 0.01$)

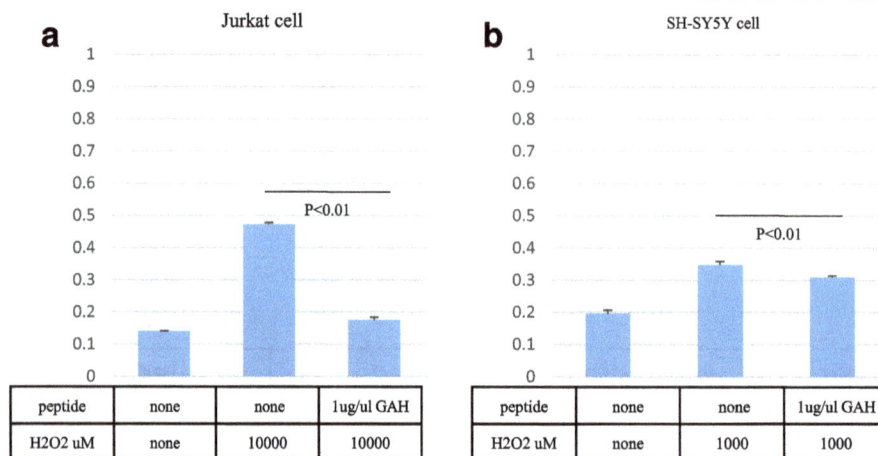

**Fig. 10** GAH protects against Jurkat and SH-SY5Y cell damage caused by hydrogen peroxide. Jurkat cells were exposed to 10,000 μM hydrogen peroxide for 4 h (**a**). SH-SY5Y cells were exposed to 1000 μM hydrogen peroxide for 24 h (**b**). Cell viability was measured by the LDH leakage assay. All data are shown as mean ± SE of three independent experiments ($p < 0.01$).

## Conclusions

The present studies showed that GAH has protective effect against cell damage determined by LDH leakage, by ethidium bromide staining, cell morphology, TUNEL assays, CCK-8 assays, and PI assays. GAH also protected Jurkat cells and SH-SY5Y cells. GAH might has a potential for cytoprotective agents.

### Acknowledgements
We acknowledge R Ishikawa for performing experiments.

### Funding
This study was partly supported by a High Technology Research Center grant and a Grant-in-Aid for exploratory research from the Ministry of Education, Culture, Sports, Science and Technology in Japan and by a Grant-in-Aid for Scientific Research from the Japan Society for the Promotion of Science (21,700,395, H. Shimura;) and from the Takeda Science Foundation (H. Shimura). This study was supported (in part) by a Grant-in-Aid (S1311011) from the Foundation of Strategic Research, Projects in Private Universities from the Ministry of Education, Culture, Sports, Science, and Technology, Japan. The funders had no role in study design, data collection and analysis, decision to publish, or preparation of the manuscript.

### Authors' contributions
HS designed and performed the experiments, participated in the interpretation of data, and wrote manuscript. RT, YS, KY performed the biological experiments. NH, TU designed the experiments, interpreted the data, and wrote the manuscript. All the authors read and approved the final manuscript.

### Competing interests
The authors declare that they have no competing interests.

### Author details
[1]Department of Neurology, Juntendo University Urayasu Hospital, 2-1-1 Tomioka, Urayasu, Chiba, Japan. [2]Department of Neurology, Juntendo University School of Medicine, Tokyo, Japan. [3]Institute for Environment and Gender Specific Medicine, Juntendo University School of Medicine, Chiba, Japan.

### References
1. Padhi A, Sengupta M, Sengupta S, Roehm KH, Sonawane A. Antimicrobial peptides and proteins in mycobacterial therapy: current status and future prospects. Tuberculosis (Edinb). 2014;94:363–73.
2. Buchwald H, Dorman RB, Rasmus NF, Michalek VN, Landvik NM, Ikramuddin S. Effects on GLP-1, PYY, and leptin by direct stimulation of terminal ileum and cecum in humans: implications for ileal transposition. Surg Obes Relat Dis. 2014;10:780–6.
3. Robinson SD, Safavi-Hemami H, McIntosh LD, Purcell AW, Norton RS, Papenfuss AT. Diversity of conotoxin gene superfamilies in the venomous snail, Conus Victoriae. PLoS One. Public Libr Sci. 2014;9:e87648.
4. Fosgerau K, Hoffmann T. Peptide therapeutics: current status and future directions. Drug Discov Today. 2014;20:122–8.
5. Yagasaki M, Hashimoto S. Synthesis and application of dipeptides; current status and perspectives. Appl Microbiol Biotechnol. 2008;81:13–22.
6. Santos S, Torcato I, Castanho MARB. Biomedical applications of dipeptides and tripeptides. Biopolymers. 2012;98:288–93.
7. Chakraborty S, Tai D-F, Lin Y-C, Chiou T-W. Antitumor and antimicrobial activity of some cyclic tetrapeptides and tripeptides derived from marine bacteria. Mar Drugs. 2015;13:3029–45.
8. Faden AI, Knoblach SM, Movsesyan VA, Lea PM, Cernak I. Novel neuroprotective tripeptides and dipeptides. Ann N Y Acad Sci. 2005;1053:472–81.
9. Turpeinen AM, Järvenpää S, Kautiainen H, Korpela R, Vapaatalo H. Antihypertensive effects of bioactive tripeptides-a random effects meta-analysis. Ann Med. 2013;45:51–6.
10. Baye E, Ukropcova B, Ukropec J, Hipkiss A, Aldini G, de Courten B. Physiological and therapeutic effects of carnosine on cardiometabolic risk and disease. Amino Acids. 2016;48:1131–49.
11. Chan KM, Decker EA. Endogenous skeletal muscle antioxidants. Crit Rev Food Sci Nutr. 1994;34:403–26.
12. Kohen R, Yamamoto Y, Cundy KC, Ames BN. Antioxidant activity of carnosine, homocarnosine, and anserine present in muscle and brain. Proc Natl Acad Sci U S A. 1988;85:3175–9.
13. Wade AM, Tucker HN. Antioxidant characteristics of L-histidine 11The work described in this manuscript was partially sponsored and funded by Cytos pharmaceuticals. LLC J Nutr Biochem Elsevier. 1998;9:308–15.
14. Kussman M, Stover PJ. Nutrigenomics and Proteomics in Health and Disease: Food Factors and Gene Interactions. John Wiley & Sons; 2009.
15. Tsuge N, Eikwa Y, Nomura Y, Yamamoto M, Sugisawa K. Antioxidative activity of peptides prepared by enzymatic hydrolysis of egg-white albumin. J Agric Chem Soc Japan. 1991;65:1635–41.

16. Chen H-M, Muramoto K, Yamauchi F, Nokihara K. Antioxidant activity of designed peptides based on the Antioxidative peptide isolated from digests of a soybean protein. J Agric Food Chem American Chemical Society. 1996;44:2619–23.

17. Saito K, Jin D-H, Ogawa T, Muramoto K, Hatakeyama E, Yasuhara T, et al. Antioxidative properties of tripeptide libraries prepared by the combinatorial chemistry. J Agric Food Chem. 2003;51:3668–74.

18. Boccellino M. Styrene-7,8-oxide activates a complex apoptotic response in neuronal PC12 cell line. Carcinogenesis. 2003;24:535–40.

19. Lindenboim L, Haviv R, Stein R. Inhibition of drug-induced apoptosis by survival factors in PC12 cells. J Neurochem 1995;64:1054–63.

20. Ivins KJ, Ivins JK, Sharp JP, Cotman CW. Multiple pathways of apoptosis in PC12 cells: CrmA INHIBITS APOPTOSIS INDUCED BY -AMYLOID. J Biol Chem American Society for Biochemistry and Molecular Biology. 1999;274:2107–12.

21. Dahl TA, Midden WR, Hartman PE. SOME PREVALENT BIOMOLECULES AS DEFENSES AGAINST SINGLET OXYGEN DAMAGE. Photochem Photobiol Blackwell Publishing Ltd. 1988;47:357–62.

22. Tamaoki T, Nomoto H, Takahashi I, Kato Y, Morimoto M, Tomita F. Staurosporine, a potent inhibitor of phospholipidCa++dependent protein kinase. Biochem Biophys Res Commun Academic Press. 1986;135:397–402.

23. Rüegg UT, Gillian B. Staurosporine, K-252 and UCN-01: potent but nonspecific inhibitors of protein kinases. Trends Pharmacol Sci Elsevier Current Trends. 1989;10:218–20.

24. Perry RP, Kelley DE. Inhibition of RNA synthesis by actinomycin D: characteristic dose-response of different RNA species. J Cell Physiol. 1970;76:127–39.

25. Hipkiss AR, Gaunitz F. Inhibition of tumour cell growth by carnosine: some possible mechanisms. Amino Acids. 2014;46:327–37.

# 15

# Characterization of sulfhydryl oxidase from *Aspergillus tubingensis*

Outi Nivala[1*], Greta Faccio[1,2], Mikko Arvas[1], Perttu Permi[3,4,5], Johanna Buchert[1,6], Kristiina Kruus[1] and Maija-Liisa Mattinen[1,7]

## Abstract

**Background:** Despite of the presence of sulfhydryl oxidases (SOXs) in the secretomes of industrially relevant organisms and their many potential applications, only few of these enzymes have been biochemically characterized. In addition, basic functions of most of the SOX enzymes reported so far are not fully understood. In particular, the physiological role of secreted fungal SOXs is unclear.

**Results:** The recently identified SOX from *Aspergillus tubingensis* (AtSOX) was produced, purified and characterized in the present work. AtSOX had a pH optimum of 6.5, and showed a good pH stability retaining more than 80% of the initial activity in a pH range 4-8.5 within 20 h. More than 70% of the initial activity was retained after incubation at 50 °C for 20 h. AtSOX contains a non-covalently bound flavin cofactor. The enzyme oxidised a sulfhydryl group of glutathione to form a disulfide bond, as verified by nuclear magnetic resonance spectroscopy. AtSOX preferred glutathione as a substrate over cysteine and dithiothreitol. The activity of the enzyme was totally inhibited by 10 mM zinc sulphate. Peptide- and protein-bound sulfhydryl groups in bikunin, gliotoxin, holomycin, insulin B chain, and ribonuclease A, were not oxidised by the enzyme. Based on the analysis of 33 fungal genomes, SOX enzyme encoding genes were found close to nonribosomal peptide synthetases (NRPS) but not with polyketide synthases (PKS). In the phylogenetic tree, constructed from 25 SOX and thioredoxin reductase sequences from IPR000103 InterPro family, AtSOX was evolutionary closely related to other Aspergillus SOXs. Oxidoreductases involved in the maturation of nonribosomal peptides of fungal and bacterial origin, namely GliT, HlmI and DepH, were also evolutionary closely related to AtSOX whereas fungal thioreductases were more distant.

**Conclusions:** AtSOX (55 kDa) is a fungal secreted flavin-dependent enzyme with good stability to both pH and temperature. A Michaelis-Menten behaviour was observed with reduced glutathione as a substrate. Based on the location of SOX enzyme encoding genes close to NRPSs, SOXs could be involved in the secondary metabolism and act as an accessory enzyme in the production of nonribosomal peptides.

**Keywords:** Secreted sulfhydryl oxidase, Dithiol oxidase, Aspergillus tubingensis, Glutathione oxidation, Nonribosomal peptide synthesis, Secondary metabolism

## Background

Sulfhydryl oxidases (SOXs) are flavin-dependent enzymes that catalyse the oxidation of free thiol groups to disulfide bonds with the concomitant reduction of molecular oxygen to hydrogen peroxide. SOXs have been isolated from animal and microbial sources [1–4]. Both intracellular and secreted enzymes have been reported [5–9]. SOX enzymes can be classified to four major families based on their structural features: intracellular single-domain proteins essential for respiration and vegetative growth / augmenter of liver regeneration (the Erv/Alr family), the endoplasmic reticulum oxidase (Ero) family, the quiescin-sulfhydryl oxidase (QSOX) family with multi-domain enzymes, and the secreted fungal SOXs family with FAD-dependent dimeric single-domain enzymes (reviewed in [10]). The secreted fungal SOXs differ from the members in other SOX families. They carry features from thioredoxin reductase and pyridine nucleotide flavin disulphide oxidoreductase sequences [10]. Many fungal species such as *Aspergillus niger, Penicillium chrysogenum, Aspergillus*

* Correspondence: outi.nivala@helsinki.fi
[1]VTT Technical Research Centre of Finland, Ltd., P.O. Box 1000, FI-02044 Espoo, Finland
Full list of author information is available at the end of the article

oryzae, *Aspergillus sojae,* and *Calodon sp.* have been reported to secrete SOXs (glutathione oxidase EC 1.8.3.3) [1, 3, 5, 6, 10, 11]. Numerous SOX-coding genes have been found in the genome of fungi. The genome of the yeast *Saccharomyces cerevisiae* contains two putative genes encoding SOX enzymes, while so far published genomes of the fungal genus *Aspergilli* typically contain from 10 to 12 putative genes encoding SOX enzymes [12]. The genome of *A. tubingensis* contains genes encoding eight putative SOXs [13].

SOXs are attractive catalysts for industrial applications. They have been tested in food applications such as in dairy and baking. SOXs have been used for ultra-high temperature (UHT)-treated milk to remove of unpleasant flavour [6, 14]. In baking, SOX enzymes have been utilized to strengthen the structure of dough and improving bread properties [15–17]. SOXs have been suggested to reduce the allergenicity of pharmaceuticals [18]. The characterized fungal SOXs are known to cross-link bonds between peptides but not between proteins [1, 3, 19]. There occurs variation in the substrate specificity within SOXs from different sources. The small thiol-containing molecules are typical substrates for secreted fungal SOXs, whereas mammalian bovine SOXs are known to oxidase protein-bound cysteine residues [1, 10, 20]. The function of intracellular *S. cerevisiae* SOXs and of some multi-domain SOXs is elucidated but the role for secreted fungal SOXs is still unclear [7, 21]. The sectered fungal SOXs do not form a mixed disulphide intermediate on the contrary to the mammalian SOXs and intracellular *S. cerevisiae* SOXs (Erv1p, Erv2p) [22–24], which implies that secreted fungal SOXs are not directly involved in the oxization of the reduced proteins. Multi-domain SOXs from the quiescin-sulfhydryl oxidase (QSOX) family have been proposed to be involved in the formation of the extracellular matrix, in the maturation of proteins along the secretory pathway, and to act as antimicrobial agent [4, 21, 25, 26]. Extracellular QSOXs are multi-domain sulfhydryl oxidases from multicellular organisms, and greatly differ from the single-domain secreted fungal SOXs. The QSOX and secreted fungal SOX proteins belong to different protein families and appear to be evolutionary unrelated. In bacteria, SOXs might be involved in the synthesis of bioactive compounds such as nonribosomal peptides [27].

This paper reports the biochemical characterization of SOX from *A. tubingensis* (AtSOX) previously identified in the screening study [28]. AtSOX was purified and its substrate specificity was further characterized and its activity on different peptide- and protein-bound sulfhydryl groups analysed. The formation of enzyme-catalysed disulphide bond in glutathione was confirmed by NMR spectroscopy. Furthermore, 33 fungal genomes were analysed to examine location of the selected genes, for example nonribosomal peptide synthetase or polyketide synthase coding genes, in relation to SOX-coding genes.

## Methods

### Production and purification of AtSOX

*Aspergillus tubingensis* strain D-85248 was obtained from the VTT Culture Collection [29] and cultivated as described in [28]. *A. tubingensis* was grown on PeptoneTM-D(+)-glucose media adapted from [3, 5] in liquid cultivation at 30 °C under shaking (250 rpm). After removal of the fungal biomass by filtration, AtSOX was purified from the cell-free extract with four chromatographic steps according to the procedure described below. The culture supernatant was concentrated and buffer exchanged to 20 mM Tris-HCl pH 7 with a PD10 column (no 17-0851-01, GE Healthcare, Uppsala, Sweden), and applied on a QSepharose fast flow column (volume = 20 mL, Amersham Biosciences, Piscataway, USA). Proteins were eluted with a linear gradient from 0 to 0.3 M NaCl in 20 mM Tris-HCl pH 7. The SOX activity of the fractions was detected using 5,5-dithiobis(2-nitrobenzoic acid) (Ellman's reagent, DTNB, no D 218200, Sigma-Aldrich, Helsinki, Finland) and oxygen consumption measurements. The SOX active fractions were pooled and applied to a Superdex 75 column (volume = 24 mL, Amersham Biosciences, USA) using 50 mM Tris-HCl pH 7 buffer containing 150 mM NaCl at a 0.1 mL/min flow rate. SOX-active fractions were pooled, concentrated, buffer exchanged to 20 mM Tris-HCl pH 7, and applied twice to a Resource Q Sepharose column (volume = 1 mL, Amersham Biosciences, USA). In the first separation using Resource Q Sepharose resin, the proteins were eluted with a linear gradient from 0 to 0.2 M NaCl in 20 mM Tris-HCl pH 7, whereas in the second separation a shallower gradient from 0 to 0.11 M NaCl was used. Prepacked anion exchange columns were connected to ÄKTA™ chromatography system and UNICORN Control Software (GE Healthcare, Uppsala, Sweden).

Protein concentration was determined using a Bio-Rad DC protein assay kit (Bio-Rad, Hercules, USA) and bovine serum albumin (BSA, no. A8022, Sigma, St. Louis, USA) as a standard. Proteins were analysed by ready-made 12% Tris-HCl SDS-PAGE gel (no 161-1156, Bio-Rad, Hercules, USA) to ensure purity of the isolated AtSOX enzyme.

### Amino acid analysis of AtSOX and strain identification

The partial amino acid sequence of AtSOX was determined from the N-terminus and internal peptide fragments. SDS-PAGE protein bands were stained with Coomassie Brilliant Blue, and the protein band of interest was excised and subjected to N-terminal sequencing. Edman degradation was performed using a Procise 494A protein sequencer from Perkin Elmer, Applied Biosystems Division (Foster City, CA, USA). The Coomassie stained protein band, assumably containing AtSOX, was cut off from the SDS-PAGE and in gel digested essentially as described by [30] for obtaining internal sequences and for peptide mass fingerprinting (PMF). In gel digestion was done by reducing proteins with dithiothreitol (DTT), alkylating with iodoacetamide and digesting with trypsin (no V5111, Promega, Madison, USA). The enzymatic cleavage occurred during overnight incubation at 37 °C. The peptides produced by enzymatic cleavage were analysed by MALDI-TOF MS after desalting using mC18 ZipTip (no ZTC18M096, Millipore, Billerica, USA). MALDI-TOF mass spectra of peptide fragments for PMF were obtained using an Ultraflex TOF/TOF instrument (Bruker-Daltonik GmbH, Bremen, Germany) using α-cyano-4-hydroxycinnamic acid (CHCA) as matrix. The sample solution was pipetted onto the sample plate together with matrix and air-dried. An electrospray ionization quadrupole time-of-flight tandem mass spectra for de novo sequencing were acquired using a Q-TOF instrument (Micromass, Manchester, UK) as described by [31]. Protein identification with the generated data, the obtained peptide masses and the partial sequences, was performed using the Mascot Peptide Mass Fingerprint and MS/MS Ion Search programmes (http://www.matrixscience.com). Identification of the production strain (VTT D-85248) based on the morphology and DNA sequencing was performed at the Identification Services CBS (Utrecht, The Netherlands).

## Peptide and protein substrate preparation

The reduced glutathione (GSH, no G4251), urinary trypsin inhibitor fragment (bikunin, no U-4751), and ribonuclease A (RNase A, no R5500) were purchased from Sigma (St. Louis, USA). Peptides, gliotoxin (no. A7665) and holomycin (no. sc-49,029), were purchased from PanReac AppliChem GmbH (Darmstadt, Germany) and Santa Cruz Biotechnology Inc. (Dallas, USA), respectively. The insulin B chain as Bunte salt derivative was kindly provided by Dr. Elisabeth Heine from Deutsches Wollforschungsinstitut (DWI, Aachen, Germany). Insulin chain B and RNaseA were reduced before use with 3% (v:v) 2-mercaptoethanol in the presence of 8 M urea according to the method used by [3]. For the reduction, insulin B was dissolved in 50 mM $(NH_4)HCO_3$ pH 8.3 containing 8 M urea and 3% (v:v) 2-mercaptoethanol was used as solvent. In the case of RNaseA, 200 mM Tris-HCl buffer (pH 7.4) with urea and 2-mercaptoethanol was used. The solutions were incubated overnight at room temperature. The final concentration of insulin chain B was 1 mM, and concentration of RNaseA was 0.5 mM. After reduction the free sulfhydryl groups were detected with 5,5-dithio-bis(2-nitrobenzoic acid). Gliotoxin and holomycin were reduced with two equivalence of Tris(2-carboxyethyl)phosphine hydrochloride (TCEP-HCl) under

nitrogen flow for 1 h prior liquid chromatography mass spectrometry (LC-MS) analysis.

## Assay of AtSOX activity and pH and temperature behaviour

Different methods were used to measure AtSOX activity. First DTNB [32] was used for the detection of free sulfhydryl groups in the peptides after the enzymatic reaction. The spectroscopic measurements were done with a Cary 100 Bio UV-vis spectrophotometer (Varian Inc., Houten, the Netherlands). Second, the oxygen consumption measurement with Fibox 3 PreSens fiber-optic oxygen meter (Presens GmbH, Regensburg, Germany) was used to measure the changes in the concentration of dissolved oxygen during enzymatic reaction as described by [28]. GSH (5 mM) was used as a substrate when determining AtSOX activity. The substrate was dissolved in phosphate buffered saline (PBS) containing 68 mM NaCl and 75 mM $KH_2PO_4$ pH 7.4. Activity of AtSOX was measured by the oxygen consumption assay in the presence of different GSH concentrations (0.25-10 mM) to determine the kinetic parameters. The Michaelis-Menten constant ($K_m$) and maximum velocity ($V_{max}$) were determined with graphing software GraphPad Prism (GraphPad Software Inc., San Diego, USA) using nonlinear curve fitting to the Michaelis-Menten equation.

The thermal stability of AtSOX was determined at 30, 40, 50, 60 and 70 °C. The enzyme preparation was incubated in McIlvaine buffer pH 6.5 in a 0.2 mg/mL protein concentration for 1, 2, 15.5 and 20 h at 30 - 70 °C and also for 15 min at 70 °C, and the residual activity was measured by the oxygen consumption assay. pH stability was determined for AtSOX in a pH range between 2.3 and 10, and the residual enzyme activity was analysed by the oxygen consumption assay after 1 and 20 h incubation. pH optimum was determined by measuring AtSOX activity with the oxygen consumption assay using glutathione in McIlvaine citrate/phosphate buffer (pH 2.3-7.5), 50 mM Tris-HCl (pH 7-9) and 50 mM Glycine-NaOH (pH 8.5-10).

## Spectroscopy measurements

UV-vis absorption spectra were measured in 20 mM Tris-HCl pH 7.5 at 20 °C using a Cary 100/300 UV-vis spectrophotometer (Varian Inc., Houten, the Netherlands). In order to release a flavin cofactor, enzyme was thermally denatured at 100 °C for 15 min followed by centrifugation (13,000 rpm, 10 min). Fluorescence was measured at 20 °C with a Cary Eclipse Fluorescence Spectrophotometer (Varian Inc., Houten, the Netherlands) using AtSOX solution in 20 mM Tris-HCl pH 7.5 in quartz cuvette with four optical faces. FAD fluorescence was recorded by exciting at 450 nm and monitoring emission between 450 and 600 nm.

## Inhibition analysis of AtSOX

The AtSOX activity was analysed in a buffer solution and in the presence of different inhibitors. The effect of possible inhibitors on the AtSOX activity was determined using 5 mM GSH as substrate. The tested potential inhibitors were DTT, ethylenediaminetetraacetic acid (EDTA), potassium iodide, magnesium sulphate, manganese sulphate, sodium sulphate, sodium chloride, zinc sulphate, and sodium dodecyl sulphate (SDS). The inhibitors were tested at 10 mM concentration in 200 mM Tris-HCl buffer pH 7.5, and zinc sulphate was tested also at 1 mM concentration. The residual SOX activity was measured by oxygen consumption measurements (Fibox 3 PreSens fiber-optic oxygen meter, PreSens GmbH, Regensburg, Germany). Inhibition by zinc sulphate was confirmed by homovanillic acid (HVA) and peroxidase coupled assay as described in [33], and the assay was done according to [34] using GSH (5 mM) as a substrate. Chemicals, HVA (Cat. no. H1252) and peroxidase type II (Cat. no. P8250), were purchased from Sigma-Aldrich (St. Louis, USA). Fluorescence from the production of a HVA dimer was measured in a black 96-well microtiter plate at 320 nm excitation and 420 nm emission wavelengths using a Varioskan spectral scanning multimode microplate reader (Thermo Electron co., Vantaa, Finland).

## Activity of AtSOX with reduced peptides

The ability of AtSOX to oxidise the peptides carrying free sulfhydryl groups was analysed by following the oxygen consumption. The activity of AtSOX (112 nkat) was analysed using reduced RNase A (0.25 mM solution in 200 mM Tris-HCl pH 7.4), and reduced GSH (5 mM solution in 75 mM $KH_2PO_4$ pH 7.4) as substrates in a total 1.86 mL reaction volume. The enzyme reactions were monitored by following the oxygen consumption of the co-substrate. The reactions were performed at room temperature and monitored for 5-20 min. For NMR spectroscopy, the enzymatically treated GSH was prepared by incubating the substrate with AtSOX for 20 min while the dissolved oxygen was totally consumed, as assessed by oxygen consumption measurements. Freshly prepared and 3 day old GSH solutions (5 mM) were used as a control samples to assess possible auto-oxidation. Reaction mixtures were analysed with $^1$H NMR as well as with $^{13}$C heteronuclear single quantum correlation ($^{13}$C-HSQC) spectroscopy. The enzyme treated sample and the control sample were prepared in 75 mM potassium phosphate pH 7.4 in Shigemi NMR tubes. The NMR spectra were recorded on Varian INOVA 600 MHz NMR spectrometer at 293 K. One dimensional $^1$H spectra were recorded with water pre-saturation at 25 °C along with the gradient enhanced $^{13}$C-HSQC spectroscopy at 20 °C [35]. Homonuclear

Total Correlation Spectroscopy (TOCSY) experiments [36] were recorded to confirm the assignment.

Besides oxygen consumption method and NMR spectroscopy, the selected peptides were analysed with a MALDI-TOF MS on an Autoflex II spectrometer (Bruker Daltonik GmbH, Bremen, Germany) using CHCA matrix for peptides and sinapic acid (SA) for proteins. The reduced substrates GSH and insulin B were dissolved in 50 mM $(NH_4)HCO_3$, bikunin in 75 mM $KH_2PO_4$, and RNase A in 200 mM Tris-HCl pH 7.4. Purified SOX was used in the experiments in a dosage of 5.6 nkat (4.1 µg protein) and in case of insulin B and RNase A also in 56 nkat (41 µg protein). The enzymatic reactions were performed at 40 °C, except for bikunin that was incubated at room temperature for 20 h. Matrix solutions were prepared by dissolving CHCA or SA in a 1:1 solution of 0.1% trifluoroacetic acid and 100% acetonitrile. The spots for MALDI plate were prepared by using 1:1 proportion of matrix and sample. Typically 1 µl matrix and 1 µl sample were used for MALDI spot, where matrix was spotted first and dried before adding sample.

Reduced gliotoxin and holomycin were used in 200 µM final concentration in 0.1 M phosphate buffer (pH 6.5) for spectroscopic measurements. Gliotoxin and holomycin were used in 440 and 680 µM final concentrations, respectively, for LC-MS analysis. UV-vis absorption spectra from 200 to 600 nm were recorded with Varioskan spectral scanning multimode microplate reader (Thermo Electron co., Vantaa, Finland). The UV-vis absorption spectra from 200 to 600 nm was recorded for 10 min before and after addition of AtSOX (100 µl reaction volume). Reduced peptides incubated (50 min at ambient temperature) with AtSOX or with denatured AtSOX were analysed with a ultra performance LC (UPLC) combined with a photodiode array detector and SYNAPT G2-S High Definition Mass Spectrometry (Waters, Milford Massachusetts, USA). One microliter of the sample was injected to a LC pre-column. LC-MS system was using a C18 Acquity UPLC VanGuard pre-column (2.1 × 5 mm, 1.7 µm,) and a C18 Acquity UPLC column (2.1 × 100 mm, 1.7 µm). All solvents used were spectral grade. Eluents were 5 mM ammonium acetate 0.1% formic acid in $H_2O$ (A) and in methanol (B). Elution was started with 10% B for 1 min, followed by a linear gradient from 10 to 100% B for 10 min and finally at 100% B for 2 min, with 0.4 mL min$^{-1}$ flow. In these conditions reduced gliotoxin eluted from the LC column after 5.23 min and gliotoxin standard after 6.02 min. Elution times for reduced holomycin and holomycin standard were 1.69 and 3.74 min, respectively.

### Analysis of SOX-coding genes in fungal genomes

The search was carried out as described in [37]. In brief, scaffolds of 33 fungal genomes from [38] were divided in windows of 16 genes that overlapped with two genes. InterPro protein annotations of the genes were then used to look for windows that contained a flavin adenine dinucleotide (FAD)-dependent pyridine nucleotide-disulfide oxidoreductase (PNDR) that recognizes also SOX enzymes (InterPro: IPR000103), nonribosomal peptide synthetases (NRPS) (InterPro: IPR000873) or polyketide synthases (PKS) (InterPro: IPR001227) and cytochrome P450 monooxygenases (P450, InterPro: IPR001128) and/or Zn2Cys6 transcription factors (Zn2, InterPro: IPR001138). Searches were carried out and the results visualised with R essentially as described by [38].

### Phylogenetic analysis

A phylogenetic analysis of the selected 25 proteins from the protein family pyridine nucleotide-disulphide oxidoreductase, class-II (InterPro: IPR000103) was carried out. The selected protein sequences were obtained from UniProtKB except AtSOX sequence was retrieved from the genome of *A. tubingensis* from Joint Genome Institute (JGI) genome portal [39]. The alignment of the sequences was done with MAFFT [40], and alignment was trimmed with trimAl [41]. The aligned sequences were from Ascomycetes species except two proteins were bacterial origin, namely, DepH from *Chromobacterium violaceum* (UniProtKB: A4ZPY8) and HlmI from *Streptomyces clavuligerus* (UniProtKB: E2PZ87). A phylogenetic tree was constructed with FastTree [42] and visualised by Geneious version 10.0 created by Biomatters.

### Results
### Purification and biochemical characterization of AtSOX from *Aspergillus tubingensis*

Purification of AtSOX was performed in four chromatographic steps from the culture cell-free medium. Fractions containing SOX activity eluted at a 110-260 mM NaCl concentration from a Q Sepharose anion exchange column. Pooled active fractions were applied to a Superdex 75 column for separation by SEC. Further purification of AtSOX was achieved with high resolution anion exchange chromatography. The SOX containing fractions eluted at a 60-175 mM and 100 mM NaCl concentration from two sequential purification steps where Resource Q columns were used. The fraction with the highest specific SOX activity was applied on a Resource Q column once again. AtSOX containing fractions started to elute at a 100 mM NaCl concentration. The purification process was monitored by SDS-PAGE (Additional file 1), and the final yield of protein was 3% of the initial activity.

AtSOX preferred GSH (3 mM) as a substrate over cysteine and DTT. AtSOX had residual activity on L-cystein $14 \pm 0.2\%$, on D-cystein $3 \pm 1.2\%$ and on DTT 9

$\pm 1.8\%$ measured with oxygen consumption assay. Oxygen consumption in AtSOX-catalysed reaction using GSH (3 mM) as a substrate is shown in Additional file 2. Michaelis-Menten behaviour was observed for AtSOX using GSH as a substrate. Michaelis-Menten constant $(K_m)$ was $0.80 \pm 0.09$ mM and maximum velocity $(V_{max})$ $(33.15 \pm 0.99) \times 108$ nkat/ml (Fig. 1a). AtSOX retained 89% of the initial activity after 20 h incubation at 40 °C (Fig. 1c). Under identical conditions at 50 and 60 °C, AtSOX showed 75% and 18% residual activity, respectively. At 70 °C, the enzyme was inactivated within 15 min. AtSOX was stable within a broad pH range as it retained 80-90% of the initial activity after 20 h incubation at pH range from 4 to 8.5 (Fig. 1d). The pH optimum of AtSOX was pH 6.5 (Fig. 1b). The enzyme activity was totally inhibited by 10 mM zinc sulphate in the assay conditions. The relative SOX activity was 4% with 1 mM zinc sulphate. The inhibition of zinc sulphate was confirmed with HVA-peroxidase coupled assay (Additional file 3). The other analysed compounds did not inhibit AtSOX activity, or had only minor effect on it, as residual activity was more than 90% in the presence of the analysed compound. For example the AtSOX retained 92% of its activity in 10 mM of potassium iodide solution and denaturant SDS did not inhibit enzyme activity. The flavoenzymatic nature of AtSOX was determined by spectral analyses using UV-vis and fluorescence spectrophotometry. The absorption spectra were recorded before and after thermal denaturation of AtSOX. The flavin cofactor was released by denaturation and detected in solution (Fig. 2). The UV-vis spectrum of AtSOX showed the characteristic peaks of flavoproteins with absorbance maxima at 275, 365 and 445 nm with a shoulder at 475 nm. AtSOX fluorescence emission spectrum had a peak at 525 nm when excited at 450 nm, which is characteristic for flavins.

The purified AtSOX was subjected to Edman degradation, and trypsin digestion to perform peptide mass fingerprinting (PMF), and to identify the protein. The PMF gave six peptides made of more than six amino acids (Fig. 3). Mining the available genome of A. tubingensis [39] with the sequences of the identified peptides allowed the identification of a single SOX-coding gene on scaffold 13. This gene coded for a 41.8 kDa secreted protein with a predicted N-terminus, after removal of the signal peptide, SSIPQ. This protein contained four of the six peptides identified from AtSOX, and the other two peptides were present but carry either an amino acid insertion or deletion. Due to the high sequence similarity with secreted SOX from A. niger (UniProtKB: A2QUK3), the strain was re-identified at the Identification Services CBS (Utrecht, the Netherlands). The identification results confirmed that the used VTT strain D-85248 was *A. tubingensis*.

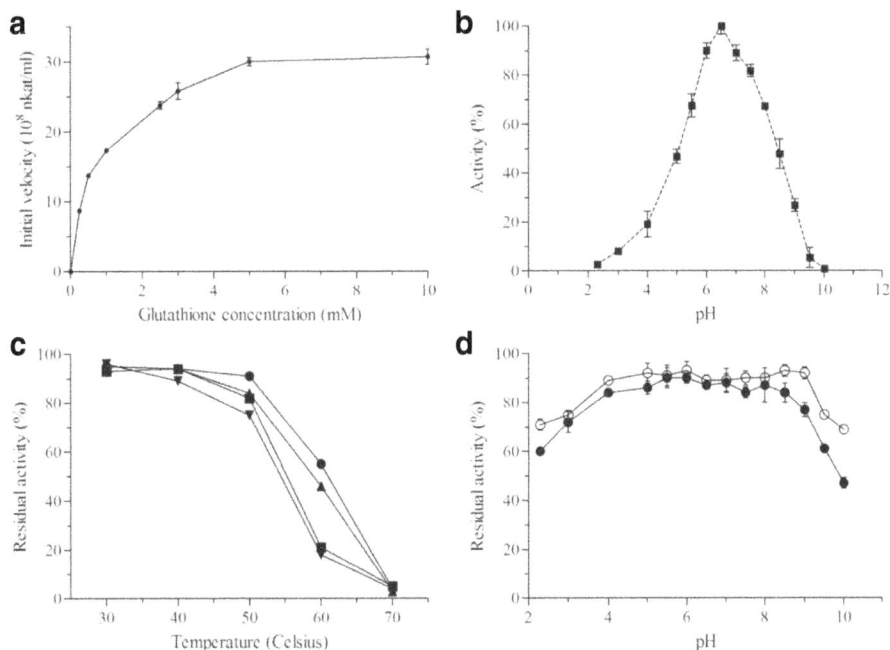

**Fig. 1** Kinetics, pH optimum and thermal and pH stability of AtSOX. AtSOX activity on GSH (0.25-10 mM) measured by the oxygen consumption assay (**a**). AtSOX activity was measured with the oxygen consumption assay on GSH (3 mM) at room temperature to determine the pH optimum (**b**) and temperature stability after 1 h (filled *circle*), 2 h (filled *triangle*), 15.5 h (filled *square*) and 20 h (filled inverted *triangle*) of incubation at different temperatures (30-70 °C) (**c**). pH stability of AtSOX as analysed by oxygen consumption measurements, when incubated within a pH range from 2.3 to 10 for 1 (empty *circles*) or 20 h (filled *circles*) (**d**). Each curve represents average of two replicates (standard deviation < 10%)

## Oxidation of peptides and proteins with AtSOX

The ability of AtSOX to oxidise peptide-bound cysteine was tested with reduced model peptides and protein namely GSH, bikunin, insulin B chain, gliotoxin, holomycin and RNase A as substrates. Reaction products were analysed by MALDI-TOF MS and NMR spectroscopy. The MALDI-TOF MS analysis of the reaction mixtures showed a dimer (613 Da) formation with GSH after overnight enzymatic treatment whereas only a peak

correspond to the monomer (308 Da) was detected in the mass spectrum of the untreated substrate. The enzymatic disulphide bond formation between GSH peptides was observed also with the spectroscopic measurements at 250 nm (Additional file 4). No dimer formation was detected with bikunin (919 Da), insulin B chain (3496 Da) or RNaseA (13,700 Da) as substrates by MALDI-TOF analysis. Other analysed small peptides – gliotoxin (326 Da) and holomycin (214 Da) – were not enzymatically oxidised as observed in the LC-MS analysis. Holomycin was auto-oxidised as oxidation was observed also with the denatured AtSOX enzyme after 50 min incubation.

Oxidation of GSH and RNase A by AtSOX was also followed. AtSOX could oxidise only the tripeptide GSH, which was the shortest tested peptide. The final product formed in the reaction of AtSOX was analysed using NMR spectroscopy. Hereby, it was possible to follow the change in the cysteine oxidation state at atomic resolution. The oxidation was confirmed by following the oxygen consumption for 20 min after addition of AtSOX to the tripeptide L-GSH. The fresh GSH and enzyme-treated sample were analysed with one dimensional [1]H NMR. The one dimensional [1]H spectra of GSH with and without AtSOX enzyme are shown in Fig. 4. Upon addition of AtSOX to GSH solution, the two degenerate protons bonded to $C_\beta$ of cysteine residue became non-degenerated due to the formation of covalent disulphide

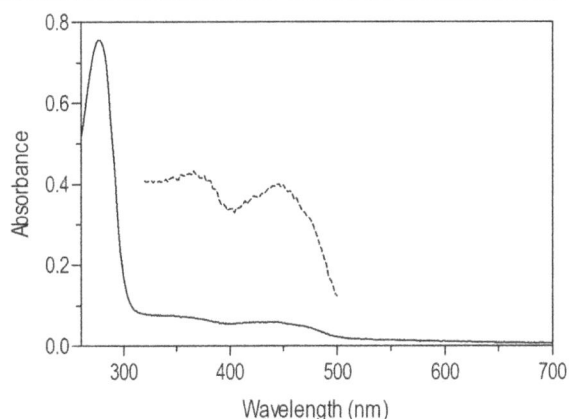

**Fig. 2** UV-visible spectrum of AtSOX enzyme. The absorbance spectra of AtSOX (continuous line) and of the released FAD (values 10×, *dashed line*) are shown

Scaffold 13  *MAPKSLFYSLFSTLSVALA*ssipqTDYDVIVVGGGPAGLSVLSSLGRMRRRTVMFDSGEYR
AtSOX                                                           TVMFDSGEYR

Scaffold 13  NSVTREMHDVLGFDGTPPAQFRGLARQQISKYNSTSVIDIKIDTITPIEDAAANSSYFRA
AtSOX                  DVLGFDGTPPAQFR

Scaffold 13  VDANGTEYTSRKVVLGTGLVDVIPDVPGLREAWGKGIWW***CPWC***DGYEHRDEPLGIL
AtSOX                               GLVDVIPDV

Scaffold 13  GSLTDVVGSVMETHTLYSDIIAFTNGTYTPANEVALAAKYPNWKEQLEAWNIGIDNRS
AtSOX                                                                              S

Scaffold 13  IASIERLQDGDDHRDDTGRQYDIFRVHFTDGSSVIRNTFITNYPTAQRSTLPEELSLVMV
AtSOX        IASIER              QYDIFRVHFTDGSSVLRNTFITNYPTAQR

Scaffold 13  DNKIDTTDY–TGMRTSLSGVYAVGDCNSDGSTNVPHAMFSGKRAGVYVHVEMSREES
AtSOX              IDTTDYY–GMR

Scaffold 13  NAAISKRDFDRRALEKQTERMVGNEMEDLWKRVLENHHRRS
AtSOX

**Fig. 3** The reconstructed AtSOX sequence. Alignment of the peptides obtained by tryptic digestion of AtSOX and the protein they identify on scaffold 13 of the genome of *A. tubingensis* [13]. The predicted signal sequence is in italic, the experimentally determined N-terminus of AtSOX is in lower case and the catalytic di-cysteine motif is bold in italic

bond. In addition, the two dimensional $^{13}$C-HSQC spectrum, exhibiting one-bond $^{1}$H-$^{13}$C connectivities (Fig. 5a) showed a significant change in the $^{13}$C chemical shift of the Cys $^{13}$C$_\beta$-$^{1}$H correlations as common for disulfide bond formation. This confirmed the fast dimerization of the GSH tripeptide through the disulfide bond upon addition of AtSOX. The control GSH substrate was checked once more 3 days after preparation with the gradient-enhanced $^{13}$C-HSQC spectrum to analyse auto-oxidation. The $^{1}$H, $^{13}$C cross peak, corresponding to the oxidised Cys C$_\beta$ form, started to appear which

refers to dimerization (Fig. 5b). The dimerization of GSH occurred also without the SOX enzyme, although at a much slower rate. Measurement of the pure GSH sample 3 days after preparation showed an evidence of both monomeric and dimeric conformations in a ratio of about 5:1.

### SOX-coding genes in fungal genomes and phylogenetics

*Aspergilli* are known for their expanded secondary metabolism in comparison to yeasts [38], while fungi in general are known for their metabolic gene clusters, which are involved in, for example, synthesis of secondary metabolites [43] and catabolism of nutrients [37]. In order to suggest physiological roles for fungal secreted SOXs, fungal genomes were analysed. In brief, 33 fungal genomes were searched with custom R scripts for short genomic regions that would contain a SOX-coding genes and type of fungal secondary metabolism genes i.e. nonribosomal peptide synthetases (NRPS), polyketide synthases (PKS) and associated genes, i.e. cytochrome P450 monooxygenases (P450), and/or Zn2Cys6 transcription factors (Zn2).

No genomic regions containing genes coding for PKS and SOX could be found, instead ten regions with SOX, NRPS and P450 or Zn2 coding genes were identified (Fig. 6). Two of these regions were the gliotoxin synthesis cluster of *A. fumigatus* [44] and the candidate gliotoxin cluster of *Trichoderma reesei* [45]. In addition, the *A. tubingensis* candidate gliotoxin cluster was detected (Fig. 6a). Based on protein clustering of SOX enzymes [38] the SOX enzymes found in the chromosomal gene clusters were divided in two groups: SOX enzymes of candidate gliotoxin gene clusters (Fig. 6a) and SOX enzyme of candidate secondary metabolism gene clusters (Fig. 6b).

AtSOX sequence retrieved from *Aspergillus tubingensis* genome project was aligned with selected 24 SOX and thioredoxin reductase sequences from IPR000103

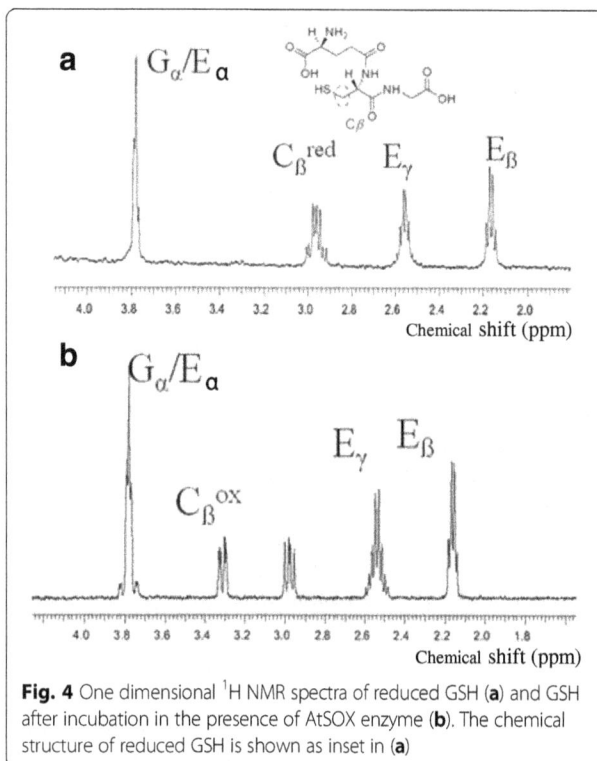

**Fig. 4** One dimensional $^{1}$H NMR spectra of reduced GSH (**a**) and GSH after incubation in the presence of AtSOX enzyme (**b**). The chemical structure of reduced GSH is shown as inset in (**a**)

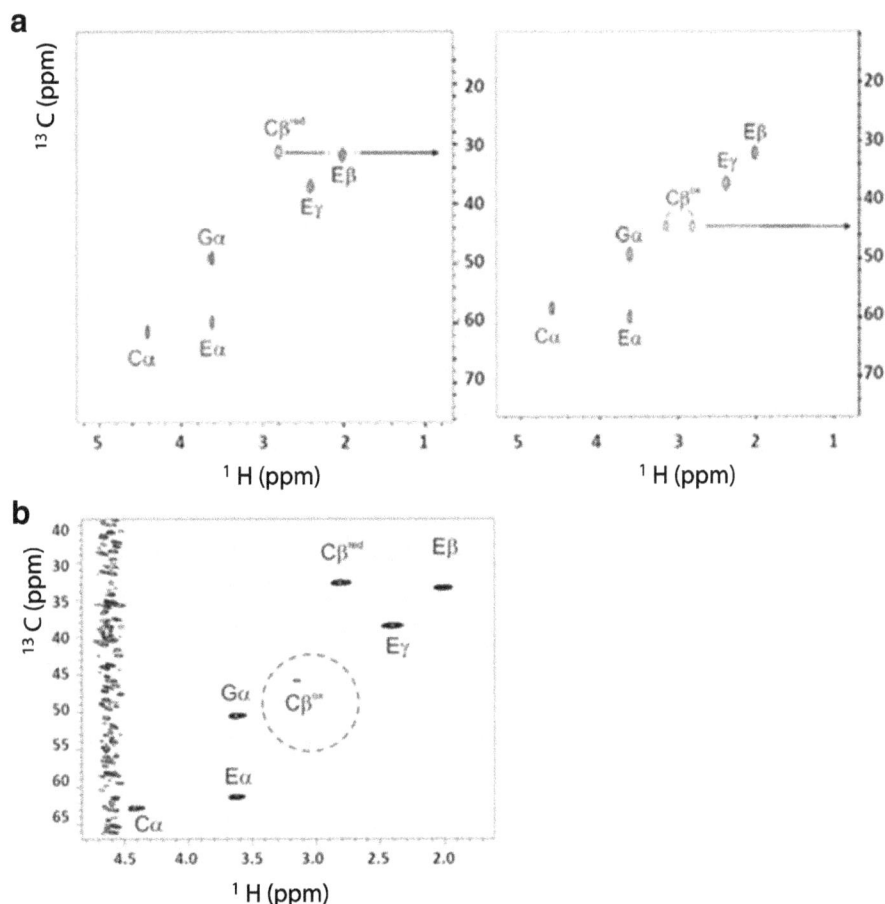

**Fig. 5** Oxidation of GSH by AtSOX as analysed by two-dimensional heteronuclear NMR spectroscopy ($^{13}$C-HSQC spectra showing one-bond $^{1}$H-$^{13}$C connectivities). The reduced substrate is shown on the left side as control and the enzyme-treated sample on the right side. The assignment of the tripeptide is indicated on the top of the cross-peaks. *Arrows* emphasize the chemical shift change in the C$_\beta$ correlation of cysteine in the two conformations i.e. in the cross-linked and open form (**a**). The gradient selected $^{13}$C-HSQC spectrum of the substrate control to evaluate auto-oxidation three days after the preparation (**b**)

InterPro family, and the phylogenetic tree was constructed based on the alignment (Fig. 7). The C-X-X-C-motif was found from the majority of the aligned sequences (Additional file 7). It was observed from the phylogenetic tree that *Aspergillus* SOXs were evolutionary closely related. Oxidoreductases with activity on nonribosomal peptides, namely GliT, DepH and HlmI, were also closely related to AtSOX.

## Discussion

Various fungi and bacteria have been shown to secrete enzymes with a sulfhydryl oxidase activity. In this work, we focused on a secreted SOX from the fungus *A. tubingensis* and report the biochemical characterization of the secreted sulfhydryl oxidase AtSOX. *A. tubingensis* was previously identified as a natural SOX-producing strain [28]. In addition, we investigated the possible physiological role of SOX enzymes by bioinformatics means, as

no clear role has yet been established for secreted fungal SOXs thus far. The SOX produced by *A. tubingensis* showed characteristics common to other secreted fungal SOXs. The small thiol group containing tripeptide GSH was a preferred substrate of AtSOX as earlier also reported for other SOXs of fungal origin [1, 3, 6, 19, 34]. The enzyme had a non-covalently bound FAD as a cofactor, which could be released by thermal denaturation of the protein, and the UV-vis spectrum of AtSOX showed the characteristic peaks of flavoproteins [3, 19, 34]. Inhibition studies with different salts showed that AtSOX activity was inhibited in the presence of zinc sulphate. Inhibition by zinc sulphate was observed also with secreted fungal SOXs from *A.oryzae* and *Penicillium* [3, 19, 34]. The catalytic center of SOX contains two reactive cysteine residues forming a C-X-X-C-motif [46, 47]. The cysteine residues are able to chelate divalent metal ions such as zinc, and the enzyme inhibition might be due to interact of zinc and cysteine as discussed in [23]. AtSOX showed

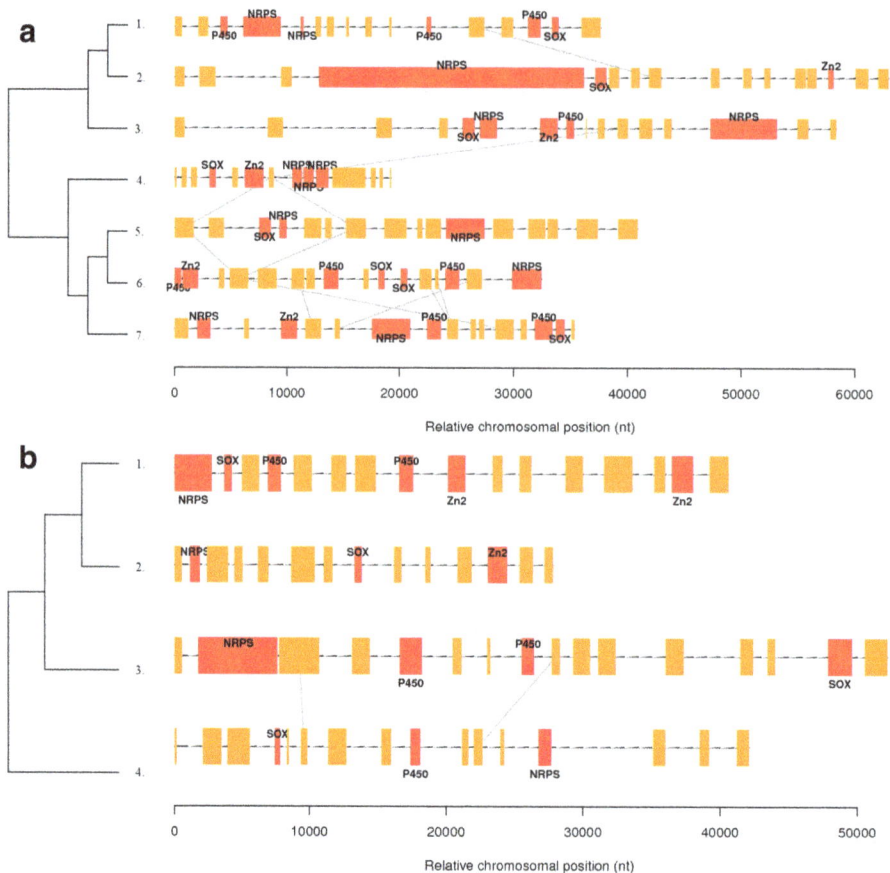

**Fig. 6** Candidate secondary metabolism clusters with SOX enzymes on fungal chromosomes. On the *left* an approximate phylogenetic tree of the species compiled from literature [36, 37]. On the *right* a stretch of a scaffold from each species containing the cluster and neighbouring genes. Genes are shown as *boxes* on the scaffold stretch. NRPS, PKS, P450 and Zn2 are indicated when present. *Grey lines* connect genes with identical protein domains on adjacent scaffolds (excluding NRPS, PKS, P450, Zn2 and SOX-coding genes) in order to reveal syntenies. Panel **a**. shows the gliotoxin clusters, while panel (**b**). shows other clusters. The strains in panel (**a**). are (1.) *Trichoderma reesei*, (2.) *Fusarium graminearum*, (3.) *Chaetomium globusum*, (4.) *Phaeosphaeria nodorum*, (5.) *A. tubingensis*, (6.) *A. oryzae* and (7.) *A. fumigatus*. The stains shown in panel (**b**). are (1.) *Fusarium graminearum*, (2.) *Magnaporthe grisea*, (3) *A. fumigatus* and (4.) *Phanerochaete chrysosporium*. Additional details are given in the Additional files 5 and 6

good thermal stability up to 50 °C retaining 75% of its activity for 20 h. The thermostability of other known fungal SOX, AoSOX2 from *A. oryzae*, is on the same range [19].

The small thiol-containing molecules, like reduced GSH, cystein and DTT, are typical substrates for secreted fungal SOXs [3, 11, 19, 34]. The protein-bound thiols groups, particularly on reduced RNase A, have also been reported to be substrates for microbial SOXs from *A. niger* and *Penicillium* [1, 3]. Later analysis have shown that reported oxidation of RNaseA was spontaneous and non-enzymatic [23]. Our results also indicated that protein-bound cysteines in RNaseA (protein size 13.7 kDa) were not oxidised by AtSOX, but instead among the analysed compounds the tripeptide GSH was the preferred substrate. It was confirmed by NMR spectroscopy studies that the oxidation reaction of GSH was enzyme-catalysed. GSH was a good substrate for AtSOX based on the oxygen consumption analysis. AtSOX preferred GSH ($K_m$ 0.8 mM) as a substrate over DTT, L-

cysteine and D-cysteine. Other characterised SOXs from *A. oryzae* preferred cysteine as a substrate over GSH. AoSOX1 had smaller $K_m$ for L-cystein (0.9 mM) than for GSH (2.78 mM), and AoSOX2 preferred D-cystein ($K_m$ 1.55 mM) over GSH ($K_m$ 3.7 mM) [19, 34]. The sulfhydryl groups in the longer peptides and protein-bound sulfhydryl groups were not oxidised suggesting preference for small substrates, which could have a better access into the active site of the enzyme. The QSOXs from an animal source, for instance bovine SOX, are known to oxidase protein-bound thiol groups and contribute in the oxidative protein folding [20], however secreted fungal SOXs are shown to have different substrate specificity and probably also a different physiological role. In this work, based on the genome analyses was suggested that secreted fungal SOXs might have role in the maturation of nonribosomal peptides.

It is not common for secondary metabolites of fungal and bacterial origin to contain disulphide bonds.

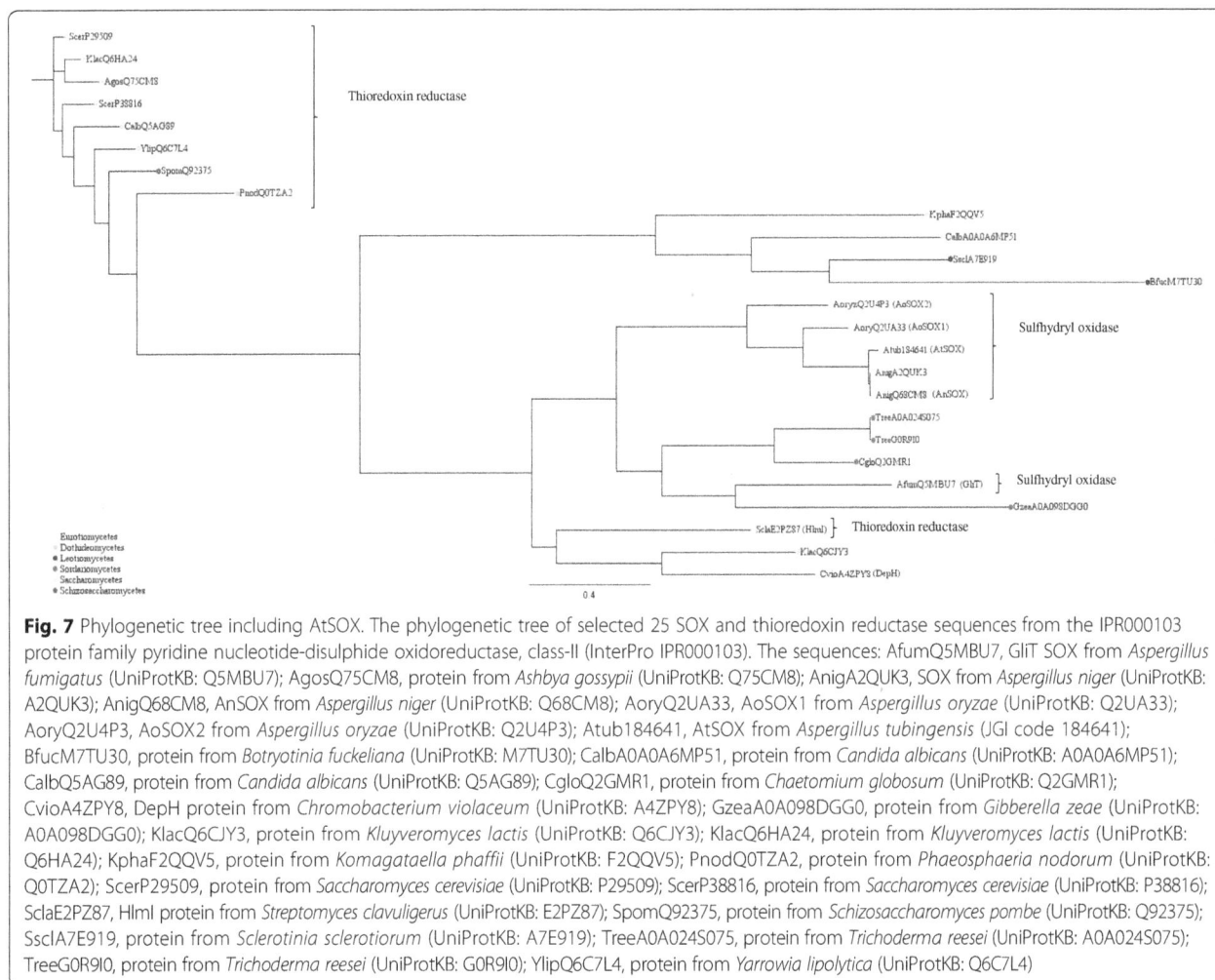

**Fig. 7** Phylogenetic tree including AtSOX. The phylogenetic tree of selected 25 SOX and thioredoxin reductase sequences from the IPR000103 protein family pyridine nucleotide-disulphide oxidoreductase, class-II (InterPro IPR000103). The sequences: AfumQ5MBU7, GliT SOX from *Aspergillus fumigatus* (UniProtKB: Q5MBU7); AgosQ75CM8, protein from *Ashbya gossypii* (UniProtKB: Q75CM8); AnigA2QUK3, SOX from *Aspergillus niger* (UniProtKB: A2QUK3); AnigQ68CM8, AnSOX from *Aspergillus niger* (UniProtKB: Q68CM8); AoryQ2UA33, AoSOX1 from *Aspergillus oryzae* (UniProtKB: Q2UA33); AoryQ2U4P3, AoSOX2 from *Aspergillus oryzae* (UniProtKB: Q2U4P3); Atub184641, AtSOX from *Aspergillus tubingensis* (JGI code 184641); BfucM7TU30, protein from *Botryotinia fuckeliana* (UniProtKB: M7TU30); CalbA0A0A6MP51, protein from *Candida albicans* (UniProtKB: A0A0A6MP51); CalbQ5AG89, protein from *Candida albicans* (UniProtKB: Q5AG89); CgloQ2GMR1, protein from *Chaetomium globosum* (UniProtKB: Q2GMR1); CvioA4ZPY8, DepH protein from *Chromobacterium violaceum* (UniProtKB: A4ZPY8); GzeaA0A098DGG0, protein from *Gibberella zeae* (UniProtKB: A0A098DGG0); KlacQ6CJY3, protein from *Kluyveromyces lactis* (UniProtKB: Q6CJY3); KlacQ6HA24, protein from *Kluyveromyces lactis* (UniProtKB: Q6HA24); KphaF2QQV5, protein from *Komagataella phaffii* (UniProtKB: F2QQV5); PnodQ0TZA2, protein from *Phaeosphaeria nodorum* (UniProtKB: Q0TZA2); ScerP29509, protein from *Saccharomyces cerevisiae* (UniProtKB: P29509); ScerP38816, protein from *Saccharomyces cerevisiae* (UniProtKB: P38816); SclaE2PZ87, HlmI protein from *Streptomyces clavuligerus* (UniProtKB: E2PZ87); SpomQ92375, protein from *Schizosaccharomyces pombe* (UniProtKB: Q92375); SsclA7E919, protein from *Sclerotinia sclerotiorum* (UniProtKB: A7E919); TreeA0A024S075, protein from *Trichoderma reesei* (UniProtKB: A0A024S075); TreeG0R9I0, protein from *Trichoderma reesei* (UniProtKB: G0R9I0); YlipQ6C7L4, protein from *Yarrowia lipolytica* (UniProtKB: Q6C7L4)

Maturation mechanisms of these low-molecular mass metabolites are still unclear [48]. Thiol oxidising enzymes, oxidoreductases GliT, DepH and HlmI, have been reported to be involved in the maturation of gliotoxin from *A. fumigatus*, of the anticancer peptide romidepsin (FK228) in *Chromobacterium violaceum* No. 968, and of the antimicrobial compound holomycin in *Streptomyces clavuligerus*, respectively [48, 49]. A FAD-dependent SOX enzyme, GliT (dimeric ~ 60 kDa), belongs to the *A. fumigatus* gliotoxin chromosomal gene cluster, and it has been shown to form the atypical intramolecular disulfide bond responsible for toxicity of gliotoxin [44, 50]. GliT utilizes molecular oxygen as terminal electron acceptor with concomitant formation of hydrogen peroxide [51]. Similarly, a dimeric FAD-dependent pyridine nucleotide-disulfide oxidoreductase DepH (dimeric ~ 70 kDa; UniProtKB: A4ZPY8) has been found responsible for the introduction of a disulfide bond in the FDA-approved anticancer peptide FK228 from the Gram-negative *C. violaceum* [27]. Functionally homologous to GliT and DepH, a dimeric FAD-dependent thiol oxidising enzyme HlmI has been proven to be involved in the maturation of holomycin from *S. clavuligerus* [49]. DepH, GliT and HlmI shared common features with secreted fungal SOXs such as AoSOX1 and AoSOX2 (UniProtKB: Q2UA33, Q2U4P3), characterized by our group, and the reconstructed AtSOX sequence of this work (Fig. 3). They all carried sequence features of FAD-dependent pyridine nucleotide-disulfide oxidoreductase (InterPro: IPR013027 and the subclass II InterPro: IPR000103 used in this genome analysis study). AoSOX1 and AoSOX2 were proven FAD-dependent, dimeric and able to oxidise small thiol-containing peptide molecules such as GSH, cysteine and DTT. The reconstructed sequence of AtSOX shared 22-96% identity to other known secreted *Aspergillus* SOXs and oxidoreductases DepH, GliT and HlmI (Table 1). The flavoenzymes GliT, DepH and HlmI have a common reaction mechanism, although they have differences in substrate specificity [48]. The substrate binding clefts of GliT, DepH and HlmI enzymes are different but they all oxidise small molecules and introduce the disulfide bond to the corresponding secondary metabolites of fungi or bacteria [48].

**Table 1** Biochemical characteristics of selected secreted flavin-dependent sulfhydryl oxidases and the enzymes DepH, GliT and HlmI reported to be involved in the secondary metabolism

| Enzyme | Biochemical properties | | | | | Identity (%) to AtSOX | Reference |
|---|---|---|---|---|---|---|---|
| | MW (kDa) | pH optimum | pH stability | Temperature stability | Cofactor, non covalent | | |
| AtSOX | 55 | 6.5 | > 80% activity after 20 h at pH 4-8.5 | > 85% activity after 20 h at 40 °C | FAD | 100 | This study |
| AnSOX | 53 (dimer) | 5.5 | n.a. | n.a. | FAD | 96.0 | [1] |
| AoSOX1 | 45 (dimer) | 8.0 | > 80% activity after 24 h at pH 5-8.5 | > 70% activity after 24 h at 40 °C | FAD | 64.6 | [34] |
| AoSOX2 | 45 (dimer) | 7.5-8.0 | > 80% activity after 24 h at pH 5-8 | > 65% activity after 1 h at 60 °C | FAD | 46.5 | [19] |
| DepH | 34.4 (dimer) | 7 (assay) | n.a. | n.a. | FAD | 22.0 | [27] |
| GliT | 30 (dimer) | 6.5 (assay) | n.a. | n.a. | FAD | 24.5 | [44, 51] |
| HlmI | 39 (dimer) | 6.5 (assay) | n.a. | n.a. | FAD | 22.3 | [49] |

*Abbreviations*: *AtSOX* secreted SOX from *Aspergillus tubingensis*, *AnSOX* secreted SOX from *Aspergillus niger* (UniProtKB: Q68CM8), *AoSOX* secreted SOX from *Aspergillus oryzae* (AoSOX1 UniProtKB: Q2UA33, AoSOX2 UniProtKB: Q2U4P3), *DepH* enzyme from *Chromobacterium violaceum* (UniProtKB: A4ZPY8), *GliT* enzyme from *Aspergillus fumigatus* (UniProtKB: Q5MBU7), *HlmI* enzyme from *Streptomyces clavuligerus* (UniProtKB: E2PZ87), *MW* molecular weight of the subunit, *SOX* sulfhydryl oxidase, *n.a.* not available

Most of the secondary metabolites are derivatives from nonribosomal peptides and polyketides, the synthesis of which is catalysed by the multidomain enzymes belonging to NRPSs and PKSs [52]. Nonribosomal peptides are a class of molecules characterised by a vast structural and functional diversity, e.g. they can have linear, cyclic, or branched structures and activities ranging from antibiotic to metal-binding, from immunosuppressive to toxic and cytostatic [53]. Most of the fungal gene clusters for the secondary metabolite biosynthesis are silent under laboratory conditions [52]. The results (Fig. 6) indicated that SOX-coding genes are associated with NRPS, but not PKS, clusters and thus SOXs are possibly to act as accessory enzymes in the production of nonribosomal peptides. Thus, based on the genome studies, a connection between nonribosomal peptides and SOXs was suggested, since SOX and the nonribosomal peptide synthetase coding genes were found in the same clusters in the analysed fungal genomes. However, in vitro enzymatic oxidation of selected nonribosomal peptides (i.e. gliotoxin and holomycin) was not detected. This could be due to a challenging location of sulfhydryl groups in the circular structure of the peptides, and hence poor availability. Moreover flavoenzymes GliT, DepH and HlmI all have different substrate specificity. The cellular localisation, ability to oxidise small thiol-containing peptides, and genome comparison of secreted SOXs support the idea that fungal secreted SOXs are involved in the biosynthesis of disulfide-containing secondary metabolites, such as nonribosomal peptides. Their role could thus be in the maturation of peptides produced nonribosomally.

## Conclusions

This paper describes the characterization of a flavin-dependent secreted fungal SOX from *Aspergillus tubingensis* (AtSOX). AtSOX was shown to have good thermal stability and the enzyme retained high activity in the broad pH range. The enzyme preferred GSH as a substrate over the tested small-thiol containing molecules (reduced cysteine and DTT). The enzyme activity was drastically reduced in the presence of zinc sulphate. The enzymatic oxidation of the tripeptide GSH and formation of a disulphide bond was verified by nuclear magnetic resonance spectroscopy. AtSOX was evolutionary closely related to other Aspergillus SOXs and the oxidoreductases GliT, HlmI and DepH of fungal and bacterial origin, whereas fungal thioreductases were evolutionary more distant. Based on the location near to NRPSs encoding genes, SOXs could be involved in the secondary metabolism and act as an accessory enzyme in the production of nonribosomal peptides.

## Additional files

**Additional file 1:** Purification of AtSOX as analysed by SDS-PAGE. Molecular weight (MW) standards are shown in lanes 1 and 9. The sample from initial crude cell-free medium is shown in lane 2. As a first purification step was used anion exchange chromatography with a Q Sepharose column. In lane 3 are the unbound proteins, and in the lanes 4–6 bound and then eluted proteins, from Q Sepharose column. Lane 6: AtSOX containing fractions selected for further purifications steps. In the lanes marked 7 and 8 are shown fractions obtained from the last purification step using anion exchange chromatography with Resource Q column (analysed in a separate SDS-PAGE gel with MW standards in lane 9). (PPTX 137 kb)

**Additional file 2:** AtSOX activity measured by oxygen consumption assay using 3 mM reduced GSH as a substrate (continuous line). The reaction occurred at the enzymatic rate (the linear area ca. 0.5 - 3.5 min). The amount of dissolved oxygen in the reduced GSH solution prior addition of enzyme is shown with a dashed line. The triplicate measurements were done. (PPTX 6691 kb)

**Additional file 3:** AtSOX activity measured by HVA-peroxidase coupled assay using reduced GSH (5 mM) as a substrate according to [33]. The enzyme reaction is at the enzymatic rate (linear area ca. 0 - 150 s). The production of the fluorescent HVA dimer was followed at excitation wavelength 320 nm and emission wavelength 420 nm. The reduced AtSOX activity with the inhibitor zinc sulphate (10 mM) is also shown (dashed line). The triplicate measurements were done. (PPTX 121 kb)

**Additional file 4:** Absorbance spectra (ca. 10 min) of 5 mM reduced GSH (a.) and 5 mM reduced GSH with AtSOX (b.). Arrow indicates the direction of increased UV adsorption due to enzymatic oxidation of the substrate. (PPTX 395 kb)

**Additional file 5:** Details to Fig. 6 Candidate secondary metabolism clusters with SOX enzymes on fungal chromosomes. On the left an approximate phylogenetic tree of the species compiled from literature [54, 55]. On the right a stretch of a scaffold from each species containing the cluster and neighbouring genes. Genes are shown as boxes on the scaffold stretch. NRPS, PKS, P450 and Zn2 are indicated when present. Grey lines connect genes with identical protein domains on adjacent scaffolds (excluding NRPS, PKS, P450, Zn2 and SOX genes) in order to reveal syntenies. Codes above the gene boxes are their identifiers and below them the Interpro protein domain identifiers found in the genes. Panel a. shows the gliotoxin clusters, while panel b. shows other clusters. The strains shown in panel a. are *Trichoderma reesei, Fusarium graminearum, Chaetomium globosum, Phaeosphaeria nodorum, A. tubingensis, A. oryzae* and *A. fumigatus*. The stains shown in panel b. are *F. graminearum, Magnaporthe grisea, A. fumigatus* and *Phanerochaete chrysosporium*. (PNG 335 kb)

**Additional file 6:** For each chromosomal cluster the table shows accession numbers for genes (Accession), scaffold identifier (Scaffold), start and end on the scaffold, direction of the gene (Direction) and Interpro protein domain identifiers found in the genes (Interpro), as details for Additional file 5. (XLSX 20 kb)

**Additional file 7:** Part of the alignment of 25 sequences from the same protein family (InterPro IPR000103). On the first line is shown AtSOX retrieved from *A. tubingensis* genome. The C-X-X-C motifs are marked with a box. The sequences: AtSOX, secreted SOX from *A. tubingensis;* AnSOX, secreted SOX from *A. niger;* AoSOX, secreted SOX from *A. oryzae;* DepH, enzyme from *C. violaceum;* GliT, enzyme from *A. fumigatus;* Hlml, enzyme from *S. clavuligerus.* The other abbreviations are shown in the legend of Fig. 7. (PPTX 95 kb)

## Abbreviations

$^{13}$C-HSQC: $^{13}$C heteronuclear single quantum correlation (spectroscopy); AtSOX: Secreted SOX enzyme from *Aspergillus tubingensis*; CHCA: α-cyano-4-hydroxycinnamic acid; DTNB: 5,5-dithio-bis(2-nitrobenzoic acid), Ellman's reagent; DTT: Dithiothreitol; EDTA: Ethylenediaminetetraacetic acid; FAD: Flavin adenine dinucleotide; GSH: The reduced glutathione; HVA: Homovanillic acid; MALDI-TOF MS: Matrix-assisted laser desorption-ionization time of flight mass spectrometry; NMR: Nuclear magnetic resonance (spectroscopy); NRPS: Nonribosomal peptide synthetases; PKS: Polyketide synthases; PMF: Peptide mass fingerprinting; PNDR: Pyridine nucleotide-disulfide oxidoreductase; QSOX: The quiescin-sulfhydryl oxidase; Q-TOF: Quadruple time-of-flight; RNase A: Ribonuclease A; SDS: Sodium dodecyl sulphate; SEC: Size-exclusion chromatography; SOX: Sulfhydryl oxidase; TOCSY: Homonuclear total correlation spectroscopy; UPLC: Ultra performance liquid chromatography

## Acknowledgements

Dr. Peter Würtz is acknowledged for NMR spectroscopy work (Finnish National Biological NMR center, Institute of Biotechnology, University of Helsinki, Finland). Docent Nisse Kalkkinen is acknowledged for Edman degradation (Institute of Biotechnology, University of Helsinki, Finland). Dr. Elisabeth Heine (DWI, Deutsches Wolforshunginstitut, Aachen, Germany) is thanked for providing Insulin B chain for the study. Dr. Heli Nygren is thanked for the guidance in the LC-MS analysis. In addition Outi Liehunen and Riitta Isoniemi are thanked for the technical assistance relating to the purification of the enzyme. Päivi Matikainen and Gunilla Rönnholm (Institute of Biotechnology, University of Helsinki, Finland) are thanked for the technical assistance related to mass spectrometry.

## Funding

The research was conducted in the Academy of Finland funded project 'Enzymatic crosslinking of food proteins: impact of food protein folding on the mode of action of crosslinking enzymes' (no: 110,965), the Finnish Funding Agency for Technology and Innovation (Tekes) funded project 'Tailored nanostabilizers for biocomponent interfaces' (no: VTT-R-06743-08) and FP6 funded project 'High performance industrial protein matrices' (no: NMP-3-CT-2003-505,790). The work of Dr. Mikko Arvas was supported by Academy Postdoctoral Researcher's fellowship (no. 127715). Dr. Greta Faccio was funded by TYROMAT, a project within the Empa Postdocs programme that is co-funded by the FP7: People Marie-Curie action COFUND. The financiers had no role in the design of the study, collection, analysis and interpretation of data, or in writing the manuscript.

## Authors' contributions

ON carried out the experimental work and drafted the manuscript. MA carried out the genome mining study and assisted in writing the manuscript. KK and M-LM designed the research and participated in writing the manuscript. GF participated in the design of the study and helped to draft and write the manuscript. PP participated in the design of NMR studies. JB participated in the design and coordination of the study. All authors read and approved the final manuscript.

## Competing interests

The authors declare that they have no competing interests.

## Author details

$^1$VTT Technical Research Centre of Finland, Ltd., P.O. Box 1000, FI-02044 Espoo, Finland. $^2$Independent scientist, St. Gallen, CH, Switzerland. $^3$Institute of Biotechnology, University of Helsinki, P.O. Box 65, FI-00014 Helsinki, Finland. $^4$Department of Biological and Environmental Sciences, Nanoscience Center, University of Jyväskylä, P.O. Box 35, FI-40014 Jyväskylä, Finland. $^5$Department of Chemistry, Nanoscience Center, University of Jyväskylä, P.O. Box 35, FI-40014 Jyväskylä, Finland. $^6$Natural resources institute Finland (Luke), P.O. Box 2, FI-00790 Helsinki, Finland. $^7$Department of Forest Products Technology, Bioproduct Chemistry, Aalto University, School of Chemical Technology, P.O. Box 16300, FI-00076 Espoo, Finland.

## References

1. de la Motte RS, Wagner FW. Aspergillus niger Sulfhydryl oxidase. Biochem Int. 1987;26:7363–71. Available from: http://dx.doi.org/10.1021/bi00397a025
2. Gerber J, Mühlenhoff U, Hofhaus G, Lill R, Lisowsky T. Yeast Erv2p is the first Microsomal FAD-linked Sulfhydryl Oxidase of the Erv1p/Alrp protein family. J Biol Chem. 2001;276:23486–91. Available from: http://dx.doi.org/10.1074/jbc.M100134200
3. Kusakabe H, Kuninaka A, Yoshino H. Purification and properties of a new enzyme, glutathione Oxidase from Penicillium sp. K-6-5. Agric Biol Chem. 1982;46:2057–67. Available from: http://dx.doi.org/10.1080/00021369.1982.10865382
4. Ostrowski MC, Kistler WS. Properties of a Flavoprotein Sulfhydryl Oxidase from rat seminal vesicle secretion. Biochemistry. 1980;19:2639–45. Available from: http://dx.doi.org/10.1021/bi00553a016
5. Kusakabe H, Midorikawa Y, Kuninaka A, Yoshino H. Distribution of extracellular oxygen related enzymes in Molds. Agric Biol Chem. 1983;47:1385–7. Available from: http://dx.doi.org/10.1080/00021369.1983.10857185
6. Starnes RL, Katkocin DM, Miller CA, Strobel Jr. RJ. Microbial sulfhydryl oxidases [internet]. United States; 1986. Available from: http://www.freepatentsonline.com/4632905.pdf.
7. Fass D. The Erv family of sulfhydryl oxidases. Biochim Biophys Acta - Mol Cell Res. 2008;1783:557–66. Available from: http://dx.doi.org/10.1016/j.bbamcr.2007.11.009
8. Kodali VK, Thorpe C. Oxidative protein folding and the Quiescin-Sulfhydryl Oxidase family of Flavoproteins. Antioxid Redox Signal. 2010;13:1217–30. Available from: http://dx.doi.org/10.1089/ars.2010.3098
9. Sevier CS, Kaiser CA. Formation and transfer of disulphide bonds in living cells. Nat Rev Mol Cell Biol. 2002;3:836–47. Available from: http://dx.doi.org/10.1038/nrm954

10. Faccio G, Nivala O, Kruus K, Buchert J, Saloheimo M. Sulfhydryl oxidases: sources, properties, production and applications. Appl Microbiol Biotechnol. 2011;91:957–66. Available from: http://dx.doi.org/10.1007/s00253-011-3440-y

11. Vignaud C, Kaid N, Rakotozafy L, Davidou S, Nicolas J. Partial purification and characterization of Sulfhydryl Oxidase from Aspergillus niger. J Food Sci. 2002;67:2016–22. Available from: http://dx.doi.org/10.1111/j.1365-2621.2002.tb09494.x

12. Hunter S, Jones P, Mitchell A, Apweiler R, Attwood TK, Bateman A, et al. InterPro in 2011: new developments in the family and domain prediction database. Nucleic Acids Res. 2012;40:D306–12. Available from: http://dx.doi.org/10.1093/nar/gkr948

13. de Vries RP, Riley R, Wiebenga A, Aguilar-Osorio G, Amillis S, Uchima CA, et al. Comparative genomics reveals high biological diversity and specific adaptations in the industrially and medically important fungal genus Aspergillus. Genome Biol. 2017;18:1–45. Available from: http://dx.doi.org/10.1186/s13059-017-1151-0

14. Swaisgood HE. Process of removing the cooked flavor from milk. 1977. Available from: http://www.patents.com/us-4053644.html.

15. Faccio G, Flander L, Buchert J, Saloheimo M, Nordlund E. Sulfhydryl oxidase enhances the effects of ascorbic acid in wheat dough. J. Cereal Sci. Elsevier Ltd; 2012;55:37–43. Available from: http://dx.doi.org/10.1016/j.jcs.2011.10.002

16. Haarasilta S, Väisanen S, Scott D. Method for improving flour dough. 1989. Available from: https://www.google.com/patents/EP0321811B1?cl=en.

17. Kaufman SP, Fennema O. Evaluation of Sulfhydryl Oxidase as a Strengthening Agent for Wheat Flour Dough [Internet]. Cereal Chem. 64AD. p. 172–6. Available from: http://www.aaccnet.org/publications/cc/backissues/1987/Documents/64_172.pdf

18. Koppelman SJ, van den Hout RHJA, Sleijster-Selis HE, Luijkx DMAM. Modification of allergens [Internet]. Netherlands; 2010. Available from: http://www.freepatentsonline.com/20100086568.pdf

19. Faccio G, Kruus K, Buchert J, Saloheimo M. Production and characterisation of AoSOX2 from Aspergillus oryzae, a novel flavin-dependent sulfhydryl oxidase with good pH and temperature stability. Appl Microbiol Biotechnol. 2011;90:941–9. Available from: http://dx.doi.org/10.1007/s00253-011-3129-2

20. Jaje J, Wolcott HN, Fadugba O, Cripps D, Yang AJ, Mather IH, et al. A Flavin-Dependent Sulfhydryl Oxidase in Bovine Milk. Biochemistry. 2007;46:13031–40. Available from: http://dx.doi.org/10.1021/bi7016975

21. Ilani T, Alon A, Grossman I, Horowitz B, Kartvelishvily E, Cohen SR, et al. A secreted disulfide catalyst controls extracellular matrix composition and function. Science. 2013;341:74–6. Available from: http://dx.doi.org/10.1126/science.1238279

22. Ang SK, Zhang M, Lodi T, Lu H. Mitochondrial thiol oxidase Erv1: both shuttle cysteine residues are required for its function with distinct roles. Biochem J. 2014;460:199–210. Available from: http://dx.doi.org/10.1042/BJ20131540

23. Janolino VG, Swaisgood HE. A comparison of sulfhydryl oxidases from bovine milk and from Aspergillus niger. Milchwissenschaft. 1992;47:143–6. Available from: http://eurekamag.com/research/006/937/006937840.php

24. Bin SK, Swaisgood HE, Horton HR. Requirement for a Sulfhydryl Group for Sulfhydryl Oxidase Activity. J Dairy Sci. 1986;69:2589–92. Available from: https://doi.org/10.3168/jds.S0022-0302(86)80705-6

25. Musard J-F, Sallot M, Dulieu P, Fraîchard A, Ordener C, Remy-Martin J-P, et al. Identification and expression of a new Sulfhydryl Oxidase SOx-3 during the cell cycle and the Estrus cycle in uterine cells. Biochem Biophys Res Commun. 2001;287:83–91. Available from: http://dx.doi.org/10.1006/bbrc.2001.5440

26. Tury A, Mairet-Coello G, Esnard-Fève A, Benayoun B, Risold P-Y, Griffond B, et al. Cell-specific localization of the sulphydryl oxidase QSOX in rat peripheral tissues. Cell Tissue Res. 2006;323:91–103. Available from: http://dx.doi.org/10.1007/s00441-005-0043-x

27. Wang C, Wesener SR, Zhang H, Cheng Y-Q. An FAD-Dependent Pyridine Nucleotide-Disulfide Oxidoreductase Is Involved in Disulfide Bond Formation in FK228 Anticancer Depsipeptide. Chem. Biol. Elsevier Ltd; 2009; 16:585–93. Available from: http://dx.doi.org/10.1016/j.chembiol.2009.05.005

28. Nivala O, Mattinen ML, Faccio G, Buchert J, Kruus K. Discovery of novel secreted fungal sulfhydryl oxidases with a plate test screen. Appl Microbiol Biotechnol. 2013;97:9429–37. Available from: http://dx.doi.org/10.1007/s00253-013-4753-9

29. Suihko M-L. VTT culture collection catalogue of strains. 4th ed. Espoo: VTT; 1999. Available from: http://culturecollection.vtt.fi/m/html?p=m

30. Shevchenko A, Wilm M, Vorm O, Mann M. Mass spectrometric sequencing of proteins from silver-stained polyacrylamide gels. Anal Chem. 1996;68: 850–8. Available from: http://dx.doi.org/10.1021/ac950914h

31. Poutanen M, Salusjärvi L, Ruohonen L, Penttilä M, Kalkkinen N. Use of matrix-assisted laser desorption/ionization time-of-flight mass mapping and nanospray liquid chromatography/electrospray ionization tandem mass spectrometry sequence tag analysis for high sensitivity identification of yeast proteins separated by tw. Rapid Commun Mass Spectrom. 2001;15: 1685–92. Available from: http://dx.doi.org/10.1002/rcm.424

32. Ellman GL. Tissue sulfhydryl groups. Arch Biochem Biophys. 1959;82:70–7. Available from: http://dx.doi.org/10.1016/0003-9861(59)90090-6

33. Raje S, Glynn NM, Thorpe C. A continuous fluorescence assay for sulfhydryl oxidase. Anal Biochem. 2002;307:266–72. Available from: http://dx.doi.org/10.1016/S0003-2697(02)00050-7

34. Faccio G, Kruus K, Buchert J, Saloheimo M. Secreted fungal sulfhydryl oxidases: sequence analysis and characterisation of a representative flavin-dependent enzyme from Aspergillus oryzae. BMC Biochem. 2010;11:31. Available from: http://dx.doi.org/10.1186/1471-2091-11-31

35. Kay LE, Keifer P, Saarinen T. Pure absorption gradient enhanced Heteronuclear single quantum correlation spectroscopy with improved sensitivity. J Am Chem Soc. 1992;114:10663–5. Available from: http://dx.doi.org/10.1021/ja00052a088

36. Braunschweiler L, Ernst RR. Coherence transfer by isotropic mixing: application to proton correlation spectroscopy. J Magn Reson. 1969;53:521–8. Available from: http://dx.doi.org/10.1016/0022-2364(83)90226-3

37. Koivistoinen OM, Arvas M, Headman JR, Andberg M, Penttilä M, Jeffries TW, et al. Characterisation of the gene cluster for L-rhamnose catabolism in the yeast Scheffersomyces (Pichia) stipitis. Gene. Elsevier B.V.; 2012;492:177–85. Available from: http://dx.doi.org/10.1016/j.gene.2011.10.031

38. Arvas M, Kivioja T, Mitchell A, Saloheimo M, Ussery D, Penttilä M, et al. Comparison of protein coding gene contents of the fungal phyla Pezizomycotina and Saccharomycotina. BMC Genomics. 2007;8:325. Available from: http://dx.doi.org/10.1186/1471-2164-8-325

39. Grigoriev IV, Nordberg H, Shabalov I, Aerts A, Cantor M, Goodstein D, et al. The genome portal of the Department of Energy Joint Genome Institute. Nucleic Acids Res. 2011;40:D26–32. Available from: http://dx.doi.org/10.1093/nar/gkr947

40. Katoh K, Standley DM. MAFFT multiple sequence alignment software version 7: improvements in performance and usability. Mol Biol Evol. 2013; 30:772–80. Available from: http://dx.doi.org/10.1093/molbev/mst010

41. Capella-Gutiérrez S, Silla-Martínez JM, Gabaldón T. trimAl: a tool for automated alignment trimming in large-scale phylogenetic analyses. Bioinformatics. 2009; 25:1972–3. Available from: http://dx.doi.org/10.1093/bioinformatics/btp348

42. Price MN, Dehal PS, Arkin AP. FastTree 2 - approximately maximum-likelihood trees for large alignments. PLoS One. 2010;5:e9490. Available from: http://dx.doi.org/10.1371/journal.pone.0009490

43. Hoffmeister D, Keller NP. Natural products of filamentous fungi: enzymes, genes, and their regulation. Nat Prod Rep. 2007;24:393–416. Available from: http://dx.doi.org/10.1039/b603084j

44. Scharf DH, Heinekamp T, Remme N, Hortschansky P, Brakhage AA, Hertweck C. Biosynthesis and function of gliotoxin in Aspergillus fumigatus. Appl Microbiol Biotechnol. 2012;93:467–72. Available from: http://dx.doi.org/10.1007/s00253-011-3689-1

45. Mukherjee PK, Horwitz BA, Kenerley CM. Secondary metabolism in Trichoderma - a genomic perspective. Microbiology. 2012;158:35–45. Available from: http://dx.doi.org/10.1099/mic.0.053629-0

46. Chivers PT, Prehoda KE, Raines RT. The CXXC motif: a rheostat in the active site. Biochemistry. 1997;36:4061–6. Available from: http://dx.doi.org/10.1021/bi9628580

47. Quan S, Schneider I, Pan J, Von Hacht A, Bardwell JCA. The CXXC motif is more than a Redox rheostat. J Biol Chem. 2007;282:28823–33. Available from: http://dx.doi.org/10.1074/jbc.M705291200

48. Scharf DH, Groll M, Habel A, Heinekamp T, Hertweck C, Brakhage AA, et al. Flavoenzyme-catalyzed formation of disulfide bonds in natural products. Angew Chem - Int Ed. 2014;53:2221–4. Available from: http://dx.doi.org/10.1002/anie.201309302

49. Li B, Walsh CT. Streptomyces clavuligerus HlmI is an Intramolecular disulfide-forming Dithiol Oxidase in Holomycin biosynthesis. Biochemistry. 2011;50:4615–22. Available from: http://dx.doi.org/10.1021/bi200321c

50. Dolan SK, O'Keeffe G, Jones GW, Doyle S. Resistance is not futile: gliotoxin biosynthesis, functionality and utility. Trends Microbiol. Elsevier Ltd; 2015;23: 419–28. Available from: http://dx.doi.org/10.1016/j.tim.2015.02.005

51. Scharf DH, Remme N, Heinekamp T, Hortschansky P, Brakhage AA, Hertweck C. Transannular Disulfide Formation in Gliotoxin Biosynthesis and Its Role in Self-Resistance of the Human Pathogen Aspergillus fumigatus. 2010;132: 10136–41. Available from: http://dx.doi.org/10.1021/ja103262m

52. Brakhage AA. Regulation of fungal secondary metabolism. Nat. Rev. Microbiol. Nat Publ Group. 2013;11:21–32. Available from: http://dx.doi.org/10.1038/nrmicro2916

53. Schwarzer D, Finking R, Marahiel MA. Nonribosomal peptides: from genes to products. Nat Prod Rep. 2003;20:275–87. Available from: http://dx.doi.org/10.1039/b111145k

54. Samson RA, Noonim P, Meijer M, Houbraken J, Frisvad JC, Varga J. Diagnostic tools to identify black aspergilli. Stud Mycol. 2007;59:129–45. Available from: http://dx.doi.org/10.3114/sim.2007.59.13

55. Wang H, Xu Z, Gao L, Hao B. A fungal phylogeny based on 82 complete genomes using the composition vector method. BMC Evol Biol. 2009;9:195. Available from: http://dx.doi.org/10.1186/1471-2148-9-195

# 16

# Evaluating the role of a trypsin inhibitor from soap nut *(Sapindus trifoliatus L. Var. Emarginatus)* seeds against larval gut proteases, its purification and characterization

V D Sirisha Gandreddi[1], Vijaya Rachel Kappala[1*], Kunal Zaveri[1] and Kiranmayi Patnala[2]

## Abstract

**Background:** The defensive capacities of plant protease Inhibitors (PI) rely on inhibition of proteases in insect guts or those secreted by microorganisms; and also prevent uncontrolled proteolysis and offer protection against proteolytic enzymes of pathogens.

**Methods:** An array of chromatographic techniques were employed for purification, homogeneity was assessed by electrophoresis. Specificity, Ki value, nature of inhibition, complex formation was carried out by standard protocols. Action of SNTI on insect gut proteases was computationally evaluated by modeling the proteins by threading and docking studies by piper using Schrodinger tools.

**Results:** We have isolated and purified Soap Nut Trypsin Inhibitor (SNTI) by acetone fractionation, ammonium sulphate precipitation, ion exchange and gel permeation chromatography. The purified inhibitor was homogeneous by both gel filtration and polyacrylamide gel electrophoresis (PAGE). SNTI exhibited a molecular weight of 29 kDa on SDS-PAGE, gel filtration and was negative to Periodic Acid Schiff's stain. SNTI inhibited trypsin and pronase of serine class. SNTI demonstrated non-competitive inhibition with a Ki value of $0.75 \pm 0.05 \times 10\text{-}10$ M. The monoheaded inhibitor formed a stable complex in 1:1 molar ratio. Action of SNTI was computationally evaluated on larval gut proteases from *Helicoverpa armigera* and *Spodoptera frugiperda*. SNTI and larval gut proteases were modeled and docked using Schrodinger software. Docking studies revealed strong hydrogen bond interactions between Lys10 and Pro71, Lys299 and Met80 and Van Der Waals interactions between Leu11 and Cys76amino acid residues of SNTI and protease from *H. Armigera*. Strong hydrogen bonds were observed between SNTI and protease of *S. frugiperda* at positions Thr79 and Arg80, Asp90 and Gly73, Asp2 and Gly160 respectively.

**Conclusion:** We conclude that SNTI potentially inhibits larval gut proteases of insects and the kinetics exhibited by the protease inhibitor further substantiates its efficacy against serine proteases.

**Keywords:** $K_i$, Monoheaded inhibitor, PAGE, SNTI, Specificity, Gut proteases, Protein-protein docking

## Background

Plants commonly exhibit structural and biochemical defense mechanisms when challenged by pathogens and herbivores. With rising incidence of destructive activities of numerous pests like fungi, weeds and insects leading to radical decrease in yields, use of pesticides has become inevitable. An ecofriendly alternative to chemical pesticides is bio-pesticides, which encompasses a broad array of microbial pesticides, bio-chemicals derived from microorganisms and other natural sources. One such compound is polypeptide that act upon the cell membrane of pathogens [1, 2]. They are capable of blocking protease mediated pathway by targeting the active sites of proteases by forming protease-inhibitor complex and results in breaking down of enzymatic activity [3] and hence these compounds are termed as protease inhibitors [PIs]. PIs are small proteins quite common in nature and these defense-

* Correspondence: rachelr68@gmail.com
[1]Assistant professor, Department of Biochemistry/Bioinformatics, Institute of Science, GITAM University, Rushikonda, Visakhapatnam 530045, Andhra Pradesh, India
Full list of author information is available at the end of the article

related proteins are often present in seeds, and also induced in certain plant tissues by herbivory or wounding [4]. The defensive capacities of plant PIs rely on inhibition of proteases present in insect guts or secreted by microorganisms and reduce the availability of amino acids necessary for their growth and development. Protease inhibitors play an endogenous role in preventing uncontrolled proteolysis and/or in protecting against the foreign proteolytic enzymes of pests or pathogens [5, 6]. Observations of their wound-inducible expression [7, 8] have led to investigations focusing on their role in plant protection against insects [9–11].

Antifungal proteins contribute defense against pathogenic fungi. A variety of antifungal proteins were isolated from the seeds of leguminous plants, including French bean, cowpea, field bean, mung bean, peanut and red kidney bean. Nearly all leguminous antifungal proteins examined were able to inhibit HIV-1 reverse transcriptase, protease and integrase to some extent [12]. There is also a wide range of herbal ingredients which have been documented to have anti pityrosporum or anti dandruff activity [13]. Understanding the molecular mechanisms of natural plant products against dermatophytes could lead to the development of safe antifungal agents for controlling human skin diseases.

One of the most diverse species among living organisms on earth are insects [14]. One fifth of the total crop production is destroyed by insects [15]. Synthetic insecticides are most widely used to control the insects, but they in turn cause harmful effects on soil and plants [16]. In order to overcome such adverse effects of synthetic insecticides, there is a need for natural insecticides which are derived from plants or microorganisms. Allelochemicals are natural plant compounds which are invariably studied on plant-insect relationship [17]. Some protease inhibitors have shown anti insecticidal activity [18, 19]. Potential natural insecticidal compounds from different plants are identified which are referred to as trypsin inhibitors. The mid gut region of insect digestive system comprises of digestive proteases which catalyses the breakdown of proteins into small molecules [20]. Protease inhibitors are proficient in interfering with digestive enzymes of insect gut and hence are able in controlling them [21]. JSTI (Jack fruit Seed Trypsin Inhibitor) has effectively shown insecticidal activity against proteases of larval mid gut [18]. SSTI (*Sapindus saponaria* Trypsin Inhibitor) from *Sapindus saponaria* L., of the family *Sapindaceae* also exhibited substantial inhibitory activity against gut proteases of rice and flour moths, velvet bean caterpillar moth and sugar borer [19].

Ever since these inhibitors are identified, their role in medicinal and agricultural fronts are being extensively investigated. Accordingly, preliminary studies on protease inhibitors are carried out by screening different plant species (Table 1) and found *Sapindus trifoliatus* seed protease inhibitor to exhibit higher inhibitor activity among the group. Soap nut tree (*Sapindus trifoliatus L.*)

**Table 1** Protease inhibitor activities in different plant seeds

| Name of the plant | % of Protease Inhibitory activity |
| --- | --- |
| Annona squamosal | 46.60 |
| Achras sapota | 58.30 |
| Mimordica charantia | 57.14 |
| Moringa sp. | 11.10 |
| Trichosanthus sp | Negligible |
| Cucurbita maxima | 64.90 |
| Termenalia sp | 27.70 |
| Vamu | 66.60 |
| Chironji | 28.20 |
| Sapindus trifoliatus | 75.40 |
| Pomegranate | 40.00 |

belongs to the family of *Sapindaceae*, which is native primarily to tropical climate. It is an evergreen plant and most commonly found in South India whose fruits are rich in saponins [1, 22] and nuts are rich in Kaempferol, quercetin and β-sitosterol and is one of the well-known plant with rich medicinal values [23]. Soap nut is used for curing eczema, treating psoriasis and removing freckles. This herb is also used for removing lice from the scalp, since they have gentle insecticidal properties. The crushed seeds are widely used for making soaps and shampoos for their antibacterial, antifungal, stomachic and spermicidal activity. With a known potential for its medicinal properties the present study is carried out to purify and characterize this protease inhibitor from the seeds of soap nut for further applications.

The broad aim of this study is to isolate, purify and characterize protease inhibitor from Soap Nuts and computationally evaluate protease inhibitory action on larval mid gut proteases.

## Materials and methods
### Extraction and Purification of SNTIS
Sapindus trifoliatus trees bearing soap nuts were selected from Annavaram, East Godavari District, India. Ripe fruits are collected from selected trees and seeds are removed and preserved for extraction of protein. The endosperm was collected from the seeds after the removal of the hard coat and 25 g of the endosperm was homogenized with 200 ml of 0.1M sodium phosphate buffer, pH 7.6 and then made up to 250 ml with the same buffer. The extract was then centrifuged at 2500 rpm for 15 minutes at 4 °C and the supernatant was used for further steps. The supernatant obtained was treated with 50 % ice cold acetone (1:5 v/v) and the resultant mixture was centrifuged at 2500 rpm for 15 minutes at 4 °C. The precipitate was then re-suspended in 0.1M sodium phosphate buffer, pH 7.6.

The inhibition spectrum of SNTI was established by first assaying the protease or esterase activity of the

**Table 2** Summary of purification of soap nut seed protease inhibitor

| Preparation | Volume (ml) | Total protein (mg) | Total Trypsin inhibitory activity (Units) TIU0 $\times 10^3$ | Specific activity TIA $\times 10^2$ (Units/mg) protein | Yield % | Fold purification |
|---|---|---|---|---|---|---|
| Crude extract | 250 | 4030 | 63.10 | 0.156 | 100 | 1.00 |
| Acetone fractionated extract | 225 | 2644 | 54.81 | 0.207 | 86.60 | 1.33 |
| Heat treated extract | 200 | 2334 | 53.10 | 0.227 | 84.15 | 1.44 |
| 50 % Ammonium sulphate | 30 | 172.5 | 23.78 | 1.370 | 37.69 | 8.78' |
| CM – Cellulose | 27 | 112 | 14.74 | 1.440 | 23.36 | 9.23 |
| Sephadex G-100 | 12 | 52 | 13.2 | 2.54 | 20.92 | 16.28 |

Yield and fold purification were calculated on the basis of TIU and TIA respectively
*TIU* trypsin inhibitory Units, *TIA* trypsin inhibitory activity

enzyme on an appropriate substrate and then incubating a fixed amount of the enzyme with various amounts of the inhibitor and assaying the residual enzyme activity. The activities of trypsin and pronase or their inhibition were assayed by the method of Kakade et al., [24] using either BAPNA or casein as the substrate. The inhibitory activity towards chymotrypsin was determined using casein [25] or ATEE [26] as the substrate. The proteolytic activity of papain was assayed using casein as substrate by the method of Arnon [27]. The Esterolytic activity of subtilisin was assayedby using ATEE [28] as the substrate. Thermolysin was assayed according to the method of Matsubara [29]. The method of Saunders and Lang [30] was employed for assaying pancreatic α –amylase. Purification of the inhibitor is carried out by 50 % Ammonium Sulfate precipitation and dialyzed against PB (Phosphate Buffer).

Further purification is achieved by ion exchange and gel permeation chromatography. Purity is assessed by native PAGE.50 % Ammonium Sulfate precipitation: The supernatant obtained after acetone fractionation was subjected to Ammonium Sulfate precipitation. Solid Ammonium Sulfate was added gradually with constant stirring at 4 °C to obtain 50 % saturation. The mixture was allowed to stand overnight at 4 °C.

Dialysis: The precipitate from 50 % Ammonium Sulfate precipitation was collected by centrifugation at 2500 rpm for 15 minutes at 4 °C, then dissolved in 0.1 M sodium phosphate buffer pH 7.6 and dialyzed against the same buffer for 12 hours at 4oC. The dialysate obtained is subjected to column chromatography. The dialysate was further purified by column chromatography for separation of the inhibitor protein from the mixture of

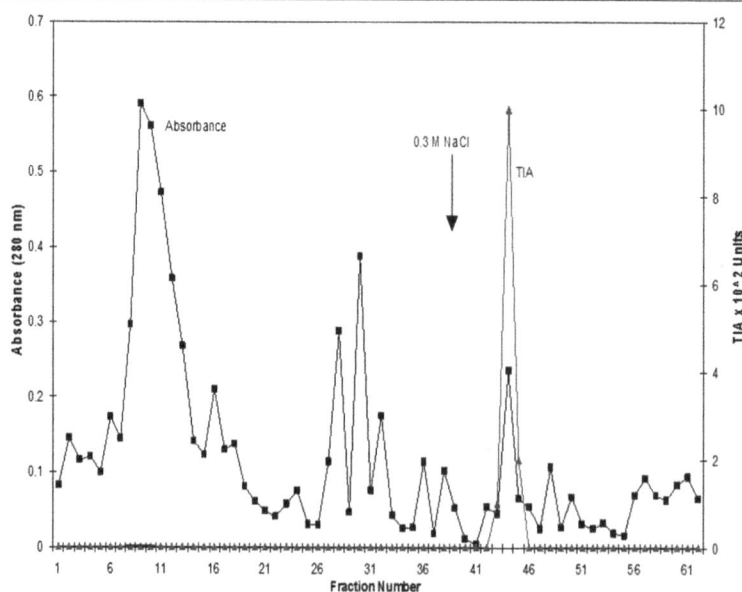

**Fig. 1** Ion exchange chromatography of SNTI on CM-Cellulose. One hundred seventy two milligram of the ammonium sulphate fractionated sample (0–50 %) was applied on to the column (2 × 80 cm) in 0.1 M sodium phosphate buffer (pH 5.8) and the adsorbed proteins were eluted with stepwise gradient in the buffer. Fractions of 5 ml were collected at a flow rate of 60 ml per hour. The protein was monitored by absorbance at 280 nm. * When the elution was done with a gradient of 0.1 to 1.0 M NaCl a single but broad peak was obtained (Results not shown. To obtain a sharp peak, the elution was performed using stepwise gradient

molecules based on charge and size using CM-cellulose and Sephadex G-100 columns.

Ion Exchange Chromatography: The dialyzed sample was loaded on a CM-Cellulose column (2×30cm) previously equilibrated with 0.1M sodium phosphate buffer pH 7.6. After washing with the equilibration buffer, stepwise elution was performed with increasing concentrations of 0.1M, 0.2M, 0.3M, 0.4M and 1.0 M NaCl and the respected fractions were collected. These fractions were monitored for protein by measuring their absorbance at 280 nm and fractions of each peak are pooled. Each pooled fraction samples were tested for the inhibitory activity against trypsin.

Gel Filtration Chromatography: The method of Andrews [31] was used to determine the molecular weight of the inhibitor by molecular sieve chromatography on Sephadex G – 100. Sephadex G-100 was swollen in 0.1M Phosphate buffer, pH 7.6 and packed in a column (2×30cm). The pooled fraction exhibiting inhibitory activity was loaded on Sephadex G- 100. The column was equilibrated and developed with the same buffer. The fractions were collected and the protein was monitored by measuring the absorbance at 280 nm". The fractions from a single peak were pooled, dialyzed against phosphate buffer at 4 °C and lyophilized.

High Performance Liquid Chromatography (HPLC): The column fraction with SNTI activity was then separated by reverse-phase HPLC, as described by Macedo et al., [19], on a C18 column(Shimadzu) that was previously equilibrated with water and >5 % acetonitrile. The SNTI active fraction was finally purified by rechromatography in a reverse phase HPLC with a flow rate of 1.0 ml/min for 35 min by isocratic elution. Proteins were monitored absorbance at 280 nm.

Polyacrylamide gel electrophoresis was carried out as described by Reisfield et al., [32] and followed by Gabriel [33]. PAGE was carried out under non-denaturing condition using 12 % slab gels. About 50 µg of protein was layered on the gel. After the electrophoretic run, proteins were fixed in 10 % TCA. Proteins were visualized using coomassie brilliant blue according to the method of Fairbanks [34].

## Characterization of SNTI

Molecular weight of the enzyme inhibitor: SDS Polyacrylamide gel electrophoresis was carried out using Phosphorylase-b (97.4 kDa), bovine serum albumin (66 kDa), ovalbumin (43 kDa), carbonic anhydrase (29 kDa), trypsin inhibitor (22.1 kDa) and lysozyme (14.3 kDa) as standard proteins for calibration. The molecular weight of the inhibitor was determined using the calibration curve.

Effect of pH on the stability of the inhibitor: Solutions of the inhibitor (1mg ml-1) in 10 mM buffers of five pH values (pH 3.0, glycine-HCl; 5.0, sodium citrate; 7.0, sodium phosphate; 9.0, Tris-HCl; 12.0, glycine-NaOH) were kept at 50C for 24h. Aliquots of the inhibitor were diluted with 0.1M phosphate buffer, pH 7.6 and assayed for trypsin inhibitory activity (TIA).

Effect of temperature on the stability of the inhibitor: The inhibitor solutions (100µg ml-1) were separately incubated in a water bath at various temperatures for 10 min and then quickly cooled in ice and appropriate aliquots were assayed for trypsin inhibitory activity (TIA).

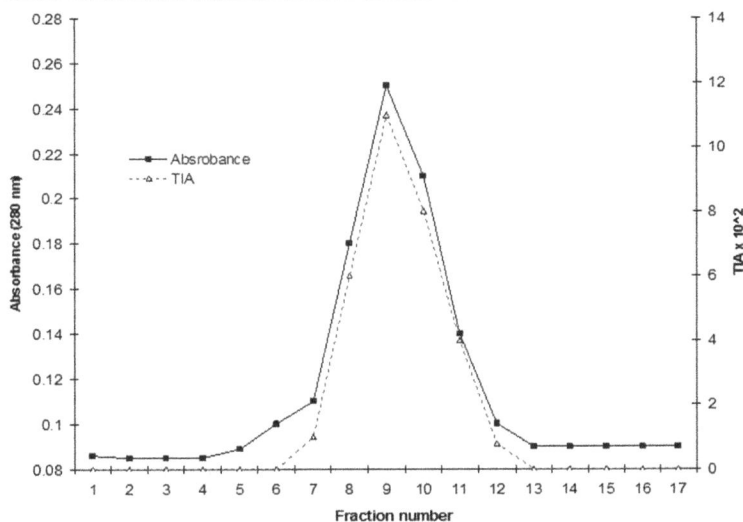

**Fig. 2** Gel filtration of SNTI on Sephadex G100. One hundred ten milligram of lyophilized preparation was applied to the Sephadex G-100 column 2 × 80 cm in 0.1 M phosphate buffer pH 7.6 and eluted with the same buffer. 2 ml fractions were collected at a flow rate of 12 ml per hour. The protein was monitored at 280 nm. Protease inhibitory activity was followed using BAPNA as the substrate

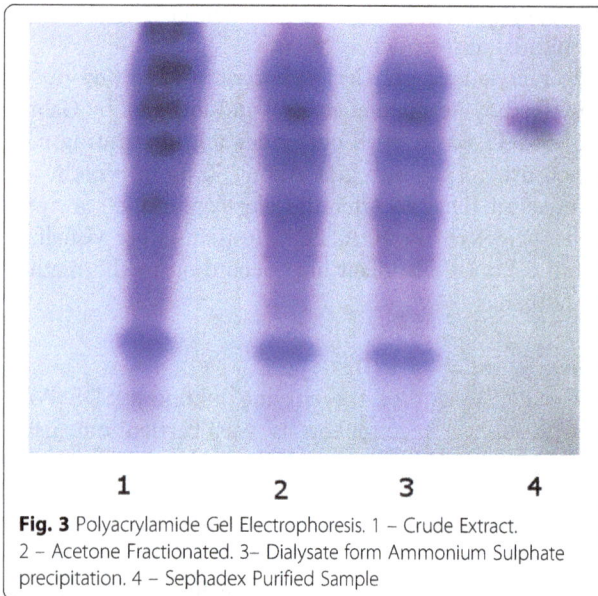

**Fig. 3** Polyacrylamide Gel Electrophoresis. 1 – Crude Extract. 2 – Acetone Fractionated. 3– Dialysate form Ammonium Sulphate precipitation. 4 – Sephadex Purified Sample

Kinetic measurements: The amidolytic activity of trypsin (50 µg) was determined with various concentrations of BAPNA (1.2to5.0µmol) in the absence of the inhibitor. The assays were then repeated in the presence of 5 and 15 µg of the inhibitor. The Ki values were calculated from Lineweaver-Burk plots.

Competition experiments: 50 µg trypsin was separately incubated with 5, 10, 15 and 20 µg of SNTI for 10 minutes at 37oC in 1ml phosphate buffer, pH7.6. Suitable aliquots of all the samples were taken and assayed for residual trypsin activity using BAPNA as a substrate.

Studies on complex formation: The trypsin-SNTI complex was isolated by gel filtration on Sephadex G-100. To form the complex, 2 mg of SNTI was incubated with 5 mg of trypsin at 37oC for 15 minutes". Excess trypsin was used to make sure that all the inhibitor is complexed, such a mixture was applied onto a column of Sephadex G-100 at 5oC previously calibrated with SNTI and trypsin run separately and the absorbance was monitored at 280 nm. Trypsin and Trypsin inhibitory activities are monitored in the fractions collected.

Fourier Transform Infra-Red Spectroscopy (FTIR): The solid state FTIR spectra are recorded in the middle infrared (4000 cm-1to400 cm-1) on Perkin Elmer. The sample for FTIR analysis are prepared by grinding the dry blended powders of trypsin inhibitor with powdered KBr and then compressed to form discs.

Database and Sequence information: To demonstrate the inhibitory activity of SNTI on mid gut proteases, computational approach has been applied where in the proteases from larval guts of two insects viz., H. armigera and S. frugiperda were considered as these organisms commonly cause damage to agricultural fields. The sequence information for SNTI was taken from MALDI-TOF analysis done by Rachel et al., (2012) [35] and the gut proteases of H. armigera and S. frugiperda were retrieved from NCBI protein data base [36].

Homology modeling of SNTI and Threading based modeling of insect gut proteases: Primarily all the three sequences are subjected to PDB BLAST for template identification. The templates are identified based on homology and the homologous template was used for

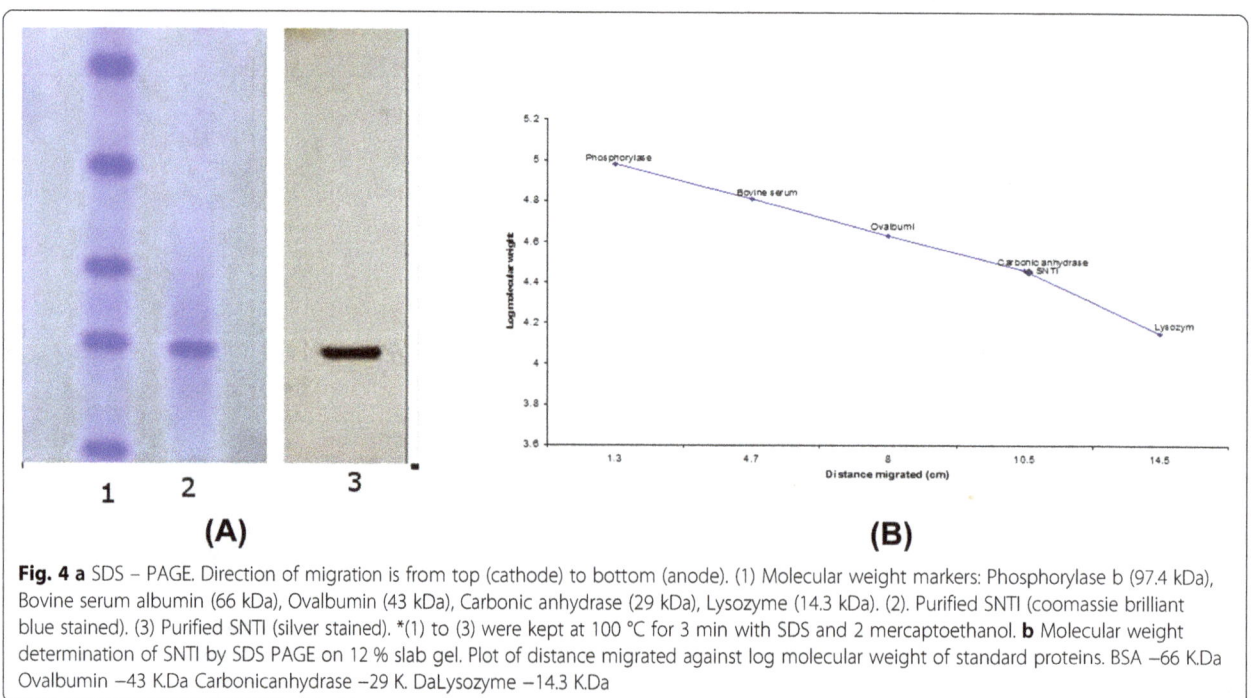

**Fig. 4 a** SDS – PAGE. Direction of migration is from top (cathode) to bottom (anode). (1) Molecular weight markers: Phosphorylase b (97.4 kDa), Bovine serum albumin (66 kDa), Ovalbumin (43 kDa), Carbonic anhydrase (29 kDa), Lysozyme (14.3 kDa). (2). Purified SNTI (coomassie brilliant blue stained). (3) Purified SNTI (silver stained). *(1) to (3) were kept at 100 °C for 3 min with SDS and 2 mercaptoethanol. **b** Molecular weight determination of SNTI by SDS PAGE on 12 % slab gel. Plot of distance migrated against log molecular weight of standard proteins. BSA –66 K.Da Ovalbumin –43 K.Da Carbonicanhydrase –29 K. DaLysozyme –14.3 K.Da

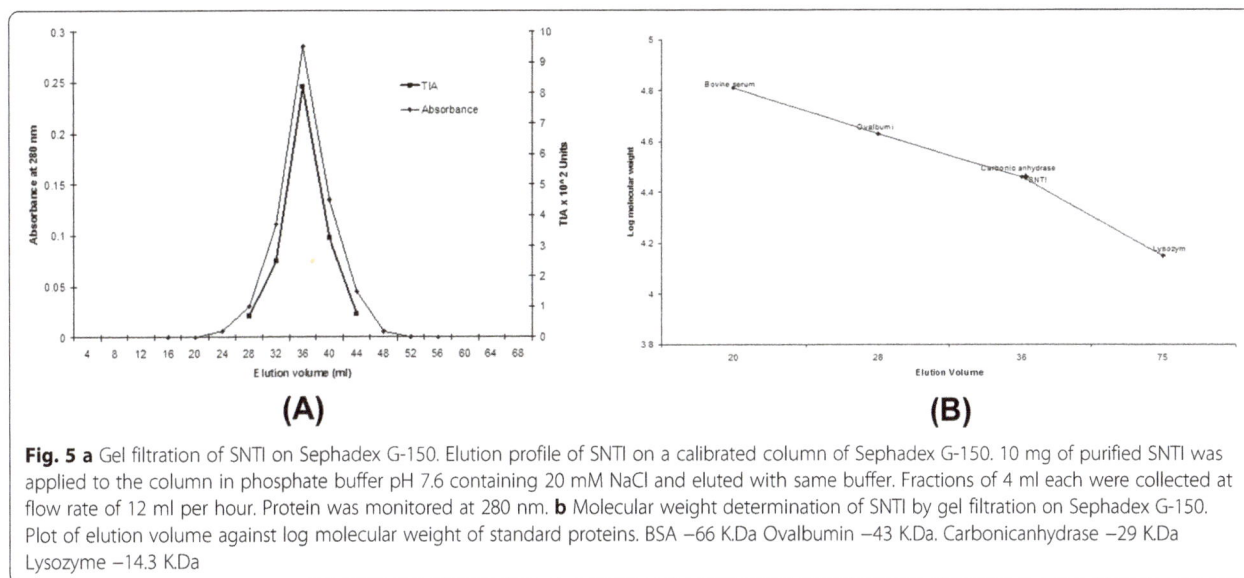

**Fig. 5 a** Gel filtration of SNTI on Sephadex G-150. Elution profile of SNTI on a calibrated column of Sephadex G-150. 10 mg of purified SNTI was applied to the column in phosphate buffer pH 7.6 containing 20 mM NaCl and eluted with same buffer. Fractions of 4 ml each were collected at flow rate of 12 ml per hour. Protein was monitored at 280 nm. **b** Molecular weight determination of SNTI by gel filtration on Sephadex G-150. Plot of elution volume against log molecular weight of standard proteins. BSA −66 K.Da Ovalbumin −43 K.Da. Carbonicanhydrase −29 K.Da Lysozyme −14.3 K.Da

homology modeling of SNTI sequence. For the insect gut proteases no homologous templates were available, hence they are subjected for modeling using threading approach. In threading approach folds (secondary structure) of the protein are considered and templates are identified from fold library. Based on the folds of template identified the target protein is modeled. Prime module from Schrodinger suite (Schrodinger 2011) was used for modeling proteins by Homology and Threading approaches.

Structure Validation: The predicted structures are subjected for validation to ERRAT and PROCHECK servers. The validations by PROCHECK were done based on the stereo chemical quality, hydrogen bonding energy and torsion angles [37]. Based on the interaction of atoms with respect to amino acid residues ERRAT validates the predicted protein structure by separating correct and incorrect determined structures [38].

Binding Site prediction: The active site of all the three modeled proteins were predicted using SiteMap [39]. SiteMap determines primary binding site on a receptor by calculating the sites on protein surface by searching the grid points called site points. Then the contour site maps are generated, producing hydrophobic and hydrophilic maps.

Protein-Protein Docking: To understand role of SNTI in inhibiting proteases, modeled SNTI and gut proteases were subjected for protein-protein docking using Piper

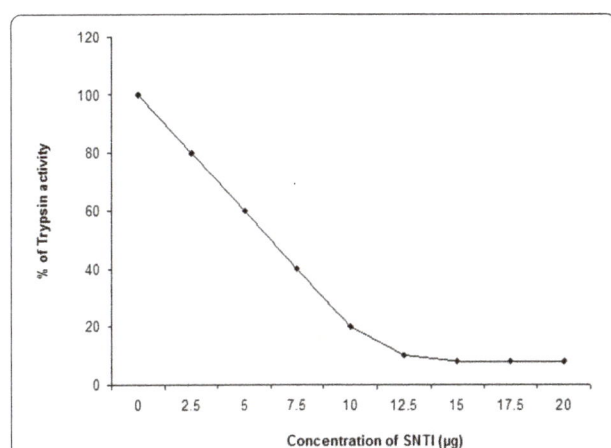

**Fig. 6** Activity of SNTI towards Bovine Trypsin. Thirty microgram of trypsin was incubated with varying amounts of SNTI for 10 min at 37 °C. The percentage residual enzyme activity was assayed using BAPNA as the substrate. The concentration of the inhibitor required to cause 50 % inhibition of the enzyme activity was determined from the graph

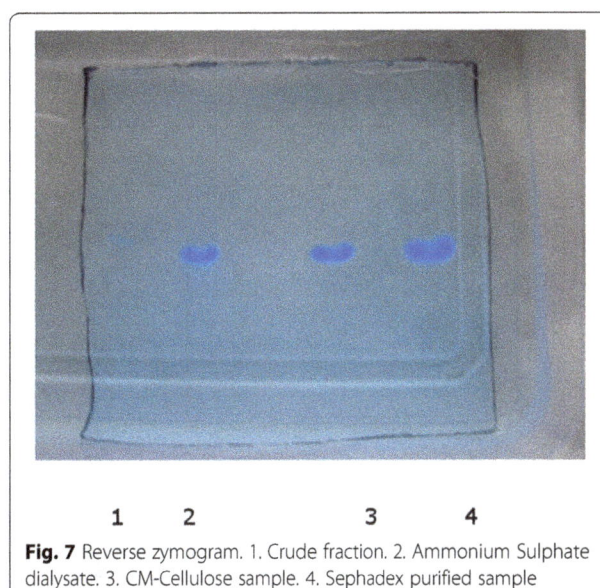

**Fig. 7** Reverse zymogram. 1. Crude fraction. 2. Ammonium Sulphate dialysate. 3. CM-Cellulose sample. 4. Sephadex purified sample

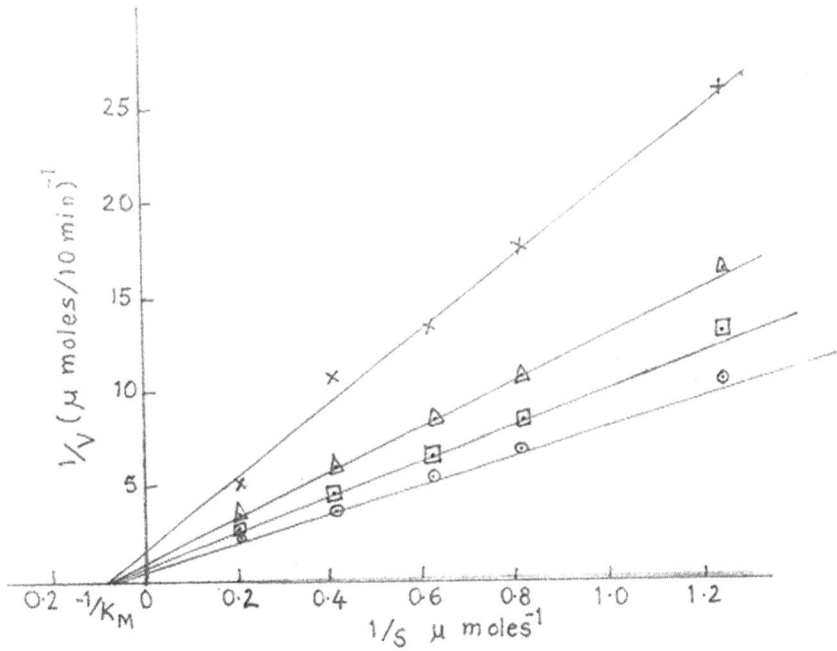

(-0-0- )  without SNTI,  (-Î--Î--)    2.5 μg SNTI,  (-Δ-Δ-) 5.0 μg SNTI, (-X-X-) 7.5μg SNTI

**Fig. 8** Mode of inhibition of trypsin by SNTI Line weaver –Burk plot. Inhibition of amidolytic activity of trypsin by SNTI was done by incubating 30 μg of trypsin and BAPNA solution (0.8 to 5.0 μ mole) with reaction system containing 2.5 to 7.5 μg of SNTI

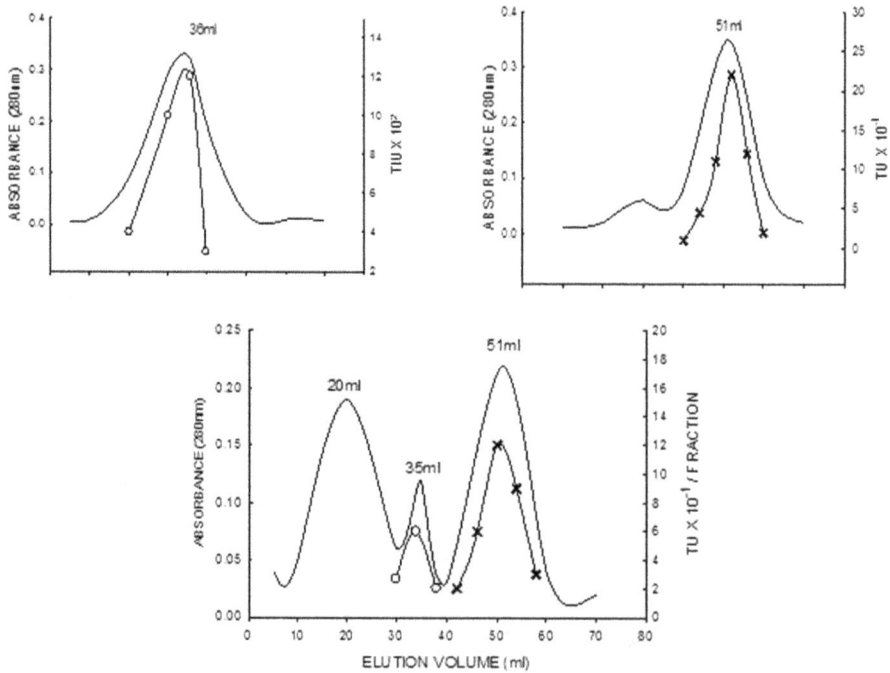

**Fig. 9** Elution patterns of SNTI, Trypsin and Trypsin-SNTI complex on Sephadex G-100 column. Elution patterns of trypsin and trypsin – SNTI complex on Sephadex G – 100. Protein was monitored at 280 nm

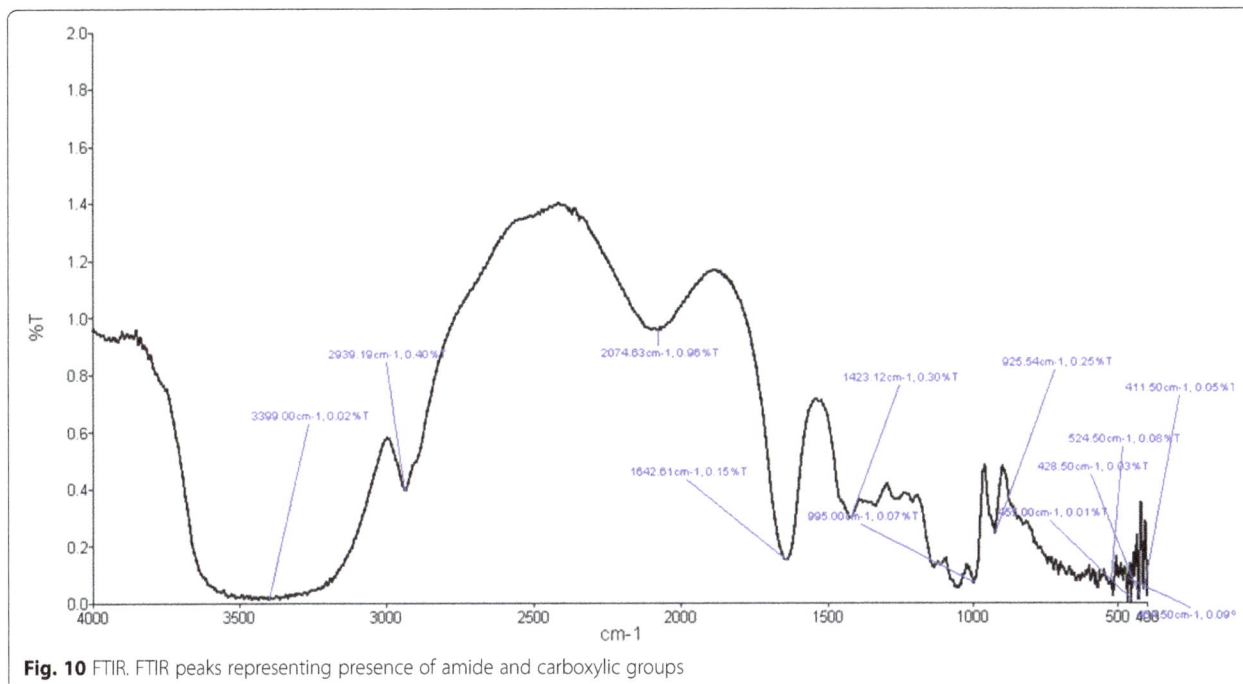

**Fig. 10** FTIR. FTIR peaks representing presence of amide and carboxylic groups

[40]. Prior to docking the proteins are subjected for protein preparation to optimize the molecule using PrepWizard. For protein-protein docking gut proteases were set as ligands and docked separately with receptor SNTI. Number of ligand rotation to probe were set for 10000 rotation and for each dock ten poses were retrieved. After docking, the best pose was selected and then these complex structures were again set for optimization by PrepWizard. These prepared complexes are set for analysis of protein-protein interaction using protein interaction option in Bioluminate [41].

## Results

### Extraction and Purification

of SNTI The Soap Nut Trypsin Inhibitor was isolated and purified from soap nut seeds (Sapindus trifoliatus L.) according to the procedure adopted by Annapurna and Siva Prasad [42] and the results are shown in the Table- 2.

The extraction procedure was carried out maintaining physiological conditions and ice cold acetone was used

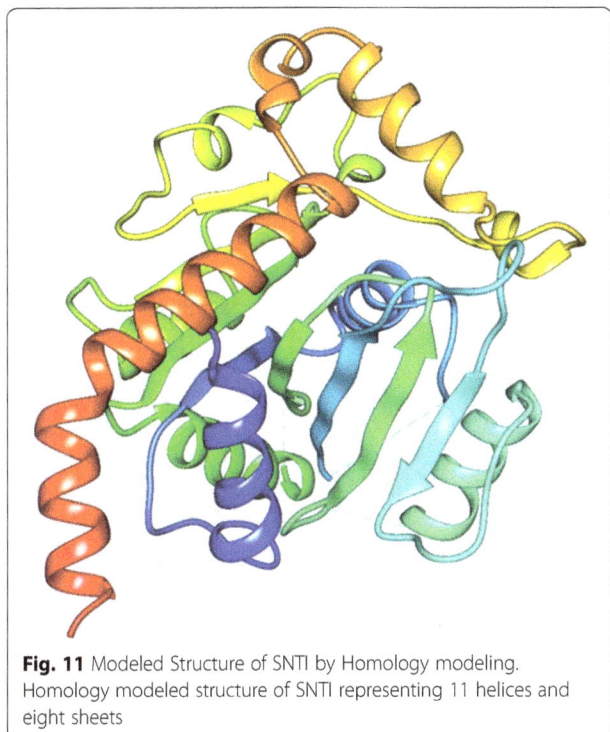

**Fig. 11** Modeled Structure of SNTI by Homology modeling. Homology modeled structure of SNTI representing 11 helices and eight sheets

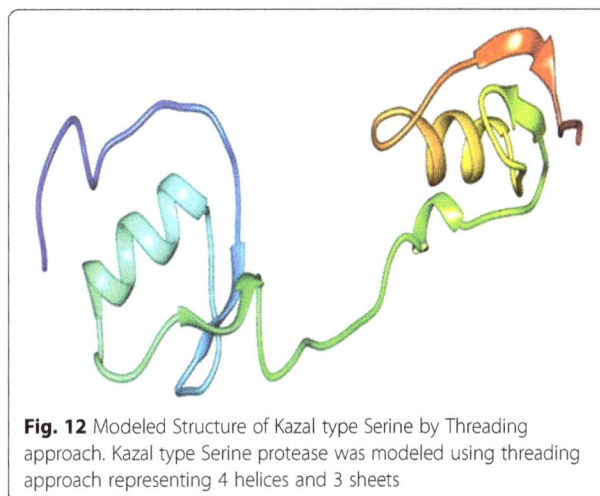

**Fig. 12** Modeled Structure of Kazal type Serine by Threading approach. Kazal type Serine protease was modeled using threading approach representing 4 helices and 3 sheets

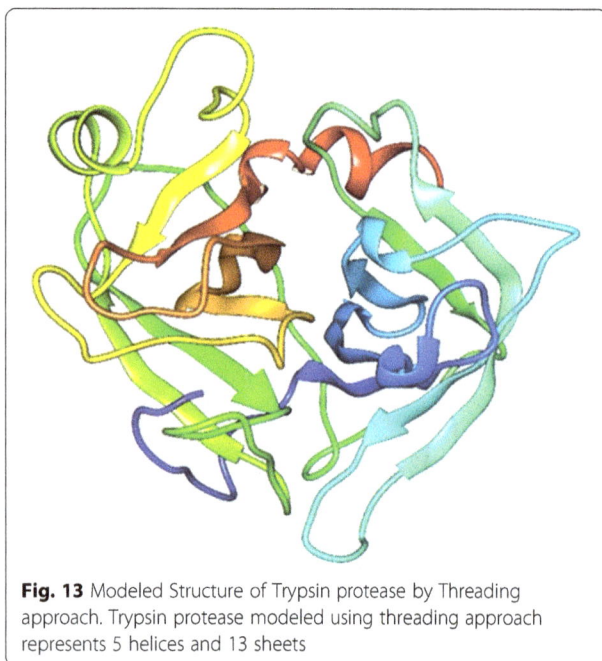

**Fig. 13** Modeled Structure of Trypsin protease by Threading approach. Trypsin protease modeled using threading approach represents 5 helices and 13 sheets

to remove lipids. The endosperm was collected from the seeds after the removal of the hard seed coat and 25 g of the endosperm was homogenized with 200 ml of 0.1M sodium phosphate buffer, pH 7.6 and then made up to 250 ml with the same buffer. The extract was then centrifuged at 2500 rpm for 15 minutes at 4 °C and the supernatant (230 ml) was used in further steps.

The supernatant (230 ml) was treated with 50 % ice cold acetone (1:5 V) and the resultant mixture was centrifuged at 2500 rpm for 15 minutes at 4 °C to remove

lipids. The resultant defatted solution was subjected to ammonium sulphate precipitation.

To the supernatant (200 ml) from acetone fractionation, solid ammonium sulphate (62.6 g) was added gradually with constant stirring at 4 °C to obtain 50 % saturation. The mixture was allowed to stand overnight at 4 °C. The precipitate was collected by centrifugation at 2500 rpm for 15 minutes at 4 °C, then dissolved in 30 ml of 0.1 M sodium phosphate buffer pH 7.6 and dialyzed against the same buffer.

Proteins have numerous functional groups that can have both positive and negative charges. Ion exchange chromatography separates proteins with regards to their net charge. If a protein has a net positive charge at pH 7, then it will bind to a column of negatively charged beads, whereas a negatively charged protein would not. By changing the pH so that the net charge on the protein is negative, it too will be eluted.

The dialyzed sample (172 mg) was loaded on a CM-Cellulose column (2×80cm) previously equilibrated with 0.1M sodium phosphate buffer pH 7.6. After washing with 250 ml of the equilibration buffer, the following stepwise elution was performed with 200 ml each of 0.1M, 0.2M, 0.3M, 0.4M and 1.0 M NaCl in 0.1 M phosphate buffer pH 7.6. Fractions of 5 ml were collected at a flow rate of 60 ml per hour. These fractions were assayed for protein by measuring their absorbance at 280 nm as well as the inhibitory activity against trypsin using BAPNA as the substrate. The elution profile of CM-Cellulose chromatography for the inhibitor is shown in Fig. - 1. The fractions containing trypsin inhibitory activity (fractions 42-48) were pooled, dialyzed against distilled water at 4 °C

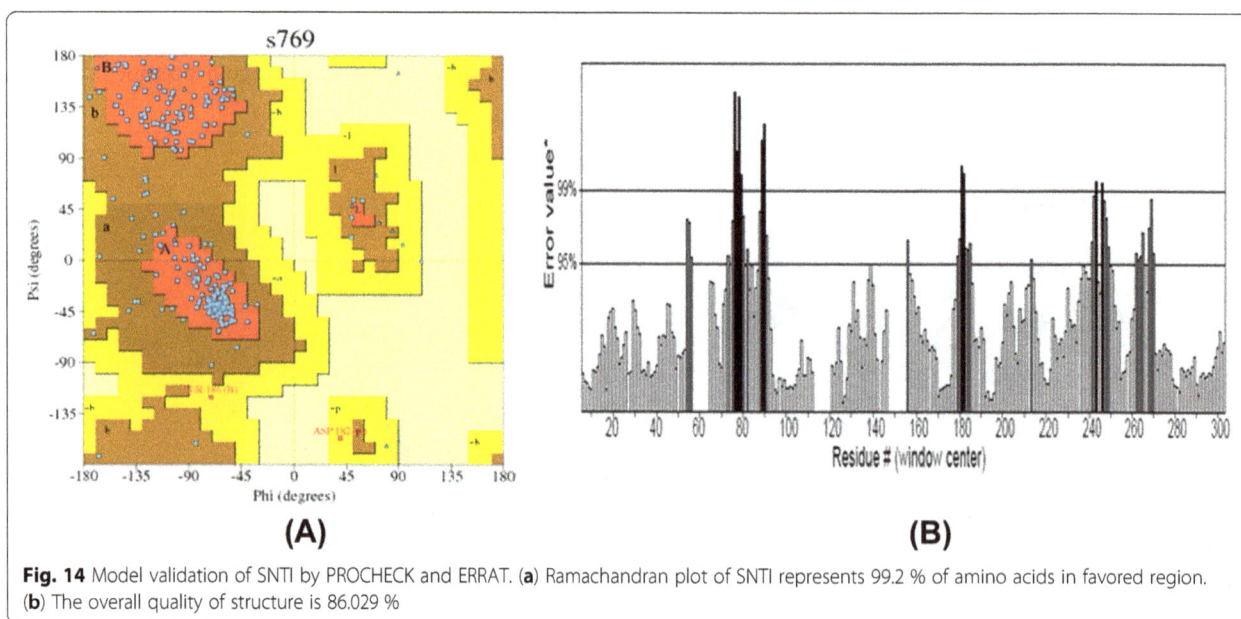

**(A)**                                                                   **(B)**

**Fig. 14** Model validation of SNTI by PROCHECK and ERRAT. (**a**) Ramachandran plot of SNTI represents 99.2 % of amino acids in favored region. (**b**) The overall quality of structure is 86.029 %

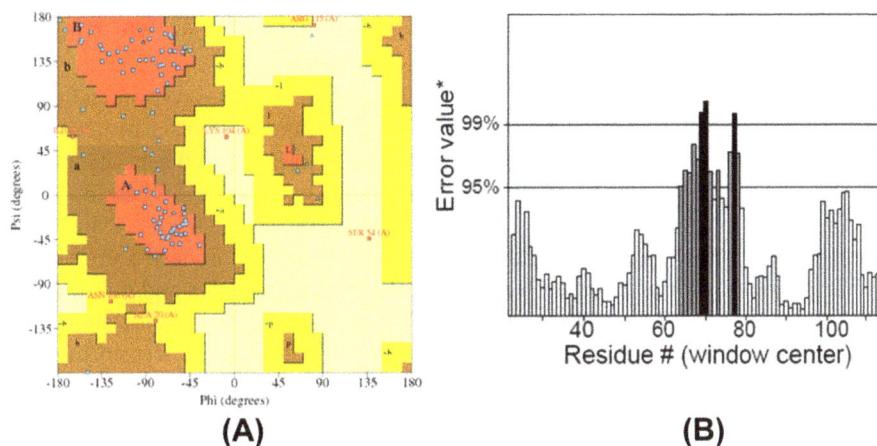

**Fig. 15** Model validation of Kazal type Serine protease by PROCHECK and ERRAT. (**a**) Ramachandran plot of Kazal type Serine protease represents 92.5 % of amino acids in favored regions. (**b**) The overall quality of structure is 86.96 %

and lyophilized. The protein yield from ion exchange chromatography was 112 mg.

The sample from ion exchange chromatography (110 mg) was dissolved in 0.1 M phosphate buffer pH 7.6 and was loaded on Sephadex G-100 column (1.8 × 30 cm) which was previously equilibrated with 0.1 M phosphate buffer, pH 7.6. The inhibitor was eluted with the same buffer. 2 ml fractions were collected at a flow rate of 12 ml per hour and the protein was monitored by measuring the absorbance at 280 nm. The trypsin inhibitory activity of the fractions was assayed using BAPNA as the substrate.

The elution profile of the gel permeation chromatography is shown in Fig. - 2. A single protein peak with corresponding trypsin inhibitory activity was observed. The fractions (8 - 12) containing the trypsin inhibitory activity were pooled, dialyzed against distilled water at 4 °C and lyophilized. The yield of protein after gel permeation chromatography was 52 mg. This preparation was stored at 0 °C. The preparation thus stored, showed full activity even after three months. By this procedure about 52 mg of the inhibitor was obtained and the final yield was about 20.9 %.

SNTI was analyzed using reverse phase HPLC to confirm its purity. HPLC analysis revealed a single peak (result not shown). The methodological procedure resulted in high purification with a 20.92 % yield.

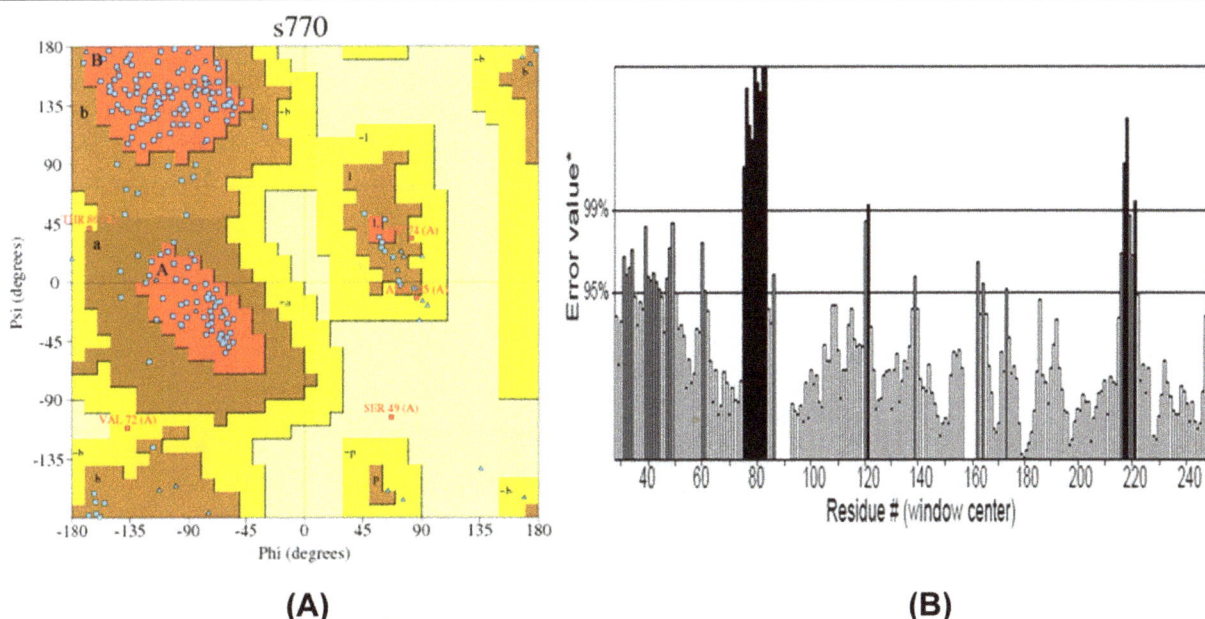

**Fig. 16** Model validation of Trypsin protease by PROCHECK and ERRAT. (**a**) Ramachandran plot of Trypsin protease represents 97.5 % of amino acids in favored regions. (**b**) The overall quality of structure is 81.04 %

**Table 3** Binding site surfaces predicted by SiteMap

| Protein | Residues present in hydrophobic surface |
|---|---|
| SNTI | Asp 2, Lys 10, Val 14, Thr 79, Lys 299, Asn 90, |
| Kazal type Serine | Leu 70, Pro 71, Gly 75, Met 80 |
| Trypsin | Gly 73, Ala 76,Arg 80, Gly 160, |

Amino acids present in the hydrophobic binding surface of SNTI, Kazal type Serine and Trypsin

A sharp band was obtained on 12 % slab gel at pH 8.3 signifying the homogeneity of the purified SNTI (Fig. - 3). SNTI did not respond to PAS (Periodic Acid Schiff's) stain suggesting it to be a non-glycoprotein.

## Characterization of SNTI

Figure – 4a shows the protein band pattern of the inhibitor on 12 % SDS slab gels when stained with coomassie brilliant blue. Silver staining of SNTI showed a sharper band on SDS-PAGE. From the plot of distance migrated in cm versus log molecular weight for standard proteins (Fig. – 4b), the inhibitor showed a molecular weight of 29 kDa. When subjected to Gel filtration on Sephadex G-150, SNTI eluted out as a single protein with a corresponding activity peak (Fig. – 5a). The plot of elution volume versus log molecular weight of the calibrating proteins is shown in Fig. – 5b. The molecular weight of SNTI calculated from the plot was 28.5 kDa.

Enzyme inhibition studies were carried out to identify the specificity of the inhibitor towards the mechanistic classes of proteases. SNTI was tested for its inhibiting capacity against bovine trypsin using both BAPNA and casein as the substrates. The inhibition patterns of the amidolytic activity of bovine trypsin by SNTI was linear up to 80 % inhibition (Fig. - 6). On extrapolation, it was found that 12 μg of the inhibitor can totally inhibit amidase activity of 30 μg of trypsin.

The activity of the SNTI against chymotrypsin, elastase and pronase (Streptomyces griseus protease)

subtilisin, papain, pepsin, thermolysin and α-amylase was tested. Except pronase, the rest of the enzymes were not affected by SNTI.

The serine proteases trypsin and pronase were inhibited by SNTI. Majority of plant protease inhibitors isolated so far have been found to be specific for serine proteases and there are some reports of these inhibitors inhibiting other classes of proteases. SNTI specifically inhibited serine proteases trypsin and pronase and it has no effect on thiol, acidic, metalloproteases and α-amylase.

Reverse zymography: Substrate containing SDS- PAGE enables visualization of trypsin inhibitor. The inhibitory activity produced by SNTI detected using trypsin and gelatin substrate in the gel is shown in Fig. - 7. SNTI showed a single inhibitory band specific to trypsin and subjected to electrophoresis.

Mode of inhibition of Trypsin: Trypsin activity in the presence and absence of SNTI was measured at different substrate concentrations. The double reciprocal plot of the kinetic data is shown in Fig. - 8. In the presence of inhibitor, there was a decrease in the Vmax and the curves met on the X –axis at a point equivalent to -1/km. The mode of inhibition of trypsin by SNTI was non-competitive. The Ki value of trypsin for SNTI calculated from Dixon plot was $0.75 + 0.05 \times 10\text{-}10$ M.

Complex Studies: SNTI was treated with excess trypsin and the mixture was pre-incubated at 37oC for 15 minutes. This mixture when applied onto a column of Sephadex G-150 at 5oC gave rise to two distinct at 280nm peaks (Fig. - 9). Peak-I had an elution volume of 20 ml which is higher than free SNTI 35 ml. The binary complex of trypsin - SNTI did not show any trypsin activity or trypsin inhibitory activity.

The molecular weight calculated for trypsin – SNTI complex on Peak-I based on the calibration curve for standard proteins (Fig. – 4b) gave a value of 68.9 kDa. This would mean a mole/mole interaction of SNTI with

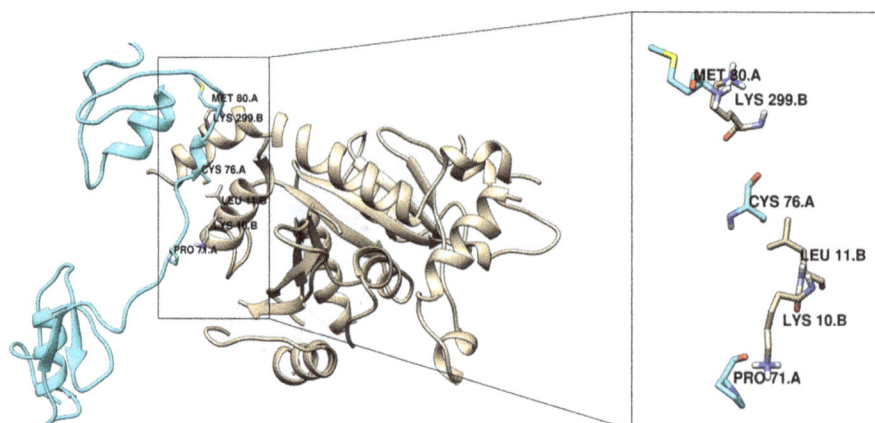

**Fig. 17** Protein-Protein interactions of SNTI with Kazal type Serine protease from *H. armigera*. Cyan color residues represents Kazal type Serine and Tan color residues represents SNTI

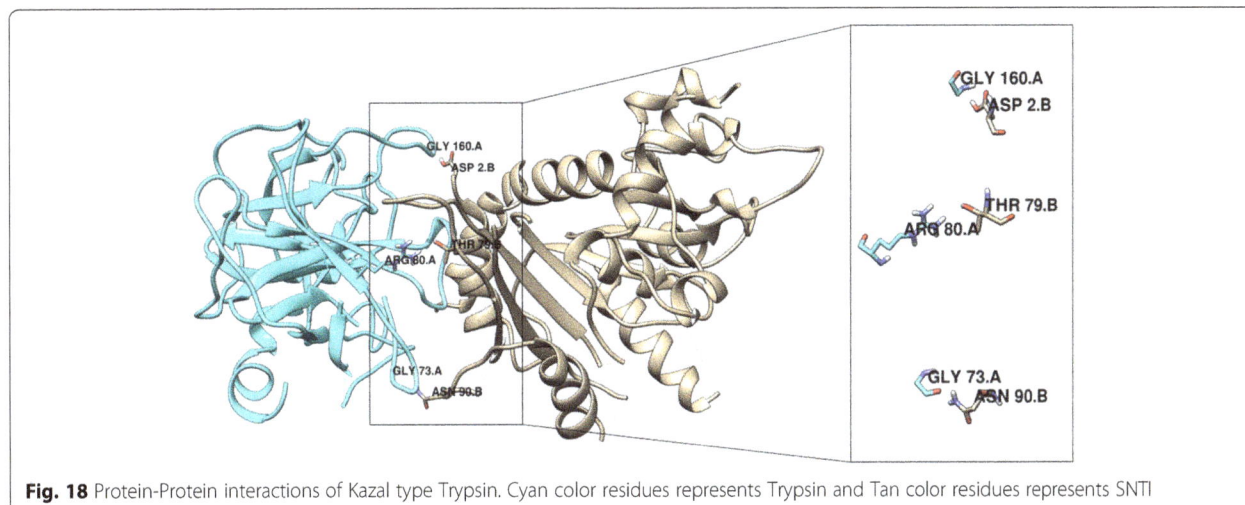

**Fig. 18** Protein-Protein interactions of Kazal type Trypsin. Cyan color residues represents Trypsin and Tan color residues represents SNTI

trypsin. Peak-II was small and represented uncomplexed SNTI with corresponding trypsin inhibitory activity. The trypsin left over after the enzyme inhibitor complex formation, was eluted out as peak-III with a corresponding elution volume of 51 ml.

Fourier Transform Infra-Red Spectroscopy (FTIR): IR spectroscopic studies elucidates functional groups in a molecule. The IR peak at 3399 (broad) and 2939 cm-1can be assigned to OH of carboxylic group and asymmetric CH3 stretching. The over ton peak can be observed at 2074 cm-1. The peak at 1642 and 1423 cm-1 could be due to amide C=O (CONH2) and CH3 bending vibrations. The peaks observed at 995 and 925 could be attributed to OH bending vibrations. The presence of amide and carboxylic groups are confirmed by the above peaks (Fig. - 10).

Database and Sequence information: Protein sequences of gut proteases of H. armigera and S. frugiperda were retrieved from protein NCBI database bearing the accession number AHX25877.1 (Kazal-type serine protease from H. armigera) and ACR25157.1 (Trypsin protease from S. frugiperda).

Homology modeling of SNTI and Threading based modeling of insect gut proteases:

**Table 4** Protein-protein interaction analysis of SNTI with Kazal type Serine protease and SNTI with Trypsin

| Protein-protein docking | Residues interacted | Distance in Å | Type of interaction |
|---|---|---|---|
| SNTI - Kazal type Serine protease | Lys 10 – Pro 71 | 3.1 | Hydrogen Binding |
| | Lys 299 – Met 80 | 3.1 | Hydrogen Binding |
| | Lue 11 – Cys 76 | 3.8 | Van Der Waals |
| SNTI – Trypsin protease | Thr 79 – Arg 80 | 3.1 | Hydrogen Binding |
| | Asn 90 – Gly 73 | 3.0 | Hydrogen Binding |
| | Asp 2 – Gly 160 | 2.9 | Hydrogen Binding |

The protein structures of all the three were modeled using Prime module from Schrodinger Suite. PDB BLAST provided a template 2C1X_A (UDP-Glucose Flavonoid 3-O Glycosyltransferase) with 42 % identities, 55 % positives and score of 192.6 for SNTI. Secondary structure of target SNTI sequence was identified using run SSP. After secondary structures for target are identified, template and target sequences are aligned and then the structure of SNTI is modelled based on the template 2C1X_A and the structure represents 11 helices and 8 beta sheets (Fig. - 11).

The target proteins Kazal type serine protease from H. armigera and trypsin protease from S. frugiperda are subjected to BLAST search for the identification of homologous template. Template structures with very low identity were retrieved so, instead of homology modeling threading or fold recognition approach was further used to model these proteases. In threading first the secondary structure of target protein sequences were predicted using run SSP option. Based on these secondary structures template is identified from the fold library and the best templates identified were crystal structure of insect derived kazal complex of serine protease (1TBQ) and crystal structure of a non-psychrophilic trypsin (1A0J) respectively.

These templates were further used for modeling Kazal type Serine and Trypsin proteases by homology modeling approach. Now, again the same first step is repeated but instead of finding the homologs, the template structure predicted by threading is used and the sequence of template and targets are aligned. Finally the model was built for Kazal type serine protease from H. armigera and was found to have 4 helices and 3 beta sheet (Fig. - 12). Similarly, Trypsin from S. frugiperda has 5 helices and 3 beta sheets (Fig. - 13). The 3D structure obtained is then validated using PROCHECK and ERRAT (Fig. - 14a, 14b). The protein structure that is modeled is satisfactory as evidenced by the validation tools. Ramachandran plot derived from PROCHECK analysis represents about 99.2 % of amino acids residues of SNTI are in favored region (Fig. - 14a) and ERRAT validates the overall structure quality to be 86.029 % (Figure - 14b). About 92.5 % of amino acids residues falling in favored regions for Kazal type Serine protease (Figure - 15a) and 97.5 % for trypsin protease (Fig.- 16a). ERRAT validates the overall structure quality of Kazal type Serine protease to be 86.96 % (Fig.- 15b) and 81.04 % for trypsin protease (Fig. - 16b).

Binding Site prediction: The predicted structures were subjected to SiteMap for binding site identification. The hydrophobic binding sites predicted by SiteMap on the surface of SNTI and gut proteases are shown in the Table - 3.

Protein-Protein Docking: By following these combinations of SNTI x Kazal type Serine protease and SNTI x Trypsin protease was performed using PIPER. All these proteins prior to docking were prepared, optimized and energy minimized. From the resultant set of 10 poses the hydrogen binding interactions and other interactions were identified (Figs. – 17, 18). They are further checked whether these interactions are present in predicted binding sites. Table – 4 represents interacting residues of SNTI x Kazal type Serine protease complex and SNTI x Trypsin protease that are involved in binding site regions.

## Discussion

The observation that the trypsin inhibitory activity in the crude extracts of these seeds is stable at 700C for 10 min has led to the use of this treatment as the first step in the purification of the inhibitor. This step also helps in the removal of the endogenous proteolytic activity present in the seed extracts [32]. About 50 % of the proteins present in the crude extract were removed by this step. When the ammonium sulphate fraction was subjected to CM-cellulose column chromatography, protease inhibitory activity was found to be associated with the protein eluted from the column by 0.3M NaCl. SNTI has been found to be homogeneous by the criteria of native PAGE and gel filtration. HPLC analysis revealed a single peak which represents the purity of the compound.

Protease inhibitors have been separated by several gel electrophoretic methods. Anionic inhibitors have been examined by the Davis method [43], cationic inhibitors by the Reisfeld method [32] and neutral inhibitors by Weber and Osborn [44] and Laemmli [45] methods. While these methods have proved useful for establishing the purity and determining the molecular weight of proteins, they cannot be used to distinguish isoinhibitors or compare the activity of the particular inhibitor against different proteolytic enzymes. SNTI in all the fractions during its purification showed a single inhibitory band

specific to trypsin on gelatin PAGE. A single inhibitory band also signifies the absence of isoforms which are common in many sources [46, 47, 48].

The molecular weight of SNTI as determined by SDS-PAGE was 29 kDa. This is close to the value 28.5 kDa obtained for the inhibitor by gel filtration on Sephadex G-100. Some protease inhibitors have exhibited anomalous behavior on Sephadex gel columns due to the existence of oligomeric forms of inhibitors arising from monomer–dimer – trimer equilibrium [49, 50, 51] or due to the presence of carbohydrate moieties. The subunit nature of SNTI has been analyzed by the SDS – PAGE technique. The inhibitor showed a single sharp band on SDS – PAGE when stained with silver supporting the monomeric nature of the protein. Further SNTI did not give positive result with PAS stain suggesting it free from carbohydrate moieties. Most of the trypsin inhibitors are non-glycoproteins. Papain inhibitor from potato tubers is a glycoprotein with a molecular weight of 80 kDa [52]. Shakuntala [53] first identified a glycoprotein trypsin inhibitor from Jack fruit seeds and to possess lectin activity.

The unusual stability of protease inhibitors, in general, is their most remarkable property. SNTI showed similarities to other protease inhibitors from soy bean [54] in their stability. The low cysteine content [35] in these inhibitors negates the possibility of the stability of the inhibitors rendered due to extensive intra-peptide cross - linking. However, the unusual stability of the inhibitor may be due to strong hydrophobic interactions forming an inner core in the protein.

The result of the investigation of the inhibitory specificity of SNTI has shown it to be a serpin and is strongly active against bovine trypsin and porcine elastase. SNTI was ineffective against other proteases such as papain (thiol), pepsin (carboxyl) and thermolysin (metallo). Majority of plant protease inhibitors isolated so far have been found to be specific for serine proteases [55]. However, there are reports of plant protease inhibitors inhibiting other classes of proteases. The trypsin/ chymotrypsin inhibitor from broad beans inhibits the sulphydryl enzyme papain [56]. Serine protease inhibitors such as barley subtilisin inhibitor [57] and wheat germ protease K inhibitor [58] are found to be active against α – amylases. The human LEKTI has 15 domains and inhibits plasmin, trypsin, elastase, subtilisin A and cathepsin G [59].

As regards the mechanism of action, SNTI has shown a non-competitive type of inhibition. Although a few like soybean trypsin inhibitor has shown the competitive type of inhibition, the majority of the inhibitors follows non-competitive inhibition kinetics [46]. Jack fruit seed protease inhibitor isolated by Annapurna et al., [42] also showed non-competitive enzyme inhibition but the one isolated by Bhat [60] exhibited uncompetitive inhibition. The Ki value of SNTI was found to be $0.75+0.05 \times 10^{-10}$

M. The low Ki value indicates high affinity of SNTI towards trypsin.

The formation of stable trypsin inhibitor complex has been demonstrated by Sephadex G-100 gel filtration studies. The results obtained suggest that the inhibitor binds to trypsin in a 1:1 molar ratio. SNTI is a mono-headed inhibitor with a site for trypsin. Double-headed inhibitors with overlapping or non-overlapping binding sites are reported from plant sources [42, 53].

Protein modelling is now widely used in docking studies which paved way for the availability of disease causing target proteins in living organisms. SNTI has a stretch of 278 amino acids and the gut proteases from H. armigera and S. frugiperda are 119 and 254 amino acid residues respectively. In the present study two different modeling approaches homology and threading were used based on the availability of the template. The protein SNTI was modeled using the homology modeling approach as the sequence similarity between the template and target is more than 40 % whereas for the gut proteases from insects the structure was modeled using threading approach. The main difference between the homology modeling and threading is that in homology modeling the structure is built based on the sequence of the template whereas in threading it uses the knowledge of folds. Upon validation of the modeled proteins the structures are found to be reliable for subjecting to docking as about more than 80 % of amino acids fall in favorable regions of Ramachandran plot. The protein-protein docking and interaction analysis have indicated that the interacting residues between the surfaces of the docked proteins are the same residues that were predicted by Site Map. As the interactions involves the binding site residues with strong hydrogen bond and Van der Walls forces, these studies indicate that SNTI have shown potent activity against the gut proteases. Further evaluation of the study using wet lab techniques could bring out a natural source of bio-pesticide.SNTI belongs to the Serine Cereal super family and was found to exhibit both anti-bacterial and anti-fungal activity [61]. SNTI was reported to have anti-bacterial and anti-fungal activity but till date no adequate literature is available on insecticidal activity of SNTI, but SSTI was shown to exhibit insecticidal activity. As SNTI was reported to have anti-bacterial and anti-fungal activity, it might also possess insecticidal activity and hence was evaluated using various in silico tools. The sequence of SNTI obtained from MALDI-TOF [35] and the gut proteases of H. armigera and S. frugiperda were subjected to modeling. The resultant structures were validated and then subjected for protein-protein docking to understand the inhibitory role of SNTI on larval gut proteases.

## Conclusion

Results demonstrate that SNTI is a very stable, purified and highly potent trypsin inhibitor. Inhibitors of proteinases have been successfully engineered for protection of plants against pests and microorganisms. Protease inhibitors are proficient in interfering with digestive enzymes of insect gut and hence are able to control them. The overall structural quality of three proteins SNTI, Kazal type serine protease and trypsin protease validated by ERRAT server was found to be 86.02, 86.96 and 81.04 %. Docking results reveal that SNTI strongly interacts by hydrogen bonds and Vander Waals forces with the gut proteases at their active sites. SNTI binds at the active sites of gut protease enzymes which renders them inactive. This blocks the process of breaking down of nutrient proteins thereby causing malnourishment of the larvae leading to lethality. Hence Protease inhibitors can be commonly used as natural bio-pesticides in controlling pests.

Further analysis of structure, protein-protein interactions and diverse biological activities of SNTI on different proteases of diverse biological origins need to be carried out to confirm the biotechnological potential of SNTI as a bio control agent and its therapeutic potentials. Compromises between increased complexity, pharmacokinetic profiles, and drug affordability will challenge biochemists to find new general methods for the simple creation of new inhibitors, which are potent, selective and bioavailable or to find better methods for efficient delivery of protein inhibitors against proteases. We hope that this endeavor can help to stimulate new efforts towards achieving such goals.

### Abbreviations
FTIR: Fourier transform infra-red spectroscopy; SNTI: Soap Nut Trypsin Inhibitor; PI: Protease inhibitor; SDS-PAGE: Sodium dodecyl sulphate – polyacrylamide gel electrophoresis; PAS: Periodic acid schiff's; PB: Phosphate buffer.

### Competing interest
The authors declare that they have no competing interests.

### Authors' contributions
GVDS and KVR performed wet lab experiments and interpreted the results. KZ and GVDS performed in silico work and analysis. KVR and PK conceived the study and drafted manuscript. All authors read and approved the final manuscript.

### Acknowledgements
Purification and characterization of this work was supported and funded by University Grants Commission (UGC) with grant number F.No. 42–659/2013 (SR) to Dr. K. Vijaya Rachel, Department of Biochemistry and Bioinformatics, Institute of Science, GITAM University. Lab facilities were provided by the Department of Biochemistry and Bioinformatics and the computational part of this work was executed using a commercial version of Prime 3.0 module from Schrodinger software which was procured under UGC-Project sanctioned to Dr. P. Kiranmayi with grant number F.No. 42–669/2013 (SR) and was used for modeling of protein. The authors acknowledge the support from project operated within the UGC-Major Research Project Program.

### Author details
[1]Assistant professor, Department of Biochemistry/Bioinformatics, Institute of Science, GITAM University, Rushikonda, Visakhapatnam 530045, Andhra Pradesh, India. [2]Department of Biotechnology, Institute of Science, GITAM University, Rushikonda, Visakhapatnam 530045, Andhra Pradesh, India.

## References

1.  Bulet P, Stocklin R, Menin L. Anti-microbial peptides: from invertebrates to vertebrates. Immunol. Rev. 2004;198:169-184.
2.  Boman H. Antibacterial peptides: basic facts and emerging concepts. J. Intern. Med. 2003;25:197-215.
3.  Syed Rakashanda, Asif Khurshid Qazi, Rabiya Majeed, Shaista Rafiq, Ishaq Mohammad Dar, et al. Antiproliferative activity of Lavatera cashmeriana-protease inhibitors towards human cancer cells. Asian Pac. J. Cancer Prev.Asian Pac J Cancer Prev. 2013;14(6):3975-8.
4.  Koiwa H, Bressan RA, Hasegava PM. Regulation of protease inhibitors and plant defense. Trends in Plant Science. 1997;2:379-384.
5.  Ryan CA. Proteinase inhibitor gene families: strategies for transformation to improve plant defenses against herbivores. BioEssays, 1989;10:20-24.
6.  Brzin J, Kidric M. Proteinases and their inhibitors in plants: role in normal growth and in response to various stress conditions. Biotechnol Genet Eng Rev. 1995;13:420–467.
7.  Pena Cortes H, Sanchez Serrano J, Rocha SosaM, Willmitzer L. Systemic induction of proteinase-inhibitor-II gene expression in potato plants by wounding. Planta. 1988;174; 84–89.
8.  Pearce G, Johnson S, Ryan CA. Purification and characterization from tobacco (Nicotiana tabacum) leaves of six small, wound-inducible, proteinase isoinhibitors of the potato inhibitor II family. Plant Physiol. 1993;102: 639–644.
9.  Johnson R, Narvaez J, An G, Ryan CA. Expression of proteinase inhibitors I and II in transgenic tobacco plants: effects on natural defense against Manduca sexta larvae. Proc. Natl. Acad. Sci. USA. 1989;86:9871-9875.
10. Duan X, Li X, Xue Q, Abo-El-Saad M, Xu D, Wu R. Transgenic rice plants harboring an introduced potato proteinase inhibitor II gene are insect resistant. Nature Biotechnol. 1996;14:494-498.
11. Klopfenstein N, Allen KK, Avila FJ, Heuchelin S, Martinez J, Carman R, et al. Proteinase Inhibitor II Gene in Transgenic Poplar: Chemical and Biological Assays. Biomass and Bioenergy. 1997;12(4): 299-311.
12. Ng TB, Au TK, Lam TL, Ye XY, Wan DC. Inhibitory effects of antifungal proteins on human immunodeficiency virus type 1 reverse transcriptase, protease and integrase. Life Sci. 2002;70(8):927-35.
13. Prabhamanju M, Shankar S Gokul, and Babu K. "Herbal vs. Chemical Actives as Antidandruff Ingredients -Which Are More Effective in the Management of Dandruff? – An Overview," Ethnobotanical Leaflets. 2009;11:5.
14. Imms AD. Outlines of Entomology. Methuen. London, UK. 1964; 5th ed: 224.
15. Roush DK and McKenzie JA. Ecological genetics of insecticide and acaricide resistance. Annual Review of Entomology. 1987;32:361-380.
16. Christos A, Damalas and Ilias G, Eleftherohorinos. Pesticide Exposure, Safety Issues, and Risk Assessment Indicators. Int. J. Environ. Res. Public Health 2011;8:1402-1419.
17. Fraenkel G S. The raison d'être of secondary plant substances. Science. 1959;129:1466-70.
18. Sudha C V. Studies on insecticidal and other biological properties of jack fruit (Artocarpus heterophyllus) seed proteinase inhibitor. Ph.D. thesis, The Andhra University: Visakhapatnam, March, 1999.
19. Maria Li´gia Macedo R, Eduardo Diz Filho BS, Mariadas Grac¸as M, Freire Maria Luiza V, Oliva Joana Sumikawa T, Marcos Toyama H, et al. A Trypsin Inhibitor from Sapindus saponaria L. Seeds: Purification, Characterization, and Activity towards Pest Insect Digestive Enzyme. Protein J. 2011;30:9–19.
20. Maarten, Jongsma, caroline bolter. The Adaptation of Insects to Plant Protease Inhibitors. J. Insect Physiol. 1997;43(10): 885–895.
21. Moloud Gholamzadeh Hitgar, Mohammad Ghadamyari and Mahbobe Sharifi. Identification and Characterisation of Gut Proteases in the Fig Tree Skeletoniser Moth, Choreutis nemorana. Plant Protect. Sci. 2013;49(1):19–26.
22. Jin-Young Kim, Seong-Cheol Park, Indeok Hwang, Hyeonsook Cheong, Jae-Woon Nah, Kyung-Soo Hahm, et al. Protease Inhibitors from Plants with Antimicrobial Activity. Int. J. Mol. Sci. 2009;10:2860-2872.
23. Rachel KV, Vimala Y, Apta Chaitanya D. A trypsin inhibitor-SNTI with antidandruff activity from Sapindus trifoliatus. Indian journal of applied research. 2013;3(3):3-5.
24. Kakade ML, Simons NR, Liener IE. An Evaluation of Natural vs. Synthetic Substrates for Measuring the Antitryptic Activity of Soybean Samples. Cereal Chem. 1969;46:518-526.
25. Sumati S, Pattabiraman TN. Natural plant enzyme inhibitors: part I. Protease inhibitors of tubers and bulbs. Indian J. Biochem Biophys. 1975;12:383–385.
26. Prabhu KS, Pattabiraman TN. Natural plant enzyme inhibitors: Part II. Protease inhibitors of seeds. Indian J. Biochem. Biophys. 1977;14:96-98.
27. Arnon R, Perlmann GE, Lorand L. Methods in Enzymology, Ed, Academic Press, New York. 1970;19:226-244.
28. Schwert GW, TakenakaY. A spectrophotometric determination of trypsin and chymotrypsin. Biochem. Biophys. Acta., 1955;16:570-575.
29. Matsubara H. Methods Enzymol. 19, 642, Nucl. Acids. Res. 1970;30:347-348.
30. Saunders RM, Lang JA. α-Amylase inhibitors in Triticum aestivum: Purification and physical chemical properties. Phytochemistry. 1973;12:1237-1241.
31. Reisfield RA, Lewis UJ, Williams DE. Disk electrophoresis of basic proteins and peptides on polyacrylamide gels. Nature. 1962;195:285.
32. Kanamori N, Sakabe K, Olasakd R. Extracellular nucleases of L. Bacillus subtlis Purification and properties. Biochim. Biophys. Acta. 1974;335:155-172.
33. Fairbanks G, Steck TL, Wallach. DFH. Electrophoretic analysis of the major polypeptides of the human erythrocyte membrane. Biochemistry. 1971;10:2606-2617.
34. Andrews P. Estimation of the molecular weights of proteins by Sephadex gel-filtration. J. Biochem. 1964;91(2):222-33.
35. Vijaya Rachel K, Sandeep Solmon K, Kiranmayi P, Bhaskar Reddy I, Siva Prasad D. In silico modeling and docking studies of Soap Nut Trypsin Inhibitor. Process Biochemistry. 2012;47:453–459.
36. Pruitt KD, Tatusova T, Maglott DR. NCBI reference sequences (RefSeq): a curated non-redundant sequence database of genomes, transcripts and proteins. Nucleic Acids Res. 2007;35(Database issue):D61-5. Epub 2006 Nov 27.
37. Morris AL, MacArthur MW, Hutchinson EG, Thornton JM. Stereo chemical quality of protein structure coordinates. Proteins. 1992;12(4):345-64.
38. Colovos C, Yeates TO. Verification of protein structures: patterns of nonbonded atomic interactions. Protein Sci. 1993;2(9):1511–1519.
39. Halgren T. Identifying and characterizing binding sites and assessing druggability. J Chem Inf Model. 2009;49:377–389.
40. Chuang G-Y, Kozakov D, Brenke R, Comeau SR, Vajda S. DARS (Decoys As the Reference State) Potentials for Protein-Protein Docking. Biophys. J. 2008;95:4217-4227.
41. Biologics Suite 2014-2: BioLuminate, version 1.5, Schrödinger, LLC, New York, NY. 2014.
42. Annapurna, Shakuntala S, Prasad DS. Purification of trypsin/chymotrypsin inhibitor from Jack fruit seeds. J Sci. Food Agric. 1991;54:399.
43. Davis BJ. Disc electrophoresis .II. Method and application to human serum proteins. Ann N Y Acad Sci. 1964;28(121):404-427.
44. Weber K, Osborn M. The Reliability of Molecular Weight Determinations by Dodecyl Sulfate-Polyacrylamide Gel Electrophoresis. J. Biol. Chem. 1969;244:4406–4412.
45. Laemmli. Cleavage of structural proteins during the assembly of the head of bacteriophage T4. Nature 227. UK. 1970; 5259:680–685.
46. Vogel R, Trautschold I, Werle E. Natural proteinase inhibitors. New York, Academic Press. 1968.
47. Richardson M. The proteinase inhibitors of plants and microorganisms. Phytochemistry. 1977;16:159-169.
48. Francoise M, Robert, Peter P, Wong. Isozymes of Glutamine Synthetase in Phaseolus vulgaris L. and Phaseolus lunatus L. Root Nodules. Plant Physiol. 1986;81:142-148.
49. Gennis LS, Canto CR. Double-headed protease inhibitors from black-eyed peas. I. Purification of two new protease inhibitors and the endogenous protease by affinity chromatography. J Biol Chem. 1976;251:734–740.
50. Gatehouse AMR, Gatehouse JA, Boulter D. Isolation and characterization of trypsin inhibitor from cowpea. Phytochemistry. 1980;19:751-756.
51. Prabhu KS, Pattabiraman TN. A colorimetric method for the estimation of the esterolytic activity of chymotrypsin. Indian J Biochem Biophys. 1977;14:96-98.
52. Rodis P, Hoff JE. Naturally occurring protein crystals in the potato inhibitor of papain, chymopapain and ficin. Plant Physiol. 1984;74:907-911.
53. Shakuntala G V, Prasad DS. Varietal Variations in Jack fruit (Artocarpus heterophyllus) seed protease inhibitor and its haemagglutinating activity. Ph.D., Thesis. 1996; Andhra University, Visakhapatnam, India.
54. Edelhoch, Harold, Steiner RF. Structural transitions of soybean trypsin inhibitor. II. The denatured state in urea. J. Biological Chemistry. 1963;238:931-8.
55. Chiche L, Heitz A, Gelly JC, Gracy J, Chau PT, Ha PT, et al. Squash inhibitors: from structural motifs to macrocyclic knottins. Curr. Protein Pept. Sci. 2004;5:341-349.
56. Warsy AS, Norton G, Stein M. Proteinase inhibitors from broad bean, Isolation and purification. Phytochem. 1974;13:2481-2486.

57.  Mundy J, Svendsen I, Hejgaard J. Barley alpha-amylase/subtilisin inhibitor, Isolation and characterization. Carlsberg Res. Commun. 1983;48:81-90.

58.  Zemke KJ, Muller-Fahrnow A, Jany K. The three-dimensional structure of the bifunctional proteinase K/alpha-amylase inhibitor from wheat (PKI3) at 2.5 Ao resolution. FEBS. 1991;397:240-242.

59.  Mitsudo K, Jayakumar A, Henderson Y, Frederick MJ, Kang Y, Wang M, et al. Inhibition ofserine proteinases plasmin, trypsin, subtilisin A, cathepsin, and elastase by LEKTI: a kinetic analysis. Biochemistry. 2003;42:3874-3881.

60.  Bhat AV. Pattabiraman TN. Protease inhibitors from jackfruit seed (Artocarpus integrifolia). J. Biosci. 1989;14(4):351–365.

61.  Dr. Hari Jagannadha Rao G, Lakshmi P. Sapindus trifoliatus: A review. International Journal of Pharmacy & Technology. 2012;4(3):2201-2214.

# 17

# Structural plasticity of green fluorescent protein to amino acid deletions and fluorescence rescue by folding-enhancing mutations

Shu-su Liu[1], Xuan Wei[1], Xue Dong[1], Liang Xu[1], Jia Liu[1,2]* and Biao Jiang[1]*

## Abstract

**Background:** Green fluorescent protein (GFP) and its derivative fluorescent proteins (FPs) are among the most commonly used reporter systems for studying gene expression and protein interaction in biomedical research. Most commercially available FPs have been optimized for their oligomerization state to prevent potential structural constraints that may interfere with the native function of fused proteins. Other approach to reducing structural constraints may include minimizing the structure of GFPs. Previous studies in an enhanced GFP variant (EGFP) identified a series of deletions that can retain GFP fluorescence. In this study, we interrogated the structural plasticity of a UV-optimized GFP variant ($GFP_{UV}$) to amino acid deletions, characterized the effects of deletions and explored the feasibility of rescuing the fluorescence of deletion mutants using folding-enhancing mutations.

**Methods:** Transposon mutagenesis was used to screen amino acid deletions in GFP that led to fluorescent and nonfluorescent phenotypes. The fluorescent GFP mutants were characterized for their whole-cell fluorescence and fraction soluble. Fluorescent GFP mutants with internal deletions were purified and characterized for their spectral and folding properties. Folding-ehancing mutations were introduced to deletion mutants to rescue their compromised fluorescence.

**Results:** We identified twelve amino acid deletions that can retain the fluorescence of $GFP_{UV}$. Seven of these deletions are either at the N- or C- terminus, while the other five are located at internal helices or strands. Further analysis suggested that the five internal deletions diminished the efficiency of protein folding and chromophore maturation. Protein expression under hypothermic condition or incorporation of folding-enhancing mutations could rescue the compromised fluorescence of deletion mutants. In addition, we generated dual deletion mutants that can retain GFP fluorescence.

**Conclusion:** Our results suggested that a "size-minimized" GFP may be developed by iterative incorporation of amino acid deletions, followed by fluorescence rescue with folding-enhancing mutations.

**Keywords:** Green fluorescent protein (GFP), Transposon mutagenesis, Amino acid deletions, Protein folding, Chromophore maturation

## Background

The discovery and application of green fluorescent proteins (GFP) have revolutionized biomedical research during the past several decades. GFP was identified in jellyfish *Aequorea victoria* in the 1960s [1] and then isolated and characterized by Prasher *et al.* in 1992 [2]. Soon after this, GFP was adapted as a fluorescent tag to unravel the details of cellular events, such as protein localization, trafficking and interaction [3]. Extensive engineering studies resulted in a panel of fluorescent protein (FP) variants with a wide range of spectral properties [4]. Other properties such as brightness, cytotoxicity and photostability have been also optimized to facilitate the application of FPs.

Notably, considerable efforts have been made to optimize the oligomerization state of FPs. Many of the naturally occurring FPs are dimeric or tetrameric [4]. Oligomerization does not limit the application of GFPs as reporters for gene expression, but may interfere with the native function of fused proteins. Tsien's group first demonstrated that tetrameric *Discosoma sp.* red fluorescent protein (DsRed) could be engineered to be monomeric [5]. In this study, they first disabled

* Correspondence: liujia@shanghaitech.edu.cn; jiangbiao@shanghaitech.edu.cn
[1]Shanghai Institute for Advanced Immunochemical Studies, ShanghaiTech University, Shanghai, China
Full list of author information is available at the end of the article

the formation of tetrameric structure by site-directed mutagenesis and then used random mutagenesis to rescue the fluorescence. It is now known that many oligomeric FPs can be converted into monomers by point mutations without appreciable deleterious effects [6].

Oligomerization rarely caused troubles for the native function of GFPs, however it may introduce structural constraints and unfavorably affect the role of GFPs as a genetic tag for biological applications such as protein-protein interaction and subcellular localization. Despite the extensive optimization, the artifacts associated with GFP oligomerization still require careful assessment in each experiment [7]. In addition to traditional strategy, an alternative approach to reducing the structural constraints associated with GFP tag is to develop a size-minimized construct. This is extremely challenging in the case of GFP because it is generally considered as a "compact" protein due to the well-shaped β-can structure [8–10].

The wild-type *A. victoria* GFP is a 27 kDa protein containing 238 amino acids. GFP and all its known variants adopt a β-can structure assembled by 11 antiparallel β-strands [3, 11]. Most strands are connected by small loops consisting of one to four amino acids. Two larger loops appear at positions 129 ~ 143 and 189 ~ 197. The chromophore of GFP is located in the central α helix, surrounded by β-strands. The top and bottom "lids" composed mainly of residues 74–91 and 128–145, respectively [11, 12].

Several early studies have been performed to understand the tolerance of GFP to amino acid deletions. GFP is well tolerant to deletions at N or C terminus [8, 9]. One study showed that the minimal domain required for the fluorescence of GFP contains residues from 2 to 232 [8]. An enhanced version of GFP (EGFP, mutation F64L/S65T) can still be fluorescent with 5 amino acids deletion from N terminus or 10 amino acid deletion from C terminus [9]. However, existing evidence suggested that GFP is very sensitive to deletions at internal positions. Deletions of the two large loops or the two small helices eliminated the fluorescence of EGFP [9]. Targeted deletion analysis of the longest loop (129 ~ 143) in SuperGlo GFP (sgGFP, mutation F64L/

S65C/I167T) showed that I128Δ and D129Δ are the only two single deletions that can retain GFP fluorescence [10].

Nevertheless, all early studies focused on analyzing deletions at the termini or large loops of GFP. In recent studies, Jones *et al.* used a transposon-based mutagenesis method [13] to investigate the global structural plasticity of EGFP to amino acid deletions [14]. A series of deletion mutants that can retain GFP fluorescence have been identified and some mutants even exhibited improved cellular fluorescence [14, 15]. These studies shed the light on the important roles of amino acid deletions on GFP fluorescence and indicated that incorporation of amino acid deletions might be a feasible approach to the development of size-minimized GFP construct. In this study, we created a set of GFP$_{UV}$ deletion mutants using a similar approach, characterized their properties in a comprehensive manner and explored the feasibility of rescuing the fluorescence of deletion mutants by introducing folding-enhancing mutations.

## Results

### Transposon-mediated deletion mutagenesis and colony screening

The transposon mutagenesis used in this study was described by Jones *et al.* [13]. It relies on an *in vitro* transposition reaction using MuA transposase and a mini-Mu transposon DNA [16] with engineered *Mly* I recognition site at each end [13]. In the presence of the donor transposon DNA, MuA transposase cleaves the acceptor DNA at a random position in a five nucleotide staggered manner. Following transposon insertion and transformation of the transposition product into bacteria, the five nucleotide overhang is repaired by bacterial DNA repair machinery, leading to duplication of the five nucleotides at the transposon insertion site. Subsequent removal of the transposon DNA using *Mly* I enzyme cleaves four nucleotide from the acceptor DNA at each end of transposon, resulting in a net 3 bp deletion from the target DNA (Fig. 1).

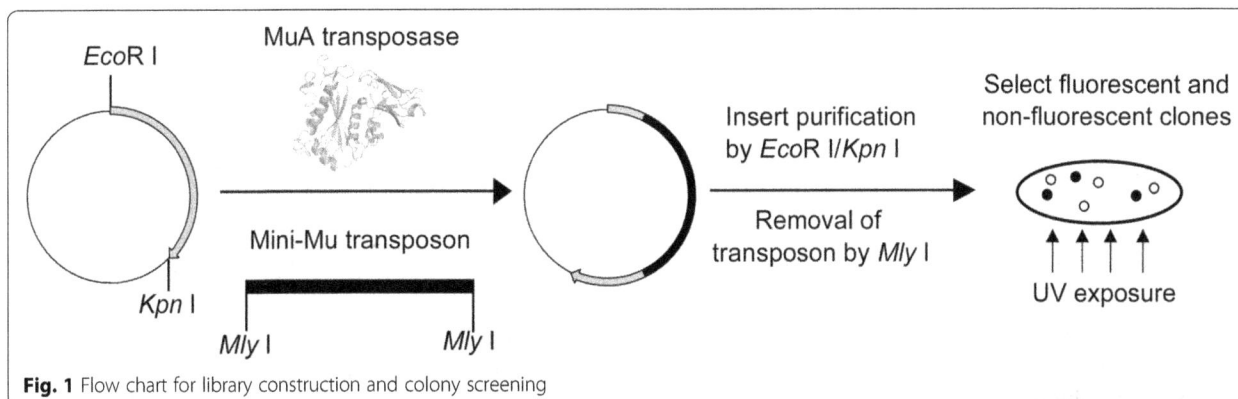

**Fig. 1** Flow chart for library construction and colony screening

In this study, we performed transposon reaction using *Mly* I Mu transposon [13] as the donor DNA and pGFP$_{UV}$ plasmid as the acceptor DNA. We collected 38,000 individual colonies from the transformation of transposition product, which were sufficient to cover more than 95 % of all possible insertion sites even considering the minor substrate preference of the transposon system [16]. We next used *EcoR* I/*Kpn* I to isolate the transposon inserts within the GFP$_{UV}$ gene (Additional file 1: Figure S1). The final deletion library was obtained by releasing the transposon DNA from pGFP$_{UV}$ plasmid using *Mly* I restriction digestion.

pGFP$_{UV}$ plasmid encodes a GFP variant with enhanced fluorescence under UV light (GFP$_{UV}$) [17]. pGFP$_{UV}$ plasmid constitutively expresses GFP$_{UV}$ protein and thus can be used directly for functional screen. In pilot experiments, we observed that some mutants exhibited considerable difference in fluorescence when grown at different temperature. We chose to perform the fluorescence screening at low temperature (20 °C). Forty fluorescent and 24 non-fluorescent colonies were identified and selected for sequencing for further analysis.

### Sequence analysis of deletion mutants

Sequencing results revealed that a small fraction of mutants contained four nucleotide deletion, which presumably resulted from excessive *Mly* I digestion (Additional file 2: Table S1). This result was consistent with previous study [18] and suggested that *Mly* I digestion might require further optimization. In total, we identified thirteen fluorescent and seven non-fluorescent mutants with unique deletions. Of the thirteen fluorescent mutants, eight contained deletions at N- or C-terminus whereas five carried deletions at internal positions. The five internal deletions were all located at the termini of α-helices or β-strands. For the seven deletions that eliminated GFP fluorescence, one was found in the N-terminal region, one was in the middle of β-strand and the rest were all found at the termini of α-helices or β-strands (Fig. 2). It is worth noting that the random nucleotide deletions may occur in two neighboring residues, leading to amino acid mutations in addition to deletions.

### Whole-cell fluorescence and fraction soluble of deletion mutants

In order to quantitatively analyse the identified fluorescent deletion mutants, we characterized their whole-cell fluorescence in the context of pGFP$_{UV}$ plasmid. As temperature may affect the fluorescence of some mutants, whole-cell fluorescence was determined under different temperature: 37, 30 and 20 °C. It was found that, for all temperature, terminal deletions generally retained higher degree of fluorescence compared with internal deletions. The fluorescence of all deletion mutants was

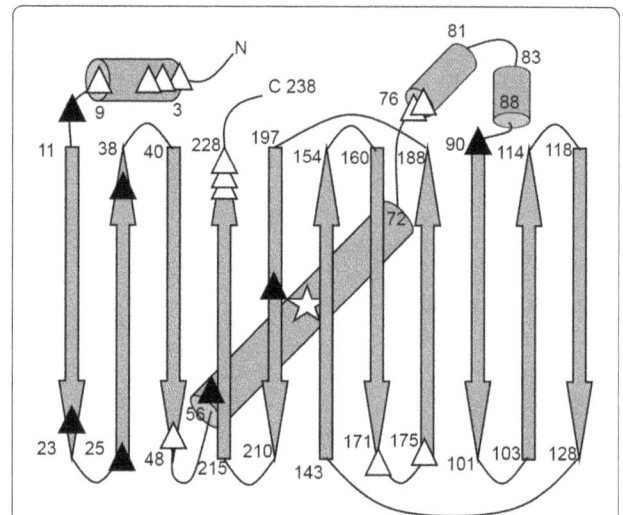

**Fig. 2** Sequence analysis of identified deletions in GFP$_{UV}$. α-helices and β-strands are shown as cylinders and arrows, respectively. The chromophore is indicated by a star. Deletions leading to fluorescent and non-fluorescent phenotypes are shown as white and black triangles, respectively. Fluorescent mutants include S2/K3→R, G4Δ, E5Δ, F8/T9→S, C48Δ, P75/D76→H, P75Δ, E172Δ, S175/V176→F, A226Δ, G228Δ and G228/I229→V (M1Δ mutant is not included due to the deletion of the start codon). Non-fluorescent mutants include G10Δ, D21Δ, G24Δ, E34Δ, P58Δ, P89/E90→Q and T203Δ

compromised at high temperature (37 °C). Decrease of temperature to 30 °C restored the fluorescence of most mutants except F8/T9→S and P75/D76→H. Expression of protein at 20 °C essentially recovered the fluorescence of all mutants (Fig. 3).

The recovery of fluorescence at low temperature suggested that these deletions might impair protein folding, which is a temperature-dependent process. In order to understand the effect of amino acid deletions on protein folding of GFP$_{UV}$, we sought to determine the fraction of soluble GFP$_{UV}$ at different temperature using a previously described method [19]. Wild-type and mutant GFP$_{UV}$ was expressed from pGFP$_{UV}$ plasmid and the bacterial cells were then lysed to assess the GFP present in the entire or soluble fraction of the cell extracts (Additional file 3: Figure S2). The soluble fraction of deletion mutants was clearly temperature dependent (Fig. 3) and plausibly correlated with the whole-cell fluorescence. For example, the mutants with no fluorescence at 37 °C (F8/T9→S, C48Δ, P75/D76→H, P75Δ and E172Δ) had dramatically reduced level of soluble fraction of proteins compared with other mutants. However, factors other than protein folding need be considered to explain the reduced fluorescence of deletion mutants. In one case, P75/D76→H mutant exhibited similar level of soluble fraction of proteins at 30 and 20 °C, but it was only fluorescent at 20 °C. It is known that both

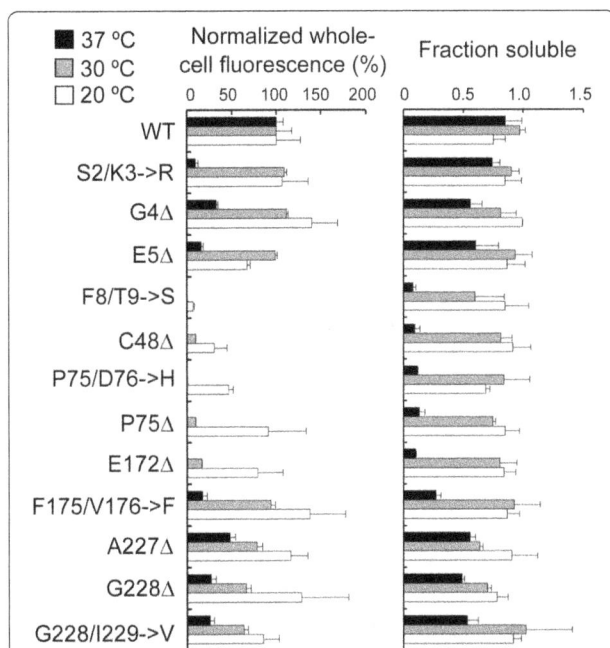

**Fig. 3** Whole-cell fluorescence and fraction soluble of deletion mutants. The fluorescence of deletion mutants are normalized to that of wtGFP$_{UV}$ under the same temperature. The arbitrary fluorescence of wtGFP$_{UV}$ are 3497, 2620 and 442 under 37, 30 and 23 °C, respectively

protein folding and chromophore maturation of GFP are temperature dependent [20]. Therefore, we intended to purify the deletion mutants and characterize their biochemical and biophysical properties. We chose to analyze the fluorescent mutants with internal deletions. Characterization of these mutants and search of feasible means to rescuing their compromised fluorescence might be the first step toward generating a size-minimized GFP.

**Protein purification and the effect of amino acid deletions on spectral properties**

Deletion mutants C48Δ, P75/D76→H, P75Δ, E172Δ, S175/V176→F were purified to more than 95 % homogeneity (Fig. 4a). The emission spectra of all variants under 397 and 475 nm excitation can be superimposed with that of wild-type GFP$_{UV}$ (wtGFP$_{UV}$) (Fig. 4b), suggesting that the chromophores of these mutants did not have structural alteration. For excitation spectra, two peaks (397 and 475 nm) were observed for all samples. These deletion mutants showed variations in the intensity ratio between the major and minor peaks. For example, P75Δ showed slightly increased minor peak whereas the minor peak in S175/V176→F mutant was considerably reduced (Fig. 4c). The 395 nm and 475 nm excitation peaks represent neutral and anionic chromophore, respectively [20]. Mutations that alter the ratio of the two chromophore species are frequently found in

GFP variants. In one case, enhanced GFP (EGFP) only has a single excitation peak at 475 nm [21] because S65T mutation transform the chromophore to be completely ionized [20]. It is interesting to observe in this study that amino acid deletions, in addition to substitutions, can have impact on the ionization state of chromophore.

We next determined the extinction coefficients at 397 and 475 nm ($\varepsilon_{397}$ and $\varepsilon_{475}$) and quantum yield of the GFP$_{UV}$ variants. Consistent with the spectra, $\varepsilon_{397}$ of all deletion mutant presented no appreciable difference from that of wtGFP$_{UV}$, while modest variation was observed for $\varepsilon_{475}$ (Table 1). We also found that all variants have very similar quantum yield to wtGFP$_{UV}$ protein, suggesting that these internal amino acid deletions did not alter the intrinsic brightness of GFP (Table 1). Taken together, these results indicated that the diminished fluorescence of deletion mutants is mainly attributed to the disrupted folding process, rather than intrinsic brightness.

**Characterization of refolding kinetics and chromophore maturation of deletion mutants**

As whole-cell lysis study indicated that amino acid deletions may affect protein folding, we next sought to determine the refolding kinetics of deletion variants using purified proteins. Previous study suggested that the refolding process of GFP$_{UV}$ could be fitted equally well with sequential or parallel model [22]. Here we performed refolding experiments using guanidine hydrochloride and recorded the recovered fluorescence as described [22]. The data were fitted into a double exponential equation with a parallel model using Prism 4.0 (Fig. 5). The $k_1$ and $k_2$ of wtGFP$_{UV}$ were generally consistent with the reported values [23]. Deletion C48Δ led to notable decrease in $k_1$ value, suggesting that this deletion was involved in the major folding pathway of GFP. It was also found that P75/D76→H and P75Δ reduced the refolding rate in the slow phase (decreased $k_2$ value) (Fig. 5). Interestingly, the *cis-trans* isomerization of P75 was proposed as a rate-limiting step in the folding process of GFP [24]. It was also noted that deletion E172Δ impacted neither fast nor slow phase of protein folding. This observation suggested that factors other than protein folding may account for the compromised fluorescence in deletion mutants.

Instructed by these results, we then analyzed the efficiency of chromophore maturation of deletion mutants. In this experiment, we denatured 7.0 μM protein of each sample and determined their absorbance at 450 nm (Fig. 6). As the extinction coefficient of matured chromophore at 450 nm has been determined in previous study [25], the concentration of matured chromophore can be calculated using Beer's law. It is evident

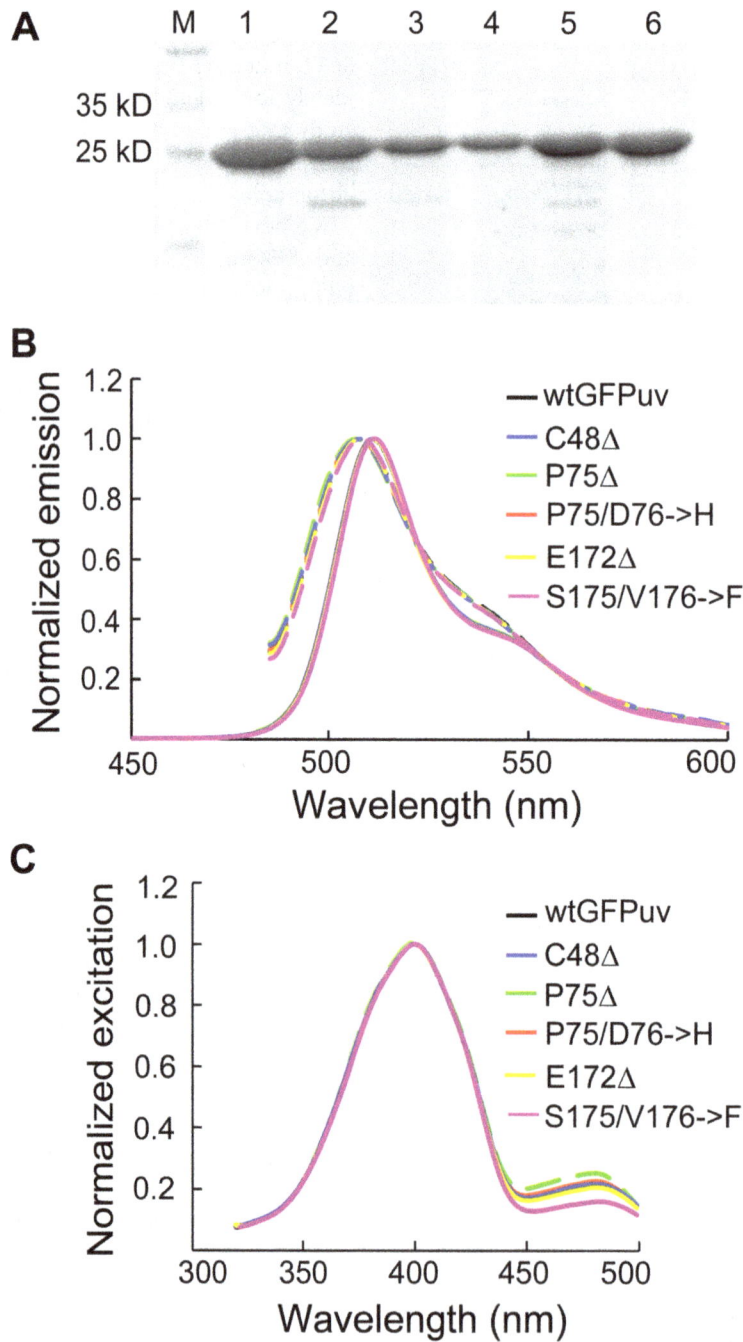

**Fig. 4** Excitation and emission spectra of purified GPF$_{UV}$ variants. **a** Purified wtGFP$_{UV}$ and deletion mutants, resolved on 12 % SDS-PAGE. M, marker. Lane 1 to 6, wtGFP$_{UV}$, C48Δ, P75Δ, P75/D76➔H, E172Δ and S175/V176➔F, respectively. **b** Normalized excitation spectra. **c** Normalized emission spectra. Note that the spectra of these variants are largely overlapped

that all deletions reduced the efficiency of chromophore maturation (Fig. 6). It is worth mentioning that the protein samples used for this experiment was purified at 20 °C. Previous analysis of the whole-cell lysis showed that the five mutant with internal deletions displayed only trace fluorescence but had considerable amount of soluble proteins when expressed at 30 °C. According to

the analysis of chromophore, this discrepancy is likely attributed to the impaired chromophore maturation of GFP proteins present in the soluble fraction.

**Fluorescence rescue by folding-enhancing mutations**

Based on the above results, the compromised fluorescence of GFP deletion mutants resulted from disrupted

**Table 1** Extinction coefficient and quantum yield of deletion mutants

| | $\varepsilon_{397}$ (M$^{-1}$ cm$^{-1}$) | $\varepsilon_{475}$ (M$^{-1}$ cm$^{-1}$) | Quantum yield ($\lambda_{ex} = 397$ nm) |
|---|---|---|---|
| wtGFP$_{UV}$ | 29600 | 7500 | 0.79[a] |
| C48Δ | 31600 | 8500 | 0.80 |
| P75/D76→H | 29700 | 6800 | 0.79 |
| P75Δ | 29400 | 7100 | 0.79 |
| E172Δ | 29300 | 7600 | 0.81 |
| S175/V176→F | 30900 | 5000 | 0.79 |

[a]Patterson et al., [26]

protein folding and chromophore maturation. As a first step toward generating a size-minimized GFP, we speculated whether the defects in these deletion mutants can be restored by folding-enhancing mutations. We chose to test the effect of two previously described folding mutations F64L [26] and S30R [19] on the mutants with internal deletions. Remarkably, F64L recovered a considerable fraction of fluorescence in all variants except P75Δ, whereas S30R only improved the fluorescence of S175/V176→F (Fig. 7a). This result illustrated that folding-enhancing mutations can rescue the compromised fluorescence of deletion mutants and that some mutations may be more effective than others. Encouraged by these results, we attempted to use F64L mutation to rescue mutants carrying multiple internal deletions. We first generated nine dual deletion mutants by combining the identified five internal deletions. As expected, GFP fluorescence was completely abolished by dual deletions, even with prolonged incubation at 4 °C. We next incorporated F64L mutation into these dual mutants. Although F64L could not recover GFP fluorescence at 37 °C, it successfully restored the fluorescence of all mutants to a considerable degree when proteins were

expressed at 20 °C (Fig. 7b). To the best of our knowledge, it has not been reported before that GFP mutants carrying multiple internal deletions can retain fluorescence.

## Discussion

Amino acid deletion, insertion and substitution are all important mechanisms for protein evolution in nature. While amino acid substitutions are responsible for the alteration of protein properties in many cases, insertions and deletions (indels) are critical for generating length variation of proteins [27, 28]. Although less commonly used, indels are efficient means for improving protein functions [29, 30]. Established methods for introducing random indels into proteins include RID mutagenesis reported by Murakami et al. [31] and RAISE mutagenesis described by Fujii et al. [32]. Notably, Jones et al. developed a facile transposon-based method for introducing random triplet nucleotide deletions within the full sequence of a given protein [13] and has gone to employ this method to investigate the structural tolerance of TEM-1 β-lactamase to amino acid deletions [33]. This method has been soon adapted for generating nucleotide substitutions with canonical [34] and non-canonical [18, 35–37] amino acids as well as domain insertions [38, 39]. Subsequent modification of this method allowed generation of multiple in-frame codon mutations [40, 41]. In this study, we used this transposon mutagenesis [13] to interrogate the global plasticity of GFP to amino acid deletions.

Early studies suggested that GFP is well tolerated to amino acid deletions at termini [8, 9] but not to those at internal positions [10]. However, recent deletion analysis of EGFP identified a set of internal deletions that can retain or even improve GFP fluorescence [14, 15]. Herein, we generated a deletion library of GFP$_{UV}$ [17] using a similar approach. This GFP variant (mutation F99S/M153T/V163A) has enhanced absorbance under UV light and thus functional screen can be performed directly by visual inspection

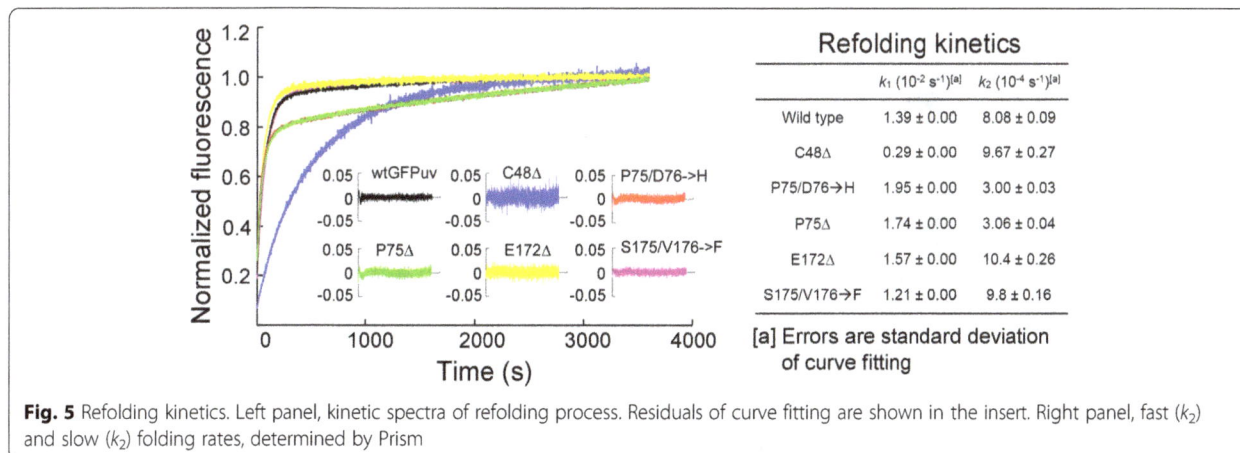

Refolding kinetics

| | $k_1$ (10$^{-2}$ s$^{-1}$)[a] | $k_2$ (10$^{-4}$ s$^{-1}$)[a] |
|---|---|---|
| Wild type | 1.39 ± 0.00 | 8.08 ± 0.09 |
| C48Δ | 0.29 ± 0.00 | 9.67 ± 0.27 |
| P75/D76→H | 1.95 ± 0.00 | 3.00 ± 0.03 |
| P75Δ | 1.74 ± 0.00 | 3.06 ± 0.04 |
| E172Δ | 1.57 ± 0.00 | 10.4 ± 0.26 |
| S175/V176→F | 1.21 ± 0.00 | 9.8 ± 0.16 |

[a] Errors are standard deviation of curve fitting

**Fig. 5** Refolding kinetics. Left panel, kinetic spectra of refolding process. Residuals of curve fitting are shown in the insert. Right panel, fast ($k_2$) and slow ($k_2$) folding rates, determined by Prism

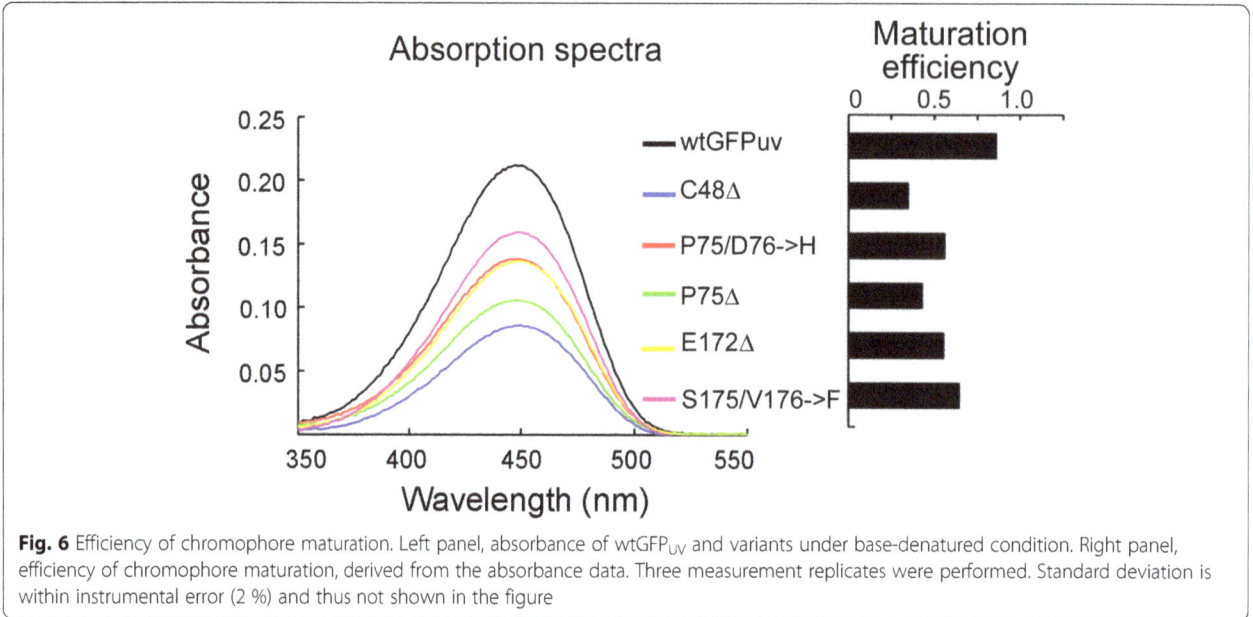

**Fig. 6** Efficiency of chromophore maturation. Left panel, absorbance of wtGFP$_{UV}$ and variants under base-denatured condition. Right panel, efficiency of chromophore maturation, derived from the absorbance data. Three measurement replicates were performed. Standard deviation is within instrumental error (2 %) and thus not shown in the figure

using UV excitation. We identified 13 fluorescent and 7 non-fluorescent mutants with unique deletions. We first analysed the whole-cell fluorescence of the mutants that retained GFP fluorescence and found that their fluorescence is strongly temperature dependent. Remarkably, several mutants (G4Δ, F175/V176➔F, A227Δ and G228Δ) that showed decreased fluorescence at high temperature (37 °C) are even more fluorescent than wtGFP$_{UV}$ when the temperature is reduced to 4 °C. This suggested that these deletions may enhance GFP fluorescence at the cost of protein stability. Further analysis of the fraction of soluble GFP in the whole-cell lysis suggested that the impaired

fluorescence of mutants at high temperature was attributed to the disrupted protein folding and that other process need be considered to explain the loss of fluorescence. Characterization of purified GFP proteins carrying internal deletions confirmed the decreased refolding rates in some variants. Additional analysis showed that all internal deletions significantly reduced the efficiency of chromophore maturation. Interestingly, these mutants have the same intrinsic brightness (quantum yield) with wtGFP$_{UV}$. This finding indicated that these mutants used the same autocatalytic process to form GFP chromophore. The present and previous [14, 15] results suggested that it

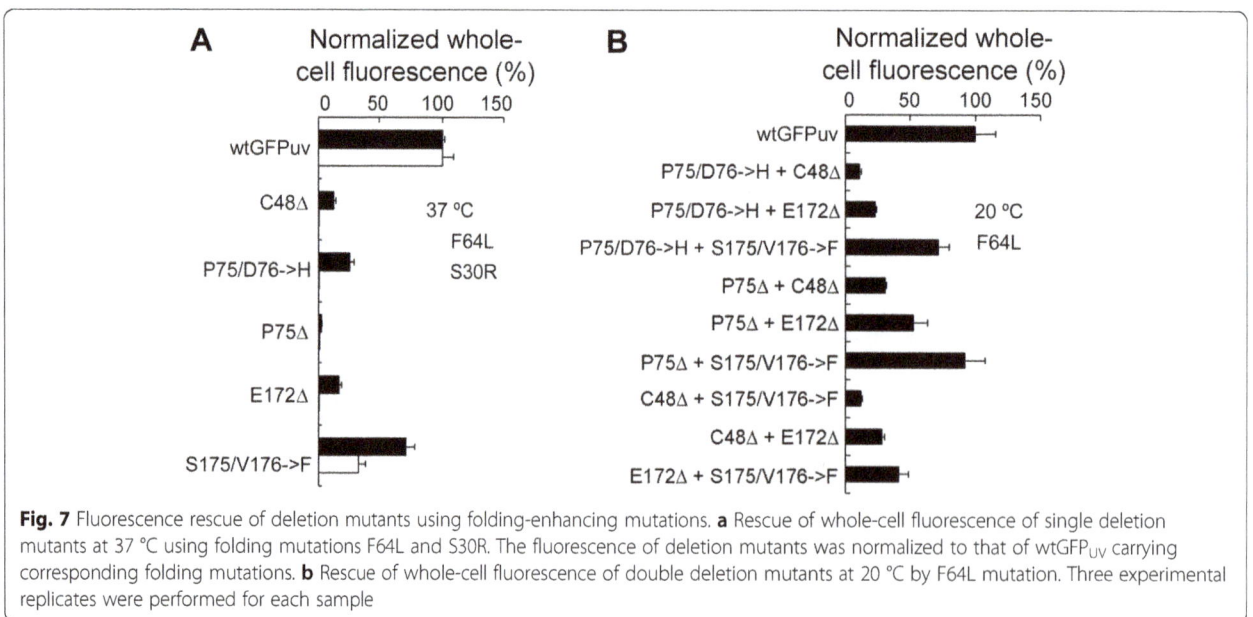

**Fig. 7** Fluorescence rescue of deletion mutants using folding-enhancing mutations. **a** Rescue of whole-cell fluorescence of single deletion mutants at 37 °C using folding mutations F64L and S30. The fluorescence of deletion mutants was normalized to that of wtGFP$_{UV}$ carrying corresponding folding mutations. **b** Rescue of whole-cell fluorescence of double deletion mutants at 20 °C by F64L mutation. Three experimental replicates were performed for each sample

is feasible to recover protein folding and chromophore maturation by introducing beneficial mutations.

Next we attempted to recover the fluorescence of deletion mutants by employing two folding mutations previously identified in GFP. It was shown that F64L mutation successfully rescued the fluorescence of most mutants whereas S30R was less effective. The different roles of F64L and S30R can be explained by their distinct functions in GFP. F64L is a central mutation in close proximity to chromophore and many studies have demonstrated its critical role in protein folding and chromophore maturation [21, 24, 26]. S30R, in contrast, is a distal mutation that enhanced protein folding in an indirect manner [19]. In addition, we analysed the effect of F64L on the non-fluorescent deletion mutants (G10$\Delta$, D21$\Delta$, G24$\Delta$, E34$\Delta$, P58$\Delta$, P89/E90$\rightarrow$Q and T203$\Delta$) and found that only G10$\Delta$ can recover trace fluorescence at 20 °C (data not shown). It is also worth noting that GFP variants containing F64L mutation can be also used as the starting template for engineering experiments. Previous studies suggested that in the presence of F64L, more deletions may be found to retain GFP fluorescence [14, 15].

Toward the goal of generating a size-minimized GFP, we next set to generate dual deletion mutants and then determined the effects of F64L on these variants. It was found that F64L was capable of rescuing their fluorescence at 20 °C but not that at 37 °C. We also generated GFP$_{UV}$ variants containing triple internal deletions and found that their fluorescence could not be restored by F64L. This result indicated that additional folding mutations or global optimization might be required for compensating the deleterious effects of multiple amino acid deletions. Recent study highlighted the importance of epistasis in protein evolution [42], therefore it may be also interesting to combine the beneficial deletions (G4$\Delta$, F175/V176$\rightarrow$F, A227$\Delta$ and G228$\Delta$) to explore their interactions.

## Conclusion
We have explored the structural plasticity of GFP to amino acid deletion on a whole-protein scale and showed that compromised fluorescence associated with deletions can be recovered by introducing folding-enhancing mutations. Our results suggested that a "size-optimized" GFP might be developed by iterative deletions of amino acids, followed by fluorescence rescue using folding mutations.

## Methods
### Construction of deletion library
pGFP$_{UV}$ vector (Clontech, Mountain View, CA) was used as the transposon target plasmid for *in vitro* transposition reaction. GFP$_{UV}$ protein carrying a 24 amino acid N-terminal tag is constitutively expressed from this vector. All of the four *Mly* I sites (249, 2334, 2836 and 3322) in pGFP$_{UV}$ vector were removed by site-directed mutagenesis without changing the amino acid sequence (see Additional file 4: Table S2 for primers). The engineered Mu transposon bearing chloramphenicol-resistant gene and *Mly* I sites was constructed as described [13]. Transposon DNA was released from pUC19 vector by *Bgl* II digestion, gel purified and then resolved on a 1 % agarose gel to determine DNA concentration and purity. Transposition reaction was performed in a 20 μL mixture containing 50 mM Tris-acetate, pH 7.5, 150 mM potassium acetate, 10 mM magnesium acetate, 4 mM spermidine, 570 ng of pGFPuv vector, 140 ng of transposon DNA (~1.3 molar excess) and 1 unit of HyperMu MuA transposase (Epicentre Biotechnologies, Madison, WI). The reaction was kept at 30 °C for 4 hrs and stopped by addition of 0.1 % SDS, followed by heat-inactivation at 70 °C for 10 min. The reaction product was transformed into chemically competent GeneHogs *Escherichia coli* cells and plated on LB agar containing 100 μg/mL ampicillin and 10 μg/mL chloramphenicol for selection of pGFP$_{UV}$ vector with transposon insertion. Approximately 38,000 colonies were collected and maxi-prepped to build the pGFP$_{UV}$-MuDel library.

pGFP$_{UV}$-transposon library DNA was digested with *EcoR* I/*Kpn* I, and the 2.0 kb DNA fragment corresponding to transposon-carrying GFP$_{UV}$ gene was re-ligated with the 2.6 kb vector backbone. This purified pGFP$_{UV}$-transposon library was then digested with *Mly* I to remove transposon DNA, leaving a three nucleotide scar at a random position of GFP$_{UV}$. The blunt-end intramolecular ligation was performed in a 20 μL reaction containing 50 mM Tris–HCl, pH7.5, 10 mM MgCl$_2$, 10 mM dithiothreitol (DTT) and 0.5 mM ATP, 300 ng DNA, 400 cohesive end units of T4 DNA ligase (NEB). The ligation product was transformed into chemically competent GeneHogs *E. coli* and the transformants were plated on LB agar containing 100 μg/mL ampicillin. In total, 10,000 colonies were collected, and maxi-prepped to build the triplet nucleotide deletion library.

### Screening for deletion mutants
The deletion library DNA was transformed into Gene-Hogs and the transformants were plated on LB agar supplemented with 100 μg/mL ampicillin at a density of 500 colonies per plate. Transformed cells were grown at 20 °C for 30 h and screened for fluorescence by visual inspection using Spectronics (Westbury, New York) model TC312E UV transilluminator under 310 nm wavelength. Forty fluorescent and 24 non-fluorescent colonies were selected and sequenced for further characterization.

## Liquid culture whole-cell fluorescence

Plasmids encoding wild-type $GFP_{UV}$ (wt$GFP_{UV}$) and deletion mutants were transformed into GeneHogs and plated on LB agar containing 100 μg/mL ampicillin at a density of ~200 colonies per plate. Single colonies with a diameter of ~0.5 mm were inoculated into 2 mL of liquid LB medium supplemented with 100 μg/mL ampicillin and grown at 37, 30 and 23 °C for 14, 18 and 25 h, respectively. Cells were harvested by centrifugation, washed with 500 μL TNG buffer (100 mM Tris–HCl, pH 7.5, 150 mM NaCl and 10 % glycerol) and then resuspended in 100 μL TNG buffer [19]. One milliliter of cell resuspension with an $OD_{600}$ of $0.150 \pm 0.003$ was prepared for fluorescence assay and the remaining cells in TNG buffer were stored at –20 °C for further experiments. Whole-cell fluorescence was determined using a model F4500 fluorescence spectrophotometer (Hitachi, Tokyo, Japan) as described [19]. The excitation and emission wavelength was set as 397 nm and 509 nm, respectively. Background fluorescence of empty GeneHogs cells was subtracted from each reading. All data were normalized to those of wt$GFP_{UV}$ under the same conditions. Three experimental replicates were performed for each sample.

## Protein expression and fraction soluble

For wild-type and mutant $GFP_{UV}$, 300 μL cell suspension in TNG buffer with an $OD_{600}$ of 0.100 was sonicated by Branson model 450 sonicator (Branson Ultrasonics, Danbury, CT) equipped with a 1/2 inch horn and a 1/8 in. tip. The power output and duty time were both set 50 %. Cell suspension was forced to two sequences of 10 pulse sonication with an interval of 3 min. Following sonication, 150 μL cell lysis was centrifuged at 12,000 $g$ for 10 min and the supernatant was transferred into a new tube. SDS loading samples were prepared in a 30 μL solution containing 15 μL crude cell lysis (whole protein) or supernatant (soluble fraction) and 15 μg bovine serum albumin (BSA) protein (Sigma, St. Louis, MO) as an internal standard. Protein samples were resolved in 12 % acrylamide SDS-PAGE gels and analyzed by Image J (http://rsbweb.nih.gov/ij/). GFP expression was quantified based on the density ratio of GFP and BSA bands as described [19]. The fraction soluble of each sample was extrapolated from the density ratio of soluble GFP present in supernatant and overall GFP present in crude cell lysis. Three experimental replicates were performed for each sample.

## Protein expression and purification and determination

wt$GFP_{UV}$ and mutants with internal deletions were subcloned into pET28b(+) vector (EMD Chemicals Inc., San Diego, CA) using primers CTAgctagcATGAGTAAAGG AGAAGAACTT (*Nhe* I site in lowercase) and CCCaa

gcttTTATTTGTAGAGCTCATC (*Hind* III site in lowercase). pET28b vector encoding wild-type and mutant $GFP_{UV}$ with N-terminal His$_6$ tags were transformed into *E. coli* BL21 (DE3) cells (Stratagene Inc., La Jolla, CA) and spread on LB agar plates supplemented with 50 μg/mL kanamycin. A single colony was inoculated into 500 mL LB media supplemented with 50 μg/mL kanamycin and grown at 37 °C to an $OD_{600}$ of 0.8. Protein expression was induced with 1 mM isopropyl-β-D-1-galactopyronaside (IPTG) for 12 h at 20 °C. The next day, cells were centrifuged and resuspended in 25 mL binding buffer (100 mM HEPES, pH 7.5, 5 mM imidazole, 1 mM phenylmethanesulfonyl fluoride, PMSF). Cells were lysed by sonication and then centrifuged at 12,000 g for 30 min. Supernatant was transferred into a new tube and then loaded on to a column pre-packed with 1 mL HisLink resins (Promega Corporation, Madison, WI). The protein-bound resins were washed with 20 mL wash buffer (100 mM HEPES, pH 7.5, 20 mM imidazole) and then eluted with 5 mL elution buffer (100 mM HEPES, pH 7.5, 300 mM imidazole). Purified proteins were concentrated and buffer-exchanged into TNG buffer (100 mM Tris–HCl, pH 7.5, 150 mM NaCl and 10 % glycerol). The concentration of purified proteins were determined using Pierce BCA Protein Assay kit (Thermo Fisher Scientific, Rockford, IL).

## Biophysical properties of deletion mutants

The excitation and emission scan were performed using 6 μg/mL purified proteins in TNG buffer. An emission wavelength of 509 nm was used for excitation scan, while emission scan was performed using excitation wavelengths of 397 nm and 495 nm.

The concentration of matured GFP was determined using "base-denatured" method [25]. Briefly, 7 μM purified proteins were denatured in 0.1 M NaOH for 5 min at 25 °C. Absorption spectra from 300 nm to 600 nm were recorded. The concentration of matured GFPs was calculated from experimentally determined $A_{447}$ and previously reported $\varepsilon_{447}$ (44,000 $M^{-1}$ $cm^{-1}$ for denatured GFPs) [25] using Beer's Law. The efficiency of chromophore maturation was calculated based on the concentrations of matured and overall GFP.

The extinction coefficients of each sample at 397 nm ($\varepsilon_{397}$) and 495 nm ($\varepsilon_{495}$) were determined using Beer's Law: $A = \varepsilon \times l \times c$, where $A$ is experimentally determined absorbance, $l$ is path length and $c$ is the concentration of matured GFPs (= total GFP concentration × efficiency of chromophore maturation).

For the measurement of quantum yield, all protein samples were diluted to an $OD_{397}$ of 0.100 and then further diluted 100-fold with water. The emission spectra from 450 nm to 600 nm were scanned using an excitation wavelength of 397 nm. The quantum yield of wt$GPF_{UV}$ has been defined in a previous study as 0.79

[7]. The quantum yield of mutants was calculated by comparing their integrated area of emission spectra with that of wtGPF$_{UV}$. Three experimental replicates were performed for each measurement. The instrumental error was estimated to be 2 %.

## Kinetic refolding experiments

To characterize the refolding ability of the deletion mutants, protein samples were denatured in the following solutions: 20 mM Tris–HCl, pH 7.5, 100 mM NaCl, 1 mM ethylenediaminetetraacetic acid (EDTA), 1 mM DTT, 0.20 mg/mL protein and 6 M guanidine hydrochloride (GdnHCl). Protein unfolding was processed at 25 °C for 24 h. Refolding process was initiated by diluting the denaturation solution 20 fold using the same buffer without GdnHCl. The fluorescence recovery was monitored at 25 °C for 60 min. The refolding data were fitted into a double exponential equation with a parallel refolding model using Prism 4.0 (GraphPad Software Inc., La Jolla, CA).

## Fluorescence rescue by folding-enhancing mutations

Mutations F64L [21] and S30R [19] were introduced into wtGFP$_{UV}$ and deletion mutants by site-directed mutagenesis (see Additional file 4: Table S2 for primers). The whole-cell fluorescence of F64L- or S30R-rescued mutants was assayed as described above. Mutants with double internal deletions were constructed using site-directed mutagenesis and mutation F64L was then introduced into the double deletion mutants. The rescued whole-cell fluorescence was assayed at 20 °C as described above.

### Abbreviations

GFP: Green fluorescent protein; EGFP: Enhanced green fluorescent protein; GFP$_{UV}$: UV-optimized GFP variant; wtGFP$_{UV}$: Wide-type GFP$_{UV}$; sgGFP: SuperGlo GFP; BSA: Bovine serum albumin; IPTG: Isopropyl-β-D-1-galactopyronaside; PMSF: Phenylmethanesulfonyl fluoride; HEPES: 2-[4-(2-Hydroxyethyl)-1-piperazinyl] ethanesulfonic acid; EDTA: Ethylenediaminetetraacetic acid; GdnHCl: Guanidine hydrochloride; LB: Lysogeny broth.

### Competing interests

The authors declare that they have no competing interests.

### Authors' contributions

SSL, JL and BJ designed experiments. SSL, XW, XD and LX performed the experiments. SSL, JL and BJ analysed the data and wrote the paper. All authors read and approved the final manuscript.

### Acknowledgements

We thank Dr. K.A. Daggett and J. Zheng for their help on library construction and Dr. T.A. Cropp for critical discussion. This work was supported by ShanghaiTech University, NIH grant R01GM084396 and University of Maryland.

### Author details

$^{1}$Shanghai Institute for Advanced Immunochemical Studies, ShanghaiTech University, Shanghai, China. $^{2}$Department of Chemistry and Biochemistry, University of Maryland, College Park, USA.

### References

1. Shimomura O, Johnson FH, Saiga Y. Extraction, purification and properties of aequorin, a bioluminescent protein from the luminous hydromedusan, Aequorea. J Cell Comp Physiol. 1962;59:223–39.
2. Prasher DC, Eckenrode VK, Ward WW, Prendergast FG, Cormier MJ. Primary structure of the Aequorea victoria green-fluorescent protein. Gene. 1992;111(2):229–33.
3. Shaner NC, Patterson GH, Davidson MW. Advances in fluorescent protein technology. J Cell Sci. 2007;120(Pt 24):4247–60.
4. Shaner NC, Steinbach PA, Tsien RY. A guide to choosing fluorescent proteins. Nat Methods. 2005;2(12):905–9.
5. Campbell RE, Tour O, Palmer AE, Steinbach PA, Baird GS, Zacharias DA, et al. A monomeric red fluorescent protein. Proc Natl Acad Sci U S A. 2002;99(12):7877–82.
6. Zacharias DA, Violin JD, Newton AC, Tsien RY. Partitioning of lipid-modified monomeric GFPs into membrane microdomains of live cells. Science. 2002;296(5569):913–6.
7. Voss U, Larrieu A, Wells DM. From jellyfish to biosensors: the use of fluorescent proteins in plants. Int J Dev Biol. 2013;57(6–8):525–33.
8. Dopf J, Horiagon TM. Deletion mapping of the Aequorea victoria green fluorescent protein. Gene. 1996;173(1 Spec No):39–44.
9. Li X, Zhang G, Ngo N, Zhao X, Kain SR, Huang CC. Deletions of the Aequorea victoria green fluorescent protein define the minimal domain required for fluorescence. J Biol Chem. 1997;272(45):28545–9.
10. Flores-Ramirez G, Rivera M, Morales-Pablos A, Osuna J, Soberon X, Gaytan P. The effect of amino acid deletions and substitutions in the longest loop of GFP. BMC Chem Biol. 2007;7:1.
11. Ormo M, Cubitt AB, Kallio K, Gross LA, Tsien RY, Remington SJ. Crystal structure of the Aequorea victoria green fluorescent protein. Science. 1996;273(5280):1392–5.
12. Yang F, Moss LG, Phillips Jr GN. The molecular structure of green fluorescent protein. Nat Biotechnol. 1996;14(10):1246–51.
13. Jones DD. Triplet nucleotide removal at random positions in a target gene: the tolerance of TEM-1 beta-lactamase to an amino acid deletion. Nucleic Acids Res. 2005;33(9):e80.
14. Arpino JA, Reddington SC, Halliwell LM, Rizkallah PJ, Jones DD. Random single amino acid deletion sampling unveils structural tolerance and the benefits of helical registry shift on GFP folding and structure. Structure. 2014;22(6):889–98.
15. Arpino JA, Rizkallah PJ, Jones DD. Structural and dynamic changes associated with beneficial engineered single-amino-acid deletion mutations in enhanced green fluorescent protein. Acta Crystallogr D Biol Crystallogr. 2014;70(Pt 8):2152–62.
16. Haapa S, Taira S, Heikkinen E, Savilahti H. An efficient and accurate integration of mini-Mu transposons in vitro: a general methodology for functional genetic analysis and molecular biology applications. Nucleic Acids Res. 1999;27(13):2777–84.
17. Crameri A, Whitehorn EA, Tate E, Stemmer WP. Improved green fluorescent protein by molecular evolution using DNA shuffling. Nat Biotechnol. 1996;14(3):315–9.
18. Daggett KA, Layer M, Cropp TA. A general method for scanning unnatural amino acid mutagenesis. ACS Chem Biol. 2009;4(2):109–13.
19. Pedelacq JD, Cabantous S, Tran T, Terwilliger TC, Waldo GS. Engineering and characterization of a superfolder green fluorescent protein. Nat Biotechnol. 2006;24(1):79–88.
20. Tsien RY. The green fluorescent protein. Annu Rev Biochem. 1998;67:509–44.
21. Cormack BP, Valdivia RH, Falkow S. FACS-optimized mutants of the green fluorescent protein (GFP). Gene. 1996;173(1 Spec No):33–8.
22. Fukuda H, Arai M, Kuwajima K. Folding of green fluorescent protein and the cycle3 mutant. Biochemistry (Mosc). 2000;39(39):12025–32.
23. Battistutta R, Negro A, Zanotti G. Crystal structure and refolding properties of the mutant F99S/M153T/V163A of the green fluorescent protein. Proteins. 2000;41(4):429–37.
24. Andrews BT, Schoenfish AR, Roy M, Waldo G, Jennings PA. The rough energy landscape of superfolder GFP is linked to the chromophore. J Mol Biol. 2007;373(2):476–90.

25. Ward WW. Bioluminescence and chemiluminescence. New York: Academic; 1981.

26. Patterson GH, Knobel SM, Sharif WD, Kain SR, Piston DW. Use of the green fluorescent protein and its mutants in quantitative fluorescence microscopy. Biophys J. 1997;73(5):2782–90.

27. Pascarella S, Argos P. Analysis of insertions/deletions in protein structures. J Mol Biol. 1992;224(2):461–71.

28. Toth-Petroczy A, Tawfik DS. Protein insertions and deletions enabled by neutral roaming in sequence space. Mol Biol Evol. 2013;30(4):761–71.

29. Afriat-Jurnou L, Jackson CJ, Tawfik DS. Reconstructing a missing link in the evolution of a recently diverged phosphotriesterase by active-site loop remodeling. Biochemistry (Mosc). 2012;51(31):6047–55.

30. Goldsmith M, Tawfik DS. Directed enzyme evolution: beyond the low-hanging fruit. Curr Opin Struct Biol. 2012;22(4):406–12.

31. Murakami H, Hohsaka T, Sisido M. Random insertion and deletion of arbitrary number of bases for codon-based random mutation of DNAs. Nat Biotechnol. 2002;20(1):76–81.

32. Fujii R, Kitaoka M, Hayashi K. RAISE: a simple and novel method of generating random insertion and deletion mutations. Nucleic Acids Res. 2006;34(4):e30.

33. Simm AM, Baldwin AJ, Busse K, Jones DD. Investigating protein structural plasticity by surveying the consequence of an amino acid deletion from TEM-1 beta-lactamase. FEBS Lett. 2007;581(21):3904–8.

34. Baldwin AJ, Busse K, Simm AM, Jones DD. Expanded molecular diversity generation during directed evolution by trinucleotide exchange (TriNEx). Nucleic Acids Res. 2008;36(13):e77.

35. Liu J, Cropp TA. Experimental methods for scanning unnatural amino acid mutagenesis. Methods Mol Biol. 2012;794:187–97.

36. Baldwin AJ, Arpino JA, Edwards WR, Tippmann EM, Jones DD. Expanded chemical diversity sampling through whole protein evolution. Mol Biosyst. 2009;5(7):764–6.

37. Arpino JA, Baldwin AJ, McGarrity AR, Tippmann EM, Jones DD. In-frame amber stop codon replacement mutagenesis for the directed evolution of proteins containing non-canonical amino acids: identification of residues open to bio-orthogonal modification. PLoS One. 2015;10(5):e0127504.

38. Edwards WR, Busse K, Allemann RK, Jones DD. Linking the functions of unrelated proteins using a novel directed evolution domain insertion method. Nucleic Acids Res. 2008;36(13):e78.

39. Arpino JA, Czapinska H, Piasecka A, Edwards WR, Barker P, Gajda MJ, et al. Structural basis for efficient chromophore communication and energy transfer in a constructed didomain protein scaffold. J Am Chem Soc. 2012;134(33):13632–40.

40. Liu J, Cropp TA. A method for multi-codon scanning mutagenesis of proteins based on asymmetric transposons. Protein Eng Des Sel. 2012;25(2):67–72.

41. Liu J, Cropp TA. Rational protein sequence diversification by multi-codon scanning mutagenesis. Methods Mol Biol. 2013;978:217–28.

42. Dellus-Gur E, Elias M, Caselli E, Prati F, Salverda ML, de Visser JA, et al. Negative Epistasis and Evolvability in TEM-1 beta-Lactamase-The Thin Line between an Enzyme's Conformational Freedom and Disorder. J Mol Biol. 2015;427(14):2396–409.

# 18

# Complex kinetics and residual structure in the thermal unfolding of yeast triosephosphate isomerase

Ariana Labastida-Polito[1], Georgina Garza-Ramos[2], Menandro Camarillo-Cadena[1], Rafael A. Zubillaga[1] and Andrés Hernández-Arana[1*]

## Abstract

**Background:** *Saccharomyces cerevisiae* triosephosphate isomerase (yTIM) is a dimeric protein that shows noncoincident unfolding and refolding transitions (hysteresis) in temperature scans, a phenomenon indicative of the slow forward and backward reactions of the native-unfolded process. Thermal unfolding scans suggest that no stable intermediates appear in the unfolding of yTIM. However, reported evidence points to the presence of residual structure in the denatured monomer at high temperature.

**Results:** Thermally denatured yTIM showed a clear trend towards the formation of aggregation-prone, β-strand-like residual structure when pH decreased from 8.0 to 6.0, even though thermal unfolding profiles retained a simple monophasic appearance regardless of pH. However, kinetic studies performed over a relatively wide temperature range revealed a complex unfolding mechanism comprising up to three observable phases, with largely different time constants, each accompanied by changes in secondary structure. Besides, a simple sequential mechanism is unlikely to explain the observed variation of amplitudes and rate constants with temperature. This kinetic complexity is, however, not linked to the appearance of residual structure. Furthermore, the rate constant for the main unfolding phase shows small, rather unvarying values in the pH region where denatured yTIM gradually acquires a β-strand-like conformation. It appears, therefore, that the residual structure has no influence on the kinetic stability of the native protein. However, the presence of residual structure is clearly associated with increased irreversibility.

**Conclusions:** The slow temperature-induced unfolding of yeast TIM shows three kinetic phases. Rather than a simple sequential pathway, a complex mechanism involving off-pathway intermediates or even parallel pathways may be operating. β-strand-type residual structure, which appears below pH 8.0, is likely to be associated with increased irreversible aggregation of the unfolded protein. However, this denatured form apparently accelerates the refolding process.

## Background

It is now accepted that many proteins fold and unfold following complex kinetic models [1]. The most detailed kinetic studies of conformational change have been performed on small monomeric proteins by means of rapid mixing or fast temperature jumps, because protein molecules of this sort usually unfold reversibly but with relaxation times ranging from less than a millisecond to a few seconds [2–4]. Previous studies have demonstrated the presence of transiently populated intermediates, apart from the native and unfolded end-states [1, 3]. Intermediate states may be found either on- or off-pathway, and their interconnections may even result in the consolidation of parallel, competing folding-unfolding pathways [5, 6]. Furthermore, the combination of experimental studies and molecular dynamics simulations has provided detailed structural descriptions of the multiple intermediates and transition states involved [7]. Recently, strong emphasis has been placed on the structural characterization of

* Correspondence: aha@xanum.uam.mx
[1]Área de Biofisicoquímica, Departamento de Química, Universidad Autónoma Metropolitana-Iztapalapa, San Rafael Atlixco 186, Iztapalapa D.F. 09340, Mexico
Full list of author information is available at the end of the article

unfolded states, because the presence of residual, native-like structure in parts of an otherwise unfolded polypeptide chain may be implicated in the speed of folding, as well as in the formation of misfolded molecules [8, 9].

However, there are examples of proteins that show very slow unfolding-refolding kinetics in the transition region (i.e., under conditions where the native and unfolded states are both significantly populated at equilibrium). Specifically, when unfolding is promoted by adding GuHCl or urea, slow-unfolding proteins take days to weeks to equilibrate, whereas for fast-unfolding proteins under similar conditions, equilibrium is reestablished in just a few hours [10–12]. Thus, if incubation times in the denaturing agent are not long enough, a slow-unfolding protein would display noncoincident unfolding and refolding profiles as the concentration of denaturing agent is varied (hysteresis). Likewise, hysteresis has been nicely demonstrated in the temperature-induced transitions of at least four proteins: an immunoglobulin light chain (monomer) [13], the Lpp-56 three-stranded α-helical coiled coil [14], and two dimeric triosephosphate isomerases [15, 16]. In these cases, thermal transitions detected by circular dichroism (CD) appear to be consistent with a two-state model with no intermediates.

Regarding trisosephosphate isomerase (TIM), many mesophilic members belonging to this enzyme family have been found to unfold slowly in chemical-denaturation studies, with one or more equilibrium or kinetic intermediates [11, 17, 18]. In contrast, thermal unfolding transitions of TIMs (in the absence of chemical denaturants) usually manifest themselves as monophasic profiles (i.e., simple sigmoidal curves with no evidence of intermediates), as recorded by CD [15, 19–21]. Unfortunately, irreversibility appears as a common feature in thermal unfolding, which has precluded the study of TIM refolding in cooling scans. Nevertheless, Benítez-Cardoza et al. [15] demonstrated that yeast TIM (yTIM) thermal unfolding is highly reversible at low protein concentration (≈0.20 μM), although the unfolding-refolding cycle displays marked hysteresis when a heating-cooling rate of 2.0 °C min$^{-1}$ is used. Attempts to achieve near-equilibrium transition profiles by decreasing the scan rate led to pronounced irreversibility [15].

At a fixed temperature, kinetic data for yTIM unfolding registered by far-UV circular dichroism (CD) in a restricted time span, are well-fitted by single exponential curves, whereas near-UV CD and fluorescence indicate biphasic kinetics. Refolding data are consistent with a second-order reaction [15]. Unlike yeast TIM, the enzyme from *Trypanosoma cruzi* (TcTIM) shows completely irreversible, temperature-induced denaturation, even at low protein concentration. Kinetic studies of this protein found that denaturation is a complex process in which two or three phases are clearly seen [19]. A common finding for both yTIM and TcTIM is that their denatured states appear to conserve some kind of residual structure based on calorimetric data [15, 19].

This work mainly focuses on determining the kinetic characteristics of temperature-induced yTIM unfolding in aqueous solution over long durations and in a wide pH range. Regardless of pH, three kinetic phases were observed, although the small-amplitude faster phase was detected at only low temperatures. The relative amplitudes of the second and third phases vary with temperature in a way that seems difficult to explain by a sequential mechanism. The results thus evidence that the kinetics of yTIM thermal unfolding is more complex than previously thought. Furthermore, residual secondary structure was found in denatured yTIM below pH 8.0. Because this residual structure appears to be associated with the loss of refolding ability, its presence may indicate that misfolded, aggregation-prone structures are formed at high temperature. Molecular dynamics simulations showed that yTIM has a tendency to suffer α-to-β transitions when unfolded at high temperature, but this method does not properly reproduce the marked effect of pH on the structure of the thermally unfolded protein.

## Materials

Overexpresion and purification of wild-type *Saccharomyces cerevisiae* TIM (yTIM) was carried out as described elsewhere [22]. Mass spectrometry (Additional file 1) and SDS-PAGE showed that the obtained enzyme was homogeneous. Enzymatic activity was determined by the coupled assay with α-glycerophosphate dehydrogenase (α-GDH), using D-glyceraldehyde 3-phosphate (DGAP) as the TIM substrate [23]. Assays were performed at 25.0 °C in 1.0 mL of 0.1 M triethanolamine buffer (pH 7.4) containing 10 mM EDTA, 0.20 mM NADH, 0.02 of α-GDH, and 2.0 mM DGAP; the reaction was started by the addition of 3.0 ng of yTIM, and NADH oxidation was followed by the change in absorbance at 340 nm. The catalytic efficiency ($k_{cat}/K_M$) of this enzyme was 5.0 x 10$^6$ s$^{-1}$ M$^{-1}$, a value similar to that reported previously [24, 25].

## Circular dichroism spectra

Circular dichroism (CD) spectra were obtained with a JASCO J-715 instrument (Jasco Inc., Easton, MD) equipped with a Peltier-type cell holder for temperature control and stirring with a magnetic bar. Cells of 1.00-cm path length were used to keep the protein concentration near 10 μg mL$^{-1}$ (0.19 μM). Although this somewhat restricted the lower wavelength limit of data registering, a low concentration is mandatory to observe reversible thermal unfolding scans [15]. CD spectral data are reported as mean residue ellipticity, [θ], which was calculated as [θ] = 100 θ/($C$ $l$); in this expression θ is the measured ellipticity in degrees, $C$ is the mean residue molar concentration (mean residue $M_r = 107.5$), and $l$ is the cell path length in centimeters.

## Thermal transitions

Conformational changes induced by heating or cooling of yTIM were continuously monitored by following the ellipticity at 220 nm while temperature was varied at 2.0 °C min$^{-1}$. Samples ($\approx$0.19 μM) were placed in a 1.00-cm cell with a magnetic stirrer, and the temperature within the cell was registered by the external probe of the Peltier-type accessory. Refolding profiles were registered immediately after the unfolding transitions had been completed.

## Kinetics studies

Unfolding kinetics tracings were registered by following ellipticity changes at 220 nm, as described previously [15, 25]. Unfolding was initiated by adding a small aliquot of concentrated TIM solution to a 1.00-cm cell containing buffer equilibrated at the temperature selected for each experiment. Within the cell, the temperature reached ±0.15 °C of the final equilibrium value in about 15 s. The final protein concentration was 0.19 μM in most cases. Essentially, the same procedure was used for monitoring changes in intrinsic fluorescence over time. In this case, experiments were carried out in a K2 spectrofluorometer (ISS, Champaign, IL), which had a Peltier accessory. Protein samples were excited at 292 nm, and the light emitted at 318 nm was collected. Kinetic data were analyzed using a triple exponential decay equation:

$$y = y_0 + A_1[\exp(-\lambda_1 t)-1] + A_2[\exp(-\lambda_2 t)-1] \\ + A_3[\exp(-\lambda_3 t)-1] \quad (1)$$

where $y$ is the physical observable monitored as a function of time $t$, and $y_0$ is the initial value of the observable. $A_i$ and $\lambda_i$ represent the observed amplitude and rate constant, respectively, for the $i$th exponential phase. In some cases, only two exponential terms were required for satisfactory curve fitting.

In refolding experiments, yTIM samples were first subjected to unfolding for 10 min at 63.0 °C. Then, the temperature control of the CD spectrometer Peltier accessory was set to a value 4.0 °C below the temperature intended for the study of the refolding reaction (42.0 °C) to allow for fast cooling of the sample ($\approx$15 °C min$^{-1}$). The final temperature value (42.0 °C) was entered into the cell-holder control when the solution in the cell was 0.5 °C above that value, and the CD signal was registered thereafter. Inside the cell, temperature came to equilibrium (±0.15 °C) in approximately 40 s.

## Molecular dynamics simulations

Molecular dynamics (MD) simulations were performed using GROMACS 4.5.4 software [26] with the GROMOS96 53A6 force-field [27]. The side-chain ionization states in the protein at the pH values simulated (6.7, 7.4, and 8.0) were established using $pK_a$ values estimated with PROPKA [28]. Dimeric yTIM (PDB ID: 1YPI) was placed in the center of a periodic dodecahedral box with 10 Å between the protein and the edge of the box. To simulate the solvent conditions at pH 6.7 (7.4; 8.0), a total of 21,763(21,757; 21 751) SPC water molecules, 12 (16; 20) sodium ions, and 7 (10; 12) chloride ions were needed to fill the box, neutralize the net protein charge, and reach the experimental ionic strength of 0.015 M (0.022 M; 0.027 M).

Prior to MD simulations, the system was relaxed by energy minimization, followed by 100 ps of thermal equilibration under the position restraints of protein heavy atoms through a harmonic force constant of 1000 kJ mol$^{-1}$ nm$^{-1}$. MD simulation was performed using an NPT ensemble at 423 K and 1.0 bar for 100 ns. A LINCS algorithm was applied to constrain the length of all covalent bonds [29], and a 2-fs time step was used. A cutoff of 1.0 nm was applied for short-range electrostatic and van der Waals interactions, while the long-range electrostatic forces were treated using the particle mesh Ewald method [30]. Two replicas were simulated at each solvent condition.

## Results and discussion

### Unfolding-refolding thermal transitions

Denaturation (unfolding) and renaturation (refolding) of yTIM were followed by continuous monitoring of the ellipticity (220 nm) under a constant heating or cooling rate of 2.0 °C min$^{-1}$. Temperature scanning profiles recorded at three different pH values are shown in Fig. 1. These profiles display the hysteresis phenomenon previously observed for yTIM [15, 16, 25], which indicates that unfolding and refolding events occur under kinetic control at the imposed scanning rate [14, 15]. It is clear that pH has an influence on the kinetic stability of the protein, because the apparent *melting temperature* is displaced to lower values at pH 8.5. Despite this pH effect, all the unfolding traces appear as sigmoid curves, with no evidence of stable intermediates. However, the total ellipticity change at pH 6.7 that takes place upon denaturation seems slightly larger (Fig. 1). It must be noted that the up-temperature scans in Fig. 1 were not allowed to proceed to higher temperatures to avoid reactions that make the process irreversible and thus decrease the extent of refolding on down-temperature scans [15].

In a different set of experiments, denatured samples of the enzyme were left to stand at 70 °C for 10 min to ensure that unfolding had been completed before their CD

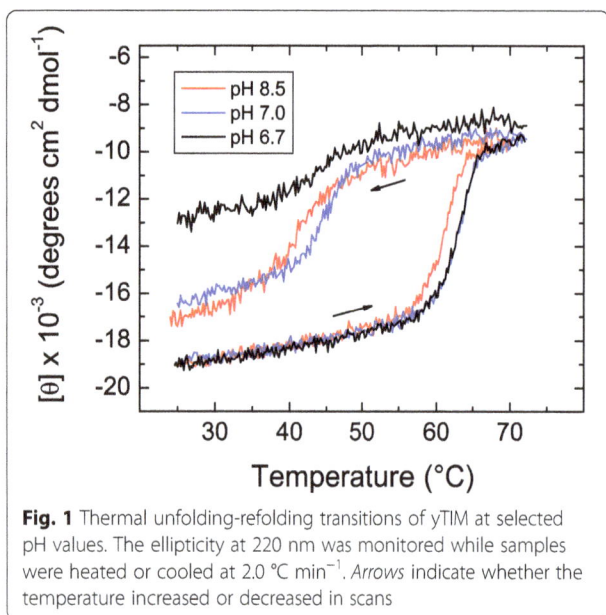

**Fig. 1** Thermal unfolding-refolding transitions of yTIM at selected pH values. The ellipticity at 220 nm was monitored while samples were heated or cooled at 2.0 °C min$^{-1}$. *Arrows* indicate whether the temperature increased or decreased in scans

spectra were recorded. Spectra shown in Fig. 2 indicate that the native α/β secondary structure of yTIM is rather insensitive to pH and is largely lost upon heating at all pH values, as judged by the decrease in magnitude in the 208–222 nm region at high temperature (Fig. 2). However, the spectrum of heat-denatured yTIM shows striking changes as pH is varied. Above pH 8.0, the spectral shape and signal magnitude of the denatured enzyme are typical of small and medium-size proteins (e.g., hen-egg lysozyme, ribonuclease A, cytochrome C, staphylococcal nuclease, cysteine proteinases) when unfolded at high

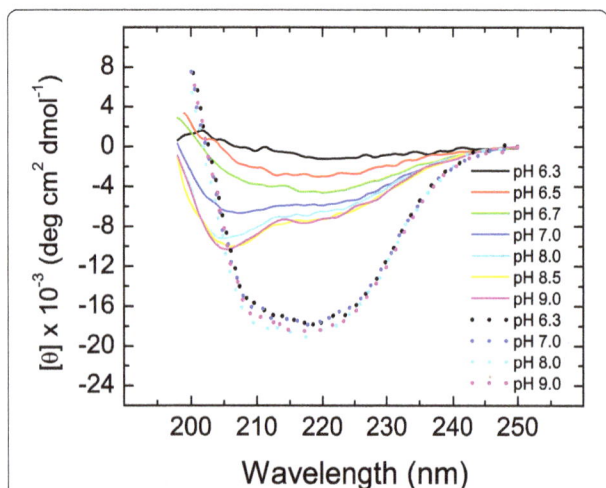

**Fig. 2** Far-UV CD spectra of thermally unfolded yTIM at different pH values. Protein samples were allowed to unfold by continuous heating (2 °C min$^{-1}$) until the end of the transition (cf. Fig. 1) and then left to stand at 70.0 °C for 10 min before recording their spectra. For comparison, spectra of native yTIM at various pH values are also shown (*dotted lines*)

temperatures in the absence of denaturant agents (see, for example, CD spectra of native and thermally unfolded lysozyme in Additional file 2). This spectral type is characterized by a negative band of approximately $10 \times 10^3$ deg cm$^2$ dmol$^{-1}$ centered at 202–204 nm, along with a broad negative shoulder with magnitude around $5 \times 10^3$ deg cm$^2$ dmol$^{-1}$ at longer wavelength [31, 32]. Below pH 8.0, the spectra of heat-denatured yTIM progressively decrease in magnitude and acquire a shape typical of all-β proteins [33], thus pointing to the presence of *residual* secondary structure in the denatured enzyme.

Regarding yTIM refolding in cooling-down scans, it is evident that this process becomes increasingly irreversible as the pH decreases below pH 7.0, as judged from the extent of recovery of native yTIM ellipticity shown in Fig. 1. To gain detailed information on the influence of pH in the unfolding and refolding events, further kinetic experiments were carried out.

### Unfolding kinetics

Kinetic studies were carried out by monitoring the time course of ellipticity at 220 nm. Experiments examining a large temperature interval were done at pH 8.0 and 6.7, where the CD spectra of denatured yTIM showed distinct features. The results at pH 8.0 indicate that at relatively high temperatures (60.0 °C and above), the loss of secondary structure shows double exponential behavior (Fig. 3), with phases well separated on the time scale. Indeed, over a restricted time interval, a single exponential-decay equation can fit the experimental data reasonably well. Only when data were recorded over a long time did a second phase become readily apparent, but this phase had small amplitude. Nevertheless, at low temperature, triple exponential behavior was observed (Fig. 3). The fastest phase, which conveys a minor ellipticity change, occurred too fast for accurate assessment of the kinetic constant by the manual-mixing method (i.e., time constant of about 20 to 100 s). This fast phase seems to be completely lost within the dead time in experiments at high temperature. Hereafter, the observed rate constants are referred to as $\lambda_1$, $\lambda_2$, and $\lambda_3$, in descending order of their magnitudes. Unfolding of yTIM at pH 6.7 showed similar behavior, with two and three kinetic phases at high and low temperature, respectively, as shown in Additional file 3.

CD spectra were recorded near the end of the unfolding process when the slowest phase was more than 98 % complete (these experiments required recordings of kinetic data for more than nine hours in the case of low temperatures). The *final* spectra appeared nearly identical, notwithstanding the temperature at which the kinetics was studied (see Additional file 4 for results obtained at pH 8.0). Furthermore, at a given pH, the spectral shape and magnitude observed at the end of unfolding were both similar to those illustrated in Fig. 2. In other

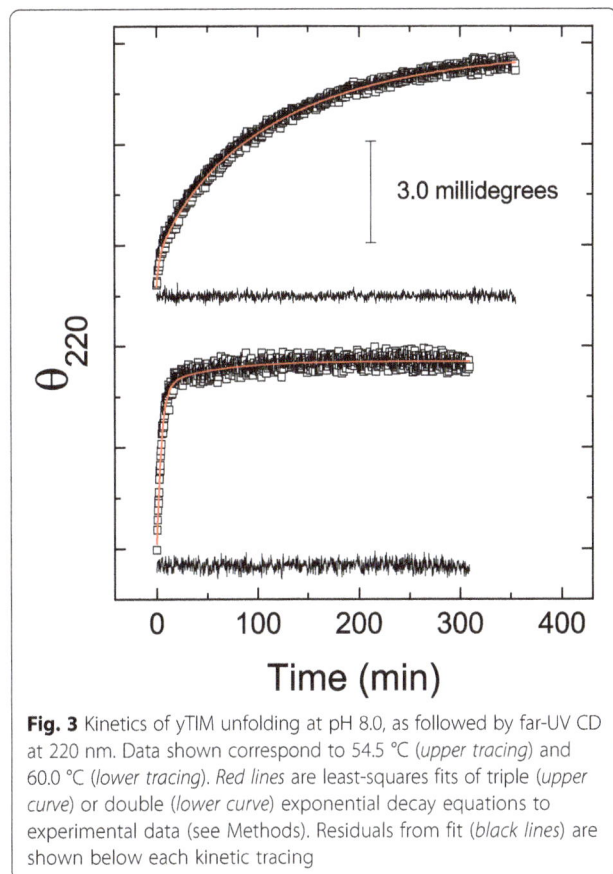

**Fig. 3** Kinetics of yTIM unfolding at pH 8.0, as followed by far-UV CD at 220 nm. Data shown correspond to 54.5 °C (*upper tracing*) and 60.0 °C (*lower tracing*). *Red lines* are least-squares fits of triple (*upper curve*) or double (*lower curve*) exponential decay equations to experimental data (see Methods). Residuals from fit (*black lines*) are shown below each kinetic tracing

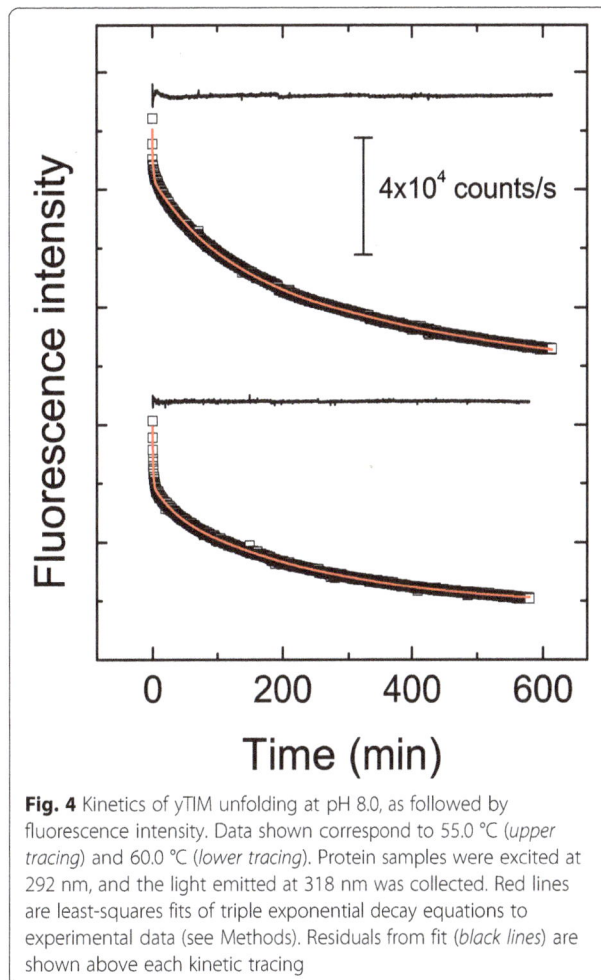

**Fig. 4** Kinetics of yTIM unfolding at pH 8.0, as followed by fluorescence intensity. Data shown correspond to 55.0 °C (*upper tracing*) and 60.0 °C (*lower tracing*). Protein samples were excited at 292 nm, and the light emitted at 318 nm was collected. Red lines are least-squares fits of triple exponential decay equations to experimental data (see Methods). Residuals from fit (*black lines*) are shown above each kinetic tracing

words, the final conformation achieved by the protein seems to be independent of the temperature (in the range studied), but is otherwise strongly affected by pH.

The voltage applied to the phototube of the CD instrument, which is proportional to the absorbance, was simultaneously recorded. The measurements indicated that changes in ellipticity associated with the first two phases are accompanied by only small changes (5.0 % or less) in the absorbance of the protein solution (see Additional file 5). Such small changes are known to occur due to alterations in the secondary and, to less extent, the tertiary structure of proteins and polypeptides [34]. However, a relatively large absorbance increment (approximately 10.0 % of the protein absorbance) was linked to the slower CD-detected kinetic phase. It is likely that this apparent increment comes from the scattering of light by aggregates of unfolded protein molecules.

Monitoring of the denaturation kinetics by changes in the fluorescence intensity also showed that this is a complex process (Fig. 4) in which there is a progressive decrease of intensity (at the wavelength of maximum emission by native yTIM). Overall, comparison of the plots shown in Figs. 3 and 4 indicates that progressive loss of secondary structure upon denaturation is accompanied by a quenching of the fluorescence signal of

tryptophan residues, which in turn likely reflects either the exposure of these residues to the aqueous solvent or less constraint by the environment [35]. Notwithstanding the temperature, three exponential terms were required to fit fluorescence data. As in CD experiments, the first fluorescence-detected phase was too fast (time constant of about 25 s) for an accurate determination of its rate constant. At low temperature (55.0 °C), the rate constant for the second phase had a value similar to that of $\lambda_2$ from CD experiments (the two values differed by 50–80 %). At 62.0 to 64.0 °C, however, it was the first fluorescence-detected rate constant that was consistent with $\lambda_2$. Furthermore, the decrease in the fluorescence intensity extended over a much longer time than the change in ellipticity (i.e., the rate constant for the slowest phase was approximately three- to fourfold smaller when determined from fluorescence than from CD). These markedly different values suggest that the slowest phase comprises several elementary steps that respond differently to the spectroscopic probes employed. For instance, formation of molecular aggregates can conceivably occur with little or no change in secondary conformation, but with an otherwise significant fluorescence quenching of tryptophan residues.

### Kinetic model for yTIM unfolding

The simplest model accounting for the results obtained from CD would be that of three sequential reactions (Scheme 1), with native and unfolded yTIM (N and U, respectively) and two intermediate species (I and X):

In this model, each of the three $\lambda$ values determined from data analysis (eqn. 1) is identical to each one of the microscopic rate constants $k_1$, $k_2$, and $k_3$. As mentioned, neither the rate constant nor the amplitude of the faster phase could be accurately determined from experiments at the lowest temperatures studied. Moreover, this phase was apparently lost within the dead time of experiments performed at high temperature. Fortunately, because $k_1$ seems to be 15–20 times larger than $k_2$, the first kinetic step occurs on a much shorter time scale than the other steps and can be regarded as kinetically separated from the other events, at least in a first approximation. This implies that amplitudes $A_2$ and $A_3$ reflect changes involved solely with steps $I \rightarrow X \rightarrow U$. Therefore, the kinetic model can be simplified to a two-step model (Scheme 2).

Equations describing the evolution in time of the fraction of each species are well known [36, 37]. By denoting the characteristic ellipticity of each species as $\theta_I$, $\theta_X$, and $\theta_U$, it can be shown that (see Additional file 6):

$$(\theta_X - \theta_I)/(\theta_U - \theta_I) = k_3/k_2 - [(k_3 - k_2)/k_2][A_2/(A_2 + A_3)] \quad (2)$$

and

$$(\theta_U - \theta_X)/(\theta_U - \theta_I) = -[(k_3 - k_2)/k_2] \times [A_3/(A_2 + A_3)] \quad (3)$$

The two equations above were used to compute $(\theta_X - \theta_I)/(\theta_U - \theta_I)$ and $(\theta_U - \theta_X)/(\theta_U - \theta_I)$, which give the ellipticity change as a fraction of the total change for each step in Scheme 2. The results indicate that the degree of unfolding occurring during the $I \rightarrow X$ step (normalized to a total unitary change) would vary from 0.35 to 0.70 over a temperature range of 11 °C (Fig. 5). For the $X \rightarrow U$ step, a concomitant decrease in the degree of unfolding would take place. Admittedly, it seems unlikely that the conformation of intermediate species would vary so drastically within such a narrow temperature range. Alternatively, these results may point to the presence of an off-pathway intermediate or even different, parallel unfolding pathways with predominance that changes with temperature.

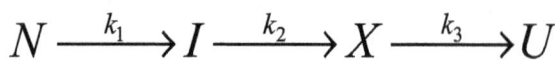

$$N \xrightarrow{k_1} I \xrightarrow{k_2} X \xrightarrow{k_3} U$$

**Scheme 1** Kinetic model for three sequential first-order reactions

$$I \xrightarrow{k_2} X \xrightarrow{k_3} U$$

**Scheme 2** Simplified kinetic model involving only two first-order steps

### Temperature dependence of unfolding rate constants

Further studies on the denaturation of the enzyme were also performed at other pH values but over a restricted temperature range to determine the *activation* parameters that control the temperature dependence of $k_2$ and $k_3$. Results for selected pH values are shown in Fig. 6 as Eyring plots, which agree with the well-known equation:

$$\ln(k/T) = \ln E + \Delta S^{\ddagger}/R - (\Delta H^{\ddagger}/R)(1/T) \quad (4)$$

where $k$ is the rate constant for an elementary reaction, $T$ is the absolute temperature, $E$ stands for a preexponential factor, and $\Delta S^{\ddagger}$ and $\Delta H^{\ddagger}$ represent the activation entropy and enthalpy, respectively. Figure 6a shows that plots corresponding to $k_2$ follow linear trends when a narrow temperature range is considered. This linearity was observed before for yTIM and has been found for a large number of other proteins [15]. However, in the cases of pH 6.7 and 8.0, at which larger temperature intervals were examined, Eyring plots appear slightly curved upwards in the low temperature region. This might be due to a shift between parallel unfolding pathways with different *activation* enthalpies [38]; that is, unfolding would switch from one predominant pathway to another as the temperature varies, in agreement with the interpretation mentioned for the change with

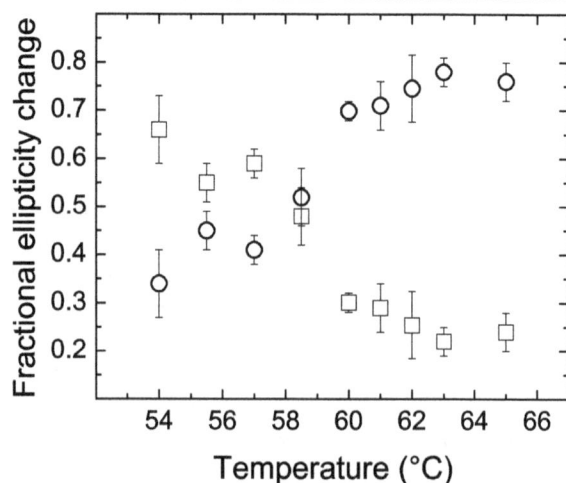

**Fig. 5** Fractional structural change for the two kinetic steps in Scheme 2 at different temperatures. Fractional changes in ellipticity were calculated from eqns. 2 and 3 from values of the rate constants and amplitudes determined at pH 6.7. Data for step $I \rightarrow X$, i.e., $(\theta_X - \theta_I)/(\theta_U - \theta_I)$, are shown as circles; data for step $X \rightarrow U$, i.e., $(\theta_U - \theta_X)/(\theta_U - \theta_I)$, are shown as *squares*

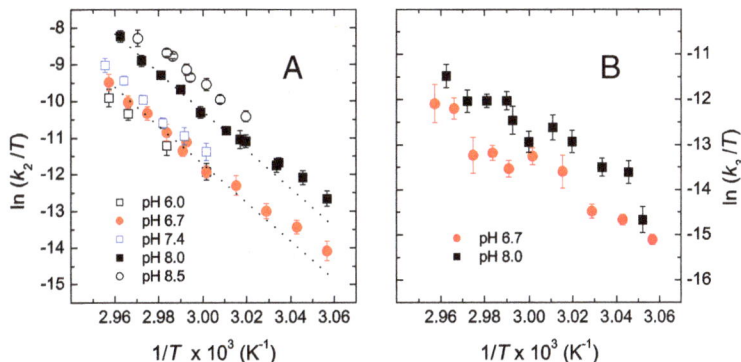

**Fig. 6** Eyring plots for the rate constants $k_2$(**a**) and $k_3$(**b**) at selected pH values. Rate constants were determined from far-UV CD kinetic experiments. Dotted lines in (**a**) are linear fits performed with data corresponding to temperatures above 60.0 °C for pH 6.7 and 8.0

temperature of the computed degree of unfolding for step I → X. However, a nonzero activation heat capacity cannot be ruled out as the origin of the curvature.

From Eyring plots, such as those in Fig. 6a, $\Delta H_2^{\ddagger}$ was determined between pH 6.0 and 8.5. It must be noted that values of $k_2$ were determined from data registered in a temperature region in which the unfolding degree accompanying step I → X remains relatively constant (i.e., from 60 to 65 °C, cf. Fig. 5). Therefore, $k_2$ can be assigned to a single predominant pathway. Overall, the value of $\Delta H_2^{\ddagger}$ was about 450 kJ mol$^{-1}$ at pH 6.0–8.0 and showed a slight decrease ($\approx$15 %) at pH 8.5 (data not shown). In contrast, Eyring plots for $k_3$ display a linear but ill-defined trend (Fig. 6b), suggesting that the slowest kinetic phase is indeed composed of several elementary steps. It is also seen that $k_3$ is much less temperature dependent than $k_2$.

The effect of pH on $k_2$ and $k_3$ was examined over a longer interval of pH values at constant temperature; 60.0 °C was chosen, because of the single apparent pathway at this temperature, and the unfolding process was slow enough to allow for determining the value of $k_2$ over an extended pH range. Results are shown in Fig. 7, which shows that pH-induced changes in $k_2$ resemble the sigmoid titration curve for an ionizable group with an approximate p$K_a$ of 8.5. Because this value of p$K_a$ is close to that of a thiol group, it may be hypothesized that a cysteine residue is responsible for the behavior observed for $k_2$. In this regard, it has been proposed that Cys126, which is a residue that is conserved with the family of TIM enzymes, plays an important role in the stability of this protein [24]. In contrast, $k_3$ values showed no defined variation with pH, again suggesting that the step X → U actually comprises multiple individual reactions.

### Refolding of yTIM

As reported previously [15, 24, 25], the kinetics of yTIM refolding at low protein concentration (0.13–0.75 μM) and in a certain temperature range is slow enough to be

monitored without resorting to fast temperature-jump techniques. By using the procedure described in the Methods section, we followed the recovery of secondary structure under two pH conditions. These studies were aimed at determining the effect of the *residual* native-like structure of unfolded yTIM (which is clearly observed at pH 6.7) on the refolding ability of the enzyme. For this purpose, yTIM samples were allowed to unfold (in the cell of the CD instrument) for 10 min at 63.0 °C. These conditions ensured ca. 85 % (pH 6.7) or 99 % (pH 8.0) unfolding, as judged by the ellipticity signal. After that, the protein solution was cooled to 42.0 °C to record the refolding reaction. Additional file 7 shows that at pH 6.7 the enzyme refolds faster than at pH 8.0. In both cases, however, refolding tracings are adequately described by a second-order kinetics equation, as determined previously [15, 24].

To explore the effect of the residual structure on the reversibility of the unfolding process, samples of yTIM

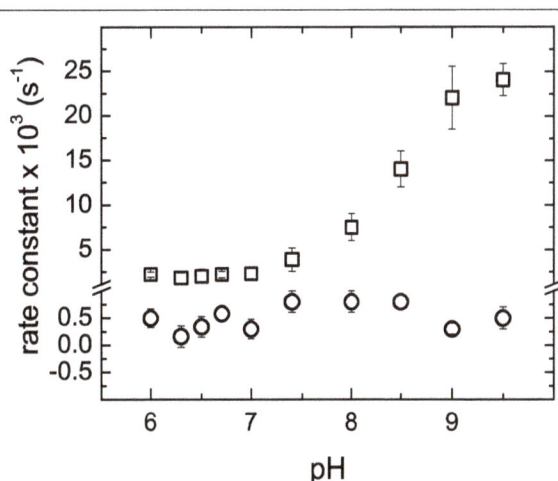

**Fig. 7** Variation of rate constants $k_2$ and $k_3$ with pH. Data for $k_2$ (squares) and $k_3$ (circles) were determined from far-UV CD kinetic experiments at 60.0 °C

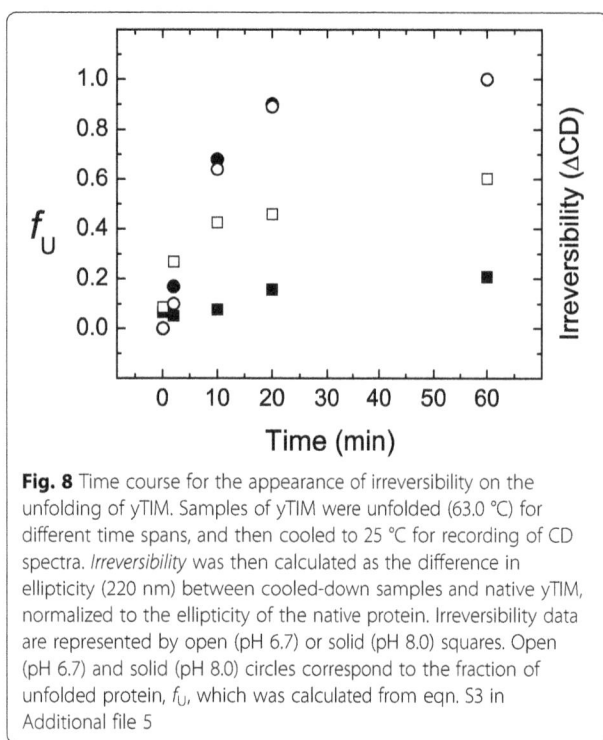

**Fig. 8** Time course for the appearance of irreversibility on the unfolding of yTIM. Samples of yTIM were unfolded (63.0 °C) for different time spans, and then cooled to 25 °C for recording of CD spectra. *Irreversibility* was then calculated as the difference in ellipticity (220 nm) between cooled-down samples and native yTIM, normalized to the ellipticity of the native protein. Irreversibility data are represented by open (pH 6.7) or solid (pH 8.0) squares. Open (pH 6.7) and solid (pH 8.0) circles correspond to the fraction of unfolded protein, $f_U$, which was calculated from eqn. S3 in Additional file 5

were unfolded for different time spans and then cooled to 25 °C to record CD spectra. As a quantitative indicator of irreversibility, the difference in ellipticity (220 nm) between cooled-down samples and native yTIM, normalized to the ellipticity of the native protein, was used. The results are shown in Fig. 8, together with the fractional values of U ($f_U$) in Scheme 2. Experimentally determined values of the kinetic constants $k_2$ and $k_3$ were used to calculate the time variation of $f_U$ according to eqn.S3 in Additional file 6. An inspection of the plots in the figure makes it evident that irreversibility is more intense at pH 6.7 than pH 8.0, as expected from the thermal scan results (cf. Fig. 1). At the lower pH, however, irreversibility begins with early unfolding times and approximately parallels the formation of $f_U$. In contrast, the onset of irreversibility at pH 8.0 is delayed and thus appears as a late event in unfolding, which takes place after the final U state becomes largely populated. This suggests again that the slowest CD-detected kinetic phase does not represent an elementary step. Reactions

that lead to irreversibility probably do not involve major changes in secondary conformation and are therefore silent in CD studies.

In summary, results from refolding studies indicate that yTIM refolds faster from the denatured state with residual structure, although such denatured state decreases the folding efficiency (i.e., the amount of native protein recovered upon refolding). Thus, it may be thought that under physiological conditions (pH near neutrality, 37 °C) the advantage of a fast folding process overcomes the difficulties posed by some degree of irreversibility. Furthermore, because irreversibility appears to be related to the time unfolded (denatured) YTIM stays at moderate to high temperatures [15], the problem of a low folding efficiency may be of less significance for mesophilic organisms such as *Saccharomyces cerevisiae*.

### Residual structure

As mentioned, thermally denatured yTIM retains a high content of β structure below pH 8.0 (see Fig. 2), which is implicated in reactions leading to irreversibility. This type of secondary structure has been found to be refractory to temperature in thermophilic and mesophilic proteins [39, 40], whereas in other instances, such as in apomyoglobin, β structure appears to be formed at elevated temperature [31] as a result of α-to-β transitions [41]. Furthermore, molecular dynamics simulations have shown that certain all-α peptides, and even full-length proteins, may be transformed to all-β structures [42, 43]. We carried out preliminary MD simulations to investigate whether this method can reproduce the structural differences in denatured yTIM that were experimentally observed when pH is varied. Simulations run at 400 K for 100 ns showed that helixes are completely lost after 75 ns, regardless of the pH value. Conversely, β-strands actually seem to be formed during the simulation, but they are slightly more abundant at pH 6.7 than at pH 8.0 (Additional file 8). Although preliminary, these results are encouraging, for they indicate that some regions in the polypeptide sequence of yTIM have a tendency to undergo α-to-β transitions. In contrast, the effect of pH does not appear to have been properly taken into account by the MD method used here, and thus deserves to be studied further.

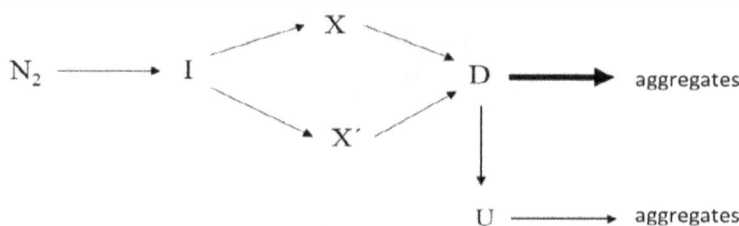

**Scheme 3** Proposed model for yTIM unfolding

## Conclusions

Two experimental approaches were used to study the influence of pH on the temperature-induced unfolding of yeast triosephosphate isomerase (yTIM). Temperature-scan experiments showed that unfolding profiles (monitored by CD) appear as monophasic transitions, with no evidence of intermediate species. pH was found to affect the *kinetic stability* of the protein based on shifts in the *melting temperature* $(T_m)$. Furthermore, below pH 8.0, CD spectra of heat-denatured yTIM gradually changed in shape to look like those for proteins rich in β-strands, but otherwise, the unfolded protein became prone to aggregate.

Despite the apparent simplicity of thermal profiles, kinetic studies performed at constant temperature clearly showed the presence of up to three kinetic phases, irrespective of pH (i.e., at high temperatures, the fastest phase was completely lost within the experimental dead time). Because the relative values of the kinetic constants suggested that the fastest phase is indeed *decoupled* from the other two, we analyzed the kinetic constants and amplitudes of the two slowest phases according to a two-step sequential mechanism. Results from the analysis, however, pointed to a more complex actual mechanism, such as one that involves parallel pathways. The temperature dependence of the rate constants appears to lend some evidence to this proposal. A simple model for yTIM unfolding that accounts for the information summarized above is shown in Scheme 3, where $N_2$ stands for the native dimer, I and X represent partially unfolded intermediates, and D and U are used to symbolize, respectively, the denatured form with β-strand residual structure and the thermally unfolded state of yTIM.

In summary, it was shown that the temperature-induced denaturation of yTIM reveals itself as a complex process when followed for a long time and over an ample temperature range. Further investigation over a wide pH range showed that the kinetic stability of yTIM responds to the titration of an ionizable group with $pK_a \approx 8.5$. Refolding studies, on the other hand, indicated that the refolding ability of the unfolded protein decreases under pH conditions that favor the formation of residual, β-strand-like structures in heat-denatured yTIM, even though refolding is faster under such conditions. Moreover, most of the reactions leading to irreversibility occur late in the unfolding process and are not detected by CD. Finally, as demonstrated in molecular dynamics simulations, yTIM unfolding shows α-to-β transition behavior, albeit with no discrimination of the experimentally observed pH effect.

## Additional files

Additional file 1: Mass Spectrum of isolated yTIM. (PDF 72 kb)

Additional file 2: Native and thermally unfolded hen-egg lysozyme CD spectra. (PDF 67 kb)

Additional file 3: Kinetics of yTIM unfolding at pH 6.7, as followed by CD. (PDF 210 kb)

Additional file 4: Far-UV CD spectra of yTIM near the end of the unfolding kinetics process. (PDF 89 kb)

Additional file 5: Kinetics of yTIM unfolding detected by light absorption. (PDF 142 kb)

Additional file 6: Derivation of eqns. 2 and 3 in text. (PDF 113 kb)

Additional file 7: Refolding kinetics of yTIM as followed by CD. (PDF 133 kb)

Additional file 8: Molecular dynamics simulations of yTIM unfolding. (PDF 113 kb)

### Abbreviations
yTIM: Triosephosphate isomerase from yeast (*Saccharomyces cerevisiae*); CD: Circular dichroism; MD: Molecular dynamics.

### Competing interests
The authors declare that they have no competing interests.

### Authors' contributions
ALP and MCC carried out most of the spectroscopic experiments, performed part of the data analysis, and participated in the experimental design and interpretation of results. GGR helped with the experiments, data analysis, and interpretation of results. RAZ performed the molecular dynamics simulations, participated in the interpretation of results, and helped in drafting the manuscript. AHA conceived of the study, participated in its design, carried out a major part of data analysis, and drafted the manuscript. All authors read and approved the final manuscript.

### Acknowledgements
This work was funded in part by CONACYT, México (SEP-CONACYT2007-80457, and SEP-CONACYT 2012-181049). ALP received a doctoral fellowship from CONACYT, México (208217). The authors thank Dr. Ponciano García-Gutiérrez (Laboratorio Interdivisional de Espectrometría de Masas, UAM-Iztapalapa) for obtaining the mass spectrum of yTIM.

### Author details
[1]Área de Biofisicoquímica, Departamento de Química, Universidad Autónoma Metropolitana-Iztapalapa, San Rafael Atlixco 186, Iztapalapa D.F. 09340, Mexico. [2]Departamento de Bioquímica, Facultad de Medicina, Universidad Nacional Autónoma de México, Coyoacán D.F. 04510, Mexico.

### References
1. Sánchez IE, Kiefhaber T. Evidence for sequential barriers and obligatory intermediates in apparent two-state protein folding. J Mol Biol. 2003;325:367–76.
2. Ferguson N, Fersht A. Early events in protein folding. Curr Op Struc Biol. 2003;13:75–81.
3. Kamagata K, Arai M, Kuwajima K. Unification of the folding mechanisms of non-two-state and two-state proteins. J Mol Biol. 2004;339:951–65.
4. Tsong TY. Detection of three kinetic phases in the thermal unfolding of ferricytochrome c. Biochemistry. 1973;12:2209–14.
5. Baldwin RL. On-pathway versus off-pathway folding intermediates. Folding & Design. 1996;1:R1–8.
6. Aghera N, Udgaonkar JB. The utilization of competing unfolding pathways of monellin is dictated by enthalpic barriers. Biochemistry. 2013;52:5770–9.
7. Travaglini-Allocatelli C, Ivarsson Y, Jemth P, Gianni S. Folding and stability of globular proteins and implications for function. Curr Opin Struct Biol. 2009;19:3–7.
8. Wong KB, Clarke J, Bond CJ, Joseˆ Luis Neira J, Freund SMV, Fersht AR, et al. Towards a complete description of the structural and dynamic properties of the denatured state of barnase and the role of residual structure in folding. J Mol Biol. 2000;296:1257–82.

9.  Pearce MC, Cabrita LD, Rubin H, Gore MG, Bottomley SP. Identification of residual structure within denatured antichymotrypsin: implications for serpin folding and misfolding. Biochem Biophys Res Commun. 2004;324:729–35.

10. Sawano M, Yamamoto H, Ogasahara K, Kidokoro S, Katoh S, Ohnuma T, et al. Thermodynamic basis for the stabilities of three CutA1s from *Pyrococcus horikoshii*, *Thermus thermophilus*, and *Oryza sativa*, with unusually high denaturation temperatures. Biochemistry. 2008;47:721–30.

11. Vázquez-Pérez AR, Fernández-Velasco DA. Pressure and denaturants in the unfolding of triosephosphate isomerase: the monomeric intermediates of the enzymes from *Saccharomyces cerevisiae* and *Entamoeba histolytica*. Biochemistry. 2007;46:8624–33.

12. Shirley BA. Urea and guanidine hydrochloride denaturation curves. In: Shirley BA, editor. Protein Stability and Folding. Theory and Practice. Totowa, NJ: Humana Press; 1995. p. 177–90.

13. Blancas-Mejía LM, Tischer A, Thompson JR, Tai J, Wang L, Auton M, et al. Kinetic control in protein folding for light chain amyloidosis and the differential effects of somatic mutations. J Mol Biol. 2014;426:347–61.

14. Dragan AI, Potekhin SA, Sivolob A, Lu M, Privalov PL. Kinetics and thermodynamics of the unfolding and refolding of the three-stranded R-helical coiled coil, Lpp-56. Biochemistry. 2004;43:14891–900.

15. Benítez-Cardoza CG, Rojo-Domínguez A, Hernández-Arana A. Temperature-induced denaturation and renaturation of triosephosphate isomerase from *Saccharomyces cerevisiae*: evidence of dimerization coupled to refolding of the thermally unfolded protein. Biochemistry. 2001;40:9049–58.

16. Samanta M, Banerjee M, Murthy MRN, Balaram H, Balaram P. Probing the role of the fully conserved Cys 126 in triosephosphate isomerase by site-specific mutagenesis – distal effects on dimer stability. FEBS Journal. 2011;278:1932–43.

17. Pan H, Raza AS, Smith DL. Equilibrium and kinetic folding of rabbit muscle triosephosphate isomerase by hydrogen exchange mass spectrometry. J Mol Biol. 2004;336:1251–63.

18. Guzman-Luna V, Garza-Ramos G. The folding pathway of glycosomal triosephosphate isomerase: Structural insights into equilibrium intermediates. Proteins. 2012;80:1669–82.

19. Mixcoha-Hernández E, Moreno-Vargas LM, Rojo-Domínguez A, Benítez-Cardoza CG. Thermal-unfolding reaction of triosephosphate isomerase from trypanosoma cruzi. Protein J. 2007;26:491–8.

20. Cabrera N, Hernández-Alcántara G, Mendoza-Hernández G, Gómez-Puyou A, Perez-Montfort R. Key residues of loop 3 in the interaction with the interface residue at position 14 in triosephosphate isomerase from *Trypanosoma brucei*. Biochemistry. 2008;47:3499–506.

21. Dhaunta N, Arora K, Chandrayan SK, Guptasarma P. Introduction of a thermophile-sourced ion pair network in the fourth beta/alpha unit of a psychophile-derived triosephosphate isomerase from Methanococcoides burtonii significantly increases its kinetic thermal stability. Biochim Biophys Acta. 1834;2013:1023–33.

22. Vázquez-Contreras E, Zubillaga RA, Mendoza-Hernández G, Costas M, Fernández-Velasco DA. Equilibrium unfolding of yeast triosephosphate isomerase: a monomeric intermediate in guanidine-HCl and two-state behavior in urea. Protein Pept Lett. 2000;7:57–64.

23. Rozacky EE, Sawyer TH, Barton RA, Gracy RW. Studies of human triosephosphate isomerase: isolation and properties of the enzyme from erythrocytes. Arch Biochem Biophys. 1971;146:312–20.

24. González-Mondragón E, Zubillaga RA, Saavedra E, Chánez-Cárdenas ME, Pérez-Montfort R, Hernández-Arana A. Conserved cysteine 126 in triosephosphate isomerase is required not for enzymatic activity but for proper folding and stability. Biochemistry. 2004;43:3255–63.

25. Reyes-López CA, González-Mondragón E, Benítez-Cardoza CG, Chánez-Cárdenas ME, Cabrera N, Pérez-Montfort R, et al. The conserved salt bridge linking two C-terminal b/a units in homodimeric triosephosphate isomerase determines the folding rate of the monomer. Proteins. 2008;72:972–9.

26. Hess B, Kutzner C, van der Spoel D, Lindahl E. GROMACS 4: Algorithms for highly efficient, load-balanced, and scalable molecular simulation. J Chem Theory Comput. 2008;4:435–47.

27. Oostenbrink C, Villa A, Mark AE, van Gunsteren WF. A biomolecular force field based on the free enthalpy of hydration and solvation: the GROMOS force-field parameter sets 53A5 and 53A6. J Comput Chem. 2004;25:1656–76.

28. Li H, Robertson AD, Jensen JH. Very fast empirical prediction and interpretation of protein pKa values. Proteins. 2005;61:704–21.

29. Hess B, Bekker H, Berendsen HJC, Fraaije JGEM. LINCS: a linear constraint solver for molecular simulations. J Comput Chem. 1997;18:1463–72.

30. Darden T, York D, Pedersen L. Particle mesh Ewald: An N*log(N) method for Ewald sums in large systems. J Chem Phys. 1993;98:10089–92.

31. Privalov PL, Tiktopulo EI, Venyaminov SY, Griko YV, Makhatadze GI, Khechinashvili NN. Heat capacity and conformation of proteins in the denatured state. J Mol Biol. 1989;205:737–50.

32. Arroyo-Reyna A, Hernández-Arana A. The thermal unfolding of stem bromelain is consistent with an irreversible two-state model. Biochim Biophys Acta. 1995;1248:123–8.

33. Manavalan P, Johnson WC. Sensitivity of circular dichroism to protein tertiary structure class. Nature. 1983;305:831–2.

34. Van Holde KE, Johnson WC, Ho PS. Principles of Physical Biochemistry. New Jersey: Prentice Hall International; 1998.

35. Campbell ID, Dwek RAR. Biological Spectroscopy. Menlo Park, CA: The Benjamin/Cummings Publishing Company; 1984.

36. Gutfreund H. Kinetics for the Llife Sciences. Receptors, Transmitters and Catalysts. Cambridge, UK: Cambridge University Press; 1995.

37. Szabo ZG. Kinetic characterization of complex reaction systems. In: Banford CH, Tipper CFH, editors. Comprehensive Chemical Kinetics. Volume 2. Amsterdam: Elsevier; 1969. p. 1–80.

38. Zaman MH, Sosnick TR, Berry RS. Temperature dependence of reactions with multiple pathways. Phys Chem Chem Phys. 2003;5:2589–94.

39. Toledo-Núñez C, López-Cruz JI, Hernández-Arana A. Thermal denaturation of a blue-copper laccase: Formation of a compact denatured state with residual structure linked to pH changes in the region of histidine protonation. Biophys Chem. 2012;167:26–32.

40. Ausili A, Scire A, Damiani E, Zolese G, Bertoli E, Tanfani F. Temperature-induced molten globule-like state in human R1-acid glycoprotein: An infrared spectroscopic study. Biochemistry. 2005;44:15997–6006.

41. Fabiani E, Stadler AM, Madern D, Koza MM, Tehei M, Hirai M, et al. Dynamics of apomyoglobin in the α-to-β transition and of partially unfolded aggregated protein. Eur Biophys J. 2009;38:237–44.

42. GC JB, Bhandari YR, Gerstman BS, Chapagain PP. Molecular dynamics investigations of the α-helix to β-barrel conformational transformation in the RfaH transcription factor. J Phys Chem B. 2014;118:5101–8.

43. Kaur H, Sasidhar YU. Molecular dynamics study of an insertion/duplication mutant of bacteriophage T4 lysozyme reveals the nature of α-β transition in full protein context. Phys Chem Chem Phys. 2013;15:7819–30.

# 19

# Solution structure and biophysical characterization of the multifaceted signalling effector protein growth arrest specific-1

Katja Rosti[1], Adrian Goldman[2,3] and Tommi Kajander[1]*

## Abstract

**Background:** The protein growth arrest specific-1 (GAS1) was discovered based on its ability to stop the cell cycle. During development it is involved in embryonic patterning, inhibits cell proliferation and mediates cell death, and has therefore been considered as a tumor suppressor. GAS1 is known to signal through two different cell membrane receptors: Rearranged during transformation (RET), and the sonic hedgehog receptor Patched-1. Sonic Hedgehog signalling is important in stem cell renewal and RET mediated signalling in neuronal survival. Disorders in both sonic hedgehog and RET signalling are connected to cancer progression. The neuroprotective effect of RET is controlled by glial cell-derived neurotrophic factor family ligands and glial cell-derived neurotrophic factor receptor alphas (GFRas). Human Growth arrest specific-1 is a distant homolog of the GFRas.

**Results:** We have produced and purified recombinant human GAS1 protein, and confirmed that GAS1 is a monomer in solution by static light scattering and small angle X-ray scattering analysis. The low resolution solution structure reveals that GAS1 is more elongated and flexible than the GFRas, and the homology modelling of the individual domains show that they differ from GFRas by lacking the amino acids for neurotrophic factor binding. In addition, GAS1 has an extended loop in the N-terminal domain that is conserved in vertebrates after the divergence of fishes and amphibians.

**Conclusions:** We conclude that GAS1 most likely differs from GFRas functionally, based on comparative structural analysis, while it is able to bind the extracellular part of RET in a neurotrophic factor independent manner, although with low affinity in solution. Our structural characterization indicates that GAS1 differs from GFRa's significantly also in its conformation, which probably reflects the functional differences between GAS1 and the GFRas.

**Keywords:** GAS1, Growth arrest specific-1, Solution X-ray scattering, Protein structure, RET, Sonic hedgehog

## Background

Growth Arrest Specific-1 gene (GAS1) was found in a screen to identify genes specifically expressed in growth-arrested mouse cells [1]. The full-length cDNA of human GAS1 was cloned [2,3] and the mature protein was found to contain 345 amino acids, a potential signal peptide, one N-glycosylation site at Asn117 and an aminated Ser318 [2,3]. The aminated Ser318 allows the mature protein to be glycophosphatidylinositol (GPI) anchored to the cell membrane [2,4].

GAS1 was found to arrest cell cycle by stopping the cells in synthesis (S) phase [1,5] and due to its ability to arrest cell proliferation in p53-dependent manner it has been considered to be a tumour suppressor protein [6,7]. Generally GAS1 might act as a tumour suppressor in adult brain, though the expression in brain leading to apoptosis has not been observed in adults [3,8]. Sequence comparison of human and murine GAS1 genes suggested that it has a conserved RGD-peptide sequence for possible RGD-dependent integrin binding at residues 306–308 [3].

Additionally GAS1 has been shown to have a significant role in development [9]. At early developmental stages GAS1 is expressed in most embryonic tissues. During development GAS1 has been reported to inhibit

* Correspondence: tommi.kajander@helsinki.fi
[1]Institute of Biotechnology, Structural Biology and Biophysics, University of Helsinki, Helsinki, Finland
Full list of author information is available at the end of the article

cell proliferation and to mediate cell death, to be involved in embryonic patterning, and to support growth of the cerebellum [3,8].

GAS1 is clearly a multifunctional protein, since it signals through at least two different kinds of transmembrane receptor proteins, Rearranged during transformation (RET) [8] and the Hedgehog receptor protein patched-1 [10,11]. The Hedgehog signalling pathway is important in development, stem cell renewal, and cancer progression. GAS1 is able to bind sonic hedgehog (SHH) and activate the signalling pathway from patched-1 [10,11]. RET, on the other hand, is a transmembrane kinase, first identified as a proto-oncogene [12]. Overactivity of RET can cause several types of cancers, and loss-of-function mutations cause varying degrees of loss in the enteric nervous system resulting in Hirschprung's disease (see *e.g.* Robertson and Mason [13]). Normally RET mediated signalling is controlled by Glial cell-derived neurotrophic factor family ligands (GFLs) and Glial cell-derived neurotrophic factor receptor alphas (GFRαs), which form a four-member protein family (GFRα1-4) [14].

Of these, GAS1 has highest (28%) similarity to GFRα1, while GAS1 and GFRα4 both have only two domains unlike GFRα1-3, which consists of three domains [15]. The secondary structure of mammalian GAS1 is predicted to be mostly α-helical separated by short β-strands and to have a long unstructured C-terminal domain [15]. By binding GFLs, GFRαs take part in controlling the survival of neurons, neuron branching, and functional recovery [14]. The most studied member of GFLs is GDNF, which was identified due to its function as a survival factor for midbrain dopaminergic neurons [14]. GDNF forms a complex with GFRα1 and promotes the survival of neurons [16]. GFLs, in general, are dimeric proteins and they are capable of binding two GFRα receptors per ligand [14]. After the formation of GFRα-GFL complex, the complex then binds to the transmembrane tyrosine kinase RET [16].

Despite the structural similarity to GFRαs, GAS1 differs from them functionally because it is able to bind to RET in a ligand independent way [8]. In addition, the intracellular signalling pathway is most probably different than for GFRα-GDNF complex, and GAS1 bound to RET blocks AKT activation, and increases ERK activation [8].

GAS1 has been suspected to be an ancestor of GFRα proteins [8,15,17]. Thus the four GFLs and GFRαs could have been generated by genome duplications at the origin of vertebrates, and at this point the gene encoding GAS1 could have diverged from GFRα-like proteins [17,18].

It has been hypothesised that the relative abundance and localization of GFRαs, GFLs and GAS1 could determine in certain conditions whether cells survive or die [15]. Furthermore, GAS1 expression is increased in neuronal cell death during early development [19]. Therefore, GAS1 could work as a switch between proliferation and differentiation in neuronal development [8]. GAS1 has been shown to colocalize to lipid rafts with RET [8]. This has led to the hypothesis that GAS1 could be a negative modulator of GDNF signalling and able to control GDNF stimulation *via* RET [8,20].

## Results

### Production and purification of human GAS1 protein

After cloning and expressing human GAS1 in *Tricoplusia Ni* cells, we purified secreted GAS1 from the insect cell growth medium using Ni-affinity chromatography (Figure 1), and the identity and size of the expressed protein was verified with a western blot (Figure 1). The purified protein is glycosylated and therefore does not run exactly according to excepted molecular weight on the SDS-PAGE, but slightly higher. Thrombin was used to cleave off the tags, and the size of the protein after cleavage was verified by SDS-PAGE and MALDI-TOF

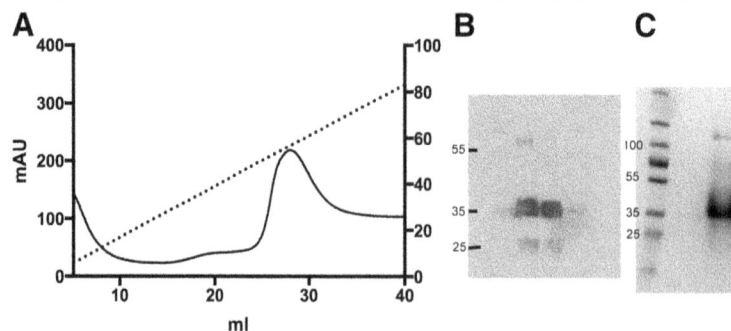

**Figure 1 Purification of recombinant GAS1 from insect cells. A)** Ni-affinity chromatogram for GAS1 purification **B)** Western blot of fractions from the Ni-affinity chromatography peak at *ca.* 28 ml. Fractions of 1 ml were collected, and fractions from peak area at 23, 27, 29 and 33 ml were tested in the western blot. **C)** SDS-PAGE analysis of GAS1 purification (left lane, molecular weight marker, with sizes indicated; right lane, purified GAS1 after gel filtration; the gel was Coomassie stained.).

mass spectroscopy. The yield of purified protein was on average ca. 1 mg/L.

Based on the primary sequence one N-glycosylation site was predicted at Asn117 located in the N-terminal domain. The corresponding site in the GFRα-structures is located at the domain interface between the two homologous disulphide rich domains, in a tightly packed two-domain structure [21,22], suggesting that the GAS1 overall conformation is quite likely very different (see below, Figure 2).

When the protein was treated with PNGase F to remove glycans, the size of the protein diminished slightly on SDS-PAGE (data not shown), and based on MALDI-TOF analysis, we observe a decrease in molecular weight of *ca.*

900 Da; the glycosylated protein had a molecular mass of 29.8 kDa and the de-glycosylated protein of 28.9 kDa, according to MALDI-TOF, while the theoretical molecular mass of the protein without glycosylation is 29.0 kD, thus matching well with the mass spectrometry results. The result obtained for glycosylated protein corresponds to approximately one N-glycan added post-translationally in the insect cells. The purified protein was functional in binding to RET *in vitro*, and found to be over 90% pure on SDS-PAGE, and monodisperse in solution after gel filtration.

### GAS1 is a monomer is solution and highly thermostable

The cleaved, non-tagged protein was found to be a monomer by analytical size exclusion chromatography and

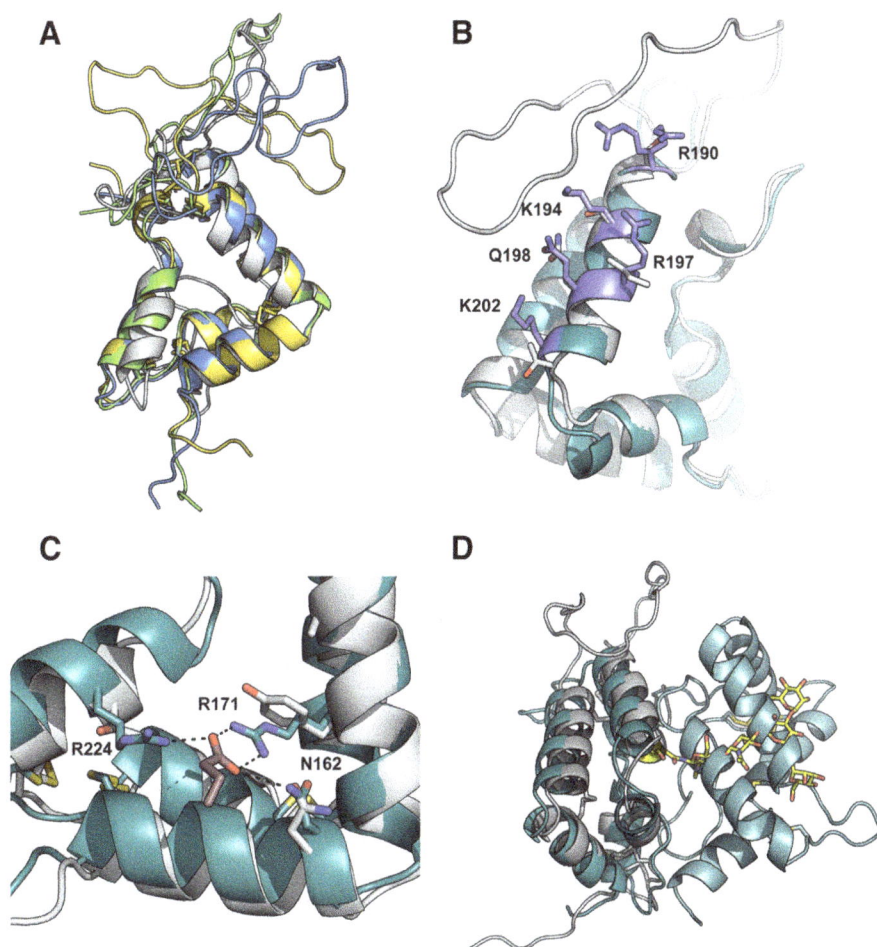

**Figure 2 Homology modelling of human GAS1 and comparison to GDNF receptor structures. A)** Four representative homology models of GAS1 N-terminal domain showing different orientations and flexibility of the extended intradomain loop of higher vertebrate proteins. The models were generated with the Raptor-X server, as mentioned in the text. The template for modelling was the GFRα1 structure (PDB: 2VE5). **B)** comparison to N-terminal domain of GFRα1 (dark cyan) and N-terminal domain (GAS1, grey) and RET/heparin binding site (grey residues, GAS, cyan residues, residues involved in heparin and indicated as putative RET binding residues). GFRα1 RET/heparin binding-site residues are labelled. **C)** The GDNF binding site residues GFRα1 (dark cyan) vs. GAS1 (grey). The GDNF Glu binding to the GFRα1 is colored brown, and hydrogen bonds are indicated with dashed lines. A Tyr and Ser residue occupy the positions in GAS1 equivalent to GFRα1 Arg171 and 224, GFRα1 GDNF binding residues are labelled. Disulphides in the figure are shown with stick presentation and atoms S atoms in yellow **D)** Position of the N-glycosylation site in GAS1 vs. GFRα1 domain interface of domains D2 and D3. N-glycan at Asn117 of GAS1 is indicated in yellow with a stick presentation, the GFRα1 D2 and D3 domains are depicted in dark cyan, GAS1 N-terminal domain is drawn in grey.

multi-angle light scattering (SEC-MALLS) (Figure 3). At 1 mg/ml and 4.5 mg/ml the SEC-MALLS runs gave a single peak with molecular mass of *ca.* 31–33 kDa (Figure 3), matching quite well to the theoretical size of monomeric GAS1 (29.0 kDa) considering the additional glycosylation at one site. Similarly, based on the small-angle X-ray scattering (SAXS) data, the molecular weight matches most closely to a monomer (Table 1). In our opinion, this is likely to reflect the oligomerization state of the lipid-anchored protein, which is unlikely to be affected by the anchor. No detectable oligomerization was observed in native PAGE or gel filtration at 4.5 mg/ml, while in SAXS

**Table 1 SAXS-derived size parameters for GAS1**

| | |
|---|---|
| I(0) (Guinier) | 25.08 |
| I(0) (Porod) | 24.8 |
| $D_{max}$ (nm) | 10.5 |
| $R_g$ (Guinier/nm) | 3.01 |
| $R_g$ (Porod/nm) | 3.00 |
| Porod volume ($V_p$) | 54.2 |
| $M_w$(theoretical) | 29 158.3 g/mol |
| $M_w$(calc) (Guinier) | 25.1 kDa |
| $M_w$(calc) (Porod vol.) | 31.9 kDa ($V_p$/1.7) |

The Guinier I(0)-value was calculated against an absolute reference (scattering of water relative to sample) [23] and the I(0) for the sample is then equal to the molecular weight. Molecular weight from the Porod volume is estimated according to Petoukhov et al. [24].

data an effect from residual aggregation was evident at higher concentrations.

Circular dichroism (CD) spectroscopy was used to verify the secondary structure content of GAS1 and, as expected, the CD spectrum was typical for an α-helical protein (Figure 3). A measured temperature denaturation curve with CD gave a result with partial melting of the structure when heated to 90°C (Figure 3). However, full temperature denaturation was not possible to obtain by CD, nor by differential scanning calorimetry (data not shown), possibly due to the high disulphide content of the protein. This suggests that the domain structure is thermally very stable. The decrease in CD signal at 222 nm did not even reach the midpoint of denaturation when heated to 90°C (Figure 3).

### Sequence analysis and evolution of GAS1

The GAS1 protein domain structure is defined by two GFRα–like domains, with 10 conserved disulphide bridge-forming cysteines in each domain [25]. GAS1 is present in all the vertebrates, and homologs are also found in lower chordates (*e.g. Ciona* and *Amphioxus* [17]). In addition, GAS1 homologs also occur in *C. elegans* and honeybee, but not, surprisingly, in *Drosophila*. The sequence identity to vertebrate proteins, however, is quite low: *ca.* 21-24% for honeybee and only 14-19% for the worm sequence (Table 2). Two conserved cysteines are missing from the *C. elegans* sequence (Figure 4), and thus the protein fold might not be fully conserved in the *C. elegans* homolog (phas-1) [26]. Alignment of the GAS1 sequences shows that, in higher vertebrates, the N-terminal domain has an insertion with low sequence complexity (Figure 4), apparently forming an extended loop structure (Figure 2). Mammals have also an RGD sequence in the C-terminal linker region. Also, a single N-glycosylation site at Asn117 is predicted to be conserved based on sequence in all vertebrates, while it is not present in the invertebrates. In the set of conserved residues beyond the structural cysteines

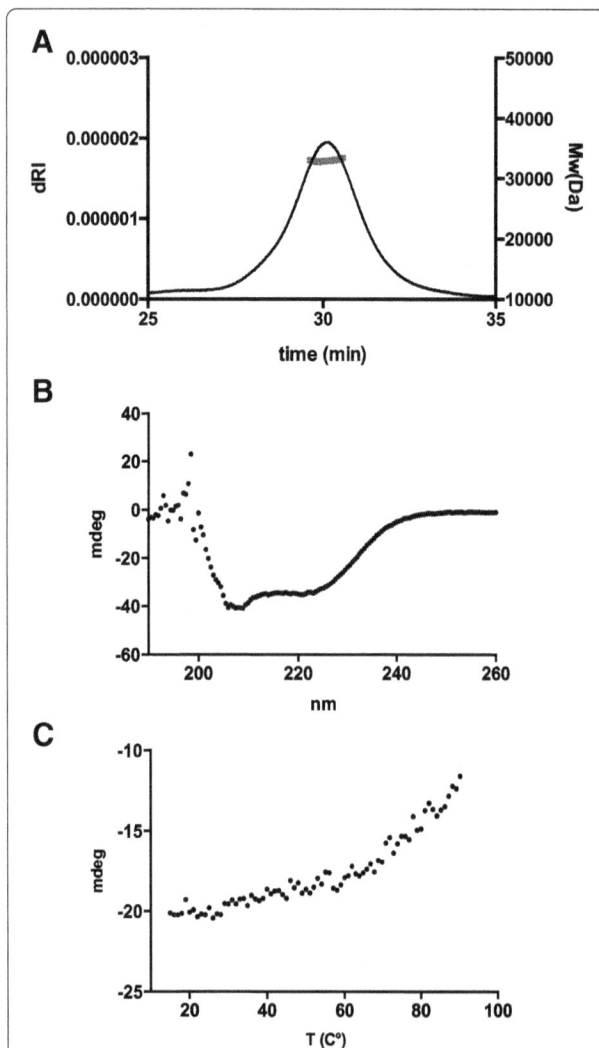

**Figure 3 GAS1 analytical gel filtration, circular dichroism and thermal unfolding. A)** GAS1 sample was run on Superdex 200 10/300 gel filtration column in TBS at 0.5 ml/min, at protein concentration of 1 mg/ml. A single major peak at 30 min (X-axis) eluted and based on multi-angle light scattering had molecular weight (right Y-axis) of *ca.* 33 kDa, matching relatively well with theoretical molecular weight of the monomer. The peak is plotted as a function of dRI signal (left Y-axis). **B)** The CD spectrum of GAS1 and **C)** the residual thermal denaturation of GAS1 as monitored by CD at 222 nm.

**Table 2 Amino acid sequence identities (%) within the GAS1 protein family**

| | Human | Sus scrofa | Bos Taurus | Canis Lupus | Mouse | Gallus gallus | Alligator | Anolis | Xenopus | Latimeria | Danio rerio | Apis | C. elegans |
|---|---|---|---|---|---|---|---|---|---|---|---|---|---|
| **Human** | 100 | | | | | | | | | | | | |
| **Sus scrofa** | **95.3** | 100 | | | | | | | | | | | |
| **Bos Taurus** | **94.1** | **94.4** | 100 | | | | | | | | | | |
| **Canis Lupus** | **91.1** | **91.4** | **90.2** | 100 | | | | | | | | | |
| **Mouse** | **85.4** | **85.4** | **84.9** | **82.9** | 100 | | | | | | | | |
| **Gallus gallus** | 61.9 | 61.7 | 60.1 | 62.1 | 59.6 | 100 | | | | | | | |
| **Alligator** | 60.3 | 60.7 | 59.9 | 59.9 | 60.1 | **70.9** | 100 | | | | | | |
| **Anolis** | 55.8 | 56.2 | 54.5 | 57 | 55.7 | 54.7 | 57.6 | 100 | | | | | |
| **Xenopus** | 51.2 | 51.7 | 50.8 | 50.6 | 50.5 | 52.2 | 55 | 48.4 | 100 | | | | |
| **Latimeria** | 47.2 | 47.4 | 46.6 | 46.3 | 45.9 | 48.4 | 50.5 | 42.1 | 49 | 100 | | | |
| **Danio rerio** | 35.2 | 34.5 | 35 | 35.6 | 35.8 | 37 | 38.7 | 34.2 | 37.6 | 43.4 | 100 | | |
| **Apis** | 22.1 | 21.7 | 21.3 | 21.6 | 21.1 | 23.1 | 23.7 | 22.6 | 22.1 | 22.8 | 23.7 | 100 | |
| **C.elegans** | 15.6 | 15.6 | 14.6 | 15.5 | 14.3 | 14.7 | 15.9 | 14.2 | 19 | 17.8 | 17.1 | 15.4 | 100 |

Pairwise identities between species that are over 70% are shown in bold.

(or positions with highly conserved mutations *e.g.* Asp to Glu) (Figure 4) a subset are present only in the vertebrate proteins, and, other than the conserved cysteines, only ten residues are conserved also in *C. elegans*.

When the conserved amino acid residues are displayed onto the surface of the modelled domains, the most conserved surface patch is found on the N-terminal domain surface formed by residues on helices 3–5, whereas the C-terminal domain surface did not reveal large patches of conservation (Figure 4). Here conservation is defined by >75% sequence similarity amongst the residue groups KHR, ED, NQSTGP, ILMVCA and FYW.

## Homology modelling of GAS1 domains and comparison to GFRαs

We constructed homology models of both GAS1 domains with the RaptorX-server (http://raptorx.uchicago. edu/) [27], designed for low sequence homology-based modelling. The models for both domains fit well to the GFRα-structure (PDB: 2VE5) [22]. As described above, the N-terminal domain of mammalian GAS1-proteins contain a large inserted loop with low sequence complexity, which based on modelling indeed appears to form a large flexible loop, but whether this region has functional significance or not remains unclear.

Modelling of GAS1 has partially been done before also by Cabrera *et al.* [8] and Schueler-Furman *et al.* [15]. Here our aim was to study possible conservation of the ligand binding regions of GFRαs vs. GAS1, to do more detailed analysis on the structure, and to provide models for the analysis of SAXS data (see below).

Although the sequence identity to the related GFRα structures is low, the cysteines involved in disulphide bridges are well conserved for the two GFRα-type

domains and make structure prediction possible. The N-terminal domain of GAS1 is equivalent of the second domain in GFRαs, which contains the growth factor binding site. We aligned our model of GAS1 with the GFRα1:GDNF complex structure [22], and, based on the structural alignment, the binding region for the GDNF is not conserved in GAS1 (Figure 2). Similarly the conserved binding residues in GFRα2:Artemin complex (PDB: 2GH0) [21], are not present in GAS1. In fact, the key ionic residues required for ligand binding are conserved in both these structure, but not present in GAS1. The conserved key residues for GDNF binding in GFRα1 are Arg171, Arg224 and Asn162. In our structural analysis GAS1 has Tyr26, Thr100, Gln17 in equivalent side chain positions; in GAS1, the ion triplet required for growth factor binding [22] is absent.

It has also been suggested by Wang et al. [21] and Parkash et al. [22] that a RET binding region would be located mostly in the second ("D2") domain on the GFRαs and would involve the GFRα1 residues Arg190, Lys194, Arg197, Gln198, Lys202, Arg257, Arg259, Glu323, and Asp324. This site forms a highly positively charged patch on the surface of GFRα1, identified also as a heparin binding site by Parkash et al. [22]. We analysed the equivalent region in the N-terminal domain of GAS1, but found no conservation between the GFRα1 structure and GAS1 (Figure 2). Overall GAS1 is not positively charged, as would be expected of a typical heparin-binding molecule. The calculated pI-value for human GAS1 is 5.0 whereas for human GFRαs the values range from 7.5-7.6 (GFRα2 – 3) to 8.4 (GFRα1) and 10.1 (GFRα4). Also, GAS1 does not contain a highly positively charged patch in the suggested RET/heparin binding region, and heparin affinity chromatography of GAS1 showed no significant binding to

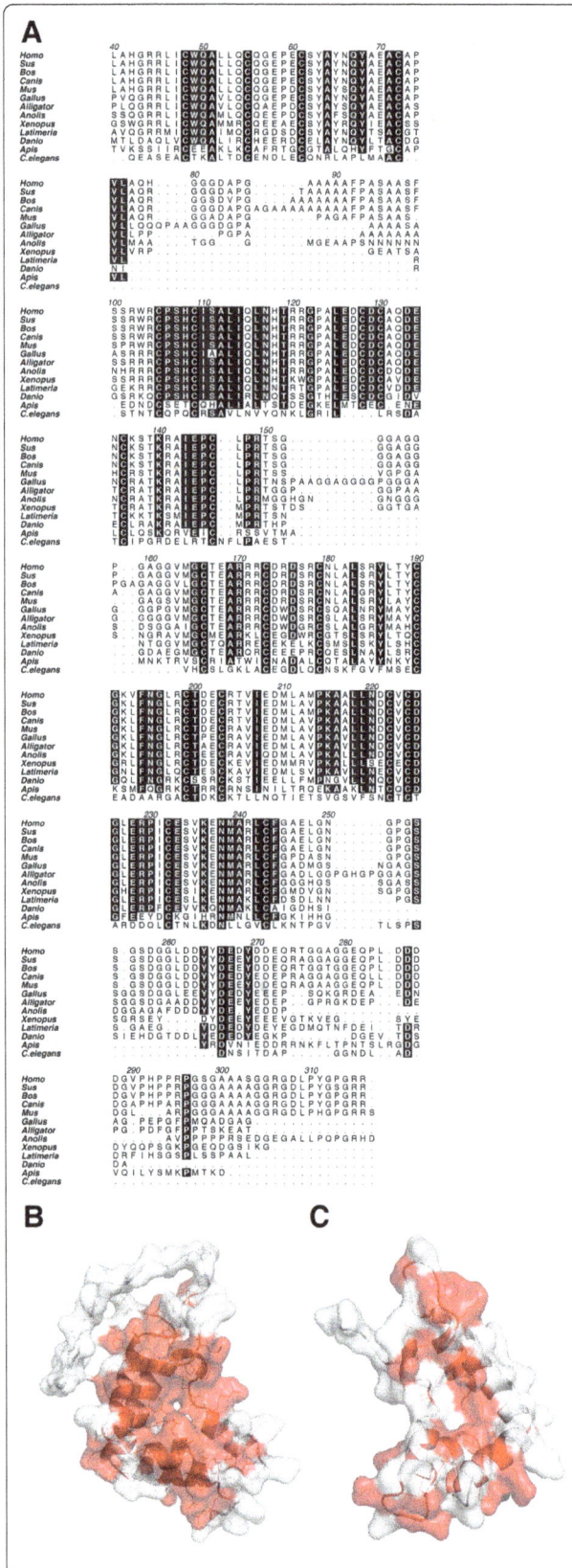

**Figure 4 The GAS1 family sequence alignment. A)** The sequences start from the beginning of the mature human GAS1 and numbered according to the human amino acid numbering. Residues over *ca.* 85% conserved (11/13) are coloured with a black-to-grey scale, in higher vertebrates (mammals) there is an extended loop in the N-terminal domain around residues 80–100 (human GAS1 numbering). The C-termini are poorly conserved (residues beyond 250), note the RGD sequence at 306–308. **B)** Conserved surface features on GAS1 displayed on the N-terminal domain. **C)** Conserved surface features on GAS1 displayed on the C-terminal domain; conserved sites in B and C are coloured in red (with >75% sequence similarity, see text).

the column (Figure 5), whereas the well-known heparin binding protein HBGAM eluted only at 1 M NaCl (Figure 5). Finally, modelling of GAS1 N- terminal domain shows that glycosylated Asn117 will be situated at the position equivalent to the domain interface between domains D2 and D3 in the GFRα structures (Figure 2).

## Structural characterization of GAS-1 by solution X-ray scattering

Solution X-ray scattering (SAXS) data indicated that GAS1 is monomeric at 0.8 mg/ml in solution based on the Porod volume and Guinier plots (Table 1): at higher concentrations the protein starts to aggregate, and the data beyond 1 mg/ml could not be analysed. Rigid body modelling of the structure was done based on homology models of the individual domains, and elongated models gave the best fits (Figure 6). We also calculated *ab initio* envelopes, which matched well with rigid body modelling of the structure (Figure 6). Both differ significantly from the compact GDNF co-receptor structures [21,22]. However, as it is clear that the structure is likely to be flexible, in particular the C-terminal long unstructured region, we also did ensemble fitting of the model against the data. This resulted in a bimodal ensemble represented by four major structures selected from the initial random pool of 10 000 structures, which fit to the data with $\chi^2 = 0.84$ (Figure 6). The selected structures represent states with extended and collapsed C-terminal linkers and variable orientations of the domains relative to each other (Figure 6). Taken together it appears from the SAXS data that the orientation of the domains of GAS1 relative to each other is not fixed; clearly the protein exists in two populations of extended and collapsed conformations.

## Binding and affinity of GAS1 to RET *in vitro*

We tested whether GAS1 is able to directly interact with RET in a ligand independent way, as previously reported [8]. For this purpose, and to determine the affinity of the interaction, the RET receptor was coupled to a chip for surface plasmon resonance assay, and binding of a concentration series of GAS1 to immobilized RET was measured. A $K_d$-value of $12.2 \pm 8.2$ μM was measured for the

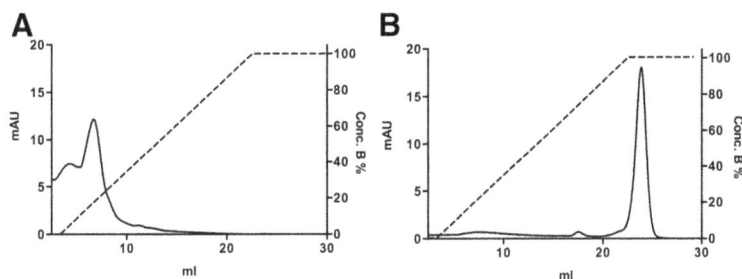

**Figure 5** Heparin affinity chromatography of GAS1. **A)** Elution of GAS1 as a function of salt concentration. **B)** Elution of HBGAM as a function of salt concentration. Chromatograms are plotted with absorbance in mAU unit (right y-axis) and salt concentration gradient to 1 M NaCl (%) (left y-axis), against volume in ml.

interaction *in vitro* (Figure 7). The kinetics of the interaction were too fast to allow for measurement of on- and off-rates, as is evident from the time scale of the binding and dissociation from the sensograms (Figure 7).

## Discussion

We have overexpressed and purified the human GAS1 protein in soluble form without the GPI-anchor, and biophysically characterized the protein. We constructed homology models for both domains of GAS1 and were able to analyse the domain structure of the protein in comparison to the structurally related GFRαs. As reported earlier by others [8], GAS1 has clearly two GFRα-like domains, but we have shown here that GAS1 differs significantly from GFRαs in both sequence and in structure.

The differences can be characterized as follows: Firstly, GAS1 has a large, 10–15 amino acid unstructured, low complexity Ala/Gly/Pro-containing loop in the N-terminal domain. This loop region is present in higher vertebrates, in mammalians and chicken, but not in fish (Figure 4). Whether this loop might have some function remains unknown. Secondly, the two-domain structure of GAS1 appears to be more flexible overall than in the characterized GFRα structures. In the GFRαs, the functional domains D2 and D3 form a compact structure, whereas GAS1 SAXS analysis reveals a flexible ensemble of structures, with the N- and C-terminal domains as independent structural units. This might reflect the location of the functional binding regions of the molecule *versus* those of the GFRαs. As expected, the protein is α-helical based on the CD spectrum. It is somewhat intriguing that we were not able to fully denature the protein; apparently the disulphide-linked arrangement of the domains is highly thermostable, and this might be a general feature of the GFRα-family.

The structural flexibility is probably a conserved feature in the protein family, as the single N-glycosylation site in the human protein is conserved in chordates. This N-glycan blocks the GFRα-equivalent domain interface,

and hence the formation of that type of compact structure. It has been also observed that this glycosylation site might have functional significance for SHH binding [28].

Our sequence analysis and that by Hätinen *et al.* [17] suggest that GAS1 is conserved during evolution, with homologs in chordates (from *Ciona* and *Amphioxus*), arthropods and roundworms, thus possibly representing an ancestral GFRα-like protein [17] However, the sequence identity from chordate to invertebrates (*e.g.* honey bee and *C. elegans*) genes is low, 14-19% for the worm phas-1 homolog of GAS1 [26], and it remains an open question whether the insect or worm genes identified as GAS1 actually share any of the functions of vertebrate GAS1/GFRα type of receptors, either in RET or the hedgehog signalling.

GAS1, as well as GFRα-like proteins, are conserved beyond vertebrates, while GFLs are not expressed in non-vertebrates. This suggests that either RET binding, independent of GFLs is conserved, or that there are alternative receptors for GAS1. In case of GAS1, this could be SHH and patched-1. Interestingly in *Drosophila*, the GRFα-like protein does not interact with RET but does interact with Drosophila NCAM analog, FasII [29]. The mammalian GFLs are known to be ligands of NCAM [30]: whether GAS1 might interact with NCAM homologs remains to be investigated.

When we compared our model with the GFRα1 structure it was clear that the crucial amino acids for GFL binding are not conserved in GAS1, and it most likely lacks the ability to bind GFL-like ligands, as they all share the same binding mode [21,22]. Indeed Cabrera et al. [8] reported that GAS1 is not able to bind GDNF. While GAS1 lacks the ability to bind GFL-type of ligands, our *in vitro* binding data support the findings by Cabrera et al. [8] that GAS1 can bind RET in a ligand independent manner, and possibly alter the intracellular signalling of RET.

The affinity of GAS1 for RET is significantly lower than that of the GFRα-GDNF ligand complex in solution ($K_d$ = 12.2 μM *versus* 0.2 nM for GFRα1-GDNF binding

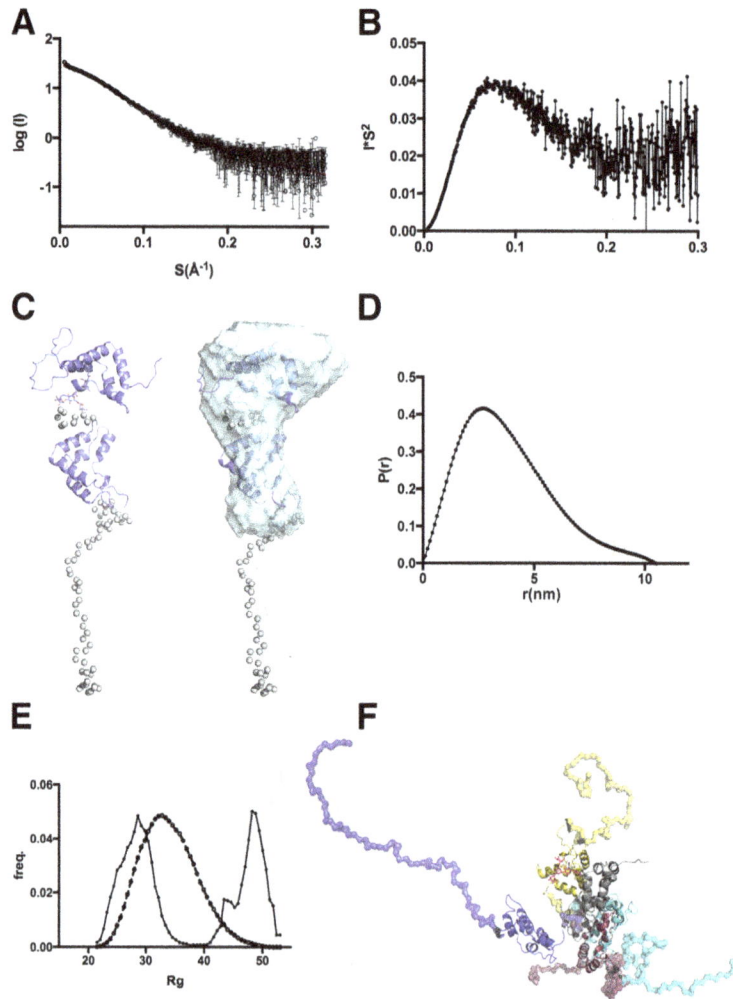

**Figure 6 Rigid body and ab initio modelling of GAS1 based on SAXS data. A)** Scattering curve and fit of CORAL rigid body model (red line) to observed data. **B)** The Kratky-plot from the experimental data, suggesting a folded structure with some flexibility. **C)** CORAL generated model with N- and C-terminal domains as rigid bodies (blue) with flexible linker regions (grey beads; left), and the ab initio model for GAS1 generated by DAMMIN (green; average of 10 calculations) fitted over the rigid body model (right). **D)** The distance distribution calculated for GAS1 SAXS data. **E)** SAXS ensemble modelling of GAS1 solution conformations shown as the statistical distribution of $R_g$-values of best fitted models (continuous line with closed circles) vs. initial random pool (dashed line with open circles) shows a bimodal distribution of GAS1 solution conformations. **F)** The selected pdb-files representing the ensemble with Chi$^2$ = 0.84 fit to the experimental data, showing extended (blue) and more collapsed models (yellow, red, cyan) in the final ensemble; the N-terminal domain (in grey) was fixed relative to the rest of the protein during the runs. The modelled glycan structure is shown as red "stick" presentation on the N-terminal domain.

RET [31,32]). However on the cell surface the affinity of GAS1 to RET is also likely to be higher as the diffusion is restricted to two-dimensions.

Another possibility is that in some cellular contexts GAS1 would be highly expressed on cell surface, which might boost the binding to RET locally. The exact mechanism of GAS1 on RET signalling remains elusive, but it seems clear that GAS1 has an effect of RET signalling, probably by inhibiting growth-factor dependent signalling [4,8].

Based on the conservation of protein surface features, as mapped on to the GAS1 models, we suggest that the N-terminal domain region defined by α – helices 3–5

might contain a functional binding site (Figure 4), whereas other possible interaction surfaces remain less clear, *e.g.* the very short RGD-peptide motif found in mammalian sequences could be functional, or exist by chance, and so far no biochemical evidence for the function exists.

GAS1 has also been reported to alter SHH-signalling through patched-1 [28,33], indicating that GAS1 has multiple functions. Related to this Pineda-Alvarez et al. [34] and Ribeiro *et al.* [35] reported missense mutations of GAS1 in holoprosencephaly (HPE) patients.In particular Thr200Arg mutation in the second domain of GAS1 Pineda-Alvarez et al. [34] was observed to result

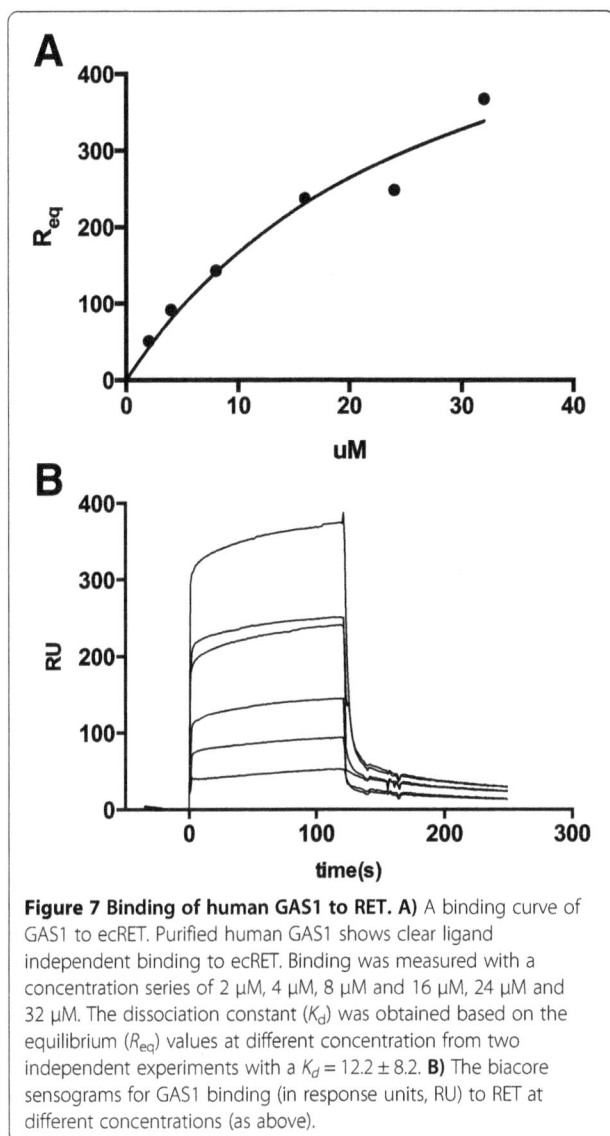

**Figure 7 Binding of human GAS1 to RET. A)** A binding curve of GAS1 to ecRET. Purified human GAS1 shows clear ligand independent binding to ecRET. Binding was measured with a concentration series of 2 µM, 4 µM, 8 µM and 16 µM, 24 µM and 32 µM. The dissociation constant ($K_d$) was obtained based on the equilibrium ($R_{eq}$) values at different concentration from two independent experiments with a $K_d = 12.2 \pm 8.2$. **B)** The biacore sensograms for GAS1 binding (in response units, RU) to RET at different concentrations (as above).

in almost complete loss of binding affinity for SHH, hence this domain could also be important for binding to SHH. Also Asn220Lys caused 20% reduction in binding according to Pineda-Alvarez et al. [34] and Ala246-Ser patient mutations are located in the same domain, while the mutations are some what scattered around the domain and do not cluster together on the surface.

## Conclusions

Our structural data reveal that GAS1 is a flexible two-domain molecule, the flexibility perhaps reflecting its multifunctional properties. The structural arrangement of the domains is clearly different form the compact GFRα structures, suggesting that it has different functional roles. In particular, neither the putative heparan sulphate proteoglycan/RET binding site [22] the known growth factor binding site are conserved in GAS1.

Thus, GAS1 must act on RET in a different way, and together with previous analysis our binding data supports the ligand-independent RET binding by GAS1, while sequence conservation analysis hints at possible sites of functional importance.

## Methods
### Ethics statement
All results of this research were based proteins expressed in cultured *Tricoplusia Ni* or *Spodoptera frugiperda* cells lines. Neither human (human subjects, human material or human data) nor animals (vertebrates or any regulated invertebrates) were used in this experimental research.

### Plasmids, reagents and cell lines
Human GAS cDNA1 in a pCR3.1 plasmid was obtained as kind gift from Prof. Mart Saarma, and the pFastBac (Invitrogen) derivative vector pK509.3 from Prof. Kari Keinänen [36]. Oligonucleotides were purchased from Sigma and Phusion polymerase and PCR reagents were from Finnzymes Inc., *E.coli* DH10Bac-cells, *Tricoplusia Ni* and *Spodoptera Frugiperda* insect cells were from Invitrogen. Serum Free insect cell culture media was purchased from HyClone, gentamycin from Dushefa. Baculovirus production was done according to Bac-to-Bac manual (Invitrogen). SDS-PAGE gels were bought from Bio-Rad. The anti-FLAG monoclonal M1 mouse antibody was from Sigma, the anti-mouse antibody from Santa Cruz biotechnology, 5 ml HisTrap crude Ni-NTA column, size exclusion column Superdex 10/300, and Thrombin protease, 3 M HyBond western-blot membrane, and the ECL reagent were all from GE Healthcare.

### PCR and cloning
Human GAS1 cDNA in pCR3.1 plasmid was used as a template for PCR. The region encoding amino acids 39–317 was amplified, thus omitting the part encoding the native secretion signal at the N-terminus and the predicted GPI-anchor in the C-terminus. The PCR product was subcloned between Not1-Hind III restriction sites to baculovirus pFastBac-derivative vector pK509.3, which has the honey bee mellitin secretion signal and a Flag-tag sequence upstream of the cloning site.

The forward PCR primer was designed to add additional amino acids at the N-terminus for a His₆-tag and a Thrombin protease cleavage site (LRPHHHHHHLVPRGS).

The PCR primer sequences used for cloning were: 5′ ACTTAACTGCGGCCGCATCATCACCATCACCATC TTGTTCCTCGTGGTTCTGCGCACGGCCGCCGCCT CATC-3′ (forward) and 5′-AGATCTTAAGCTTACCT GCGCCCAGGCCCATAG-3′ (reverse). The template was PCR amplified with 5% DMSO to optimize it for a high GC-rich template (here, 81.2 %). PCR cycling conditions were as recommended by manufacturer (Finnzymes Inc). Cleaved and agarose gel purified vector and insert were ligated using T4 ligase (New England Biolabs).

## Virus propagation and Western-blots

The GAS1 construct was transformed to DH10Bac-cells to transpose it as a part of baculovirus shuttle-vector. The resulting DNA was isolated as described in the Bac-to-Bac manual (Invitrogen). Baculoviruses were multiplied by transfecting Sf9 cells on Cellstar (GreinerBio-one) six-well plate at 70 % confluency according to the manufacturer's instructions (Mirus. USA). In short, 200 µl of serum free HyQ-SFX medium, lacking antibiotics, were placed in microcentrifuge tubes with 6 microliters of TransIt reagent (Mirus), and incubated 20 minutes at room temperature. Two micrograms of bacmid DNA was added to the reactions and incubation was continued for another 20 minutes. Cells were washed with phosphate buffered saline (PBS) pH 7.4, and the medium changed to fresh HyQ-SFX. The transfection mixture was then added to the cells drop-wise. Cells were incubated at +27°C for five hours, after which the medium was changed to serum free SFX medium supplemented with 50 µg/ml Gentamycin. Cells were incubated for five days. The virus production efficacy was estimated by comparing the wells with non-infected control wells; properly infected cells stopped dividing, grew in size, and finally lysed. Virus was passaged typically by infecting 70-90% confluent plates. Passage one was done by adding 2 ml of virus from transfected cells to 70% confluent plate, in a total culture volume of 5 ml. For passages two and three, 90 % confluent plates were made by infecting cells with three to four millilitres of virus from the previous passage in a total volume of 25 ml. Virus propagation was estimated by visual analysis, as described, and by detecting the presence of the flag-tagged GAS1 protein by western blot. Virus propagation was typically continued to at least passage four, in order to get sufficient amount of virus to infect the culture used for protein production.

## Protein production and purification

The GAS1 protein was produced by infecting 200 ml of Tn5 cells, typically at $2 \times 10^6$ cell density, with 5 ml of high titer virus, typically from passage three or four. 72 h post infection the cells were harvested by centrifugation and the supernatant was collected.

The secreted GAS1 protein was purified from the supernatant by Ni-affinity chromatography. The column was equilibrated with binding buffer containing 20 mM sodium phosphate pH 7.4, 150 mM sodium chloride and 5 mM imidazole, and the protein was eluted with linear 5–500 mM imidazole gradient with the binding buffer.

GAS1 protein was detected by SDS-PAGE and identity confirmed by a FLAG-tag Western blot; based on this 1 ml fractions from the peak area were collected (Figure 1).

GAS1 protein containing fractions were detected from the major peak. These pooled and concentrated with a 30 kDa cut-off Amicon spin concentrator (Millipore) for 4000 rpm at +4°C, typically up to 500 µl volume. The buffer was exchanged, to phosphate buffered saline (PBS), pH 7.4, by diluting to 15 ml and repeating the centrifugation step as described.

After buffer exchange, the tags were cleaved off with Thrombin protease at a ratio of ten units per milligram of protein, the cleaved protein was concentrated to a 350 ul final volume, and further purified using size exclusion chromatography with Superdex 10/300 GL column in a buffer containing Tris-buffered saline (TBS) pH 7.4, (25 mM Tris pH 7.4; 150 mM NaCl; 2 mM KCl, pH adjusted with HCl), supplemented with 250 mM NaCl at a 0.5 ml/min flow rate. Fractions of 2 ml were collected and analyzed by SDS-PAGE. The fractions containing correct, approximately 35 kDa protein were pooled and concentrated as previously described.

## Surface plasmon resonance assay

Surface plasmon resonance (Biacore™, GE Healthcare) was used to determine the binding affinity of GAS1 protein to the ectodomain of RET protein (ecRET; R&D Systems, catalog no. 1168-CR-050/CF).

For this purpose ecRET was coupled to a CM5 chip (GE Healthcare) by amide coupling. The chip was ctivated according to manufacturers instructions with 1-ethyl-3-(3-dimethylaminopropyl carbodiimine (EDC)- N-hydroxysuccimide-(NHS) solution (Amine coupling kit, GE Healthcare). The ecRET at 0.25 mg/ml in PBS was diluted 1:10 to 10 mM Na-acetate,pH 5.0 and coupled to the chip at 4000 RU level.

After ecRET was coupled to the chip, the remaining free activated carboxyl groups on the surface were inactivated with 1 M ethanolamine (GE Healthtcare). The buffer used for binding experiment was 10 mM Hepes pH 7.4, 150 mM NaCl (HBS) supplemented with 1 mM $CaCl_2$, 0.01% Triton X-100. First flow channel from the chip was used as a blank control channel showing the possible non-specific binding to a non-coated surface. A GAS1 sample concentration series from 2 µM to 32 µM was injected at 20 ul/min for 2 minutes, and after each experiment the chip surface was regenerated with 1 M $MgCl_2$ with two 10 µl injections to release the bound GAS1 from RET. The dissociation constant for GAS1 to RET binding was calculated from binding curve calculated fitted from the equilibrium response ($R_{eq}$) values for binding at each concentration (Figure 7).

## Small-angle X-ray scattering studies and multi-angle laser light scattering

The GAS1 protein *ab initio* solution structure was obtained with 0.8 mg/ml protein in 50 mM Tris pH 7.5, 50 mM NaCl buffer. Small-Angle-X-Ray scattering data collection (ESFR, France) was performed with 1 s exposure time per image, and 10 repeats per sample, and these averaged and subtracted from similarly averaged buffer baseline. The measured data was analysed using PRIMUS [37] software and the *ab initio* modelling of the protein was done by DAMMIF/DAMAVER software [38]. The model of the protein with flexible linkers was obtained by using rigid body homology modelling against collected data. Original homology modelling of the domains to GFRα1 was done with Raptor-X server and rigid body modelling of the two-domain structure and modelling of flexible linkers was done with CORAL and BUNCH within the ATSAS software package [39]. Ensemble modelling of SAXS data was done using EOM 2.0 [38,40] via the ATSAS-online server (http://www.embl-hamburg.de/biosaxs/atsas-online/).

## Multi-angle laser light scattering

SEC-MALLS measurements were run at 0.5 ml/min over an S-200 Superdex 10/300 column (GE Healtcare) in 20 mM TRIS pH 7.4, 150 mM NaCl with a Schimadzu HPLC system and MiniDAWN TREOS light scattering detector and Optilab rEX refractive index detector (Wyatt Technology Corporation). Data was analysed with ASTRA 6 software (Wyatt Technology Corporation).

## Circular Dichroism and thermal stability

Circular Dichoism (CD) spectrum at 190–260 nm was collected on a JASCO J-720 instrument. For this experiment the protein was dialyzed against 20 mM Na phosphate pH 7.4, 50 mM NaCl. The GAS1 sample was diluted to 6.5 µM concentration in the same buffer. Measurement was done with a capped 350 µl 1 mm light path quartz cuvette (Hellma-Analytics). Data for thermal denatural analysis was collected at 222 mm wavelength from 20 to 90°C, with one degree steps and 30 second incubation at each temperature.

## Abbreviations
GAS1: Growth arrest specific-1; SAXS: Small-angle X-ray scattering; CD: Circular dichroism; RET: Rearranged during transformation; SEC-MALLS: Size exclusion chromatography and multi-angle light scattering; GFRα: Glial cell-derived neurotrophic factor receptor alpha; GFL: Glial cell-derived neurotrophic factor family ligand; MALDI-TOF: Matrix Assisted Laser Desorption Ionization - Time of Flight; PBS: Phosphate buffer saline; TBS: Tris-buffered saline; SHH: Sonic Hedgehog; GDNF: Glial cell line-derived neurotrophic factor; NCAM: Neural cell adhesion molecule; HBGAM: Heparin-binding growth-associated molecule.

## Competing interests
The authors declare that they have no competing interests.

## Authors' contributions
TK, KR, and AG designed experiments. TK and KR performed experiments. TK, KR and AG wrote the paper. All authors read and approved the final manuscript.

## Acknowledgements
We thank Dr. Jaana Jurvansuu for advice and help in the initial stages of this project and Prof. Mart Saarma and Dr. Jaan-Olle Andressoo for insightful discussions and Prof. Saarma for the human GAS1 cDNA. Biacore assays were done with the help of Thomas Strandin and Maria Aatonen. SAXS data were collected at the ID14-3 beam line at ESRF. The MALDI-TOF measurements were done in the Institute of Biotechnology proteomics unit. TK and KR were funded by Academy of Finland grants 256049 and 251700 (to TK), and AG by the Academy of Finland (12522061), by the 7th European Union Framework (CADMAD), and by the Sigrid Juselius Foundation.

## Author details
[1]Institute of Biotechnology, Structural Biology and Biophysics, University of Helsinki, Helsinki, Finland. [2]Astbury Centre for Structural Molecular Biology, School of Biomedical Sciences, University of Leeds, Leeds, UK. [3]Department of Biosciences, Division of Biochemistry, University of Helsinki, Helsinki, Finland.

## References
1. Schneider C, King RM, Philipson L. Genes specifically expressed at growth arrest of mammalian cells. Cell. 1988;54(6):787–93.
2. Stebel M, Vatta P, Ruaro ME, Del Sal G, Parton RG, Schneider C. The growth suppressing gas1 product is a GPI-linked protein. FEBS Lett. 2000;481(2):152–8.
3. Del Sal G, Collavin L, Ruaro ME, Edomi P, Saccone S, Valle GD, et al. Structure, function, and chromosome mapping of the growth-suppressing human homologue of the murine gas1 gene. Proc Natl Acad Sci U S A. 1994;91(5):1848–52.
4. Ruaro ME, Stebel M, Vatta P, Marzinotto S, Schneider C. Analysis of the domain requirement in Gas1 growth suppressing activity. FEBS Lett. 2000;481(2):159–63.
5. Del Sal G, Ruaro ME, Philipson L, Schneider C. The growth arrest-specific gene, gas1, is involved in growth suppression. Cell. 1992;70(4):595–607.
6. Del Sal G, Ruaro EM, Utrera R, Cole CN, Levine AJ, Schneider C. Gas1-induced growth suppression requires a transactivation-independent p53 function. Mol Cell Biol. 1995;15(12):7152–60.
7. Derry WB, Bierings R, van Iersel M, Satkunendran T, Reinke V, Rothman JH. Regulation of developmental rate and germ cell proliferation in Caenorhabditis elegans by the p53 gene network. Cell Death Differ. 2007;14(4):662–70.
8. Cabrera JR, Sanchez-Pulido L, Rojas AM, Valencia A, Manes S, Naranjo JR, et al. Gas1 is related to the glial cell-derived neurotrophic factor family receptors alpha and regulates Ret signaling. J Biol Chem. 2006;281(20):14330–9.
9. Lee CS, Fan CM. Embryonic expression patterns of the mouse and chick Gas1 genes. Mech Dev. 2001;101(1–2):293–7.
10. Allen BL, Tenzen T, McMahon AP. The Hedgehog-binding proteins Gas1 and Cdo cooperate to positively regulate Shh signaling during mouse development. Genes Dev. 2007;21(10):1244–57.
11. Martinelli DC, Fan CM. The role of Gas1 in embryonic development and its implications for human disease. Cell Cycle. 2007;6(21):2650–5.
12. Santoro M, Melillo RM, Carlomagno F, Vecchio G, Fusco A. Minireview: RET: normal and abnormal functions. Endocrinology. 2004;145(12):5448–51.
13. Robertson K, Mason I. The GDNF-RET signalling partnership. Trends Genet. 1997;13(1):1–3.
14. Sariola H, Saarma M. Novel functions and signalling pathways for GDNF. J Cell Sci. 2003;116(Pt 19):3855–62.
15. Schueler-Furman O, Glick E, Segovia J, Linial M. Is GAS1 a co-receptor for the GDNF family of ligands? Trends Pharmacol Sci. 2006;27(2):72–7.
16. Airaksinen MS, Titievsky A, Saarma M. GDNF family neurotrophic factor signaling: four masters, one servant? Mol Cell Neurosci. 1999;13(5):313–25.
17. Hatinen T, Holm L, Airaksinen MS. Loss of neurturin in frog–comparative genomics study of GDNF family ligand-receptor pairs. Mol Cell Neurosci. 2007;34(2):155–67.

18. Airaksinen MS, Holm L, Hatinen T. Evolution of the GDNF family ligands and receptors. Brain Behav Evol. 2006;68(3):181–90.

19. Mellstrom B, Cena V, Lamas M, Perales C, Gonzalez C, Naranjo JR. Gas1 is induced during and participates in excitotoxic neuronal death. Mol Cell Neurosci. 2002;19(3):417–29.

20. Lopez-Ramirez MA, Dominguez-Monzon G, Vergara P, Segovia J. Gas1 reduces Ret tyrosine 1062 phosphorylation and alters GDNF-mediated intracellular signaling. Int J Dev Neurosci. 2008;26(5):497–503.

21. Wang X, Baloh RH, Milbrandt J, Garcia KC. Structure of artemin complexed with its receptor GFRalpha3: convergent recognition of glial cell line-derived neurotrophic factors. Structure. 2006;14(6):1083–92.

22. Parkash V, Leppanen VM, Virtanen H, Jurvansuu JM, Bespalov MM, Sidorova YA, et al. The structure of the glial cell line-derived neurotrophic factor-coreceptor complex: insights into RET signaling and heparin binding. J Biol Chem. 2008;283(50):35164–72.

23. Mylonas E, Svergun D. Accuracy of molecular mass determination of proteins in solutionby small-angle X-ray scattering. J Appl Cryst. 2007;40(Supplement):245–9.

24. Petoukhov MV, Frank D, Shkumatov AV, Tria A, Kikhney AG, Gajda M, et al. New developments in the ATSAS program package for small-angle scattering data analysis. J Appl Cryst. 2012;45:342–50.

25. Leppanen VM, Bespalov MM, Runeberg-Roos P, Puurand U, Merits A, Saarma M, et al. The structure of GFRalpha1 domain 3 reveals new insights into GDNF binding and RET activation. EMBO J. 2004;23(7):1452–62.

26. Agostoni E, Gobessi S, Petrini E, Monte M, Schneider C. Cloning and characterization of the C. elegans gas1 homolog: phas-1. Biochim Biophys Acta. 2002;1574(1):1–9.

27. Kallberg M, Wang H, Wang S, Peng J, Wang Z, Lu H, et al. Template-based protein structure modeling using the RaptorX web server. Nat Protoc. 2012;7(8):1511–22.

28. Martinelli DC, Fan CM. Gas1 extends the range of Hedgehog action by facilitating its signaling. Genes Dev. 2007;21(10):1231–43.

29. Kallijarvi J, Stratoulias V, Virtanen K, Hietakangas V, Heino TI, Saarma M. Characterization of Drosophila GDNF receptor-like and evidence for its evolutionarily conserved interaction with neural cell adhesion molecule (NCAM)/FasII. PLoS One. 2012;7(12):e51997.

30. Paratcha G, Ledda F, Ibanez CF. The neural cell adhesion molecule NCAM is an alternative signaling receptor for GDNF family ligands. Cell. 2003;113(7):867–79.

31. Kjaer S, Ibanez CF. Identification of a surface for binding to the GDNF-GFR alpha 1 complex in the first cadherin-like domain of RET. J Biol Chem. 2003;278(48):47898–904.

32. Trupp M, Arenas E, Fainzilber M, Nilsson AS, Sieber BA, Grigoriou M, et al. Functional receptor for GDNF encoded by the c-ret proto-oncogene. Nature. 1996;381(6585):785–9.

33. Seppala M, Depew MJ, Martinelli DC, Fan CM, Sharpe PT, Cobourne MT. Gas1 is a modifier for holoprosencephaly and genetically interacts with sonic hedgehog. J Clin Invest. 2007;117(6):1575–84.

34. Pineda-Alvarez DE, Roessler E, Hu P, Srivastava K, Solomon BD, Siple CE, et al. Missense substitutions in the GAS1 protein present in holoprosencephaly patients reduce the affinity for its ligand. SHH Hum Genet. 2012;131(2):301–10.

35. Ribeiro LA, Quiezi RG, Nascimento A, Bertolacini CP, Richieri-Costa A. Holoprosencephaly and holoprosencephaly-like phenotype and GAS1 DNA sequence changes: Report of four Brazilian patients. Am J Med Genet A. 2010;152A(7):1688–94.

36. Keinanen K, Jouppila A, Kuusinen A. Characterization of the kainate-binding domain of the glutamate receptor GluR-6 subunit. Biochem J. 1998;330(Pt 3):1461–7.

37. Konarev PV, Volkov VV, Sokolova AV, Koch MHJ, Svergun DI. PRIMUS: a Windows PC-based system for small-angle scattering data analysis. J Appl Cryst. 2003;36(5):1277–82.

38. Svergun DI. Restoring low resolution structure of biological macromolecules from solution scattering using simulated annealing. Biophys J. 1999;76(6):2879–86.

39. Petoukhov MV, Svergun DI. Applications of small-angle X-ray scattering to biomacromolecular solutions. Int J Biochem Cell Biol. 2013;45(2):429–37.

40. Bernado P, Svergun DI. Analysis of intrinsically disordered proteins by small-angle X-ray scattering. Methods Mol Biol. 2012;896:107–22.

# Permissions

The contributors of this book come from diverse backgrounds, making this book a truly international effort. This book will bring forth new frontiers with its revolutionizing research information and detailed analysis of the nascent developments around the world.

We would like to thank all the contributing authors for lending their expertise to make the book truly unique. They have played a crucial role in the development of this book. Without their invaluable contributions this book wouldn't have been possible. They have made vital efforts to compile up to date information on the varied aspects of this subject to make this book a valuable addition to the collection of many professionals and students.

This book was conceptualized with the vision of imparting up-to-date information and advanced data in this field. To ensure the same, a matchless editorial board was set up. Every individual on the board went through rigorous rounds of assessment to prove their worth. After which they invested a large part of their time researching and compiling the most relevant data for our readers.

The editorial board has been involved in producing this book since its inception. They have spent rigorous hours researching and exploring the diverse topics which have resulted in the successful publishing of this book. They have passed on their knowledge of decades through this book. To expedite this challenging task, the publisher supported the team at every step. A small team of assistant editors was also appointed to further simplify the editing procedure and attain best results for the readers.

Apart from the editorial board, the designing team has also invested a significant amount of their time in understanding the subject and creating the most relevant covers. They scrutinized every image to scout for the most suitable representation of the subject and create an appropriate cover for the book.

The publishing team has been an ardent support to the editorial, designing and production team. Their endless efforts to recruit the best for this project, has resulted in the accomplishment of this book. They are a veteran in the field of academics and their pool of knowledge is as vast as their experience in printing. Their expertise and guidance has proved useful at every step. Their uncompromising quality standards have made this book an exceptional effort. Their encouragement from time to time has been an inspiration for everyone.

The publisher and the editorial board hope that this book will prove to be a valuable piece of knowledge for researchers, students, practitioners and scholars across the globe.

# List of Contributors

Li Ma and Jiangqin Liu
Key Laboratory for Medical Molecular Diagnostics of Guangdong Province, Guangdong Medical College, Xincheng Road, Dongguan 523808, P R China

Liping Li, Youxiang Sun, Xiaodan Wu, Lijun Chen and Pengfei Wu
Key Laboratory for Medical Molecular Diagnostics of Guangdong Province, Guangdong Medical College, Xincheng Road, Dongguan 523808, P R China
Department of Biochemistry, School of Basic Medicine, Guangdong Medical College, Dongguan 523808, P R China

Rosa Viana, Pablo Lujan and Pascual Sanz
Instituto de Biomedicina de Valencia, CSIC, and Centro de Investigación Biomédica en Red de Enfermedades Raras (CIBERER), Jaime Roig 11, 46010 Valencia, Spain

Marco Aurelio Pardo-Galván
Instituto de Investigaciones Químico-Biológicas, Universidad Michoacana de San Nicolás de Hidalgo, Edificio B-3 Ciudad Universitaria Avenida Francisco J. Múgica S/N, Morelia, Michoacán 58030, México

Mario Javier Gutiérrez-Fernández
Instituto de Investigaciones Químico-Biológicas, Universidad Michoacana de San Nicolás de Hidalgo, Edificio B-3 Ciudad Universitaria Avenida Francisco J. Múgica S/N, Morelia, Michoacán 58030, México
Present address: Universidad Tecnológica de Morelia, Morelia, Michoacán 58200, México

César Adrián Gómez-Correa
Present address: Universidad Tecnológica de Morelia, Morelia, Michoacán 58200, México
División de Estudios de Posgrado de la Facultad de Ciencias Médicas y Biológicas "Dr. Ignacio Chávez", Universidad Michoacana de San Nicolás de Hidalgo, Morelia, Michoacán 58020, México

Edith Higareda-Mendoza
División de Estudios de Posgrado de la Facultad de Ciencias Médicas y Biológicas "Dr. Ignacio Chávez", Universidad Michoacana de San Nicolás de Hidalgo, Morelia, Michoacán 58020, México

Concetta De Santi, Bjørn Altermark, Marcin Miroslaw Pierechod and Nils-Peder Willassen
NorStruct, Department of Chemistry, Faculty of Science and Technology, UiT The Arctic University of Norway, Tromsø, Norway

Luca Ambrosino and Donatella de Pascale
Institute of Protein Biochemistry, National Research Council, Naples, Italy

Sheena Jiang and Xianqiang Li
Signosis Inc., 1700 Wyatt Drive, Suite #10-12, Santa Clara, CA 95054, USA

Eric Zhang and Rachel Zhang
Saratoga High School, 20300 Herriman Ave, Saratoga, CA 95070, USA

Luis F. Plenge-Tellechea
Departamento de Ciencias Químico Biológicas, Laboratorio de Biología Molecular y Bioquímica (Edif. T-216), Instituto de Ciencias Biomédicas, Universidad Autónoma de Ciudad Juárez, Plutarco Elías Calles #1210 Fovissste Chamizal, Ciudad Juárez, Chihuahua C.P. 32310, Mexico

Jorge A. Sierra-Fonseca
Department of Biological Sciences, University of Texas at El Paso, El Paso, TX 79968, USA

Javier Vargas-Medrano
Department of Biomedical Sciences, Center of Emphasis for Neurosciences, Texas Tech University Health Science Center, El Paso, TX 79905, USA

Robert Marmulla, Barbara Šafarić and Jens Harder
Department of Microbiology, Max Planck Institute for Marine Microbiology, Celsiusstr. 1, D-28359 Bremen, Germany

**Stephanie Markert and Thomas Schweder**
Institute for Pharmacy, Department of Pharmaceutical Biotechnology, University of Greifswald, Felix-Hausdorff-Str. 3, D-17487 Greifswald, Germany

**Yasunori Fukuda, Osamu Sano, Kenichi Kazetani, Koji Yamamoto, Hidehisa Iwata and Junji Matsui**
Pharmaceutical Research Division, Takeda Pharmaceutical Company Limited, 2-26-1, Muraokahigashi, Fujisawa, Kanagawa, Japan

**Thao Thi Nguyen, Hanh Van Vu, Nhung Thi Hong Nguyen, Tuyen Thi Do and Thanh Sy Le Nguyen**
Institute of Biotechnology, Vietnam Academy of Science and Technology, 18 Hoang Quoc Viet Road, Distr. Caugiay, 10600 Hanoi, Vietnam

**Roberto Alampi, Flavia Biundo, Giovanni Toscano and Maria Rosa Felice**
Department of Chemical, Biological, Pharmaceutical, and Environmental Sciences, University of Messina, Viale F. Stagno D'Alcontres, 31, 98166 Messina, Italy.

**Luca Marco Di Bella**
Department of Chemical, Biological, Pharmaceutical, and Environmental Sciences, University of Messina, Viale F. Stagno D'Alcontres, 31, 98166 Messina, Italy Inter University National Group of Marine Sciences (CoNISMa), Piazzale Flaminio, 9, 00196 Rome, Italy

**Carolyn J. Adamski and Timothy Palzkill**
Department of Biochemistry and Molecular Biology, Baylor College of Medicine, Houston, TX, USA Department of Pharmacology, Baylor College of Medicine, Houston, TX, USA

**Annegret Ulke-Lemée, David Hao Sun and Justin A. MacDonald**
Department of Biochemistry & Molecular Biology, University of Calgary, Cumming School of Medicine, 3280 Hospital Drive NW, Calgary, AB T2N 4Z6, Canada

**Hiroaki Ishida and Hans J. Vogel**
Biochemistry Research Group, Department of Biological Sciences, University of Calgary, 2500 University Drive NW, Calgary, AB T2N 1 N4, Canada

**Xiaowen Fei**
School of Science, Hainan Medical College, Haikou 571101, China

**Junmei Yu, Yajun Li and Xiaodong Deng**
Institute of Tropical Bioscience and Biotechnology, Chinese Academy of Tropical Agricultural Science, Key Laboratory of Tropical Crop Biotechnology, Ministry of Agriculture, Haikou 571101, China

**Yoshiaki Shimada and Takao Urabe**
Department of Neurology, Juntendo University Urayasu Hospital, 2-1-1 Tomioka, Urayasu, Chiba, Japan

**Hideki Shimura**
Department of Neurology, Juntendo University Urayasu Hospital, 2-1-1 Tomioka, Urayasu, Chiba, Japan
Institute for Environment and Gender Specific Medicine, Juntendo University School of Medicine, Chiba, Japan

**Ryota Tanaka, Kazuo Yamashiro and Nobutaka Hattori**
Department of Neurology, Juntendo University School of Medicine, Tokyo, Japan

**Outi Nivala, Kristiina Kruus and Mikko Arvas**
VTT Technical Research Centre of Finland, Ltd., P.O. Box 1000, FI-02044 Espoo, Finland

**Greta Faccio**
VTT Technical Research Centre of Finland, Ltd., P.O. Box 1000, FI-02044 Espoo, Finland
Independent scientist, St. Gallen, CH, Switzerland

**Johanna Buchert**
VTT Technical Research Centre of Finland, Ltd., P.O. Box 1000, FI-02044 Espoo, Finland
Natural resources institute Finland (Luke), P.O. Box 2, FI-00790 Helsinki, Finland

**Maija-Liisa Mattinen**
VTT Technical Research Centre of Finland, Ltd., FI-02044 Espoo, Finland
Department of Forest Products Technology, Bioproduct Chemistry, Aalto University, School of Chemical Technology, P.O. Box 16300, FI-00076 Espoo, Finland

**Perttu Permi**
Institute of Biotechnology, University of Helsinki, P.O. Box 65, FI-00014 Helsinki, Finland
Department of Biological and Environmental Sciences, Nanoscience Center, University of Jyväskylä, P.O. Box 35, FI-40014 Jyväskylä, Finland
Department of Chemistry, Nanoscience Center, University of Jyväskylä, P.O. Box 35, FI-40014 Jyväskylä, Finland

**V D Sirisha Gandreddi, Vijaya Rachel Kappala and Kunal Zaveri**
Assistant professor, Department of Biochemistry/ Bioinformatics, Institute of Science, GITAM University, Rushikonda, Visakhapatnam 530045, Andhra Pradesh, India

**Kiranmayi Patnala**
Department of Biotechnology, Institute of Science, GITAM University, Rushikonda, Visakhapatnam 530045, Andhra Pradesh, India

**Shu-su Liu, Xuan Wei, Xue Dong, Liang Xu and Biao Jiang**
Shanghai Institute for Advanced Immunochemical Studies, ShanghaiTech University, Shanghai, China

**Jia Liu**
Shanghai Institute for Advanced Immunochemical Studies, ShanghaiTech University, Shanghai, China
Department of Chemistry and Biochemistry, University of Maryland, College Park, USA

**Ariana Labastida-Polito, Menandro Camarillo-Cadena, Rafael A. Zubillaga and Andrés Hernández-Arana**
Área de Biofisicoquímica, Departamento de Química, Universidad Autónoma Metropolitana-Iztapalapa, San Rafael Atlixco 186, Iztapalapa D.F. 09340, Mexico

**Georgina Garza-Ramos**
Departamento de Bioquímica, Facultad de Medicina, Universidad Nacional Autónoma de México, Coyoacán D.F. 04510, Mexico

**Katja Rosti and Tommi Kajander**
Institute of Biotechnology, Structural Biology and Biophysics, University of Helsinki, Helsinki, Finland

**Adrian Goldman**
Astbury Centre for Structural Molecular Biology, School of Biomedical Sciences, University of Leeds, Leeds, UK
Department of Biosciences, Division of Biochemistry, University of Helsinki, Helsinki, Finland

# Index